Ordinary

Microwave and RF Circuits: Analysis, Synthesis and Design

For a complete listing of the *Artech House Microwave Library*,
turn to the back of this book

Microwave and RF Circuits: Analysis, Synthesis and Design

Max W. Medley

Artech House
Boston • London

Library of Congress Cataloging-in-Publication Data

Medley, M. W. (Max W.), 1942-
Microwave and RF Circuits: Analysis, Synthesis, and Design / Max W. Medley.
p. cm.
Includes bibliographical references and index.
ISBN 0-89006-546-2
1. Microwave circuits—Design and construction—Data processing.
2. Radio cirucits—Design and construction—Data processing
3. Computer-aided design. I. Title.

TK7876.M389 1992 92-5359
621.381'32—dc20 CIP

British Library Cataloguing in Publication Data

Medley, Max W.
Microwave and RF Circuits: Analysis, Synthesis, and Design
I. Title
621.3815

ISBN 0-89006-546-2

685 Canton Street
Norwood, MA 02062

International Standard Book Number: 0-89006-546-2
Library of Congress Catalog Card Number: 92-5359

10 9 8 7 6 5 4

To Alice, Todd, and Michael

Contents

Preface

The contemporary RF and microwave circuit designer has a number of very powerful computer-aided design (CAD) tools at his or her disposal. The first CAD programs, which became commercially available just over two decades ago, had very rigid circuit file structures, contained a limited number of ideal elements, and were restricted to circuit topologies that could be created from five basic two-port interconnections. Most programs were only available through commercial time share computer networks, which were expensive and not at all conducive to creative circuit design. A complex circuit design could take several days. The circuit model was created and analyzed to ensure that there were no obvious errors. Then, unless the company's profit margin needed trimming, the file was submitted as a batch process to run optimization overnight at reduced cost and priority. Assuming the CPU limit had not been reached and the optimization targets were correct, the results were available the next morning.

By contrast, modern CAD programs run on desktop workstations or even personal computers easily accessible to most designers. These programs are the results of hundreds of man-years of effort, and have evolved from a fragmented collection of specialized programs into what can best be characterized as a total design and development environment. Entire systems or subsystems can be modeled based on ideal specifications for each component, and then updated with measured or predicted data as the design progresses. Designs can be created by simply drawing a schematic on the screen and then analyzing and optimizing the linear or nonlinear circuit as necessary. Once the electrical performance is acceptable, a physical layout modified to suit the physical constraints can be generated automatically, and then a final optimization can be performed to account for the effects of layout parasitics. Finally, effects of component and layout tolerances can be investigated, the design can be optimized to maximize production yields, and artwork can be generated for assembly.

The obvious benefit of an integrated environment is enhanced accuracy and efficiency. Schematic capture provides a clear picture of the circuit topology, reducing modeling errors, and the link between the schematic and the layout further reduces the chance of a mistake, which, especially in the case of a monolithic microwave integrated circuit (MMIC), can be costly and time consuming. At higher microwave or millimeter wave frequencies, where layout parasitics have a pronounced effect on circuit performance, physical models provide excellent correlation between predicted and measured performance, dramatically reducing the number of design iterations.

The scope of modern CAD systems extends beyond analysis, optimization, and layout. An attached network analyzer can be used to measure active and passive components, a database can be created from the measured data, then used to improve the accuracy of the circuit models. The network analyzer can also be indispensable when the hardware does not perform as expected. The measured response can be compared with the response of the circuit model and used to identify the probable cause. Design documentation is essential, although often created as an afterthought. However, CAD systems can create schematics, assembly drawings, and parts lists almost automatically, and when changes are made anywhere in the design, all references are updated instantly.

Numerous articles, application notes, and books have been written on the subject of CAD techniques. Documentation for the CAD software is another excellent source of information. An extensive library of design examples is usually included, which show step-by-step design procedures for a variety of different circuits. However, the vast majority of this material deals with the design only after a topology has been defined. The crucial step of creating a network that meets the performance requirements and can actually be built is too often left as an exercise for the designer.

This book addresses the subject of RF and microwave circuit design and, as the title suggests, is concerned with the practical aspects of synthesizing networks that achieve specified design objectives. Although modern CAD tools can be successfully applied with only minimal knowledge of the principles involved, an understanding of the basic concepts enhances the efficiency of the designer and is often reflected in the hardware performance. This book is written for senior- and graduate-level engineering students, as well as practicing design engineers, and is intended to provide a reference source for the underlying theory, along with practical design techniques. Much of the rigorous theory and mathematical derivations is available elsewhere, although it is spread out among numerous sources. Most engineers and students do not have the time to devote to extensive research not directly related to the task at hand, and the intent of this book is to present the material as concisely as possible and in sufficient detail to explain basic concepts and derivations. Two exceptions to this approach are the multiport s-parameter interconnections and mapping, which are covered in some detail. The multiport interconnection method has not been previously published, and mapping is often completely overlooked as a powerful design tool. Extensive reference lists are provided for the reader interested in pursuing a particular topic.

A significant portion of the material in this book deals with network synthesis, which has not been fully exploited with regard to RF and microwave circuit design. Computer-aided synthesis has been available almost as long as analysis and optimization programs and, although several excellent synthesis programs have been written, has not found the same widespread acceptance as traditional CAD. Network analysis is a very important part of the engineering curriculum, and designers eagerly accept CAD programs based on familiar concepts. However, network synthesis is rarely offered as more than an elective, and even then emphasis is usually placed on filter design. This trend is reflected in the literature as well. Few papers describe design applications of network synthesis, and those that do often require the reader to have a foundation in synthesis theory to be able to understand them. Synthesis offers several advantages over other design techniques. The efficiency of a matching network can be predicted before the network is designed, and synthesis automatically provides one or more circuit topologies and associated element values. Synthesis is certainly not the perfect answer to all the designer's problems. However, it is sincerely hoped that this book will provide the engineer with alternative design methods that prove useful for practical solutions.

This book is divided into three main parts: Chapters 1, 2, and 3 cover network analysis and optimization; Chapters 4, 5, and 6 discuss network synthesis; and Chapter 7 provides some practical applications of design methods covered in the other chapters. Most major topics include numerical examples that illustrate the practical aspects of applying the procedures to specific problems. Chapter 1 is a review of the parameter sets used to represent networks and the basic types of interconnections. *S*-parameters are discussed in some detail since they turn out to be the most convenient for RF and microwave circuit design. Chapter 2 covers the representation of circuit elements, multiport analysis, and the characterization of active and passive networks. Chapter 3 addresses mapping methods, including the Smith chart and multiport networks, along with direct and indirect optimization techniques. Chapter 4 provides an introduction to modern network synthesis and presents techniques applicable to the design of lumped-element networks. Chapter 5 extends the lumped-element methods to the design of distributed-element networks and provides synthesis techniques that have no lumped-element counterpart. Chapter 6 applies the lumped and distributed synthesis methods to the design of broadband impedance-matching networks and presents techniques useful for determining the impedances that provide the specified circuit performance and can be matched with practical circuits. Chapter 7 illustrates modern development techniques by demonstrating the use of commercially available CAD tools in conjunction with design methods from preceding chapters The balanced amplifier configuration is analyzed, and the steps involved in realizing the circuit using microstrip fabrication methods are examined from inception to artwork generation. The advantages and disadvantages of feedback and resistive compensated amplifiers are discussed, and the technique is demonstrated through the design of an octave band low-noise MMIC circuit.

I would like to express my appreciation to Steven March for his many helpful comments, suggestions, and significant contributions to the reference lists in this book. I also wish to thank EEsof, Inc., of Westlake Village, California, for supplying the CAD software used to verify many of the numerical examples in this book, including the CAD applications in Chapter 7.

Max W. Medley, Jr.
Largo, Florida
April, 1992

Chapter 1

Network Characterization

Chapter 1 provides a summary of network characterization techniques, which, for the most part, will be familiar to the reader. The techniques of this chapter are based on linear time-invariant networks, although, as a practical matter, the methods can be applied to mildly nonlinear networks with some degree of success.

1.1 NETWORK REPRESENTATIONS

Circuit elements can be characterized as having two or more *terminals*, or end points. Elements that are connected together in some fashion are collectively known as *networks*. Network analysis, as the term implies, is concerned with the characteristics of the complete network rather than the individual components. Figure 1.1 shows a simple resistive voltage divider partitioned into internal circuit elements connected together at *nodes*, and external terminals which provide access to the network. We know that this circuit is a voltage divider and can easily characterize it by applying a voltage across terminals 1 and 1′, then measuring the voltage across terminals 2 and 2′. Similarly, we can apply the voltage source at 2-2′ and then measure the voltage across 1-1′. In this case, we may only be interested in the voltage divider ratio; however, we are usually interested in a complete mathematical representation that can be determined from the external terminals and used to predict the response for any source or load connected to the network, or several smaller networks combined into one large network.

The pair of external terminals used to connect a stimulus or termination is known as a *port*, and a network with n ports is called an *n-port* network. The n-

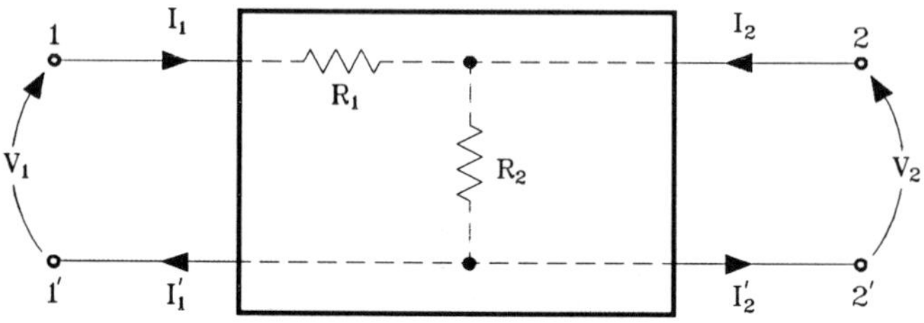

Figure 1.1 General four-terminal network.

port network does not necessarily have $2n$ terminals; a terminal can be common to more than one port and, in fact, may not even be part of the network. For example, a transistor usually has three terminals but is characterized as a two-port network formed from the base-emitter and collector-emitter terminals. Occasionally we require access to the emitter and form a three-port by referencing the base, emitter, and collector terminals to a common ground. The voltage and current definition at each port is shown in Figure 1.1. Noting that the two currents I and I' flow through the source or termination and must be the same, we see there are two parameters per port, or a total of four parameters for the two-port network. We can select any two of the four parameters as independent variables, and, since there are six combinations of voltage or current taken two at a time, we have six different representations of a network.

1.1.1 *Z*-Parameter Representation

Selecting voltage as the dependent variable, we can write

$$V_1 = z_{11}I_1 + z_{12}I_2$$
$$V_2 = z_{21}I_1 + z_{22}I_2 \tag{1.1a}$$

or, in matrix notation,

$$\begin{bmatrix} V_1 \\ V_2 \end{bmatrix} = \begin{bmatrix} z_{11} & z_{12} \\ z_{21} & z_{22} \end{bmatrix} \begin{bmatrix} I_1 \\ I_2 \end{bmatrix} \tag{1.1b}$$

The coefficients z_{ij}, which relate the dependent and independent variables, must be determined in order to characterize the network. If we set I_2 to zero, we can solve (1.1a) for z_{11} and z_{21}:

$$z_{11} = \left.\frac{V_1}{I_1}\right|_{I_2=0} \qquad z_{21} = \left.\frac{V_2}{I_1}\right|_{I_2=0} \tag{1.1c}$$

and when I_1 is set to zero,

$$z_{12} = \left.\frac{V_1}{I_2}\right|_{I_1=0} \qquad z_{22} = \left.\frac{V_2}{I_2}\right|_{I_1=0} \tag{1.1d}$$

We see that the coefficients, or *network parameters*, have units of voltage divided by current, or impedance, and are known as *z-parameters* or *impedance parameters*. From Figure 1.1, it is apparent that the port current is zero when terminated with an open circuit; hence, the z-parameters are also known as *open circuit parameters*. We can easily measure the z-parameters of any two-port network by connecting a known voltage source at port 1 while port 2 is left open, measuring the current flowing into the network at port 1 and the open circuit voltage at port 2, and then computing z_{11} as (V_1/I_1), and z_{21} as (V_2/I_1). We can reverse the ports, measure voltage and current again, and then compute z_{12} and z_{22}. Current must flow in the network when one port is open, otherwise I_1 and I_2 are both zero and the z-parameters are not defined. For example, if we remove R_2 in Figure 1.1, current cannot flow when either port is open and the z-parameters do not exist. The ideal transformer is another example of a simple element that has no z-parameter representation.

When using z-parameters for network analysis, it is often convenient to work with *normalized parameters*; that is, parameters referenced to a common impedance. Since z-parameters have units of impedance, we can normalize by dividing all elements by a reference impedance Z_o. The normalized **Z** matrix is given by

$$\mathbf{Z}_n = \begin{bmatrix} \dfrac{z_{11}}{Z_o} & \dfrac{z_{12}}{Z_o} \\[2ex] \dfrac{z_{21}}{Z_o} & \dfrac{z_{22}}{Z_o} \end{bmatrix}$$

Normalized parameters can help improve numerical stability and are usually much easier to work with for manual calculations.

Z-parameters do not provide much physical insight into the behavior of a network. The input and output parameters, z_{11} and z_{22}, are the input and output impedances, respectively. However, the impedance is measured under the condition of an open circuit at the opposite port and does not usually reflect the actual operating conditions. We can determine some characteristics of the network directly from its z-parameters. When $z_{11} = z_{22}$, the network is unchanged when ports 1 and 2 are interchanged, or *symmetrical*; and when $z_{12} = z_{21}$, the response is identical when the excitation is applied at port 1 or port 2, and the network is said to be *reciprocal.*

1.1.2 *Y*-Parameter Representation

Selecting current as the dependent variable, we can write

$$I_1 = y_{11}V_1 + y_{12}V_2$$
$$I_2 = y_{21}V_1 + y_{22}V_2 \tag{1.2a}$$

or, in matrix notation,

$$\begin{bmatrix} I_1 \\ I_2 \end{bmatrix} = \begin{bmatrix} y_{11} & y_{12} \\ y_{21} & y_{22} \end{bmatrix} \begin{bmatrix} V_1 \\ V_2 \end{bmatrix} \tag{1.2b}$$

We can solve for y_{ij} by forcing the port voltage to zero. When V_2 is set to zero and (1.2a) is rearranged,

$$y_{11} = \left.\frac{I_1}{V_1}\right|_{V_2=0} \qquad y_{21} = \left.\frac{I_2}{V_1}\right|_{V_2=0} \tag{1.2c}$$

and when V_1 is zero,

$$y_{12} = \left.\frac{I_1}{V_2}\right|_{V_1=0} \qquad y_{22} = \left.\frac{I_2}{V_2}\right|_{V_1=0} \tag{1.2d}$$

These coefficients have units of current divided by voltage, or admittance, and are known as *y-parameters* or *admittance parameters.* We can force the zero voltage condition by short-circuiting the port. Hence, the y-parameters are also known as *short circuit parameters.*

The y-parameters are measured in almost the same manner as the z-parameters, except the port opposite the voltage source is shorted and the measured current at each port is divided by the input voltage. A network consisting solely of a shunt admittance does not have a defined **Y** matrix since a short circuit at either port shorts out the element as well as the voltage source.

The y-parameters have units of admittance, and we can either divide by the reference admittance Y_o or multiply by the reference impedance Z_o to normalize:

$$\mathbf{Y}_n = \begin{bmatrix} y_{11}Z_o & y_{12}Z_o \\ y_{21}Z_o & y_{22}Z_o \end{bmatrix} = \begin{bmatrix} \dfrac{y_{11}}{Y_o} & \dfrac{y_{12}}{Y_o} \\ \dfrac{y_{21}}{Y_o} & \dfrac{y_{22}}{Y_o} \end{bmatrix}$$

A network characterized by y-parameters is symmetrical when $y_{11} = y_{22}$, and reciprocal when $y_{12} = y_{21}$.

1.1.3 *H*-Parameter Representation

We can use the voltage at the input port and the current at the output port as the dependent variables, then

$$\begin{aligned} V_1 &= h_{11}I_1 + h_{12}V_2 \\ I_2 &= h_{21}I_1 + h_{22}V_2 \end{aligned} \tag{1.3a}$$

or

$$\begin{bmatrix} V_1 \\ I_2 \end{bmatrix} = \begin{bmatrix} h_{11} & h_{12} \\ h_{21} & h_{22} \end{bmatrix} \begin{bmatrix} I_1 \\ V_2 \end{bmatrix} \tag{1.3b}$$

We can set V_2 to zero and solve (1.3a) for h_{11} and h_{21}:

$$h_{11} = \left.\frac{V_1}{I_1}\right|_{V_2=0} \qquad h_{21} = \left.\frac{I_2}{I_1}\right|_{V_2=0} \tag{1.3c}$$

and then solve for h_{12} and h_{22} by forcing I_1 to zero:

$$h_{12} = \left.\frac{V_1}{V_2}\right|_{I_1=0} \qquad h_{22} = \left.\frac{I_2}{V_2}\right|_{I_1=0} \tag{1.3d}$$

The h_{11} coefficient has units of impedance and h_{22} units of admittance. The two transfer coefficients, h_{12} and h_{21}, are voltage and current ratios which are dimensionless. These coefficients are known as *hybrid* or *h-parameters*, and are determined by shorting the output port, forcing V_2 to zero for h_{11} and h_{21} and terminating the input in an open circuit, $I_1 = 0$, for h_{12} and h_{22}.

The normalized h-parameter matrix is

$$\mathbf{H}_n = \begin{bmatrix} \dfrac{h_{11}}{Z_o} & h_{12} \\ h_{21} & h_{22}Z_o \end{bmatrix}$$

The network is reciprocal when $h_{21} = -h_{12}$, and symmetrical when $\Delta_h = 1$. Δ_h is the determinant of $\mathbf{H}$, or $h_{11}h_{22} - h_{12}h_{21}$.

1.1.4 *G*-Parameter Representation

We can create a set of parameters that are very similar to the h-parameters by interchanging the voltage and current in (1.3):

$$\begin{aligned} I_1 &= g_{11}V_1 + g_{12}I_2 \\ V_2 &= g_{21}V_1 + g_{22}I_2 \end{aligned} \tag{1.4a}$$

or

$$\begin{bmatrix} I_1 \\ V_2 \end{bmatrix} = \begin{bmatrix} g_{11} & g_{12} \\ g_{21} & g_{22} \end{bmatrix} \begin{bmatrix} V_1 \\ I_2 \end{bmatrix} \tag{1.4b}$$

Setting I_2 equal to zero and rewriting (1.4a),

$$g_{11} = \left.\frac{I_1}{V_1}\right|_{I_2=0} \qquad g_{21} = \left.\frac{V_2}{V_1}\right|_{I_2=0} \tag{1.4c}$$

and when V_1 is set to zero,

$$g_{12} = \left.\frac{I_1}{I_2}\right|_{V_1=0} \qquad g_{22} = \left.\frac{V_2}{I_2}\right|_{V_1=0} \tag{1.4d}$$

The g_{11} coefficient has units of admittance and g_{22} units of impedance. The two transfer coefficients, g_{12} and g_{21}, are dimensionless current and voltage ratios, respectively. These coefficients are known as *g-parameters* and are determined by using an open short circuit to force I_2 to zero, and a short circuit to force V_1 to zero.

The normalized *g*-parameter matrix is

$$\mathbf{G}_n = \begin{bmatrix} g_{11}Z_o & g_{12} \\ g_{21} & \dfrac{g_{22}}{Z_o} \end{bmatrix}$$

The network is reciprocal when $g_{21} = -g_{12}$, and symmetrical when $\Delta_g = 1$.

1.1.5 *ABCD*-Parameter Representation

The *z*-, *y*-, *h*-, and *g*-parameter sets are derived by defining the dependent variables as combinations of voltage and current from both ports. We can create another set of parameters by using the voltage and current at the same port as the dependent variable. We can define the port voltage and current in any manner as long as we are consistent, or at least keep track of differences, and it will simplify matters later if we make one minor change in the output current definition. The current at port 2 is defined as flowing out of the positive terminal, or $-I_2$ as shown in Figure 1.1, and we can describe the network with

$$\begin{aligned} V_1 &= AV_2 - BI_2 \\ I_1 &= CV_2 - DI_2 \end{aligned} \tag{1.5a}$$

or

$$\begin{bmatrix} V_1 \\ I_1 \end{bmatrix} = \begin{bmatrix} A & B \\ C & D \end{bmatrix} \begin{bmatrix} V_2 \\ -I_2 \end{bmatrix} \tag{1.5b}$$

Setting I_2 equal to zero and rewriting (1.5a) in terms of A and C,

$$A = \left.\frac{V_1}{V_2}\right|_{I_2=0} \qquad C = \left.\frac{I_1}{V_2}\right|_{I_2=0} \tag{1.5c}$$

and when V_2 is zero, we have

$$B = -\left.\frac{V_1}{I_2}\right|_{V_2=0} \qquad D = -\left.\frac{I_1}{I_2}\right|_{V_2=0} \tag{1.5d}$$

The input parameter A is a dimensionless voltage ratio, and the output parameter D is a current ratio. The transfer coefficient B has units of impedance, and C units of admittance. These coefficients are known as *ABCD-* or *chain parameters*, and are computed in the same manner as elements of the other parameter sets.

The normalized **ABCD** matrix is

$$\mathbf{Ch}_n = \begin{bmatrix} A & \dfrac{B}{Z_o} \\ CZ_o & D \end{bmatrix}$$

The network is symmetrical when $A = D$, and reciprocal when $\Delta = 1$; that is, $AD - BC = 1$.

The sixth network representation is similar to the *ABCD*-parameter set, except V_2 and I_2 are used as the dependent variables. These parameters are sometimes known as *inverse chain parameters* and are not as useful as the other parameters.

1.1.6 Parameter Conversions

Except for a few special cases, we can represent a network using any of the five parameter sets. For example, when we short-circuit one port of an ideal transformer, V_1 and V_2 are zero, and both currents equal zero when we open-circuit either port. Thus, we cannot determine the z- or y-parameters for this element, although we can easily determine the other parameter sets. Depending on the task at hand, one parameter set often has advantages over other sets. The next section illustrates the ease with which we can interconnect elements using certain parameters. Consequently, we find it necessary to convert from one parameter set to another.

We can premultiply both sides of (1.1b) by the inverse of the **Z** matrix:

$$\begin{bmatrix} z_{11} & z_{12} \\ z_{21} & z_{22} \end{bmatrix}^{-1} \begin{bmatrix} V_1 \\ V_2 \end{bmatrix} = \begin{bmatrix} z_{11} & z_{12} \\ z_{21} & z_{22} \end{bmatrix}^{-1} \begin{bmatrix} z_{11} & z_{12} \\ z_{21} & z_{22} \end{bmatrix} \begin{bmatrix} I_1 \\ I_2 \end{bmatrix}$$

which can be simplified:

$$\begin{bmatrix} z_{11} & z_{12} \\ z_{21} & z_{22} \end{bmatrix}^{-1} \begin{bmatrix} V_1 \\ V_2 \end{bmatrix} = \begin{bmatrix} I_1 \\ I_2 \end{bmatrix}$$

and then compared with (1.2b) to show that

$$\mathbf{Y} = \mathbf{Z}^{-1} \tag{1.6}$$

Thus, we can invert the z-parameters to obtain the y-parameters, or invert y-parameters to obtain z-parameters. However, a network that has a valid representation using one parameter set may not using another. This problem was discussed in Sections 1.1.1 and 1.1.2, and is indicated by a zero determinate for either the **Z** or **Y** matrix.

Similarly, we can premultiply both sides of (1.3b) by the inverse of the **H** matrix:

$$\begin{bmatrix} h_{11} & h_{12} \\ h_{21} & h_{22} \end{bmatrix}^{-1} \begin{bmatrix} V_1 \\ I_2 \end{bmatrix} = \begin{bmatrix} h_{11} & h_{12} \\ h_{21} & h_{22} \end{bmatrix}^{-1} \begin{bmatrix} h_{11} & h_{12} \\ h_{21} & h_{22} \end{bmatrix} \begin{bmatrix} I_1 \\ V_2 \end{bmatrix}$$

which can be simplified:

$$\begin{bmatrix} h_{11} & h_{12} \\ h_{21} & h_{22} \end{bmatrix}^{-1} \begin{bmatrix} V_1 \\ I_2 \end{bmatrix} = \begin{bmatrix} I_1 \\ V_2 \end{bmatrix}$$

and then compared with (1.4b) to conclude that

$$\mathbf{G} = \mathbf{H}^{-1} \tag{1.7}$$

and we see that conversion between the g- and h-parameters can also be accomplished by inverting the appropriate matrix.

Conversion between the other parameter sets is not quite as simple and requires comparison on a term-by-term basis for similar voltage and current conditions. For example, to convert from z- to h-parameter, we note that when $V_2 = 0$ we can use (1.3c) and (1.2c) to write

$$h_{11} = \left. \frac{V_1}{I_1} \right|_{V_2=0} \qquad y_{11} = \left. \frac{I_1}{V_1} \right|_{V_2=0}$$

or $h_{11} = 1/y_{11}$. Inverting the **Z** matrix to form y-parameters, we have $y_{11} = z_{22}/\Delta_z$, and

$$h_{11} = \frac{\Delta_z}{z_{22}} \tag{1.8a}$$

When the input current is zero, we can compare (1.3d) and (1.1d):

$$h_{22} = \left.\frac{I_2}{V_2}\right|_{I_1=0} \qquad z_{22} = \left.\frac{V_2}{I_2}\right|_{I_1=0}$$

which reveals that

$$h_{22} = \frac{1}{z_{22}} \tag{1.8b}$$

For $I_1 = 0$, we can also use (1.3d) to write

$$h_{12} = \left.\frac{V_1}{V_2}\right|_{I_1=0}$$

then solve (1.1d) for V_1 and V_2:

$$V_1 = z_{12} I_2 \qquad V_2 = z_{22} I_2$$

and substitute to show that

$$h_{12} = \frac{z_{12}}{z_{22}} \tag{1.8c}$$

Similarly, we can write h_{21} from (1.3c), solve for I_1 and I_2 from (1.2c), and then substitute, which reduces to $h_{21} = y_{21}/y_{11}$. Replacing y_{11} and y_{21} with the equivalent z-parameter expression results in the desired relationship:

$$h_{21} = -\frac{z_{21}}{z_{22}} \tag{1.8d}$$

Figure 1.2 shows the conversions between all five parameter sets, as well as the normalization. To use this figure, locate the column containing the original parameters and the row containing the converted parameters. The intersection of the row and column gives the necessary formulas expressed in matrix form. For example, to convert from **Y** to **H**, locate the second column and fourth row; h_{11} is given by the expression in the upper left corner, h_{12} by the formula in the upper right corner, and so forth. The symbol Δ_x represents the determinate of the parameter matrix. Normalization is given in the cell located at the row and column intersection of the same parameter set. Once a parameter set is normalized, all conversions result in normalized parameters as well.

1.2 NETWORK CONNECTIONS

Network analysis often requires the interconnection of several networks into one large circuit representing the structure we wish to analyze. We can connect the four-terminal network shown in Figure 1.1 to another four-terminal network in five different configurations: *cascade*, *series*, *parallel*, *series-parallel*, and *parallel-series*. The output terminals of one network are attached to the input terminals of a second network for the cascade connection, and the other four interconnections represent various connections between corresponding ports of the two networks. It is possible to conceive of other connections by interchanging the ports of one of the networks (e.g., cascade the output port of one network with the output port of a second network). However, this is easily accomplished by reversing the parameter matrix (e.g., exchange z_{11} with z_{22} and z_{12} with z_{21}), and then using one of the five basic interconnections.

1.2.1 The Cascade Connection

The cascade connection shown in Figure 1.3 is used to connect the output port of one network to the input port of a second network and is the most common type. At the interface between the two networks in Figure 1.3(b), we see that

$$\begin{bmatrix} V_{2\alpha} \\ I_{2\alpha} \end{bmatrix} = \begin{bmatrix} V_{1\beta} \\ I_{1\beta} \end{bmatrix} \tag{1.9a}$$

The *ABCD*-parameters relate the voltage and current at one port to the voltage and current at the other port, and we can write

$$\begin{bmatrix} V_{1\alpha} \\ I_{1\alpha} \end{bmatrix} = \begin{bmatrix} A_\alpha & B_\alpha \\ C_\alpha & D_\alpha \end{bmatrix} \begin{bmatrix} V_{2\alpha} \\ I_{2\alpha} \end{bmatrix} \tag{1.9b}$$

From → / To ↓	Z		Y		G		H		ABCD	
Z	$\frac{z_{11}}{Z_0}$	$\frac{z_{12}}{Z_0}$	$\frac{y_{22}}{\Delta_y}$	$-\frac{y_{12}}{\Delta_y}$	$\frac{1}{g_{11}}$	$-\frac{g_{12}}{g_{11}}$	$\frac{\Delta_h}{h_{22}}$	$\frac{h_{12}}{h_{22}}$	$\frac{A}{C}$	$\frac{\Delta}{C}$
	$\frac{z_{21}}{Z_0}$	$\frac{z_{22}}{Z_0}$	$-\frac{y_{21}}{\Delta_y}$	$\frac{y_{11}}{\Delta_y}$	$\frac{g_{21}}{g_{11}}$	$\frac{\Delta_g}{g_{11}}$	$-\frac{h_{21}}{h_{22}}$	$\frac{1}{h_{22}}$	$\frac{1}{C}$	$\frac{D}{C}$
Y	$\frac{z_{22}}{\Delta_z}$	$-\frac{z_{12}}{\Delta_z}$	$y_{11} Z_0$	$y_{12} Z_0$	$\frac{\Delta_g}{g_{22}}$	$\frac{g_{12}}{g_{22}}$	$\frac{1}{h_{11}}$	$-\frac{h_{12}}{h_{11}}$	$\frac{D}{B}$	$-\frac{\Delta}{B}$
	$-\frac{z_{21}}{\Delta_z}$	$\frac{z_{11}}{\Delta_z}$	$y_{21} Z_0$	$y_{22} Z_0$	$-\frac{g_{21}}{g_{22}}$	$\frac{1}{g_{22}}$	$\frac{h_{21}}{h_{11}}$	$\frac{\Delta_h}{h_{11}}$	$-\frac{1}{B}$	$\frac{A}{B}$
G	$\frac{1}{z_{11}}$	$-\frac{z_{12}}{z_{11}}$	$\frac{\Delta_y}{y_{22}}$	$\frac{y_{12}}{y_{22}}$	$g_{11} Z_0$	g_{12}	$\frac{h_{22}}{\Delta_h}$	$-\frac{h_{12}}{\Delta_h}$	$\frac{C}{A}$	$-\frac{\Delta}{A}$
	$\frac{z_{21}}{z_{11}}$	$\frac{\Delta_z}{z_{11}}$	$-\frac{y_{21}}{y_{22}}$	$\frac{1}{y_{22}}$	g_{21}	$\frac{g_{22}}{Z_0}$	$-\frac{h_{21}}{\Delta_h}$	$\frac{h_{11}}{\Delta_h}$	$\frac{1}{A}$	$\frac{B}{A}$
H	$\frac{\Delta_z}{z_{22}}$	$\frac{z_{12}}{z_{22}}$	$\frac{1}{y_{11}}$	$-\frac{y_{12}}{y_{11}}$	$\frac{g_{22}}{\Delta_g}$	$-\frac{g_{12}}{\Delta_g}$	$\frac{h_{11}}{Z_0}$	h_{12}	$\frac{B}{D}$	$\frac{\Delta}{D}$
	$-\frac{z_{21}}{z_{22}}$	$\frac{1}{z_{22}}$	$\frac{y_{21}}{y_{11}}$	$\frac{\Delta_y}{y_{11}}$	$-\frac{g_{21}}{\Delta_g}$	$\frac{g_{11}}{\Delta_g}$	h_{21}	$h_{22} Z_0$	$-\frac{1}{D}$	$\frac{C}{D}$
ABCD	$\frac{z_{11}}{z_{21}}$	$\frac{\Delta_z}{z_{21}}$	$-\frac{y_{22}}{y_{21}}$	$-\frac{1}{y_{21}}$	$\frac{1}{g_{21}}$	$\frac{g_{22}}{g_{21}}$	$-\frac{\Delta_h}{h_{21}}$	$-\frac{h_{11}}{h_{21}}$	A	$\frac{B}{Z_0}$
	$\frac{1}{z_{21}}$	$\frac{z_{22}}{z_{21}}$	$-\frac{\Delta_y}{y_{21}}$	$-\frac{y_{11}}{y_{21}}$	$\frac{g_{11}}{g_{21}}$	$\frac{\Delta_g}{g_{21}}$	$-\frac{h_{22}}{h_{21}}$	$-\frac{1}{h_{21}}$	$C Z_0$	D

Figure 1.2 Parameter set conversions and normalization.

for network α, and

$$\begin{bmatrix} V_{1\beta} \\ I_{1\beta} \end{bmatrix} = \begin{bmatrix} A_\beta & B_\beta \\ C_\beta & D_\beta \end{bmatrix} \begin{bmatrix} V_{2\beta} \\ I_{2\beta} \end{bmatrix} \tag{1.9c}$$

for network β. Using the equality in (1.9a), we can substitute (1.9c) into (1.9b), resulting in the expression that relates the input voltage and current to the output voltage and current:

$$\begin{bmatrix} V_{1\alpha} \\ I_{1\alpha} \end{bmatrix} = \begin{bmatrix} A_\alpha & B_\alpha \\ C_\alpha & D_\alpha \end{bmatrix} \begin{bmatrix} A_\beta & B_\beta \\ C_\beta & D_\beta \end{bmatrix} \begin{bmatrix} V_{2\beta} \\ I_{2\beta} \end{bmatrix} \tag{1.9d}$$

Thus, the cascade connection is simply the product of both **ABCD** matrices, explaining why this representation is sometimes called chain parameters.

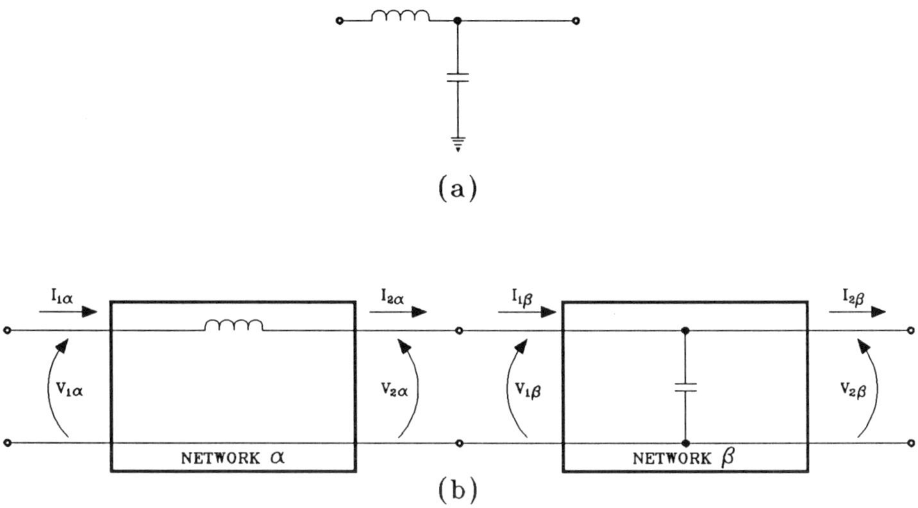

Figure 1.3 Cascade connection: (a) example; (b) voltage and current definition.

1.2.2 The Parallel Connection

The connection shown in Figure 1.4 is used to form an equivalent for two networks that are interconnected by placing each terminal of one network in parallel with the corresponding terminal of another network. The figure shows an example and the voltage and current definitions. Inspecting the input side, we see that V_1 is equal to the port voltage at both networks, and I_1 is the sum of the current flowing into the two networks, or

$$\begin{aligned} V_1 &= V_{1\alpha} = V_{1\beta} \\ I_1 &= I_{1\alpha} + I_{1\beta} \end{aligned} \tag{1.10a}$$

and from the output side,

$$\begin{aligned} V_2 &= V_{2\alpha} = V_{2\beta} \\ I_2 &= I_{2\alpha} + I_{2\beta} \end{aligned} \tag{1.10b}$$

The y-parameters are defined using the current as the dependent variable, and we can write

$$\begin{bmatrix} I_{1\alpha} \\ I_{2\alpha} \end{bmatrix} = \begin{bmatrix} y_{11\alpha} & y_{12\alpha} \\ y_{21\alpha} & y_{22\alpha} \end{bmatrix} \begin{bmatrix} V_{1\alpha} \\ V_{2\alpha} \end{bmatrix} \tag{1.10c}$$

for network α, and

$$\begin{bmatrix} I_{1\beta} \\ I_{2\beta} \end{bmatrix} = \begin{bmatrix} y_{11\beta} & y_{12\beta} \\ y_{21\beta} & y_{22\beta} \end{bmatrix} \begin{bmatrix} V_{1\beta} \\ V_{2\beta} \end{bmatrix} \tag{1.10d}$$

for network β. Using (1.10a) through (1.10d), we can express the current at the input and output ports as

$$\begin{bmatrix} I_1 \\ I_2 \end{bmatrix} = \begin{bmatrix} I_{1\alpha} \\ I_{2\alpha} \end{bmatrix} + \begin{bmatrix} I_{1\beta} \\ I_{2\beta} \end{bmatrix} = \begin{bmatrix} y_{11\alpha} & y_{12\alpha} \\ y_{21\alpha} & Y_{22\alpha} \end{bmatrix} \begin{bmatrix} V_{1\alpha} \\ V_{2\alpha} \end{bmatrix} + \begin{bmatrix} y_{11\beta} & y_{12\beta} \\ y_{21\beta} & y_{22\beta} \end{bmatrix} \begin{bmatrix} V_{1\beta} \\ V_{2\beta} \end{bmatrix} \tag{1.10e}$$

and then, since the voltage at the input port of both networks is equal to V_1 and the output port voltage is equal to V_2, we can collect terms:

$$\begin{bmatrix} I_1 \\ I_2 \end{bmatrix} = \left\{ \begin{bmatrix} y_{11\alpha} & y_{12\alpha} \\ y_{21\alpha} & y_{22\alpha} \end{bmatrix} + \begin{bmatrix} y_{11\beta} & y_{12\beta} \\ y_{21\beta} & y_{22\beta} \end{bmatrix} \right\} \begin{bmatrix} V_1 \\ V_2 \end{bmatrix} \tag{1.10f}$$

Therefore, two networks can be connected in parallel by simply adding the y-parameters of each. Mathematically, it does not matter which network is represented as α and which is β since the y-parameters add on a term-for-term basis.

1.2.3 The Series Connection

We can form a series connection by connecting the negative terminals of one network to the corresponding positive terminals of another network as shown in Figure 1.5. The port voltages for the overall network is the sum of the individual port voltages, and since the current flowing into and out of a terminal pair must be the same, port current for the combined network is equal to the current flowing in the corresponding port of each network. We can express these conditions as

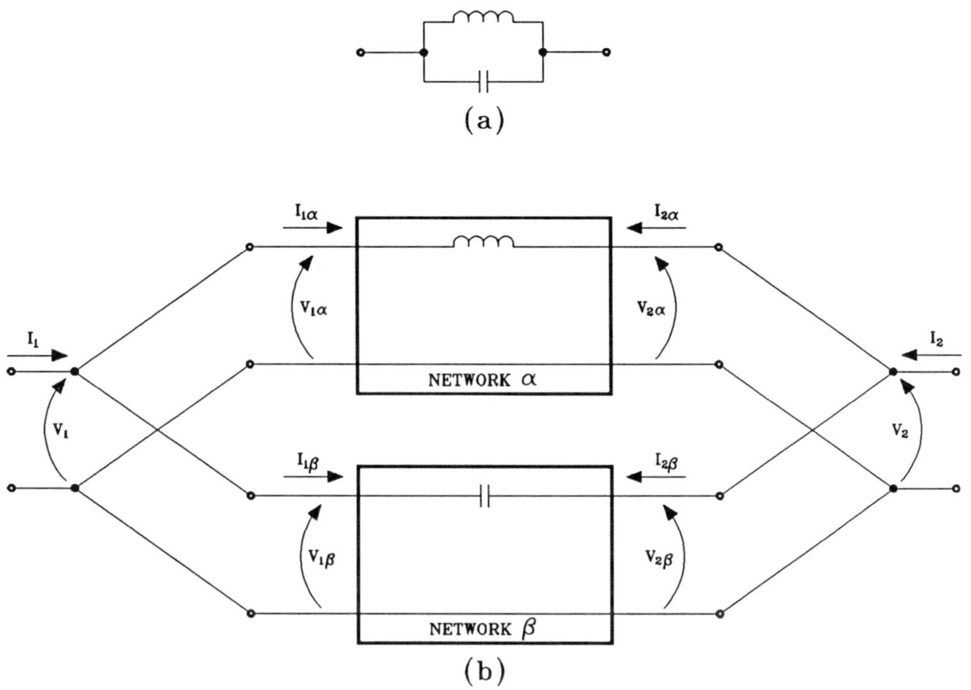

Figure 1.4 Parallel connection: (a) example; (b) voltage and current definition.

$$
\begin{aligned}
V_1 &= V_{1\alpha} + V_{1\beta} \\
I_1 &= I_{1\alpha} = I_{1\beta}
\end{aligned}
\tag{1.11a}
$$

for the input side, and

$$
\begin{aligned}
V_2 &= V_{2\alpha} + V_{2\beta} \\
I_2 &= I_{2\alpha} = I_{2\beta}
\end{aligned}
\tag{1.11b}
$$

for the output side. The *z*-parameters define the voltage in terms of current, and we can write

$$
\begin{bmatrix} V_{1\alpha} \\ V_{2\alpha} \end{bmatrix} = \begin{bmatrix} z_{11\alpha} & z_{12\alpha} \\ z_{21\alpha} & z_{22\alpha} \end{bmatrix} \begin{bmatrix} I_{1\alpha} \\ I_{2\alpha} \end{bmatrix}
\tag{1.11c}
$$

for network α, and

$$\begin{bmatrix} V_{1\beta} \\ V_{2\beta} \end{bmatrix} = \begin{bmatrix} z_{11\beta} & z_{12\beta} \\ z_{21\beta} & z_{22\beta} \end{bmatrix} \begin{bmatrix} I_{1\beta} \\ I_{2\beta} \end{bmatrix} \tag{1.11d}$$

for network β. Using (1.11a) through (1.11d), we can write the voltage and current relationship for the combined network as

$$\begin{bmatrix} V_1 \\ V_2 \end{bmatrix} = \begin{bmatrix} V_{1\alpha} \\ V_{2\alpha} \end{bmatrix} + \begin{bmatrix} V_{1\beta} \\ V_{2\beta} \end{bmatrix} = \begin{bmatrix} z_{11\alpha} & z_{12\alpha} \\ z_{21\alpha} & z_{22\alpha} \end{bmatrix} \begin{bmatrix} I_{1\alpha} \\ I_{2\alpha} \end{bmatrix} + \begin{bmatrix} z_{11\beta} & z_{12\beta} \\ z_{21\beta} & z_{22\beta} \end{bmatrix} \begin{bmatrix} I_{1\beta} \\ I_{2\beta} \end{bmatrix} \tag{1.11e}$$

Then, using the current equality from (1.11a) and (1.11b), we can write

$$\begin{bmatrix} V_1 \\ V_2 \end{bmatrix} = \left\{ \begin{bmatrix} z_{11\alpha} & z_{12\alpha} \\ z_{21\alpha} & z_{22\alpha} \end{bmatrix} + \begin{bmatrix} z_{11\beta} & z_{12\beta} \\ z_{21\beta} & z_{22\beta} \end{bmatrix} \right\} \begin{bmatrix} I_1 \\ I_2 \end{bmatrix} \tag{1.11f}$$

and we see that the series connection is simply the sum of the z-parameters.

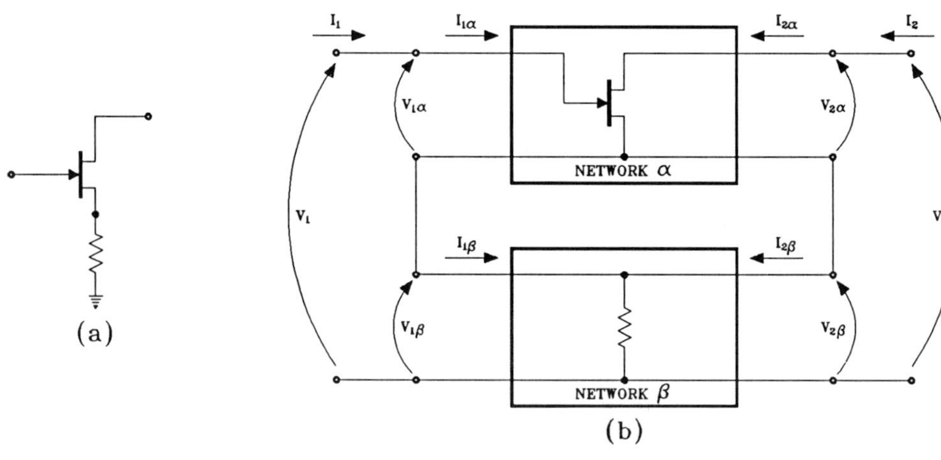

Figure 1.5 Series connection: (a) example; (b) voltage and current definition.

Since matrix addition is commutative, it does not matter which network is defined as α and which is β. However, this statement seems to be contradicted by the two networks shown in the upper half of Figure 1.6. The two resistors connected between the negative terminals of network α in Figure 1.6(a) is shorted, as is the field-effect transistor (FET) shown in Figure 1.6(b). The parallel connection in Figure 1.6(a) is based on the *y*- or short circuit parameters. If we short the output which forces V_2 to zero for both networks, then attach a source to the input port, we find that the current flowing in the positive and negative terminal pair of networks α and β is not equal, which violates one of the assumptions made during the derivation. The same is true of the network in Figure 1.6(b); however, we can augment the connections with an ideal one-to-one transformer, as shown in Figures 1.6(c) and (d), which takes care of this problem. The ideal transformer does not alter the network parameters, but it does provide the necessary isolation when computing the voltage and current for the combined network.

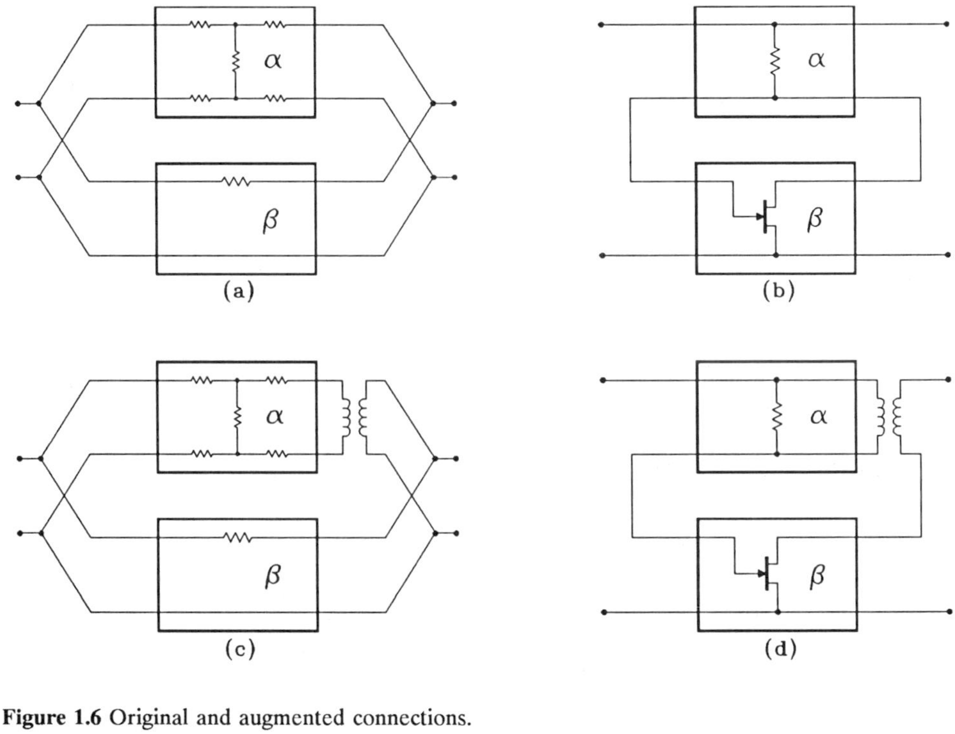

Figure 1.6 Original and augmented connections.

1.2.4 The Series-Parallel Connection

The arrangement shown in Figure 1.7 is known as a series-parallel connection and is formed by connecting the input terminals in series, and the output terminals in parallel. The example in Figure 1.7(a), which consists of a transmission line and resistor (usually 0Ω), is known as a Ruthroff transformer [1], which has excellent impedance-transforming properties across very broad bandwidths. The input current into both networks is identical, and the voltage at the input port of the combined network is the sum of the individual voltages, or

$$\begin{aligned} V_1 &= V_{1\alpha} + V_{1\beta} \\ I_1 &= I_{1\alpha} = I_{1\beta} \end{aligned} \tag{1.12a}$$

and at the output port,

$$\begin{aligned} V_2 &= V_{2\alpha} = V_{2\beta} \\ I_2 &= I_{2\alpha} + I_{2\beta} \end{aligned} \tag{1.12b}$$

We can write the input voltage and output current as

$$\begin{bmatrix} V_{1\alpha} \\ I_{2\alpha} \end{bmatrix} = \begin{bmatrix} h_{11\alpha} & h_{12\alpha} \\ h_{21\alpha} & h_{22\alpha} \end{bmatrix} \begin{bmatrix} I_{1\alpha} \\ V_{2\alpha} \end{bmatrix} \tag{1.12c}$$

for network α, and

$$\begin{bmatrix} V_{1\beta} \\ I_{2\beta} \end{bmatrix} = \begin{bmatrix} h_{11\beta} & h_{12\beta} \\ h_{21\beta} & h_{22\beta} \end{bmatrix} \begin{bmatrix} I_{1\beta} \\ V_{2\beta} \end{bmatrix} \tag{1.12d}$$

for network β. Then, from (1.12a) through (1.12d),

$$\begin{bmatrix} V_1 \\ I_2 \end{bmatrix} = \begin{bmatrix} V_{1\alpha} \\ I_{2\alpha} \end{bmatrix} + \begin{bmatrix} V_{1\beta} \\ I_{2\beta} \end{bmatrix} = \begin{bmatrix} h_{11\alpha} & h_{12\alpha} \\ h_{21\alpha} & h_{22\alpha} \end{bmatrix} \begin{bmatrix} I_{1\alpha} \\ V_{2\alpha} \end{bmatrix} + \begin{bmatrix} h_{11\beta} & h_{12\beta} \\ h_{21\beta} & h_{22\beta} \end{bmatrix} \begin{bmatrix} I_{1\beta} \\ V_{2\beta} \end{bmatrix} \tag{1.12e}$$

Using the voltage and current equality in (1.12a) and (1.12b), we can substitute and then combine terms:

$$\begin{bmatrix} V_1 \\ I_2 \end{bmatrix} = \left\{ \begin{bmatrix} h_{11\alpha} & h_{12\alpha} \\ h_{21\alpha} & h_{22\alpha} \end{bmatrix} + \begin{bmatrix} h_{11\beta} & h_{12\beta} \\ h_{21\beta} & h_{22\beta} \end{bmatrix} \right\} \begin{bmatrix} I_1 \\ V_2 \end{bmatrix} \tag{1.12f}$$

Equation (1.12f) shows that the series-parallel connection can be realized by adding the h-parameters of the individual two-port networks.

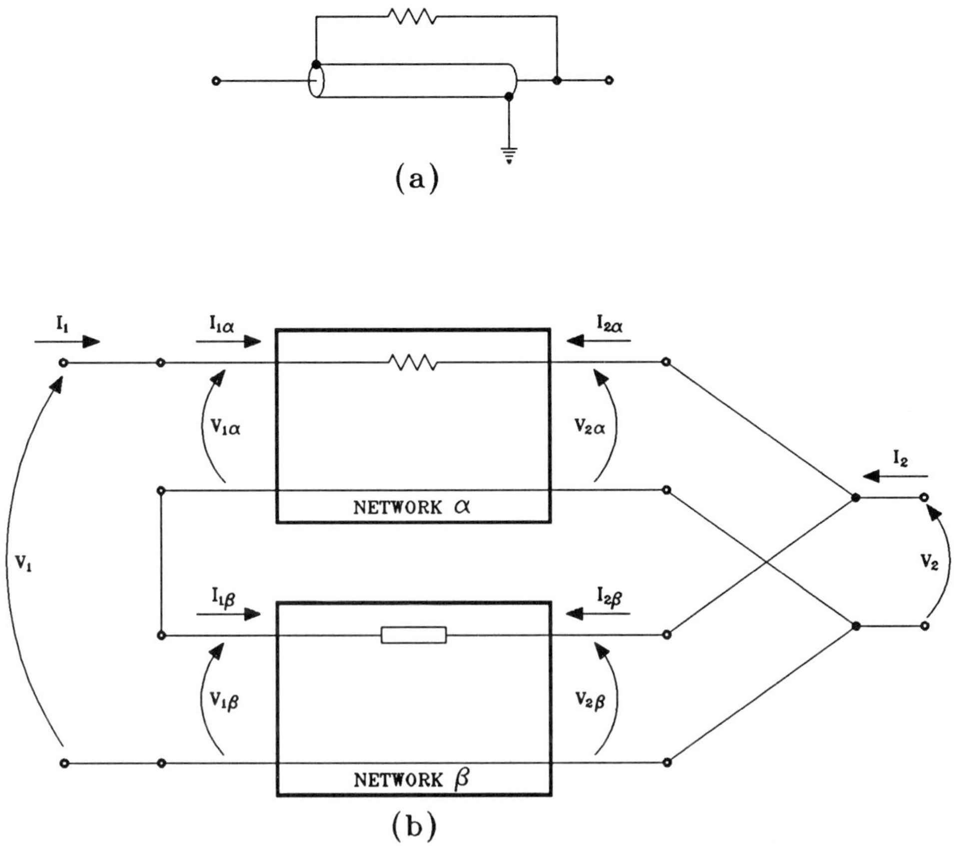

Figure 1.7 Series-parallel connection: (a) example; (b) voltage and current definitions.

1.2.5 The Parallel-Series Connection

The parallel-series connection shown in Figure 1.8 is simply the reverse of the series-parallel connection discussed in Section 1.2.4. The input terminals are connected in parallel, and the output terminals in series. We can write

$$V_1 = V_{1\alpha} = V_{1\beta}$$
$$I_1 = I_{1\alpha} + I_{1\beta} \tag{1.13a}$$

for the parallel input, and

$$\begin{aligned} V_2 &= V_{2\alpha} + V_{2\beta} \\ I_2 &= I_{2\alpha} = I_{2\beta} \end{aligned} \tag{1.13b}$$

for the series output connection. The input voltages and output currents are equal, and we can follow the same procedure as in Section 1.2.4, except that we use *g*-parameters instead of *h*-parameters. Writing equations for networks α and β, substituting into (1.13a) and (1.13b), and then simplifying the expression, gives us

$$\begin{bmatrix} I_1 \\ V_2 \end{bmatrix} = \left\{ \begin{bmatrix} g_{11\alpha} & g_{12\alpha} \\ g_{21\alpha} & g_{22\alpha} \end{bmatrix} + \begin{bmatrix} g_{11\beta} & g_{12\beta} \\ g_{21\beta} & g_{22\beta} \end{bmatrix} \right\} \begin{bmatrix} V_1 \\ I_2 \end{bmatrix} \tag{1.13c}$$

which shows that the *g*-parameters for two networks connected in parallel at the input ports and in series at the output ports are the sum of the *g*-parameters of the individual networks.

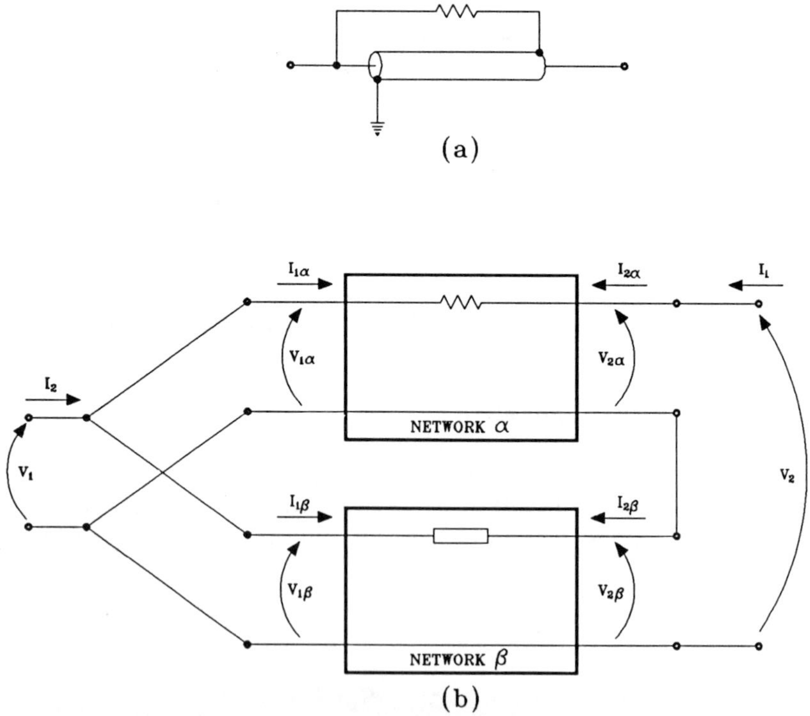

Figure 1.8 Parallel-series connection: (a) example; (b) voltage and current definitions.

1.3 *S*-PARAMETERS

The network parameters in Section 1.1 require open and short circuits to evaluate the coefficients. As we move higher in frequency, it becomes difficult to realize an open or short circuit, and the accuracy of any measurements is dependent on how well we can characterize the terminations. In addition, many active devices react to open and short circuit terminations by oscillating, and any measurements we make under these conditions are meaningless for linear circuit design.
One particularly useful network representation, which was developed to characterize microwave circuits, is *scattering*, or *s-parameters* [2-10]. The term *scattering* comes from the similarity between the behavior of light in the field of optics and microwave energy. The *s*-parameters can be measured using any convenient termination and, unlike other parameter sets, exist for most practical networks. Perhaps the most important feature is the ease and accuracy with which these parameters can be measured at very high frequencies and the direct physical interpretation of the coefficients.

Figure 1.9 shows a two-port network driven at port 1 by a $Z_1\Omega$ voltage source and terminated at port 2 by a $Z_2\Omega$ load. Z_1 and Z_2, or Z_i, in general, are known as the *reference impedances* and can be any value, although 50Ω is by far the most common. The voltage and current definitions are shown in the figure, and we can define two additional parameters[1] that are functions of V_i, I_i, and Z_i as

$$a_i = \frac{V_i + Z_i I_i}{2 \cdot \sqrt{|\mathrm{Re} Z_i|}} \tag{1.14a}$$

and

$$b_i = \frac{V_i - Z_i^* I_i}{2 \cdot \sqrt{|\mathrm{Re} Z_i|}} \tag{1.14b}$$

where Z_i^* is the complex conjugate of Z_i, and Re Z_i is the real component of the reference impedance. Essentially, the operation in (1.14) changes variables from the voltage and current parameters associated with other parameter sets. Selecting b_i as the dependent variable and a_i as the independent variable, we can write

$$\begin{aligned} b_1 &= s_{11}a_1 + s_{12}a_2 \\ b_2 &= s_{21}a_1 + s_{22}a_2 \end{aligned} \tag{1.15}$$

---

[1] Some authors define the impedance in the numerator as the square root of Z_i or R_i. However, this author prefers the definition after Bodway [11], Kurokawa [2], and Penfield [3].

for the two-port in Figure 1.9, or, in matrix notation,

$$\mathbf{b} = \mathbf{S}\mathbf{a} \tag{1.16}$$

for any n-port network.

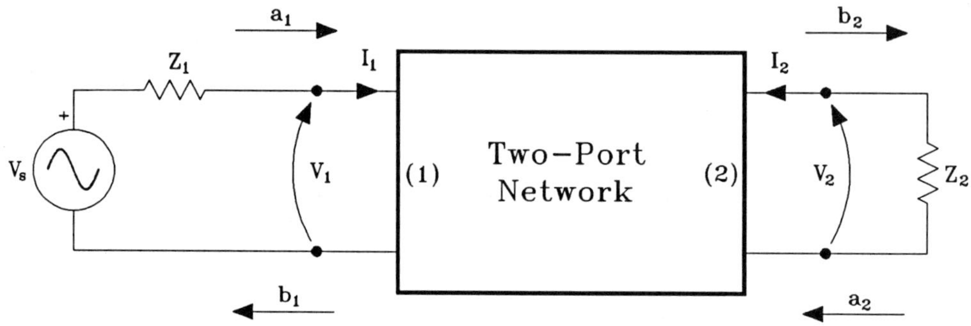

Figure 1.9 General two-port network.

We can evaluate the coefficients of the **S** matrix by setting $a_2 = 0$ and solving for s_{11} and s_{21}, then solving for s_{12} and s_{22} when a_1 is zero. How do we set a_i to zero? Using Figure 1.9, we see that the output voltage is $(-I_2Z_2)$, which can be substituted into (1.14a):

$$a_2 = \frac{V_2 + Z_2 I_2}{2 \cdot \sqrt{|\mathrm{Re} Z_2|}} = \frac{-I_2 Z_2 + Z_2 I_2}{2 \cdot \sqrt{|\mathrm{Re} Z_2|}} = 0$$

and the answer is that a_i is always zero at any port not connected to a source and terminated with the reference impedance. Thus, we can easily measure the s-parameters of any size network by connecting a source to one port at a time. However, we are still faced with the problem of actually measuring a_i and b_i, which requires further examination of these parameters.

From transmission line theory [12], we know that

$$V_i = V_{iI} + V_{iR}$$

and

$$I_i = \frac{V_{iI}}{Z_i} - \frac{V_{iR}}{Z_i}$$

where the subscripts I and R denote the incident and reflected components of voltage, respectively. Assuming Z_i is real, we can substitute into (1.14a)

$$a_i = \frac{V_i + Z_i I_i}{2 \cdot \sqrt{|\mathrm{Re} Z_i|}} = \frac{(V_{iI} + V_{iR}) + Z_i \left(\frac{V_{iI}}{Z_i} - \frac{V_{iR}}{Z_i} \right)}{2 \cdot \sqrt{|\mathrm{Re} Z_i|}} = \frac{V_{iI}}{\sqrt{|\mathrm{Re} Z_i|}}$$

and into (1.14b)

$$b_i = \frac{V_i - Z_i^* I_i}{2 \cdot \sqrt{|\mathrm{Re} Z_i|}} = \frac{(V_{iI} + V_{iR}) - Z_i^* \left(\frac{V_{iI}}{Z_i} - \frac{V_{iR}}{Z_i} \right)}{2 \cdot \sqrt{|\mathrm{Re} Z_i|}} = \frac{V_{iR}}{\sqrt{|\mathrm{Re} Z_i|}}$$

which reveals that a_i is a function of the incident voltage and b_i is a function of reflected voltage. Both parameters have units that are the square root of power, and

$$|a_i|^2 = \frac{|V_{iI}|^2}{|\mathrm{Re} Z_i|}$$

$$|b_i|^2 = \frac{|V_{iR}|^2}{|\mathrm{Re} Z_i|}$$

Thus, a_i is an incident wave, $|a_i|^2$ is incident power, b_i is a reflected wave, and $|b_i|^2$ is reflected power. From (1.15), we see that the reflected wave at each port is the sum of the incident waves from all ports modified by coefficients of the s-parameter matrix.

Using Figure 1.9, we can write $|a_1|^2$ as

$$|a_1|^2 = \left| \frac{V_1 + Z_1 \left(\frac{V_s - V_1}{Z_1} \right)}{2 \cdot \sqrt{|\mathrm{Re} Z_1|}} \right|^2 = \frac{|V_s|^2}{4 \cdot |\mathrm{Re} Z_1|}$$

and we find that $|a_1|^2$ is the power available from the source, or exchangeable power if Re Z_1 is negative. If we subtract the power reflected by the network from the power available from the source, we have

$$|a_i|^2 - |b_i|^2 = a_i a_i^* - b_i b_i^*$$

$$= \frac{(V_1 + Z_1 I_1)(V_1^* + Z_1^* I_1^*)}{4|\mathrm{Re}\, Z_1|} - \frac{(V_1 - Z_1^* I_1)(V_1^* - Z_1 I_1^*)}{4|\mathrm{Re}\, Z_1|}$$

$$= \frac{2 Z_1 (V_1 I_1^* + V_1^* I_1)}{4|\mathrm{Re}\, Z_1|} = \frac{Z_1}{|\mathrm{Re}\, Z_1|} \mathrm{Re}(V_1 I_1^*)$$

which is the power delivered to the network when Z_1 is positive real. When the source is connected to port 1, $|a_2|^2$ is zero and we can write $|b_2|^2$ as

$$|b_2|^2 = \left| \frac{V_2 - Z_2^* I_2}{2 \cdot \sqrt{|\mathrm{Re}\, Z_2|}} \right|^2 = |\mathrm{Re}\, Z_2| \cdot |I_2|^2$$

which is the power delivered to the load.

The s_{ij} coefficients of the s-parameter matrix are all ratios of reflected-to-incident waves, which turns out to be a very convenient representation for microwave circuits. When a generator with available power $|a_i|^2$ is connected to port i, we can measure a at port i, and b at all ports. At port i we have

$$s_{ii} = \frac{b_i}{a_i} = \frac{V_i - Z_i^* I_i}{V_i + Z_i I_i} = \frac{Z_{in} I_i - Z_i^* I_i}{Z_{in} I_i + Z_i I_i} = \frac{Z_{in} - Z_i^*}{Z_{in} + Z_i}$$

where Z_{in} is the impedance looking into port i. Thus,

$$s_{ii} = \Gamma_{in} = \textit{Reflection coefficient at port } i$$

and

$$|s_{ii}|^2 = \frac{|b_i|^2}{|a_i|^2} = \frac{\textit{Power reflected from the input}}{\textit{Power available from the generator}} = \textit{Return loss at port } i$$

At any port j, where $i \neq j$,

$$|s_{ji}|^2 = \frac{|b_j|^2}{|a_i|^2} = \frac{\textit{Power delivered to the load}}{\textit{Power available from the generator}} = \textit{Transducer power gain}$$

Mathematically, s-parameters are easily computed using techniques similar to those of the other parameter sets. A source is connected to port 1, voltages and currents are computed at all ports, and then s_{j1} is evaluated for $j = 1,2,...,n$. The source is moved to port 2, s_{j2} is evaluated, and so forth until n^2 parameters have been computed. The s-parameters are usually measured using a *vector network analyzer*. The basic network analyzer consists of a signal source, a directional coupler used to separate incident and reflected energy, and a means to measure the ratio and phase difference of reflected and incident power. Most modern network analyzers are computer controlled, use error correcting techniques to improve the accuracy of the measurements, and provide real-time displays of Smith, polar, and rectangular charts. Figure 1.10 shows a photograph of a microwave to millimeter wave automatic network analyzer (ANA). The upper section contains the display and data entry portion, the center section is used to select various display and measurement options using a menu-driven interface, and the lower section contains RF circuitry. The signal source, usually synthesized, is mounted underneath the small table since it is completely controlled by the ANA and does not require manual settings.

We can use the incident and reflected waves to define another network representation called the *scattering transfer* or *t-parameters*. These parameters are similar to the *ABCD*-parameters that use voltage and current from the same port. The t-parameters use either a_1 and b_1 or a_2 and b_2 as the dependent variables, and two networks can be cascaded by simply multiplying the two **T** matrices together. However, depending on the exact definition, the denominator is either s_{12} or s_{21}, which results in numerical problems for some practical networks. By the time we convert from **S** to **T**, multiply, and then convert back to **S**, there is little advantage in using t-parameters and they will not be discussed further.

1.3.1 *S*-Parameter Conversions

Although the s-parameters are the most convenient representation for microwave networks, it is sometimes necessary to convert to other parameter sets. Assuming the reference impedance is Z_o, we can add a_i and b_i:

$$2 \cdot \sqrt{|\mathrm{Re} Z_o|}\,(a_i + b_i) = (V_i + Z_o I_i) + (V_i - Z_o^* I_i) = 2V_i + I_i(Z_o - Z_o^*) \tag{1.17a}$$

and then subtract a_i and b_i:

$$2 \cdot \sqrt{|\mathrm{Re} Z_o|}\,(a_i - b_i) = (V_i + Z_o I_i) - (V_i - Z_o^* I_i) = I_i(Z_o + Z_o^*) \tag{1.17b}$$

Solving (1.17b) for I_i, we have

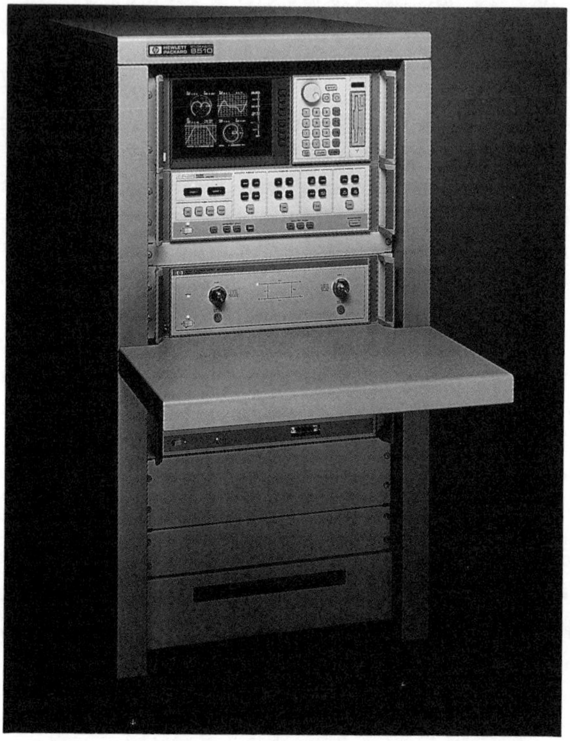

Figure 1.10 Photograph of Automatic Network Analyzer(courtesy of the Hewlett-Packard Company).

$$I_i = \frac{2 \cdot \sqrt{|\mathrm{Re}\, Z_o|}(a_i - b_i)}{(Z_o + Z_o^*)} \tag{1.17c}$$

and then substituting into (1.17a),

$$V_i = \frac{2 \cdot \sqrt{|\mathrm{Re}\, Z_o|}(a_i Z_o^* + b_i Z_o)}{(Z_o + Z_o^*)} \tag{1.17d}$$

When Z_o is real, we can write (1.17c) as

$$I_i = \frac{\sqrt{|Z_o|}}{Z_o} \cdot (a_i - b_i) \tag{1.17e}$$

and (1.17d) as

$$V_i = \sqrt{|Z_o|} \cdot (a_i + b_i) \tag{1.17f}$$

If we also assume $Z_i = Z_o$ for all i (i.e., the reference impedance is the same at all ports) and Z_o is positive real, we can write the voltage and current relationship for an n-port network in terms of the a and b variables:

$$\mathbf{V} = \sqrt{Z_o}(\mathbf{a} + \mathbf{b})$$

$$\mathbf{I} = \frac{1}{\sqrt{Z_o}}(\mathbf{a} - \mathbf{b})$$

The z-parameters are

$$\mathbf{V} = \mathbf{Z}\mathbf{I}$$

or, in terms of incident and reflected waves,

$$\sqrt{Z_o}(\mathbf{a} + \mathbf{b}) = \mathbf{Z} \cdot \frac{1}{\sqrt{Z_o}}(\mathbf{a} - \mathbf{b})$$

Dividing by the square root of the reference impedance and rearranging to move b terms to the left side,

$$\mathbf{b} + \frac{1}{Z_o}\mathbf{Z}\mathbf{b} = \frac{1}{Z_o}\mathbf{Z}\mathbf{a} - \mathbf{a}$$

$$\left(\frac{1}{Z_o}\mathbf{Z} + \mathbf{U}\right)\mathbf{b} = \left(\frac{1}{Z_o}\mathbf{Z} - \mathbf{U}\right)\mathbf{a}$$

where **U** is the unit matrix which has elements equal to one on the main diagonal and zero elsewhere. We can premultiply both sides by

$$\left(\frac{1}{Z_o}\mathbf{Z} + \mathbf{U}\right)^{-1}$$

which yields

$$\mathbf{b} = \left(\frac{1}{Z_o}\mathbf{Z} + \mathbf{U}\right)^{-1}\left(\frac{1}{Z_o}\mathbf{Z} - \mathbf{U}\right)\mathbf{a}$$

Comparing this equation with (1.16), we see that

$$\mathbf{S} = \left(\frac{1}{Z_o}\mathbf{Z} + \mathbf{U}\right)^{-1}\left(\frac{1}{Z_o}\mathbf{Z} - \mathbf{U}\right) \tag{1.18a}$$

Using a similar approach, we can convert from **S** to **Z**,

$$\frac{1}{Z_o}\mathbf{Z} = (\mathbf{U} - \mathbf{S})^{-1}(\mathbf{U} + \mathbf{S}) \tag{1.18b}$$

Y to **S**,

$$\mathbf{S} = \left(\mathbf{U} + Z_o\mathbf{Y}\right)^{-1}\left(\mathbf{U} - Z_o\mathbf{Y}\right) \tag{1.18c}$$

and **S** to **Y**,

$$Z_o\mathbf{Y} = (\mathbf{U} + \mathbf{S})^{-1}(\mathbf{U} - \mathbf{S}) \tag{1.18d}$$

The *z*- and *y*-parameters are defined using either the voltage or current from each port as the dependent variable, and they are valid for *n*-port networks. However, the other parameter sets are defined using both voltage and current, which is generally not as useful for *n*-port representation. For example, *h*-parameters are defined using V_1 and I_2 as dependent variables for a two-port. How would we write (1.13a) for a three-port network? Nevertheless, most networks are two-ports and conversion between the *s*-parameters and other sets can be useful. We can use either (1.18) with the parameter conversions in Figure 1.2 or the relationship in (1.17), along with the definition of each parameter set to derive the necessary equations. Figure 1.11 shows the conversions to and from *s*-parameters and the other parameter sets.

Except for *ABCD*-parameters, there is significant commonality between all conversions shown in the figure, and it is possible to develop a simple algorithm that can be implemented in compact form and used with a programmable calculator or manual calculations. If we define Δ as

$$\Delta = \frac{-2}{\left(1 + s_{11}\right)\left(1 + s_{22}\right) - s_{12}s_{21}}$$

	From S Parameters		To S Parameters	
Z	$z_{11} = \frac{(1+s_{11})(1-s_{22})+s_{12}s_{21}}{(1-s_{11})(1-s_{22})-s_{12}s_{21}}$	$z_{12} = \frac{2s_{12}}{(1-s_{11})(1-s_{22})-s_{12}s_{21}}$	$s_{11} = \frac{(z_{11}-1)(1+z_{22})-z_{12}z_{21}}{(1+z_{11})(1+z_{22})-z_{12}z_{21}}$	$s_{12} = \frac{2z_{12}}{(1+z_{11})(1+z_{22})-z_{12}z_{21}}$
	$z_{21} = \frac{2s_{21}}{(1-s_{11})(1-s_{22})-s_{12}s_{21}}$	$z_{22} = \frac{(1-s_{11})(1+s_{22})+s_{12}s_{21}}{(1-s_{11})(1-s_{22})-s_{12}s_{21}}$	$s_{21} = \frac{2z_{21}}{(1+z_{11})(1+z_{22})-z_{12}z_{21}}$	$s_{22} = \frac{(1+z_{11})(z_{22}-1)-z_{12}z_{21}}{(1+z_{11})(1+z_{22})-z_{12}z_{21}}$
Y	$y_{11} = \frac{(1-s_{11})(1+s_{22})+s_{12}s_{21}}{(1+s_{11})(1+s_{22})-s_{12}s_{21}}$	$y_{12} = \frac{-2s_{12}}{(1+s_{11})(1+s_{22})-s_{12}s_{21}}$	$s_{11} = \frac{(1-y_{11})(1+y_{22})+y_{12}y_{21}}{(1+y_{11})(1+y_{22})-y_{12}y_{21}}$	$s_{12} = \frac{-2y_{12}}{(1+y_{11})(1+y_{22})-y_{12}y_{21}}$
	$y_{21} = \frac{-2s_{21}}{(1+s_{11})(1+s_{22})-s_{12}s_{21}}$	$y_{22} = \frac{(1+s_{11})(1-s_{22})+s_{12}s_{21}}{(1+s_{11})(1+s_{22})-s_{12}s_{21}}$	$s_{21} = \frac{-2y_{21}}{(1+y_{11})(1+y_{22})-y_{12}y_{21}}$	$s_{22} = \frac{(1+y_{11})(1-y_{22})+y_{12}y_{21}}{(1+y_{11})(1+y_{22})-y_{12}y_{21}}$
G	$g_{11} = \frac{(1-s_{11})(1-s_{22})-s_{12}s_{21}}{(1+s_{11})(1-s_{22})+s_{12}s_{21}}$	$g_{12} = \frac{-2s_{12}}{(1+s_{11})(1-s_{22})+s_{12}s_{21}}$	$s_{11} = \frac{(1-g_{11})(1+g_{22})+g_{12}g_{21}}{(1+g_{11})(1+g_{22})-g_{12}g_{21}}$	$s_{12} = \frac{-2g_{12}}{(1+g_{11})(1+g_{22})-g_{12}g_{21}}$
	$g_{21} = \frac{2s_{21}}{(1+s_{11})(1-s_{22})+s_{12}s_{21}}$	$g_{22} = \frac{(1+s_{11})(1+s_{22})-s_{12}s_{21}}{(1+s_{11})(1-s_{22})+s_{12}s_{21}}$	$s_{21} = \frac{2g_{21}}{(1+g_{11})(1+g_{22})-g_{12}g_{21}}$	$s_{22} = \frac{(1+g_{11})(g_{22}-1)-g_{12}g_{21}}{(1+g_{11})(1+g_{22})-g_{12}g_{21}}$
H	$h_{11} = \frac{(1+s_{11})(1+s_{22})-s_{12}s_{21}}{(1-s_{11})(1+s_{22})+s_{12}s_{21}}$	$h_{12} = \frac{2s_{12}}{(1-s_{11})(1+s_{22})+s_{12}s_{21}}$	$s_{11} = \frac{(h_{11}-1)(1+h_{22})-h_{12}h_{21}}{(1+h_{11})(1+h_{22})-h_{12}h_{21}}$	$s_{12} = \frac{2h_{12}}{(1+h_{11})(1+h_{22})-h_{12}h_{21}}$
	$h_{21} = \frac{-2s_{21}}{(1-s_{11})(1+s_{22})+s_{12}s_{21}}$	$h_{22} = \frac{(1-s_{11})(1-s_{22})-s_{12}s_{21}}{(1-s_{11})(1+s_{22})+s_{12}s_{21}}$	$s_{21} = \frac{-2h_{21}}{(1+h_{11})(1+h_{22})-h_{12}h_{21}}$	$s_{22} = \frac{(1+h_{11})(1-h_{22})+h_{12}h_{21}}{(1+h_{11})(1+h_{22})-h_{12}h_{21}}$
ABCD	$A = \frac{(1+s_{11})(1-s_{22})+s_{12}s_{21}}{2s_{21}}$	$B = \frac{(1+s_{11})(1+s_{22})-s_{12}s_{21}}{2s_{12}}$	$s_{11} = \frac{A+B-C-D}{A+B+C+D}$	$s_{12} = \frac{2(AD-BC)}{A+B+C+D}$
	$C = \frac{(1-s_{11})(1-s_{22})-s_{12}s_{21}}{2s_{21}}$	$D = \frac{(1-s_{11})(1+s_{22})+s_{12}s_{21}}{2s_{12}}$	$s_{21} = \frac{2}{A+B+C+D}$	$s_{22} = \frac{-A+B-C+D}{A+B+C+D}$

Figure 1.11 Two-port *S*-parameter conversions.

then the s to y conversion can be written as

$$\mathbf{Y} = \begin{bmatrix} -\{\Delta(1 + s_{22}) + 1\} & \Delta \cdot s_{12} \\ \Delta \cdot s_{21} & -\{\Delta(1 + s_{11}) + 1\} \end{bmatrix}$$

Examining Figure 1.11, we see that by making minor changes and appropriate substitutions, this one set of equations can be used for other conversions:

Y ⇒ **S**:	use **S** ⇒ **Y**.
S ⇒ **Z**:	replace **S** with **-S**, then use **S** ⇒ **Y**.
Z ⇒ **S**:	use **S** ⇒ **Y**, then replace **Y** with **-Y**.
S ⇒ **H**:	replace s_{11} with $-s_{11}$ and s_{12} with $-s_{12}$, then use **S** ⇒ **Y**.
H ⇒ **S**:	use **S** ⇒ **Y**, then replace s_{11} with $-s_{11}$ and s_{12} with $-s_{12}$.
S ⇒ **G**:	replace s_{22} with $-s_{22}$ and s_{21} with $-s_{21}$, then use **S** ⇒ **Y**.
G ⇒ **S**:	use **S** ⇒ **Y**, then replace s_{22} with $-s_{22}$ and s_{21} with $-s_{21}$.

We can convert between any of the parameter sets using s-parameters as an intermediate step (e.g., **Z** to **S** then **S** to **Y**). However, by using matrix inversion, we can add four additional conversions:

Z ⇒ **Y**:	invert.
Y ⇒ **Z**:	invert.
H ⇒ **G**:	invert.
G ⇒ **H**:	invert.

1.3.2 Generalized *S*-Parameters

Network analyzers are usually constructed using 50Ω components; therefore, most of the s-parameter data we obtain from measurements or furnished by manufacturers is normalized to 50Ω at all ports. We found in Section 1.3 that s_{ii} is the input reflection coefficient and s_{ji} is the transducer gain from port i to port j. However, the s-parameters are only valid when all ports are terminated with their respective reference impedance. Using 50Ω as an example, s_{11} is the input reflection coefficient of a two-port network when the output is terminated in 50Ω, and s_{21} is the transducer gain when driven at port 1 from a 50Ω source and measured into a 50Ω load connected at port 2. Very often we need to know the s-parameters when the network is connected to complex terminations, which may in fact be different at each port.

Figure 1.12(a) shows port k of an n-port network. All ports are referenced to Z_o, and we can express the incident and reflected waves as

$$a_k = \frac{V_k + Z_o I_k}{2 \cdot \sqrt{|\text{Re}\, Z_o|}} \tag{1.19a}$$

$$b_k = \frac{V_k - Z_o^* I_k}{2 \cdot \sqrt{|\text{Re}\, Z_o|}} \tag{1.19b}$$

We can analyze an arbitrary termination Z_k by connecting a source with internal impedance Z_o to the one-port, as shown in Figure 1.12(b). When the termination is connected to port k, $V_1 = V_k$, $I_1 = -I_k$, and we can write the incident and reflected waves as

$$a_1 = \frac{V_k - Z_o I_k}{2 \cdot \sqrt{|\text{Re}\, Z_o|}} \tag{1.20a}$$

$$b_1 = \frac{V_k + Z_o^* I_k}{2 \cdot \sqrt{|\text{Re}\, Z_o|}} \tag{1.20b}$$

Comparing (1.20) with (1.19), we see that when $Z_o = Z_o^*$, $a_1 = b_k$ and $b_1 = a_k$, or the reflected wave from port k enters the termination as an incident wave, and the reflected wave from the termination is sent back into port k as an incident wave. The condition $Z_o = Z_o^*$ requires Z_o to be real; however, if we normalize the one-port in Figure 1.12(b) to Z_o^*, then $a_1 = b_k$ and $b_1 = a_k$ for any Z_o. Defining the reflection coefficient at port k,

$$\Gamma_k = \frac{Z_k - Z_o}{Z_k + Z_o^*} \tag{1.21a}$$

we can use (1.19) and (1.20) to relate the incident and reflected waves at port k by

$$a_k = \Gamma_k b_k \tag{1.21b}$$

We assume the two-port network shown in Figure 1.9 is initially terminated at both ports by a $Z_o\Omega$ reference impedance, and wish to compute the input reflection coefficient when port 2 is terminated with Z_L. From (1.21) we have $a_2 = \Gamma_L b_2$, and substituting a_2 into (1.15) gives us

$$b_1 = s_{11} a_1 + s_{12}(\Gamma_L b_2) \tag{1.22a}$$

$$b_2 = s_{21}a_1 + s_{22}(\Gamma_L b_2) \tag{1.22b}$$

Solving (1.22b) for b_2, we have

$$b_2 = \frac{s_{21}}{1 - s_{22}\Gamma_L} a_1$$

which can be substituted into (1.22a)

$$\frac{b_1}{a_1} = s'_{11} = s_{11} + \frac{s_{12}s_{21}\Gamma_L}{1 - s_{22}\Gamma_L} \tag{1.23}$$

S-parameters referenced to arbitrary terminations at each port are known as *generalized scattering parameters*. The expression used to convert from one set of reference impedances to another was derived by Youla and Paterna [13] for positive real Z_i, and later by Kurokawa [2] for any Z_i. The generalized form of the transformation is

$$\mathbf{S}' = \mathbf{A}^{-1}(\mathbf{S} - \mathbf{\Gamma}^{*T})(\mathbf{U} - \mathbf{\Gamma}\mathbf{S})^{-1}\mathbf{A}^{*T} \tag{1.24a}$$

where $\mathbf{A}$ and $\mathbf{\Gamma}$ are diagonal, and

$$A_i = \left(1 - \Gamma_i^*\right)\frac{\sqrt{1 - |\Gamma_i|^2}}{|1 - \Gamma_i|} \tag{1.24b}$$

$$\Gamma_i = \frac{Z'_i - Z_i}{Z'_i + Z_i^*} \tag{1.24c}$$

The new set of reference impedance values and transformed set of s-parameters are symbolized by the primed variables in (1.24). The notation $\mathbf{A}^{*T}$ indicates the transpose of a matrix formed by replacing each element of $\mathbf{A}$ with its corresponding complex conjugate. The reflection coefficient Γ_i in (1.24c) is referenced to the complex conjugate of Z_i in contrast with s_{ii}, which is referenced to Z_i. For a two-port network, (1.24) reduces to

$$s'_{11} = \frac{A_1^*}{A_1} \cdot \frac{\left(1 - \Gamma_L s_{22}\right)\left(s_{11} - \Gamma_S^*\right) + \Gamma_L s_{12}s_{21}}{\left(1 - \Gamma_S s_{11}\right)\left(1 - \Gamma_L s_{22}\right) - \Gamma_S\Gamma_L s_{12}s_{21}} \tag{1.25a}$$

$$s'_{12} = \frac{A_2^*}{A_1} \cdot \frac{s_{12}\left(1 - |\Gamma_S|^2\right)}{\left(1 - \Gamma_S s_{11}\right)\left(1 - \Gamma_L s_{22}\right) - \Gamma_S\Gamma_L s_{12}s_{21}} \tag{1.25b}$$

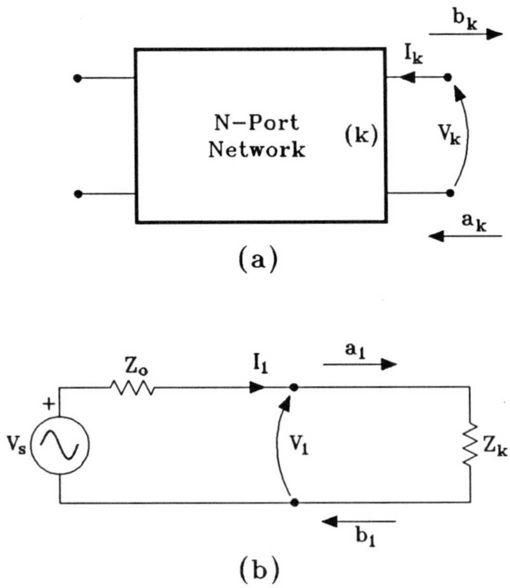

Figure 1.12 (a) Port k of n-port network; (b) one-port termination.

$$s'_{21} = \frac{A_1^*}{A_2} \cdot \frac{s_{21}\left(1 - |\Gamma_L|^2\right)}{(1 - \Gamma_S s_{11})(1 - \Gamma_L s_{22}) - \Gamma_S \Gamma_L s_{12} s_{21}} \tag{1.25c}$$

$$s'_{22} = \frac{A_2^*}{A_1} \cdot \frac{(1 - \Gamma_S s_{11})(s_{22} - \Gamma_L^*) + \Gamma_S s_{12} s_{21}}{(1 - \Gamma_S s_{11})(1 - \Gamma_L s_{22}) - \Gamma_S \Gamma_L s_{12} s_{21}} \tag{1.25d}$$

where the subscript S refers to the source and L to the load termination, or ports 1 and 2, respectively.

As an example, we can use (1.25) to compute the input reflection coefficient for an arbitrary load termination. Using (1.24c), $\Gamma_S = 0$ and Γ_L is the reflection coefficient of Z_L referenced to Z_o^*, and from (1.24b), $A_1 = 1$. Substituting into (1.25a), we have

$$s'_{11} = \frac{1}{1} \cdot \frac{(1 - \Gamma_L s_{22})(s_{11} - 0) + \Gamma_L s_{12} s_{21}}{(1 - 0 \cdot s_{11})(1 - \Gamma_L s_{22}) - 0 \cdot \Gamma_L s_{12} s_{21}} = s_{11} + \frac{s_{12} s_{21} \Gamma_L}{1 - s_{22} \Gamma_L}$$

which is the same result shown by (1.23).

When Z_o is positive real, we can write the normalized port voltage and currents from (1.17) as

$$V_{iN} = \frac{V_i}{\sqrt{Z_o}} \qquad\qquad I_{iN} = \sqrt{Z_o}\, I_i$$

Then, if we renormalize port i to Z', V_i and I_i remain unchanged, but the normalized values become

$$V'_{iN} = \frac{V_i}{\sqrt{Z'}} \qquad\qquad I'_{iN} = \sqrt{Z'}\, I_i$$

or

$$V'_{iN} = \sqrt{\frac{Z_o}{Z'}}\, V_{iN} \qquad\qquad I'_{iN} = \sqrt{\frac{Z'}{Z_o}}\, I_{iN}$$

which we recognize as the equation for an ideal transformer. Thus, at least for a positive real reference impedance, we can visualize normalization as ideal transformers connected at each port with turns ratios such that the original reference is transformed into the new value, as shown in Figure 1.13.

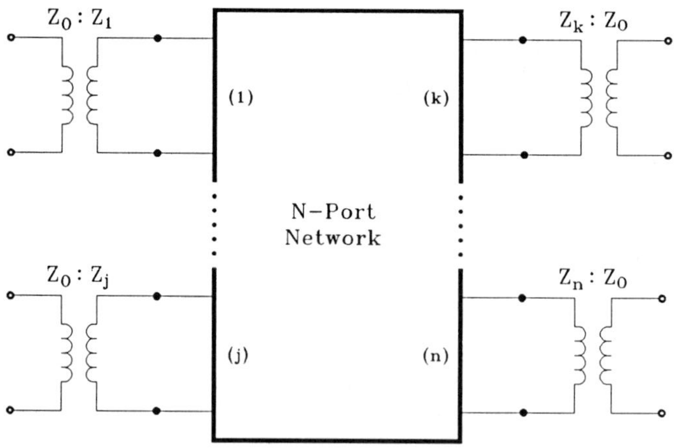

Figure 1.13 Physical n-port normalization.

1.3.3 Signal Flow Graphs

A signal flow graph is a visual aid for writing and solving network equations that can be used to simplify *s*-parameter calculations. Mason [14,15] developed flow graphs for linear circuit analysis, and later other workers [16] showed the usefulness of the technique applied to scattering parameter analysis.

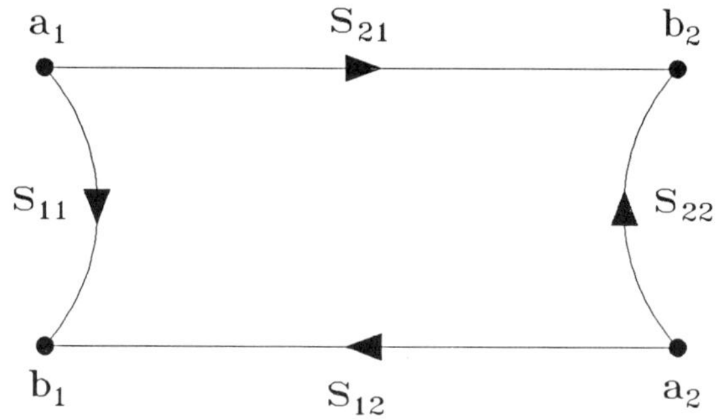

Figure 1.14 Two-port signal flow graph.

Signal flow graphs are constructed using one simple rule: *each variable becomes a node and each constant becomes a directed branch running from an independent to a dependent node.* Figure 1.14 shows the signal flow graph for a two-port network. The flow graph has two independent nodes (a_1 and a_2), two dependent nodes (b_1 and b_2), and four directed branches running from a_i to b_i. Each branch is assigned a constant that relates the dependent and independent nodes. For example, s_{11} is the ratio of b_1 to a_1 when a_2 is zero, and we see that the branch running from a_1 to b_1 is assigned a value of s_{11}. We interpret the branches as paths for the incident and reflected waves, and *the value of a wave at any node can be found by summing the adjusted wave coming through each of its branches from other nodes.*

The network shown in Figure 1.14 has two ports formed by nodes a_1 - b_1 and a_2 - b_2. A wave incident at port 1 flows into node a_1 and is split between two branches, s_{11} and s_{21}. The wave arriving at b_1 has a value of $s_{11}a_1$, and emerges from port 1 as a reflected wave. The wave following the path from a_1 to b_2 emerges from port 2 as a reflected wave, $s_{21}a_1$. Summing the adjusted waves arriving at b_1, we have

$$b_1 = s_{11}a_1 + s_{12}a_2$$

and at b_2 we have

$$b_2 = s_{21}a_1 + s_{22}a_2$$

which is equation (1.15).

We use signal flow graphs to solve for network parameters in exactly the same manner as the more traditional methods. That is, we usually connect a source at one port, terminate the other ports, and then solve for the ratio of reflected to incident waves. However, the source and termination must be represented by an equivalent flow graph. Figure 1.15(a) shows a Thevenin equivalent voltage source, which consists of an ideal voltage source V_s and internal impedance Z_s. Using Z_o as the reference, we can write the port voltage as

$$V_i = V_s + Z_s I_i$$

Substituting (1.17c) and (1.17d) for I_i and V_i,

$$\frac{2\cdot\sqrt{|\mathrm{Re}Z_o|}\left(a_i Z_o^* + b_i Z_o\right)}{\left(Z_o + Z_o^*\right)} = V_s + \frac{2Z_s\sqrt{|\mathrm{Re}Z_o|}\left(a_i - b_i\right)}{\left(Z_o + Z_o^*\right)}$$

then solving for b_i,

$$b_i = \frac{Re\,Z_o\cdot V_s}{\sqrt{|\mathrm{Re}Z_o|}\left(Z_s + Z_o\right)} + \frac{\left(Z_s - Z_o^*\right)}{\left(Z_s + Z_o\right)}a_i \tag{1.26a}$$

or

$$b_i = b_s + \Gamma_s a_i \tag{1.26b}$$

We see that the generator has two branches as shown in Figure 1.15(a). One branch connects an independent node called b_s, which is a function of V_s and Z_s, to the dependent node b_i, and has a coefficient of unity. The second branch connects independent node a_i to dependent node b_i and has a coefficient of Γ_S.

The current flowing through the load shown in Figure 1.15(b) is (V_i/Z_L), which we can substitute into (1.14):

$$2\cdot\sqrt{|\mathrm{Re}Z_o|}\,a_i = V_i\left(1 + \frac{Z_o}{Z_L}\right)$$

$$2\cdot\sqrt{|\mathrm{Re}Z_o|}\,b_i = V_i\left(1 - \frac{Z_o^*}{Z_L}\right)$$

Solving for V_i and then equating terms, we have

$$b_i = \frac{Z_L - Z_o^*}{Z_L + Z_o} a_i \qquad (1.27a)$$

or

$$b_i = \Gamma_L a_i \qquad (1.27b)$$

which results in the signal flow graph shown in Figure 1.15(b).

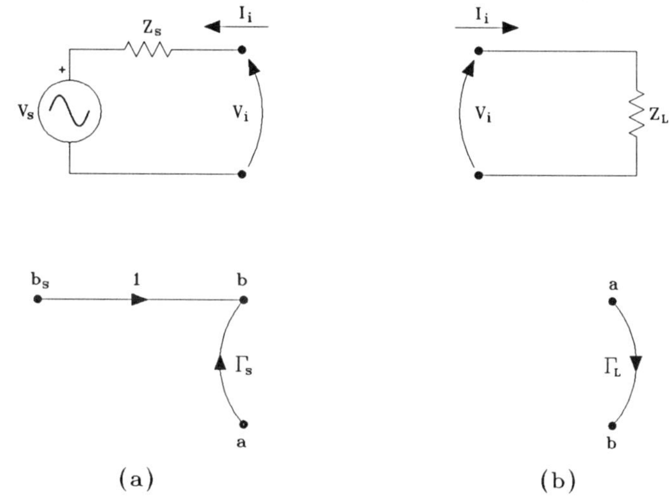

Figure 1.15 Circuit and signal flow graph for (a) generator; (b) load.

The *nontouching loop rule* [14,15] provides a means of expressing the ratio of dependent variables at any node to independent variables at any node. The ratio of output to input variable T is given by

$$T = \frac{\sum_k (P_k \cdot \Delta_k)}{\Delta} = \frac{\sum_k \left(P_k \cdot \sum_{i=0}^{n} (-1)^i \cdot L_i^k \right)}{\sum_{i=0}^{n} (-1)^i \cdot L_i} \qquad (1.28)$$

$$T = \frac{P_1\{1 - \sum L_1^1 + \sum L_2^1 - \sum L_3^1 + \cdots\} + P_2\{1 - \sum L_1^2 + \sum L_2^2 - \sum L_3^2 + \cdots\} + \cdots + P_k \cdot \Delta_k}{1 - \sum L_1 + \sum L_2 - \sum L_3 + \sum L_4 - \sum L_5 + \cdots}$$

where

P_k = the product of coefficients in a forward path from the independent to dependent variable.
L_0 = 1.
L_1 = first-order loops.
L_2 = second-order loops.
L_i = ith-order loops.
L_0^k = 1.
L_1^1 = first-order loops not touching path 1.
L_1^2 = first-order loops not touching path 2.
L_i^k = i^{th}-order loops not touching path k.
n = highest-order loop.

A forward path is formed by following a series of directed branches from the input node to the output node such that no node is encountered more than once and the gain is the product of branch coefficients in the path. A first-order loop is a forward path that begins and ends at the same node, where no other nodes may be encountered more than once, and where loop gain is the product of each branch in the loop. A second-order loop is the product of two first-order loops that do not touch or have nodes in common, a third-order loop is the product of three first-order nontouching loops, and so forth. The nontouching loop rule is not as formidable as it may appear, and an example will help illustrate its usefulness.

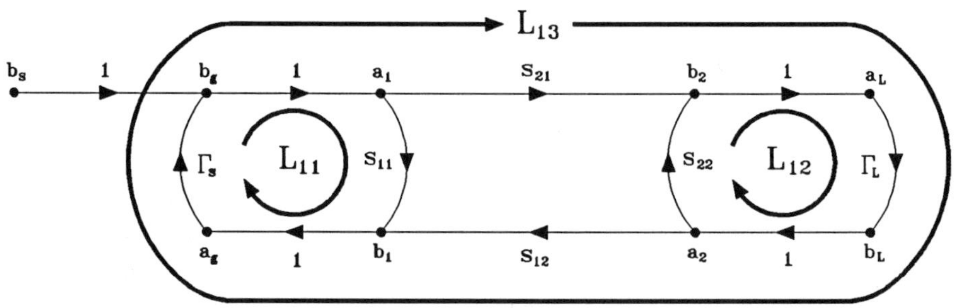

Figure 1.16 Flow graph of generator, two-port, and load.

Figure 1.16 shows the signal flow graph equivalent of Figure 1.9, which is a two-port with a generator connected at port 1 and a load connected at port 2. As shown, the generator and load are connected through unity branches, which simply means that the nodes at either end of the branch are equal, or the two nodes are connected. However, in Section 1.3.2 we found that the incident and reflected waves can only be equal when $Z_o = Z_o^*$ or Z_o is real. Rather than renormalize portions of the flow graph or use nonunity branches, we will reference all coefficients to a positive real Z_o and use (1.24) when necessary to compute generalized scattering parameters.

Examining Figure 1.16, we see there are three first-order loops:

$$L_{11} = s_{11}\Gamma_s$$

$$L_{12} = s_{22}\Gamma_L$$

$$L_{13} = s_{12}s_{21}\Gamma_s\Gamma_L$$

The three loops describe the mechanism whereby the waves are reflected and re-reflected at various interfaces. Loop L_{11} is formed by the wave emerging from the generator, reflecting back into the generator from port 1, and then re-reflecting into the network, where it is added to the incident wave. L_{12} is the wave emerging from port 2 and reflecting off the load back into port 2, where a portion is re-reflected into the termination. The large loop L_{13} results when the wave emerging from the generator passes through the network into the load, where it is reflected back into port 2, again passes through the network, then back into the generator, where it is re-reflected into the network. There are only two first-order loops that do not touch; thus we have only one second-order loop:

$$L_{21} = L_{11}\cdot L_{12} = s_{11}s_{22}\Gamma_s\Gamma_L$$

We can write the input reflection coefficient as

$$s_{11}' = \frac{b_1}{a_1}$$

and we need only use the nontouching loop rule to compute a_1 and b_1 from the flow graph. Using b_s as the input node, we follow the path to b_g then a_1. There is no other path from b_s to a_1 where a node is encountered only once. L_{12} is the only loop that does not touch at least part of this path. By inspection we can write the ratio of a_1 to b_s as

$$\frac{a_1}{b_s} = \frac{P_1\left\{1 - \sum L_1^1\right\}}{\Delta} = \frac{P_1\{1 - L_{12}\}}{\Delta} = \frac{1\cdot\{1 - s_{22}\Gamma_L\}}{\Delta} = \frac{1 - s_{22}\Gamma_L}{\Delta}$$

where Δ is defined in (1.28) and is one minus the sum of all odd-order loops plus the sum of all even-order loops. Turning our attention to b_1, we see that there are two paths from b_s. We can follow a path from b_s to a_1 and then down the s_{11}

branch to b_1. This path has a gain of s_{11}, and only first-order loop L_{12} does not touch. Thus, the first term in the numerator of (1.28) is

$$P_1 \cdot \Delta_1 = s_{11}(1 - L_{12}) = s_{11}(1 - s_{22}\Gamma_L)$$

The second path begins at b_s, continues through the network to a_L, emerges from the load at b_L, then back through the network to b_1. This path has a gain of $s_{12} \bullet s_{21} \bullet \Gamma_L$, and all loops touch at least a part of the path, resulting in a second term,

$$P_2 \cdot \Delta_2 = s_{12}s_{21}\Gamma_L(1 - 0) = s_{12}s_{21}\Gamma_L$$

which is added to the first, and

$$\frac{b_1}{b_s} = \frac{P_1 \cdot \Delta_1 + P_2 \cdot \Delta_2}{\Delta} = \frac{s_{11}(1 - s_{22}\Gamma_L) + s_{12}s_{21}\Gamma_L}{\Delta}$$

Dividing (b_1/b_s) by (a_1/b_s), we have

$$\frac{(b_1/b_s)}{(a_1/b_s)} = \frac{b_1}{a_1} = s_{11}' = \frac{s_{11}(1 - s_{22}\Gamma_L) + s_{12}s_{21}\Gamma_L}{1 - s_{22}\Gamma_L} = s_{11} + \frac{s_{12}s_{21}\Gamma_L}{1 - s_{22}\Gamma_L}$$

which is the same result obtained in Section 1.3.2 for positive real Z_o. Note that the denominator of (1.28) is a characteristic of the total network, common to all ratios, and does not need to be computed because it cancels during the final division.

It is also possible to use topological manipulation of the flow graph in such a manner that all loops are eliminated, and there is only one path from the input to output variable. Kuhn [17] presents four simple rules for the reduction of flow graphs and gives examples showing that it is possible to arrive at a solution without using the nontouching loop formula of (1.28).

1.3.4 The *S*-Parameter Cascade Connection

We can use the signal flow graph to develop the equations for a cascade connection of two separate two-port networks, as shown in Figure 1.17, where A_{ij} is an *s*-parameter coefficient of two-port A, and B_{ij} is associated with the coefficients of two-port B. As shown, there are a total of six first-order loops:

$$L_{11} = A_{11}\Gamma_s$$

$$L_{12} = A_{22}B_{11}$$

$$L_{13} = B_{22}\Gamma_L$$

$$L_{14} = B_{11}A_{12}A_{21}\Gamma_s$$

$$L_{15} = A_{22}B_{12}B_{21}\Gamma_L$$

$$L_{16} = A_{12}A_{21}B_{12}B_{21}\Gamma_s\Gamma_L$$

three second-order loops:

$$L_{21} = L_{11}L_{12} = A_{11}A_{22}B_{11}\Gamma_s$$

$$L_{22} = L_{11}L_{13} = A_{11}B_{22}\Gamma_s\Gamma_L$$

$$L_{23} = L_{12}L_{13} = A_{22}B_{11}B_{22}\Gamma_L$$

and one third-order loop:

$$L_{31} = L_{11}L_{12}L_{13} = A_{11}A_{22}B_{11}B_{22}\Gamma_s\Gamma_L$$

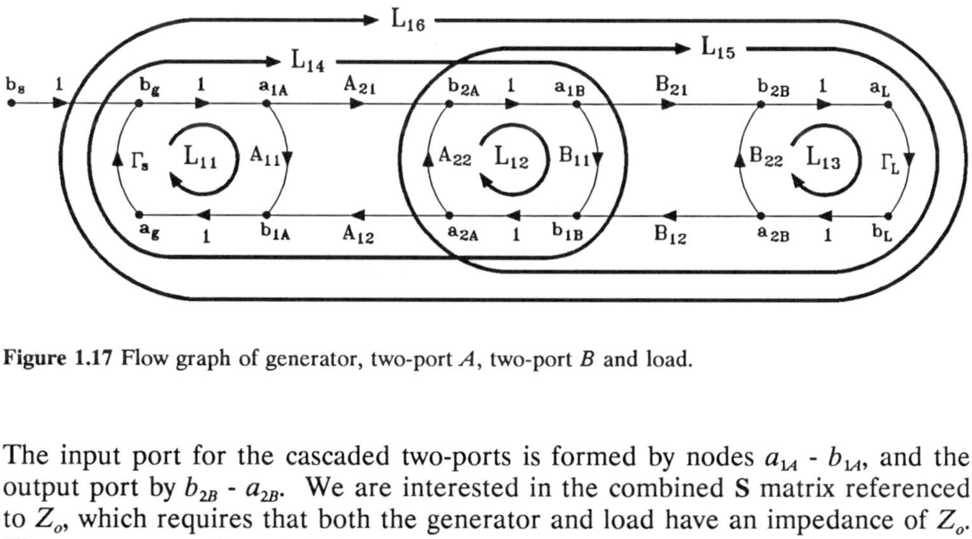

Figure 1.17 Flow graph of generator, two-port *A*, two-port *B* and load.

The input port for the cascaded two-ports is formed by nodes a_{1A} - b_{1A}, and the output port by b_{2B} - a_{2B}. We are interested in the combined **S** matrix referenced to Z_o, which requires that both the generator and load have an impedance of Z_o. Thus, $\Gamma_s = \Gamma_L = 0$, and all loops except L_{12} are zero. The signal flow graph can become complex at times, and it often helps to eliminate zero branches and use different colors for various loops. When Figure 1.17 is drawn without the Γ_s and Γ_L branches, it is readily apparent that L_{12} is the only closed loop.

When the generator and load are connected as shown, $a_2 = 0$, and we can solve for s_{11} and s_{21}. There is only one path from b_s to a_{1A}, and we can write

$$\frac{a_{1A}}{b_s} = \frac{P_1 \cdot (1 - L_{12})}{\Delta} = \frac{1 - A_{22}B_{11}}{\Delta}$$

We can follow three paths from b_s to b_{1A}. The first path is through A_{11}:

$$P_1 \cdot \Delta_1 = A_{11}(1 - L_{12}) = A_{11}(1 - A_{22}B_{11})$$

The second is through A_{21}, B_{11}, and A_{12}:

$$P_2 \cdot \Delta_2 = A_{21}B_{11}A_{12}(1 - 0) = A_{12}A_{21}B_{11}$$

The third path is through Γ_L; however, $\Gamma_L = 0$ and we need only sum the first two terms:

$$\frac{b_{1A}}{b_s} = \frac{A_{11}(1 - A_{22}B_{11}) + A_{12}A_{21}B_{11}}{\Delta}$$

We have only one path from b_s to b_{2B}, and all loops touch at least a part of this path:

$$\frac{b_{2B}}{b_s} = \frac{P_1 \cdot (1 - 0)}{\Delta} = \frac{A_{21}B_{21}}{\Delta}$$

Dividing (b_{1A}/b_s) by (a_{1A}/b_s), we have

$$s_{11} = \frac{A_{11}(1 - A_{22}B_{11}) + A_{12}A_{21}B_{11}}{1 - A_{22}B_{11}} = A_{11} + \frac{A_{12}A_{21}B_{11}}{1 - A_{22}B_{11}} \tag{1.29a}$$

and when (b_{2B}/b_s) is divided by (a_{1A}/b_s),

$$s_{21} = \frac{A_{21}B_{21}}{1 - A_{22}B_{11}} \tag{1.29b}$$

We can set a_1 to zero by exchanging the generator and load, and then solve for s_{12} and s_{22}:

$$s_{12} = \frac{A_{12}B_{12}}{1 - A_{22}B_{11}} \tag{1.29c}$$

$$s_{22} = B_{22} + \frac{A_{22}B_{12}B_{21}}{1 - A_{22}B_{11}} \tag{1.29d}$$

Thus, direct *s*-parameter cascade is not complex and is certainly more efficient than converting to *ABCD* or the transfer scattering parameters, multiplying matrices, and then converting back.

1.3.5 *S*-Parameter Parallel and Series Connections [18]

Figure 1.18(a) shows the configuration we used in Section 1.2.2 to connect two-port networks in parallel.

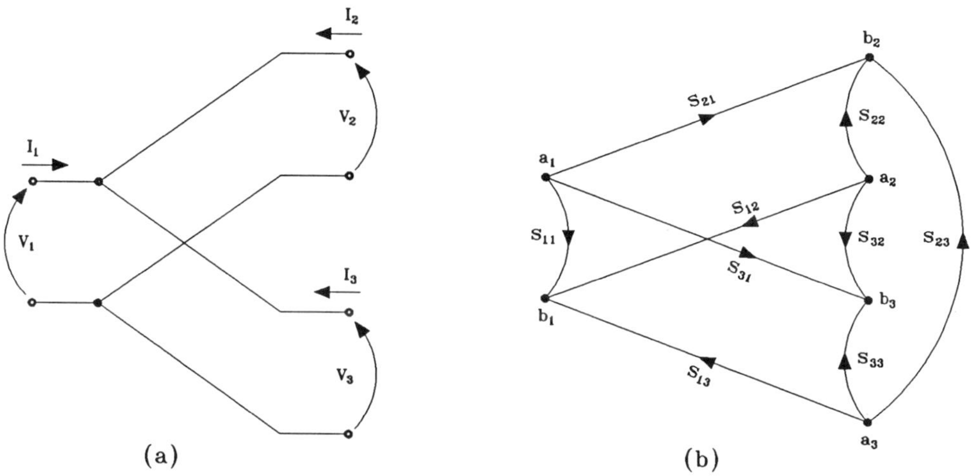

Figure 1.18 Parallel connection: (a) voltage and current definition; (b) signal flow graph.

We used the equality $V_1 = V_2 = V_3$ to determine that the *y*-parameters of the combined network are the sum of the *y*-parameters of the individual networks. Unfortunately, there is not a similar equality for the incident and reflected waves. The *parallel Y* or *tee* is a three-port network that has a signal flow graph as shown in Figure 1.18(b). As shown, a wave incident at any port is split between the other two ports, and the reflected wave from any port is the adjusted sum of the incident waves at the other two ports. The coefficients of the three-port **S** matrix must be determined before we can construct a flow graph or use other methods to connect the two-port networks in parallel.

Figure 1.19 shows the equivalent circuit of the parallel tee when port 1 is connected to a $Z_o\Omega$ voltage source, and ports 2 and 3 are terminated with $Z_o\Omega$

loads. The voltage at all three ports is equal and, noting that the circuit is a voltage divider, we have

$$V_1 = V_2 = V_3 = V_s \cdot \frac{(Z_o/2)}{(Z_o + Z_o/2)} = \frac{V_s}{3}$$

The current flowing into port 1 is

$$I_1 = \frac{(V_s/3)}{(Z_o/2)} = \frac{2V_s}{3Z_o}$$

and the current flowing into ports 2 and 3 is

$$I_2 = I_3 = -\frac{V_2}{Z_o} = -\frac{(V_s/3)}{Z_o} = -\frac{V_s}{3Z_o}$$

Using (1.14), we can write the incident and reflected waves as

$$a_1 = \frac{(V_s/3) + (2V_s/3Z_o)(Z_o)}{2 \cdot \sqrt{Z_o}} = \frac{V_s}{2 \cdot \sqrt{Z_o}}$$

$$b_1 = \frac{(V_s/3) - (2V_s/3Z_o)(Z_o)}{2 \cdot \sqrt{Z_o}} = -\frac{(V_s/3)}{2 \cdot \sqrt{Z_o}}$$

$$b_2 = b_3 = \frac{(V_s/3) - (-V_s/3Z_o)(Z_o)}{2 \cdot \sqrt{Z_o}} = \frac{(2V_s/3)}{2 \cdot \sqrt{Z_o}}$$

then solve for the first three coefficients:

$$S_{11} = \left.\frac{b_1}{a_1}\right|_{a_2=a_3=0} = \frac{(-1/3)}{1} = -\frac{1}{3}$$

$$S_{21} = \left.\frac{b_2}{a_1}\right|_{a_2=a_3=0} = \frac{(2/3)}{1} = \frac{2}{3}$$

$$S_{31} = \left.\frac{b_3}{a_1}\right|_{a_2=a_3=0} = \frac{(2/3)}{1} = \frac{2}{3}$$

We can solve for the other six coefficients by inspection. The complete **S** matrix for the parallel tee connection is

$$\mathbf{S_Y} = \begin{bmatrix} -1/3 & 2/3 & 2/3 \\ 2/3 & -1/3 & 2/3 \\ 2/3 & 2/3 & -1/3 \end{bmatrix} \tag{1.30}$$

and we find that the input reflection coefficient at all three ports is (-1/3) and all other coefficients are (2/3). The parallel tee can easily be extended to an n-port when necessary to connect more than three ports:

$$s_{ii} = \frac{2 - n}{n}$$
$$s_{ij} = \frac{2}{n} \tag{1.31}$$

where

$$i,j = 1,2,...,n$$

and

$$i \neq j$$

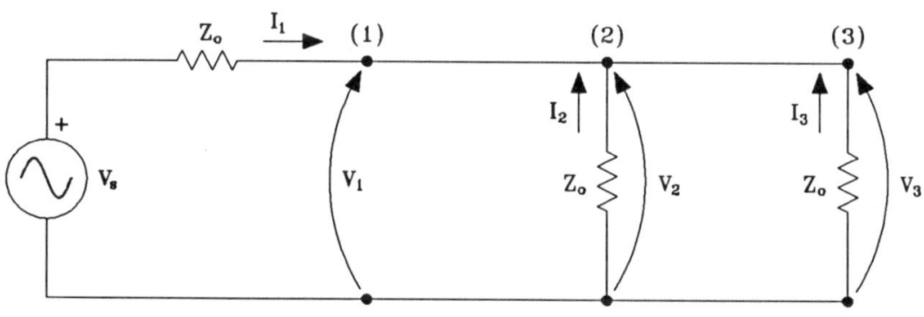

Figure 1.19 Parallel connection equivalent circuit.

The *series tee* shown in Figure 1.20(a) is useful for connecting two networks in series, as shown in Figure 1.5, and has other applications, as well. The series tee has the same flow graph as the parallel tee shown in Figure 1.18(b), and we can compute the s-parameters in the same manner.

Figure 1.20(b) shows the equivalent circuit when a $Z_o\Omega$ source is connected to port 1 and the other ports are terminated with $Z_o\Omega$ loads. Except for the sign, current is equal at all three ports:

$$I_1 = -I_2 = -I_3 = \frac{V_s}{3Z_o}$$

The voltage at port 1 is

$$V_1 = \left(\frac{V_s}{3Z_o}\right)(2Z_o) = \frac{2V_s}{3}$$

and at ports 2 and 3 is

$$V_2 = V_3 = \left(\frac{V_s}{3Z_o}\right)(Z_o) = \frac{V_s}{3}$$

We can compute the incident and reflected waves at each port

$$a_1 = \frac{(2V_s/3) + (V_s/3Z_o)(Z_o)}{2 \cdot \sqrt{Z_o}} = \frac{V_s}{2 \cdot \sqrt{Z_o}}$$

$$b_1 = \frac{(2V_s/3) - (V_s/3Z_o)(Z_o)}{2 \cdot \sqrt{Z_o}} = \frac{(V_s/3)}{2 \cdot \sqrt{Z_o}}$$

$$b_2 = b_3 = \frac{(V_s/3) - (-V_s/3Z_o)(Z_o)}{2 \cdot \sqrt{Z_o}} = \frac{(2V_s/3)}{2 \cdot \sqrt{Z_o}}$$

and then three of the nine *s*-parameters

$$s_{11} = \left.\frac{b_1}{a_1}\right|_{a_2=a_3=0} = \frac{(1/3)}{1} = \frac{1}{3}$$

$$s_{21} = \left.\frac{b_2}{a_1}\right|_{a_2=a_3=0} = \frac{(2/3)}{1} = \frac{2}{3}$$

$$s_{31} = \left.\frac{b_3}{a_1}\right|_{a_2=a_3=0} = \frac{(2/3)}{1} = \frac{2}{3}$$

The other two columns of the **S** matrix can be filled in by moving the generator to port 2 and then port 3, which yields

$$\mathbf{S_T} = \begin{bmatrix} 1/3 & 2/3 & 2/3 \\ 2/3 & 1/3 & -2/3 \\ 2/3 & -2/3 & 1/3 \end{bmatrix} \tag{1.32}$$

Although the signal flow graph can be used to derive equations for the *s*-parameter series and parallel connection, the results are not simple. It is perhaps easier to convert from *s*- into *z*- or *y*-parameters, add, and then convert back. However, there are circumstances where these connections are remarkably simple; the next two sections illustrate two special cases.

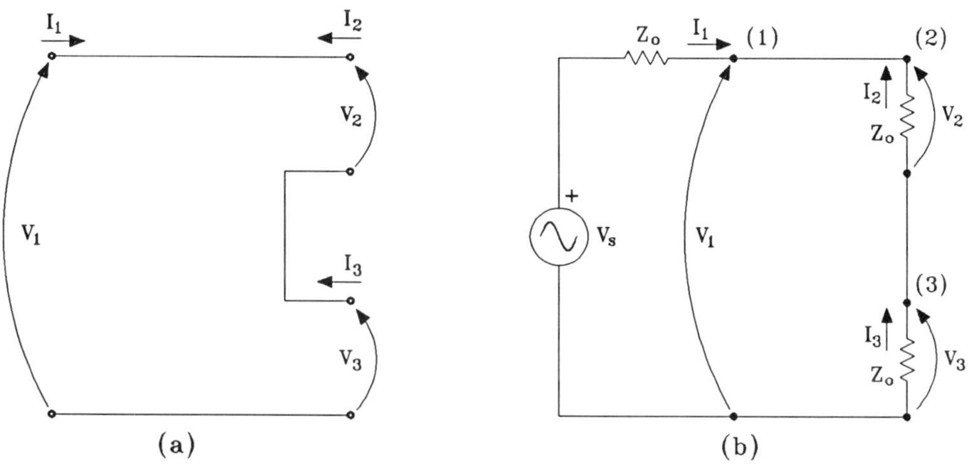

Figure 1.20 (a) Series connection; (b) equivalent circuit.

1.3.6 Indefinite *S*-Parameters

Figure 1.21(a) shows a transistor with an impedance connected from the source to ground and a second connection from the source to another point in the circuit. The three-terminal transistor is usually characterized as a two-port or three-port representation, as shown in Figure 1.21(b). However, we need access to each terminal. We can define three ports by referencing each terminal to a common external node. Although the figure illustrates the external node as ground, in fact the node needs no definite reference, and *s*-parameters defined in this manner are called *indefinite scattering parameters* [19].

Given a two-port representation of a network, we can create a third port by connecting the two-port in series with a three-port parallel tee, which results in the three-port circuit shown in Figure 1.22.

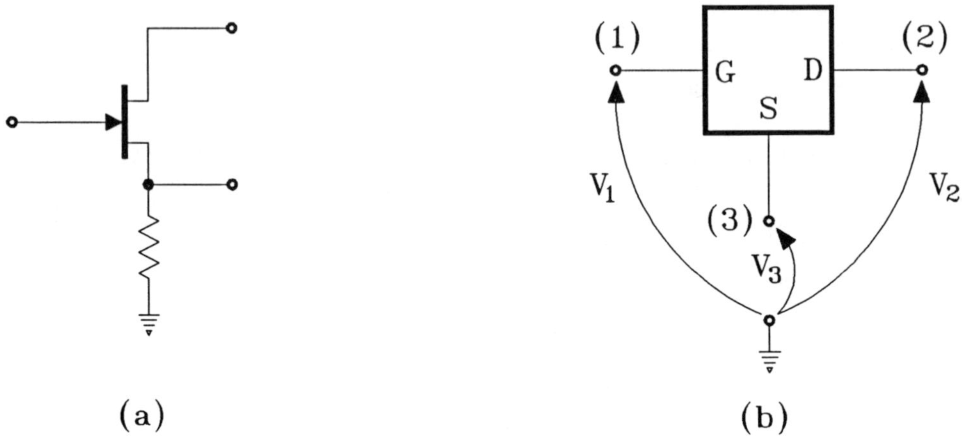

Figure 1.21 (a) Circuit; (b) port reference.

The original two-port has been transformed into a three-port with the gate, connected between the input and output series tee. Rather than use individual connections, we can create one multiport which will transform the n-port into a network with $n + 1$ ports referenced to a common external node.

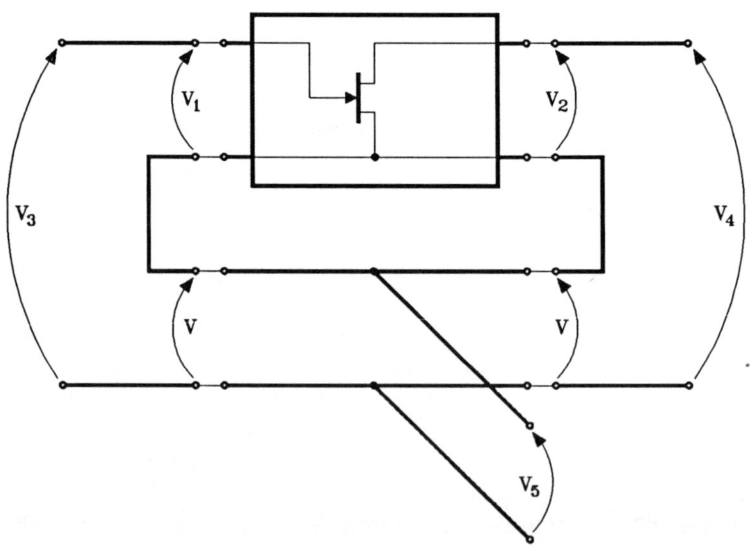

Figure 1.22 Two-port to three-port connection diagram.

We can separate the connection network into internal ports which are connected to the original network and external ports which are available after the transformation. Examining Figure 1.22, we see that each positive terminal of an internal port is connected to the corresponding positive terminal of an external port. The negative terminal of each internal port is connected to a common positive terminal of the new external port, and each external negative terminal is connected to the negative terminal of the new port. The multiport series tee is illustrated in Figure 1.23 for a network where the internal ports are numbered from 1 to k - 1, external ports from k to n - 1, and where the new port is n where n = $2k$ - 1.

We can solve for the s-parameters of the multiport series tee by applying the same techniques used to solve the three-port parallel tee and series tee. We connect a generator to port 1, terminate all other ports, solve for the voltage and currents at each port, compute the incident and reflected waves at each port, and then calculate the coefficients of the first column of the **S** matrix. We then move the generator to port 2, solve for the second column coefficients, and so forth, until n columns have been computed. The results of this exercise are

$$\begin{aligned} s_{ij}\Big|_{\substack{i=1 \text{ to } k-1 \\ j=1 \text{ to } k-1}} &= \frac{2}{n+3} \\ s_{in} = s_{ni}\Big|_{\substack{i=1 \text{ to } k-1 \\ j=1 \text{ to } k-1}} &= \frac{-4}{n+3} \end{aligned} \tag{1.33a}$$

$$\begin{aligned} s_{ij}\Big|_{\substack{i=k \text{ to } n-1 \\ j=k \text{ to } n-1}} &= \frac{2}{n+3} \\ s_{in} = s_{ni}\Big|_{\substack{i=k \text{ to } n-1 \\ j=k \text{ to } n-1}} &= \frac{4}{n+3} \end{aligned} \tag{1.33b}$$

$$s_{ij}\Big|_{\substack{i=k \text{ to } n-1 \\ j=1 \text{ to } n-1 \\ i=j+(k-1)}} = \frac{n+1}{n+3} \tag{1.33c}$$

$$s_{ij}\Big|_{\substack{i=k \text{ to } n-1 \\ j=1 \text{ to } n-1 \\ i\neq j+(k-1)}} = \frac{-2}{n+3} \tag{1.33d}$$

$$s_{ij}\Big|_{\substack{i=1 \text{ to } k-1 \\ j=k \text{ to } n-1 \\ i=j-(k-1)}} = \frac{n+1}{n+3} \tag{1.33e}$$

$$s_{ij}\Big|_{\substack{i=1 \text{ to } k-1 \\ j=k \text{ to } n-1 \\ i\neq j-(k-1)}} = \frac{-2}{n+3} \tag{1.33f}$$

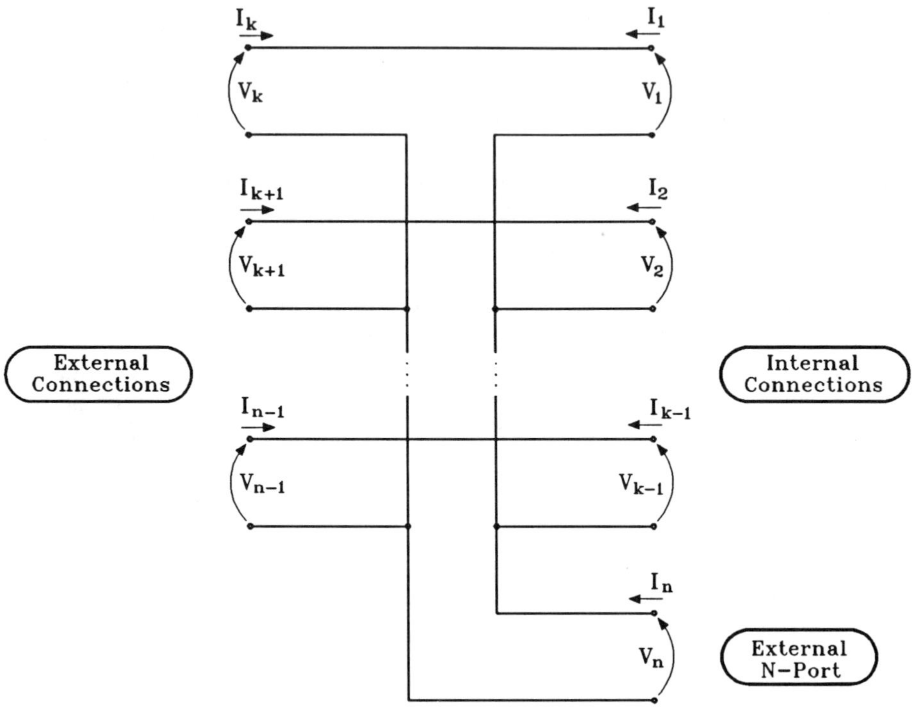

Figure 1.23 Multiport series tee connection.

$$s_{nn} = -\frac{n - 5}{n + 3} \tag{1.33g}$$

We can investigate the characteristics of the indefinite scattering matrix by analyzing the transformed three-port shown in Figure 1.24(a), where the ports are numbered as shown. The relationship between incident and reflected waves is

$$\begin{bmatrix} b_1 \\ b_2 \\ b_3 \end{bmatrix} = \begin{bmatrix} s_{11} & s_{12} & s_{13} \\ s_{21} & s_{22} & s_{23} \\ s_{31} & s_{32} & s_{33} \end{bmatrix} \begin{bmatrix} a_1 \\ a_2 \\ a_3 \end{bmatrix}$$

For positive real Z_o, the current flowing at each port is the difference between the incident and reflected currents:

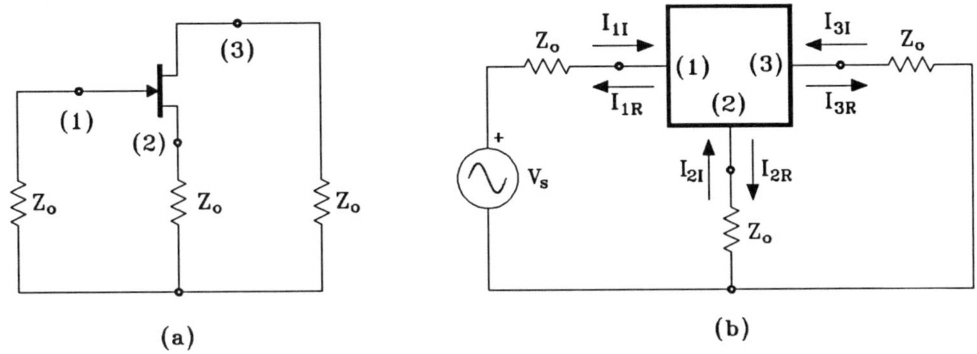

Figure 1.24 (a) FET circuit; (b) three-port equivalent.

$$I_i = I_{iI} - I_{iR}$$

and

$$a_i = \sqrt{Z_o}\, I_{iI}$$

$$b_i = \sqrt{Z_o}\, I_{iR}$$

When we connect the generator to port 1, as shown in Figure 1.24(b), the incident waves at ports 2 and 3 are zero, and

$$b_1 = s_{11} a_1$$

$$b_2 = s_{21} a_1$$

$$b_3 = s_{31} a_1$$

which can be written as

$$\sqrt{Z_o}\, I_{1R} = s_{11} \sqrt{Z_o}\, I_{1I}$$

$$\sqrt{Z_o}\, I_{2R} = s_{21} \sqrt{Z_o}\, I_{1I}$$

$$\sqrt{Z_o}\, I_{3R} = s_{31} \sqrt{Z_o}\, I_{1I}$$

Applying Kirchhoff's current law to the common node,

$$I_1 + I_2 + I_3 = (I_{1I} - I_{1R}) + (I_{2I} - I_{2R}) + (I_{3I} - I_{3R}) = 0$$

Since $I_{2I} = I_{3I} = 0$,

$$I_{1I} = I_{1R} + I_{2R} + I_{3R}$$

or

$$I_{1I} = s_{11}I_{1I} + s_{21}I_{1I} + s_{31}I_{1I}$$

Thus,

$$1 = s_{11} + s_{21} + s_{31} \tag{1.34a}$$

Moving the generator to port 2, we have

$$1 = s_{12} + s_{22} + s_{32} \tag{1.34b}$$

and to port 3,

$$1 = s_{13} + s_{23} + s_{33} \tag{1.34c}$$

Therefore, *the sum of each column of the indefinite scattering matrix is equal to one.*

We can connect identical generators between each terminal and the common node, as shown in Figure 1.25. Since all three terminals are at the same potential, no current can flow, or

$$I_{1I} = I_{1R}$$

$$I_{2I} = I_{2R}$$

$$I_{3I} = I_{3R}$$

and for equal generators, $a_1 = a_2 = a_3$, which can be expressed as

$$I_{1I} = I_{2I} = I_{3I}$$

The reflected wave from port 1 is

$$b_1 = s_{11}a_1 + s_{12}a_2 + s_{13}a_3$$

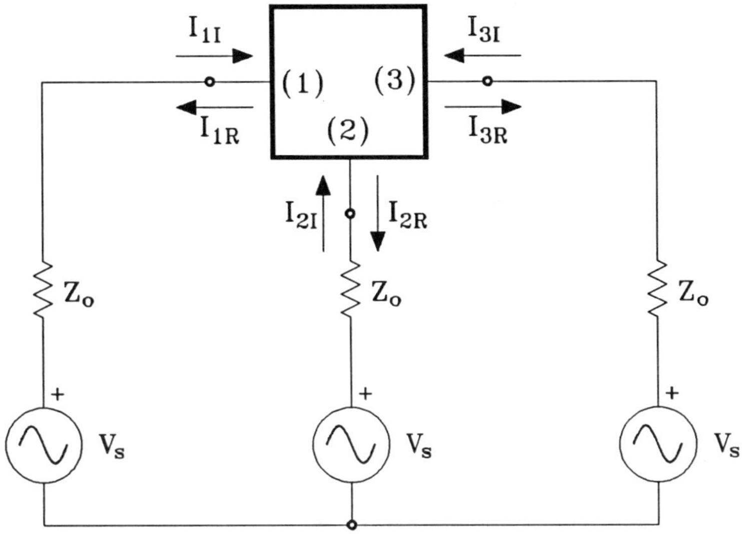

Figure 1.25 Generators connected to ports 1, 2, and 3.

or in terms of incident and reflected currents,

$$\sqrt{Z_o}\,I_{1R} = s_{11}\sqrt{Z_o}\,I_{1I} + s_{12}\sqrt{Z_o}\,I_{2I} + s_{13}\sqrt{Z_o}\,I_{3I}$$

However, all currents are equal and

$$I_{1R} = s_{11}\,I_{1R} + s_{12}\,I_{1R} + s_{13}\,I_{1R}$$

or

$$1 = s_{11} + s_{12} + s_{13} \tag{1.34d}$$

In the same manner, the second row is

$$1 = s_{21} + s_{22} + s_{23} \tag{1.34e}$$

the third row is

$$1 = s_{31} + s_{32} + s_{33} \tag{1.34f}$$

and we conclude that *the sum of elements in each row of the indefinite scattering matrix is also equal to one.*

Therefore, only two of the three elements in each row or column is independent. The third element can be calculated from any two coefficients in a row or column. We can extend the analysis to an n-port with the same results; that is, the sum of any row or column is one.

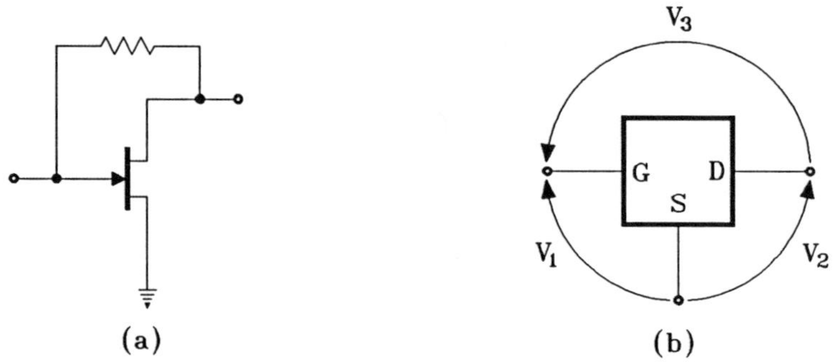

Figure 1.26 (a) Parallel feedback; (b) port reference.

1.3.7 *N*-Port Networks With *N*-Terminals

We usually define the ports of a network by referencing one terminal to another such that an n-port network has $2n$ external terminals or nodes. The indefinite scattering matrix is created by referencing each terminal to a common external node, which results in an n-port network with $n + 1$ terminals. We can also define ports by referencing terminals to each other or an n-port network formed from n terminals. Figure 1.26(a) shows an FET with parallel feedback connected from drain to gate. This is simply a parallel connection of the device and feedback element. However, if we intend to try several different values of feedback or connect a complex network, it is often more convenient to create a third port. Figure 1.26(b) illustrates a three-port network where each terminal is referenced to an adjacent terminal. Once we compute the three-port s-parameters, we can terminate port 3 with the feedback element or network.

We can easily transform the two-port device parameters into a three-port network using two parallel tees and one series tee, as shown in Figure 1.27. This connection is similar to the one used to create the indefinite s-parameter matrix, except that the two series tees have been replaced by parallel tees and the parallel tee replaced by a series tee. We found that there are only n - 1 independent parameters in each row or column of an n-port defined as shown in Figure 1.21, and a similar relationship holds for the port definitions shown in Figure 1.26.

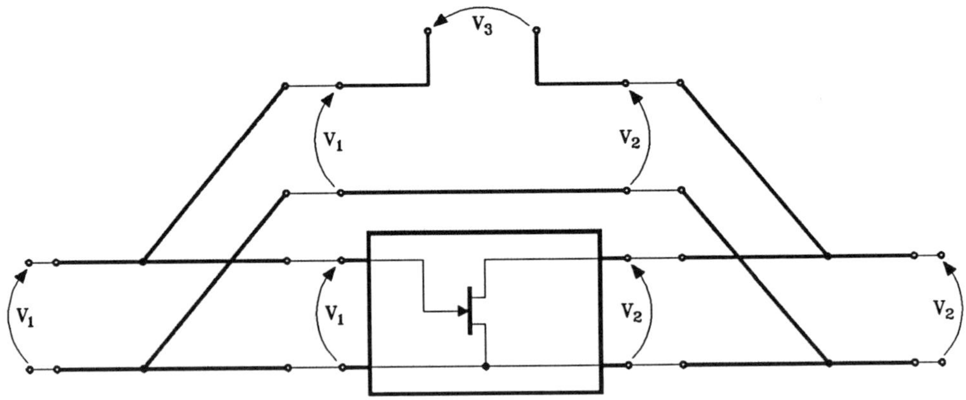

Figure 1.27 Three-terminal to three-port connection.

Figure 1.28(a) shows the FET with a $Z_o\Omega$ impedance connected between adjacent terminals, and Figure 1.28(b) shows the voltage and current definitions when a voltage source is connected to port 1 of the three-port network. The port voltage is the sum of incident and reflected components:

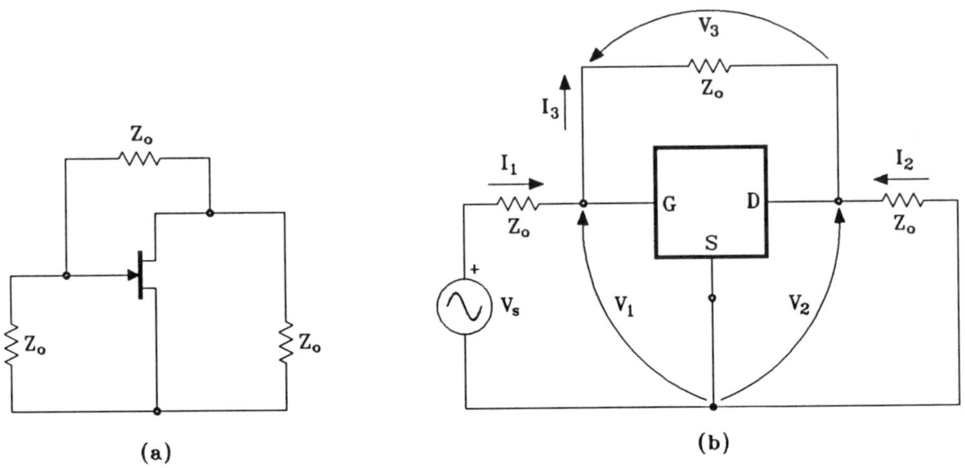

Figure 1.28 (a) Parallel feedback; (b) three-port equivalent.

$$V_1 = V_{1I} + V_{1R}$$

$$V_2 = V_{2I} + V_{2R}$$

$$V_3 = V_{3I} + V_{3R}$$

or, in terms of current,

$$V_1 = Z_o I_{1I} + Z_o I_{1R}$$

$$V_2 = Z_o I_{2I} + Z_o I_{2R}$$

$$V_3 = Z_o I_{3I} + Z_o I_{3R}$$

From Kirchhoff's voltage law,

$$V_1 = V_2 + V_3$$

or

$$\left(Z_o I_{1I} + Z_o I_{1R}\right) = \left(Z_o I_{2I} + Z_o I_{2R}\right) + \left(Z_o I_{3I} + Z_o I_{3R}\right)$$

However, since there are no generators at ports 2 and 3, $a_2 = a_3 = 0$ and $I_{2I} = I_{3I} = 0$, or

$$I_{1I} + I_{1R} = I_{2R} + I_{3R}$$

Following the procedure used in Section 1.3.6, we can solve for the three *s*-parameters in terms of incident and reflected waves, express the incident and reflected waves in terms of incident and reflected current, and then substitute, which yields

$$I_{1I} + s_{11} I_{1I} = s_{21} I_{1I} + s_{31} I_{1I}$$

or

$$1 = -s_{11} + s_{21} + s_{31} \tag{1.35a}$$

Moving the generator to port 2, the second column of the **S** matrix is

$$1 = s_{12} - s_{22} - s_{32} \tag{1.35b}$$

and the third column is

$$1 = s_{13} - s_{23} - s_{33} \tag{1.35c}$$

when the generator is moved to port 3.

We can determine the rows of the **S** matrix by connecting a generator at all three ports, as shown in Figure 1.29. The three source voltages V_{S1}, V_{S2}, and V_{S3} are selected such that the port voltages V_1, V_2, and V_3 are equal, and, in fact, zero. Since the three port voltages are zero, no current can flow into the three terminals, or

$$I_G = I_D = I_S = 0$$

and

$$V_1 = V_2 = V_3 = 0$$

Since current is not flowing in the three terminals, we can remove the three-terminal network, which results in a single path for current flow; thus,

$$I_1 = -I_2 = -I_3$$

The port voltage is zero, and we can write the incident waves as

$$a_1 = \frac{V_1 + Z_o I_1}{2 \cdot \sqrt{Z_o}} = \frac{\sqrt{Z_o} I_1}{2}$$

$$a_2 = \frac{V_2 + Z_o I_2}{2 \cdot \sqrt{Z_o}} = \frac{\sqrt{Z_o} I_2}{2} = -\frac{\sqrt{Z_o} I_1}{2}$$

$$a_3 = \frac{V_3 + Z_o I_3}{2 \cdot \sqrt{Z_o}} = \frac{\sqrt{Z_o} I_3}{2} = -\frac{\sqrt{Z_o} I_1}{2}$$

the reflected waves as

$$b_1 = \frac{V_1 - Z_o I_1}{2 \cdot \sqrt{Z_o}} = -\frac{\sqrt{Z_o} I_1}{2}$$

$$b_2 = \frac{V_2 - Z_o I_2}{2 \cdot \sqrt{Z_o}} = -\frac{\sqrt{Z_o} I_2}{2} = \frac{\sqrt{Z_o} I_1}{2}$$

$$b_3 = \frac{V_3 - Z_o I_3}{2 \cdot \sqrt{Z_o}} = -\frac{\sqrt{Z_o} I_3}{2} = \frac{\sqrt{Z_o} I_1}{2}$$

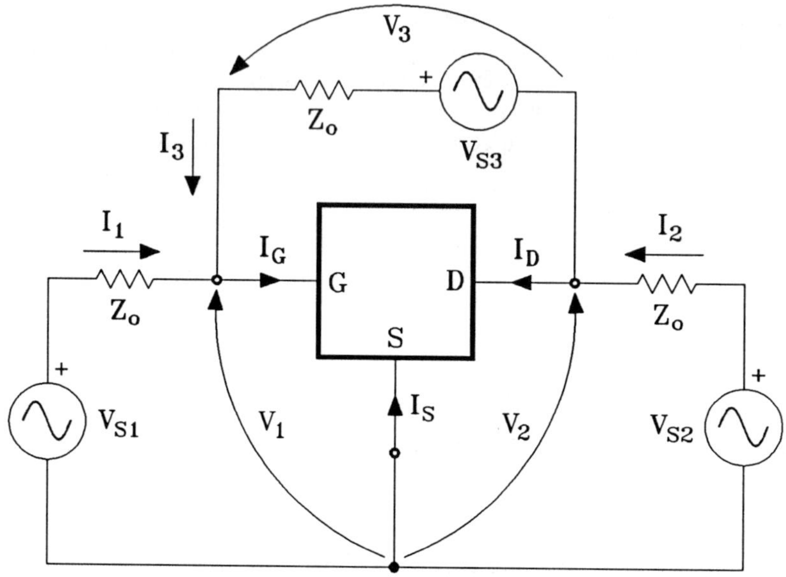

Figure 1.29 Generators connected to ports 1, 2 and 3.

and then the relationship between incident and reflected waves:

$$b_1 = -a_1 = a_2 = a_3$$

$$b_2 = a_1 = -a_2 = -a_3$$

$$b_3 = a_1 = -a_2 = -a_3$$

From (1.16),

$$b_1 = s_{11}a_1 + s_{12}a_2 + s_{13}a_3$$

$$b_2 = s_{21}a_1 + s_{22}a_2 + s_{23}a_3$$

$$b_3 = s_{31}a_1 + s_{32}a_2 + s_{33}a_3$$

or

$$b_1 = -s_{11}b_1 + s_{12}b_1 + s_{13}b_1$$

$$b_2 = s_{21}b_2 - s_{22}b_2 - s_{23}b_2$$

$$b_3 = s_{31}b_3 - s_{32}b_3 - s_{33}b_3$$

Therefore,

$$1 = -s_{11} + s_{12} + s_{13} \tag{1.35d}$$

$$1 = s_{21} - s_{22} - s_{23} \tag{1.35e}$$

and

$$1 = s_{31} - s_{32} - s_{33} \tag{1.35f}$$

Although the sum of the rows and columns does not equal one, there are only two independent variables in each row or column of the scattering matrix.

REFERENCES

[1] C.L. Ruthroff, "Some Broad-Band Transformers," *Proc. IRE*, Vol. 47, No. 8, August 1959, pp. 1337-1342.

[2] K. Kurokawa, "Power Waves and the Scattering Matrix," *IEEE Trans. on Microwave Theory and Techniques*, Vol. MTT-13, No. 3, March 1965, pp. 194-202.

[3] P. Penfield, Jr., "Noise in Negative Resistance Amplifiers," *IRE Trans. on Circuit Theory*, Vol. CT-7, No. 2, June 1960, pp. 166-170.

[4] H.J. Carlin, "The Scattering Matrix in Network Theory," *IRE Trans. on Circuit Theory*, Vol. CT-3, No. 2, June 1956, pp. 88-97.

[5] E.W. Mathews, "The Use of Scattering Matrices in Microwave Circuits," *IRE Trans. on Microwave Theory and Techniques*, Vol. MTT-3, April 1955, pp. 21-26.

[6] R.W. Anderson, "S-Parameter Techniques for Faster, More Accurate Network Design," *Hewlett-Packard Journal*, Vol. 18, No. 6, February 1967, pp. 13-24.

[7] W.H. Froehner, "Quick Amplifier Design with Scattering Parameters," *Electronics*, Vol. 40, October 16, 1967, pp. 100-109.

[8] *S-Parameters...Circuit Analysis and Design*, Hewlett-Packard Company, Application Note 95, September 1968.

[9] *S-Parameter Design*, Hewlett-Packard Company, Application Note 154, April 1972, Revised May 1973.

[10] L. Weinberg, "Fundamentals of Scattering Matrices," *Electro-Technology*, July 1967, pp. 55-72.

[11] G.E. Bodway, "Two-Port Power Flow Analysis Using Generalized Scattering Parameters," *Microwave Journal*, Vol. 10, No. 5, May 1967, pp. 61-69.

[12] S. Ramo, J. Whinnery, and T. Van Duzer, *Fields and Waves in Communication Electronics*, John Wiley & Sons, Inc., New York, 1965.

[13] D.C. Youla and P.M. Paterno, "Realizable Limits of Error for Dissipationless Attenuators in Mismatched Systems," *IEEE Trans. on Microwave Theory and Techniques*, Vol. MTT-12, No. 3, May 1964, pp. 289-299.

[14] S.J. Mason, "Feedback Theory - Some Properties of Signal Flow Graphs," *Proc. IRE*, Vol. 41, No. 9, September 1953, pp. 1144-1156.

[15] S.J. Mason, "Feedback Theory - Further Properties of Signal Flow Graphs," *Proc. IRE*, Vol. 44, No. 7, July 1956, pp. 920-926.

[16] J.K. Hunton, "Analysis of Microwave Measurement Techniques by Means of Signal Flow Graphs," *IEEE Trans. on Microwave Theory and Techniques*, Vol. MTT-8, No. 2, March 1960, pp. 206-212.

[17] N. Kuhn, "Simplified Signal Flow Graph Analysis," *Microwave Journal*, Vol. 6, No. 11, November 1963, pp. 59-66.

[18] L. Besser and C. Hsieh, "Series and Parallel Addition of Two-Ports Via S-Parameters," *Microwave Journal*, Vol. 18, No. 4, April 1975, pp. 53-56.

[19] G.E. Bodway, "Circuit Design and Characterization of Transistors by Means of Three-Port Scattering Parameters," *Microwave Journal*, Vol. 11, May 1968, pp. 55-63.

Chapter 2

Network Analysis

Network analysis is encountered during almost every phase of the design process. For example, an amplifier design usually begins by analyzing a number of transistors and selecting one that meets specifications. The device is then analyzed in more detail to determine the necessary source and load impedance. Once the impedance is defined, input and output matching networks can be designed and the complete amplifier analyzed to determine its characteristics. Assuming the design meets specifications, the ideal circuit model is modified to include nonideal components and parasitics due to the physical layout, and then the complete model is analyzed and perhaps optimized. The final step is to determine production yields based on component and layout tolerances, and then center the design to maximize the number of production circuits that meet specifications. Optimization and design centering both rely on network analysis and may require several thousands of analyses.

Techniques discussed in Chapter 1 can be used to analyze a large number of circuits. Most components can be represented as a two-port network, and many circuits can be modeled by interconnecting these networks using a combination of the cascade, series, parallel, series-parallel, and parallel-series connections. However, not all circuits can be modeled in this manner, and we must consider other techniques that are appropriate for the analysis of multiport networks. This chapter describes two methods that can be used for the analysis of multiport networks, as well as other techniques that are indispensable for the analysis and design of modern RF and microwave circuits.

2.1 CIRCUIT ELEMENTS

The most common task associated with network analysis is computing the matrix representation for circuit elements. Once individual elements have been computed, they can be interconnected to form the complete network. We can represent most elements using any of the parameter sets from Chapter 1. However, it is usually more efficient to choose a representation that offers some advantage at a later stage of the analysis. For example, if two parallel elements are to be cascaded with another, we can compute the *y*-parameters of the parallel elements, add, convert the results to *ABCD*-parameters, and then multiply with the **ABCD** matrix of the cascaded element. The final results may require another conversion (e.g., *ABCD*- to *s*-parameters), but we have avoided the extra steps of transforming the two parallel elements into *y*-parameters. Using this approach, we must develop an element library that includes expressions in each of the six parameter sets.

2.1.1 Series and Shunt Impedance

We can represent a number of elements with an equivalent series or shunt impedance. The resistor (R), inductor (L), and capacitor (C) are two-terminal elements that can be represented by an equivalent impedance or admittance. In addition, series and parallel combinations of elements such as RL, RC, LC, and RLC can also be reduced to a two-terminal equivalent, and we can generate complex combinations that represent loss and parasitic elements. The open and shorted transmission line can also be modeled in this manner, even when we include loss and termination parasitics.

Figure 2.1(a) shows a two-port network consisting of a series impedance connected between the positive terminals of ports 1 and 2. We can derive the

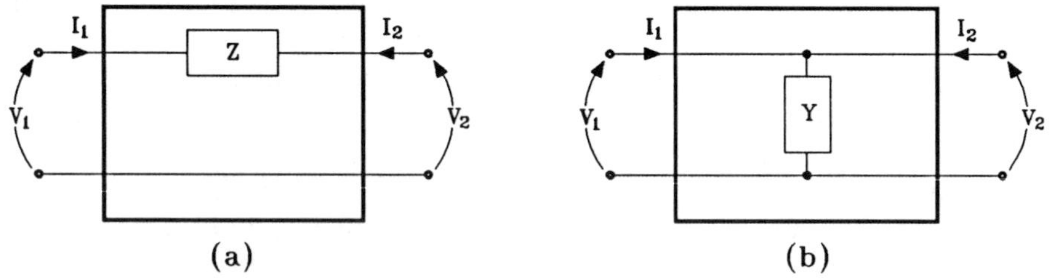

Figure 2.1 (a) Series impedance; (b) shunt admittance.

ABCD-parameters of this network by referring to Section 1.1.5. Connecting a voltage source to port 1, we can force I_2 to zero by open-circuiting the output port. Since I_2 is zero, I_1 must also be zero and $V_1 = V_2 = V_S$. From (1.5c),

$$A = \left.\frac{V_1}{V_2}\right|_{I_2=0} = \frac{V_S}{V_S} = 1$$

and

$$C = \left.\frac{I_1}{V_2}\right|_{I_2=0} = \frac{0}{V_S} = 0$$

Short-circuiting the output, we have $V_1 = V_S$, $V_2 = 0$, $I_1 = -I_2 = (V_S/Z)$; then, from (1.5d),

$$B = -\left.\frac{V_1}{I_2}\right|_{V_2=0} = -\frac{V_S}{(-V_S/Z)} = Z$$

and

$$D = -\left.\frac{I_1}{I_2}\right|_{V_2=0} = -\frac{(V_S/Z)}{(-V_S/Z)} = 1$$

Thus, the **ABCD** matrix for the series impedance is

$$Ch_Z = \begin{bmatrix} 1 & Z \\ 0 & 1 \end{bmatrix} \tag{2.1}$$

and we can use a similar approach to show that the **ABCD** matrix for the shunt admittance shown in Figure 2.1(b) is

$$Ch_Y = \begin{bmatrix} 1 & 0 \\ Y & 1 \end{bmatrix} \tag{2.2}$$

which turns out to be a very simple expression for both networks in Figure 2.1.

We can compute the *s*-parameters by connecting the appropriate source and load, as discussed in Section 1.3. Figure 2.2 shows a 1V $Z_o\Omega$ voltage source connected to port 1 of the series impedance, and a $Z_o\Omega$ termination connected at port 2. The actual value of the voltage source cancels when we compute the coefficients of the **S** matrix as ratios of reflected to incident waves, and it simplifies the mathematics somewhat if we use 1V, as shown in the figure. There is only one path for current flow, and

$$I_1 = -I_2 = \frac{1}{Z + 2Z_o}$$

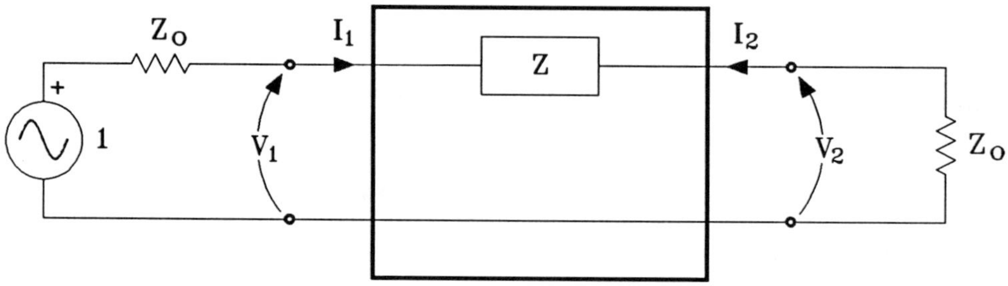

Figure 2.2 Series impedance.

Then computing port voltages using the voltage divider ratios,

$$V_1 = 1 \cdot \frac{Z + Z_o}{Z + 2Z_o}$$

$$V_2 = 1 \cdot \frac{Z_o}{Z + 2Z_o}$$

Assuming the reference impedance is positive real, we can use (1.14) to compute the incident and reflected waves as

$$a_1 = \frac{V_1 + Z_o I_1}{2\sqrt{Z_o}} = \frac{\left(\frac{Z + Z_o}{Z + 2Z_o}\right) + Z_o\left(\frac{1}{Z + 2Z_o}\right)}{2\sqrt{Z_o}} = \frac{1}{2\sqrt{Z_o}}$$

$$b_1 = \frac{V_1 - Z_o^* I_1}{2\sqrt{Z_o}} = \frac{\left(\frac{Z + Z_o}{Z + 2Z_o}\right) - Z_o\left(\frac{1}{Z + 2Z_o}\right)}{2\sqrt{Z_o}} = \frac{Z}{2\sqrt{Z_o}(Z + 2Z_o)}$$

$$b_2 = \frac{V_2 - Z_o^* I_2}{2\sqrt{Z_o}} = \frac{\left(\frac{Z_o}{Z + 2Z_o}\right) - Z_o\left(-\frac{1}{Z + 2Z_o}\right)}{2\sqrt{Z_o}} = \frac{2Z_o}{2\sqrt{Z_o}(Z + 2Z_o)}$$

and then two of the four s-parameters:

$$s_{11} = \left.\frac{b_1}{a_1}\right|_{a_2=0} = \frac{Z/\left[2\sqrt{Z_o}(Z + 2Z_o)\right]}{1/\left[2\sqrt{Z_o}\right]} = \frac{Z}{Z + 2Z_o}$$

$$s_{21} = \left.\frac{b_2}{a_1}\right|_{a_2=0} = \frac{2Z_o/\left[2\sqrt{Z_o}(Z + 2Z_o)\right]}{1/\left[2\sqrt{Z_o}\right]} = \frac{2Z_o}{Z + 2Z_o}$$

We could reverse the source and load to compute the other two parameters; however, we note that the network is symmetrical and write the complete **S** matrix by inspection:

$$\mathbf{S}_Z = \begin{bmatrix} \frac{Z}{Z + 2Z_o} & \frac{2Z_o}{Z + 2Z_o} \\ \frac{2Z_o}{Z + 2Z_o} & \frac{Z}{Z + 2Z_o} \end{bmatrix} \tag{2.3a}$$

A closer examination of the four s-parameters reveals that the expressions can be simplified somewhat:

$$s_{12} = s_{21} = \frac{2Z_o}{Z + 2Z_o} \tag{2.3b}$$

$$s_{11} = s_{22} = 1 - s_{12}$$

When we replace the series impedance by a shunt admittance and follow the same procedure,

$$\mathbf{S}_Y = \begin{bmatrix} \frac{-Z_oY}{2 + Z_oY} & \frac{2}{2 + Z_oY} \\ \frac{2}{2 + Z_oY} & \frac{-Z_oY}{2 + Z_oY} \end{bmatrix} \tag{2.4a}$$

or

$$s_{12} = s_{21} = \frac{2}{2 + Z_o Y} \tag{2.4b}$$

$$s_{11} = s_{22} = s_{12} - 1$$

Equations (2.3) and (2.4) are convenient when we know the impedance or admittance. However, there are times when a one-port reflection coefficient must be represented as a two-port for analysis. For example, a complex bias network can be modeled as a one-port network and then converted into a shunt equivalent, as shown in Figure 2.3(b). Measured data from a network analyzer is usually in the form of a reflection coefficient and must be converted to impedance or admittance before (2.3) or (2.4) can be used.

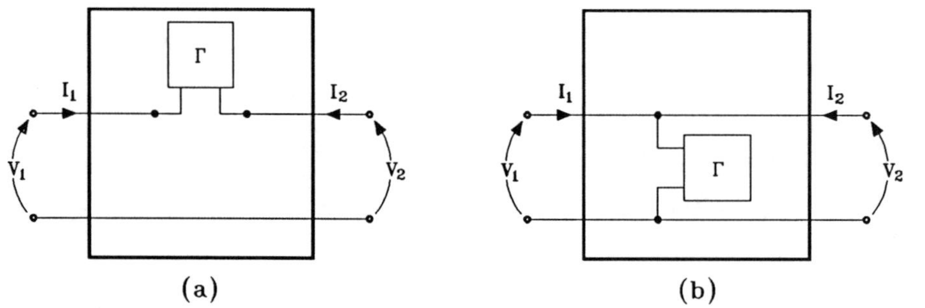

Figure 2.3 (a) Series; (b) shunt one-port.

When the reference impedance is real, the relationship between the reflection coefficient and impedance is

$$\Gamma = \frac{Z - Z_o}{Z + Z_o} \tag{2.5a}$$

which can be rearranged:

$$Z = Z_o\left(\frac{1 + \Gamma}{1 - \Gamma}\right) \tag{2.5b}$$

and then substituted into (2.3b):

$$s_{12} = s_{21} = \frac{2(1 - \Gamma)}{3 - \Gamma} \tag{2.6a}$$

$$s_{11} = s_{22} = 1 - s_{12}$$

which provides the expression for the series one-port in Figure 2.3(a). The shunt one-port shown in Figure 2.3(b) is derived by substituting the admittance as a function of reflection coefficient into (2.4b):

$$s_{12} = s_{21} = \frac{2(1 + \Gamma)}{3 + \Gamma} \tag{2.6b}$$

$$s_{11} = s_{22} = s_{12} - 1$$

2.1.2 Transmission Lines

Parameters for the transmission line element can easily be derived using the equations that describe the voltage and currents as a function of distance along the line. We can connect a source at one end of the line, terminate the other end, and then use the equations to compute the voltage and current at both ports. However, Holmes and Kuhn [1] use the signal flow graph to provide a unique method for determining the *s*-parameters of a transmission line.

Figure 2.4(a) shows the signal flow graph of a transmission line with characteristic impedance Z_c, which is embedded in a $Z_o\Omega$ system. The s'_{ij} parameters account for the mismatch between the $Z_c\Omega$ transmission line and the $Z_o\Omega$ source and load terminations. The two branches with coefficients $e^{-j\theta}$ represent phase shift along the transmission line. When the transmission line is lossless, θ is equal to $(\omega l/v)$, where ω is the radian frequency, l is the transmission line length, and v is the velocity of propagation along the transmission line. When the line is lossy, θ will be complex and the reader is referred to transmission line theory [2] for a complete discussion of this parameter.

We can use the nontouching loop rule to determine *s*-parameters of the complete transmission line shown in Figure 2.4(a):

$$s_{11} = s_{22} = s'_{11} + \frac{s'_{22}\, s'_{12}\, s'_{21}\, e^{-j2\theta}}{1 - \left(s'_{22}\right)^2 e^{-j2\theta}} \tag{2.7a}$$

$$s_{12} = s_{21} = \frac{s'_{12}\, s'_{21}\, e^{-j\theta}}{1 - \left(s'_{22}\right)^2 e^{-j2\theta}} \tag{2.7b}$$

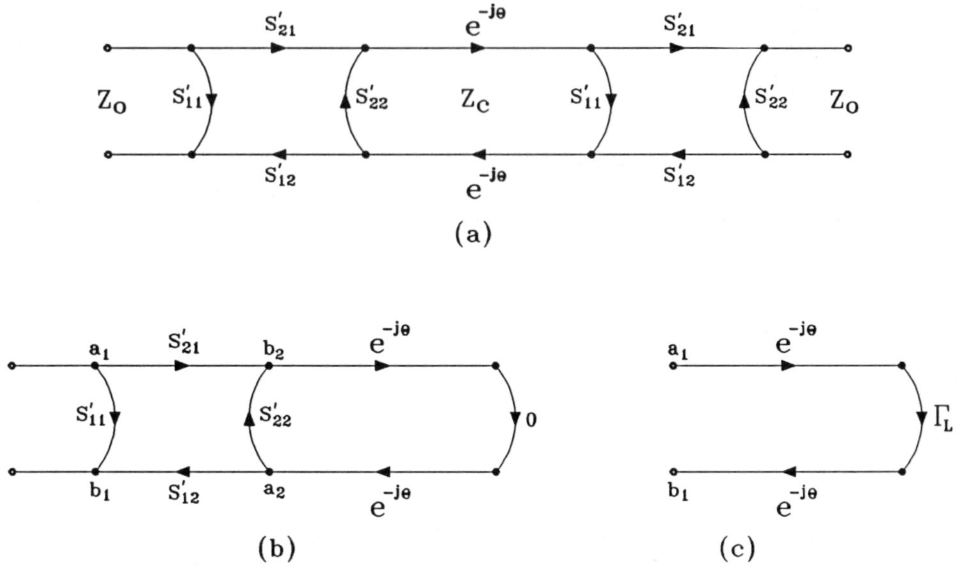

Figure 2.4 (a) Flow graph of the transmission line; (b) line terminated with Z_c; (c) line terminated with Γ_L.

The transmission line is shortened until it has infinitesimal length and θ is zero. This condition is the same as no transmission line at all, or no reflections at either port and perfect transmission from input to output. Substituting θ = 0 into (2.7a) and (2.7b), we have

$$s_{11} = s_{22} = s'_{11} + \frac{s'_{22}\, s'_{12}\, s'_{21}}{1 - \left(s'_{22}\right)^2} = 0 \tag{2.7c}$$

and

$$s_{12} = s_{21} = \frac{s'_{12}\, s'_{21}}{1 - \left(s'_{22}\right)^2} = 1 \tag{2.7d}$$

Using equation (2.7d) in (2.7c),

$$s'_{11} + s'_{22} = 0 \tag{2.7e}$$

and from (2.7d),

$$s'_{12}s'_{21} = 1 - \left(s'_{11}\right)^2 = \left(1 - s'_{11}\right)\left(1 + s'_{11}\right) \tag{2.7f}$$

The transmission line flow graph did not change as the length was made shorter; hence (2.7e) and (2.7f) are valid for any length.

We can terminate the output port with an impedance equal to Z_c, the characteristic impedance of the transmission line. The output port is no longer mismatched and we have the flow graph shown in Figure 2.4(b). When the line is again shortened to zero length, we have the one-port load shown in Figure 1.15(b) and the relationship between incident and reflected waves given by (1.27). Using (2.7e) and Figure 2.4(b), showing that (b_1/a_1), the input reflection coefficient, is s'_{11}, we have

$$s'_{11} = \Gamma = \frac{Z_c - Z_o}{Z_c + Z_o} = -s'_{22} \tag{2.7g}$$

Substituting this relationship along with (2.7f) into (2.7a) and (2.7b), we have the scattering parameters for the transmission line element

$$s_{11} = s_{22} = \frac{\Gamma\left(1 - e^{-j2\theta}\right)}{1 - \Gamma^2 e^{-j2\theta}} \tag{2.8a}$$

and

$$s_{12} = s_{21} = \frac{\left(1 - \Gamma^2\right)e^{-j\theta}}{1 - \Gamma^2 e^{-j2\theta}} \tag{2.8b}$$

We can express (2.8a) and (2.8b) in a slightly different form by substituting (2.5a) for Γ, and Euler's trigonometry identity for $e^{-j\theta}$; that is, $e^{-j\theta} = \text{Cos}\theta - j\text{Sin}\theta$:

$$s_{11} = s_{22} = \frac{z_c - 1/z_c}{\left(z_c + 1/z_c\right) - j2\,\text{Cot}\,\theta} \tag{2.8c}$$

$$s_{12} = s_{21} = \frac{1}{\text{Cos}\,\theta + j\frac{1}{2}\left(z_c + 1/z_c\right)\text{Sin}\,\theta} \tag{2.8d}$$

where z_c is normalized to Z_o, $z_c = Z_c/Z_o$. In the case of a lossy transmission line where θ is complex, (2.8c) and (2.8d) can be written in terms of hyperbolic functions.

We often find it useful to know the input reflection coefficient or impedance when a transmission line is terminated with an arbitrary load. The simplest approach is to reference the s-parameters to the characteristic impedance of the line, Z_c, which results in the signal flow graph shown in Figure 2.4(c). There is no mismatch at either port, $s_{11} = s_{22} = 0$, and the transmission coefficients are $e^{-j\theta}$, which accounts for the phase shift. We can write s_{11} by inspection as

$$s_{11} = \Gamma_L e^{-j2\theta} \tag{2.9a}$$

where

$$\Gamma_L = \frac{Z_L - Z_C}{Z_L + Z_C} \tag{2.9b}$$

In other words, the reflection coefficient of the load is referenced to the characteristic impedance of the transmission line. Using (2.5b), we can write the input impedance as

$$Z_{in} = Z_C \frac{1 + s_{11}}{1 - s_{11}} = Z_c \frac{1 + \Gamma_L e^{-j2\theta}}{1 - \Gamma_L e^{-j2\theta}} \tag{2.9c}$$

Two of the more useful elements resulting from this derivation are the shorted transmission line, $\Gamma_L = -1$:

$$Z_{SC} = Z_C \frac{1 - e^{-j2\theta}}{1 + e^{-j2\theta}} = jZ_C \mathrm{Tan}\,\theta \tag{2.9d}$$

and the open circuit line, $\Gamma_L = +1$:

$$Z_{OC} = Z_C \frac{1 + e^{-j2\theta}}{1 - e^{-j2\theta}} = -jZ_C \mathrm{Cot}\,\theta \tag{2.9e}$$

where Euler's identity is used to write the equivalent trigonometric form. Any parasitics associated with the nonideal terminations, such as fringing capacitance, ground inductance, and radiation, can be accounted for in Γ_L and (2.9c), used to compute the input impedance.

2.1.3 The Ideal Transformer

The ideal transformer shown in Figure 2.5 is an element which is difficult to realize at RF and microwave frequencies. Nevertheless, the element is useful for network analysis. The voltage and currents at the two ports are related through the number of turns in the primary and secondary:

$$n_1 \cdot V_2 = n_2 \cdot V_1$$

$$n_1 \cdot I_1 = -n_2 \cdot I_2$$

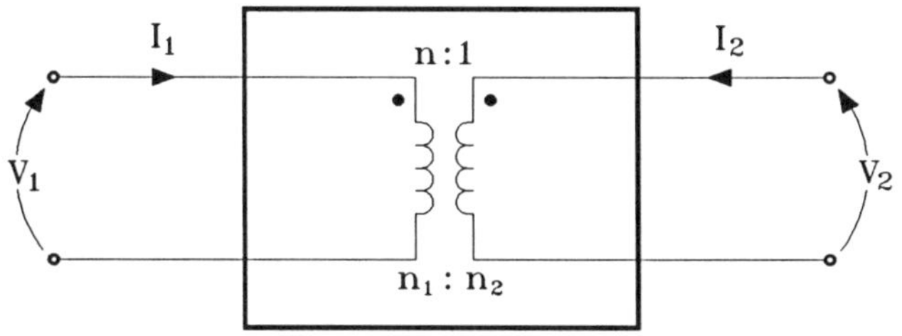

Figure 2.5 Ideal transformer.

and we can express the voltage, current, and impedance in terms of the turns ratio:

$$n = \frac{n_1}{n_2} = \frac{V_1}{V_2} = -\frac{I_2}{I_1} = \sqrt{\frac{Z_1}{Z_2}}$$

We can use these expressions in conjunction with equation (1.5) to compute the *ABCD*-parameters of the ideal transformer. When we open-circuit the secondary, $I_2 = 0$, and we can apply (1.5c):

$$A = \frac{V_1}{V_2} = \frac{n V_2}{V_2} = n$$

$$C = \frac{I_1}{V_2} = \frac{(-I_2/n)}{V_2} = 0$$

then short-circuit the secondary, $V_2 = 0$, and

$$B = -\frac{V_1}{I_2} = -\frac{n V_2}{I_2} = 0$$

$$D = -\frac{I_1}{I_2} = -\frac{(-I_2/n)}{I_2} = \frac{1}{n}$$

or

$$Ch_T = \begin{bmatrix} n & 0 \\ 0 & \frac{1}{n} \end{bmatrix} \tag{2.10}$$

The s-parameters can be computed by connecting a $Z_o \Omega$ voltage source to the primary, and terminating the secondary with a $Z_o \Omega$ load. The $Z_o \Omega$ termination reflects to the primary as an equivalent $n^2 Z_o \Omega$ impedance, and the total impedance in the input loop is $Z_o(n^2 + 1)$. Assuming a 1V source,

$$I_1 = \frac{1}{Z_o(n^2 + 1)}$$

$$I_2 = -n \cdot I_1 = \frac{-n}{Z_o(n^2 + 1)}$$

and

$$V_1 = 1 \cdot \frac{n^2 Z_o}{Z_o(n^2 + 1)} = \frac{n^2}{(n^2 + 1)}$$

$$V_2 = \frac{V_1}{n} = \frac{n}{n^2 + 1}$$

When Z_o is positive real, we can use (1.14) to compute a_1, b_1, and b_2, and then apply (1.15) to determine s_{11} and s_{21}. The termination can be connected to the primary, the voltage source to the secondary, and then a similar procedure can be used to compute the remaining two s-parameters. The **S** matrix for the ideal transformer is

$$S_T = \begin{bmatrix} \frac{n^2 - 1}{n^2 + 1} & \frac{2n}{n^2 + 1} \\ \frac{2n}{n^2 + 1} & -\frac{n^2 - 1}{n^2 + 1} \end{bmatrix} \tag{2.11a}$$

The ideal transformer in Figure 2.5 is shown with dots at the positive terminal of the primary and secondary side, indicating that there is no phase reversal between the input and output voltage. When the dots are shown at opposite ends, the sense of the output voltage and current is reversed, and we must compute another set of parameters:

$$S_T = \begin{bmatrix} \dfrac{n^2 - 1}{n^2 + 1} & \dfrac{-2n}{n^2 + 1} \\ \\ \dfrac{-2n}{n^2 + 1} & -\dfrac{n^2 - 1}{n^2 + 1} \end{bmatrix} \tag{2.11b}$$

which shows that we can simply replace *n* with *-n*. The reader is encouraged to verify these results by reversing the polarity of the output voltage and current and then repeat the above derivation.

2.1.4 Voltage and Current Sources

The voltage and current sources shown in Figures 2.6 and 2.7 are often used to model active devices such as field-effect transistors (FET) and bipolar transistors. The ideal device model should represent the device characteristics in the linear as well as nonlinear regions of operation. The techniques of this chapter are restricted to small-signal linear analysis. However, these methods can be used to fit *s*-parameter data and used as a starting point for nonlinear analysis.

Parameters for all four sources shown in the figures can be computed in a similar manner. Figure 2.6(a) shows a current-dependent voltage source where the output voltage is proportional to the input current, and 2.6(b) shows a source that provides an output voltage proportional to input voltage. The proportionality constants α and μ usually contain exponential terms that provide a time delay

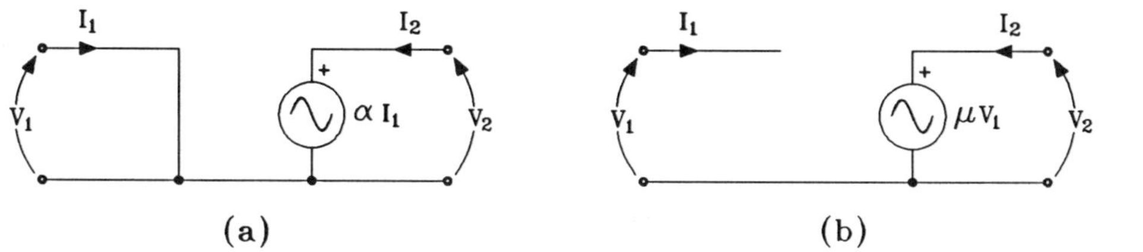

Figure 2.6 Voltage source; (a) current-dependent; (b) voltage-dependent.

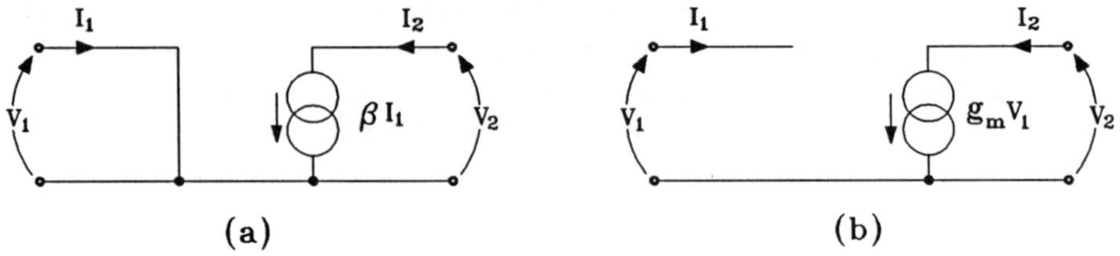

Figure 2.7 Curent source: (a) current-dependent; (b) voltage-dependent.

through the source (e.g., $\alpha e^{-j\omega\tau}$). When we connect a 1V $Z_o\Omega$ source to port 1 of Figure 2.6(a) and terminate port 2 with a $Z_o\Omega$ load, the voltages and currents are

$$V_1 = 0 \qquad I_1 = \frac{1}{Z_o}$$

$$V_2 = \alpha I_1 = \frac{\alpha}{Z_o} \qquad I_2 = -\frac{V_2}{Z_o} = -\frac{\alpha}{Z_o^2}$$

the incident and reflected waves are

$$a_1 = \frac{1}{2\sqrt{Z_o}} \qquad b_1 = \frac{-1}{2\sqrt{Z_o}}$$

$$a_2 = 0 \qquad b_2 = \frac{\alpha}{Z_o\sqrt{Z_o}}$$

and the first column of the s-parameter matrix is

$$s_{11} = \frac{b_1}{a_1} = -1$$

$$s_{21} = \frac{b_2}{a_1} = \frac{2\alpha}{Z_o}$$

Reversing the source and load, the voltages and currents become

$$V_1 = 0 \qquad I_1 = \frac{V_1}{Z_o} = 0$$

$$V_2 = \alpha I_1 = 0 \qquad\qquad I_2 = \frac{1 - V_2}{Z_o} = \frac{1}{Z_o}$$

the second column of the *s*-parameter matrix is

$$s_{12} = \frac{b_1}{a_2} = 0$$

$$s_{22} = \frac{b_2}{a_2} = -1$$

and we can write the complete *s*-parameter matrix as

$$S_{CDVS} = \begin{bmatrix} -1 & 0 \\ \dfrac{2\alpha}{Z_o} & -1 \end{bmatrix} \tag{2.12}$$

Parameters for other elements can be determined using similar techniques. We attach a source at one port, terminate the other port with an appropriate load which eliminates all but one independent variable, compute port voltages and currents and then one column of the parameter matrix. The remaining columns are filled by moving the source to each port and then computing the corresponding matrix column. Figures 2.8 and 2.9 show the *z*-, *y*-, *h*-, *g*-, *ABCD*-, and *s*-parameters for a number of common elements. The impedance and admittance of the series and shunt elements are reciprocal, that is, $Z = 1/Y$ and $Y = 1/Z$. The propagation constant of the transmission lines is discussed in Section 2.1.2, and the reader is referred to transmission line theory [2] for a more detailed description of γ and β.

2.2 NETWORK ANALYSIS

Although the techniques discussed in Chapter 1 are useful for the analysis of many circuits, we can easily find practical networks that cannot be analyzed using these methods. Analysis techniques applying to networks with arbitrary topologies are more complex than those considered in Chapter 1, usually requiring considerably more computer resources and longer execution times. However, a combination of methods can sometimes be used to lessen the impact. We might analyze as much of the network as possible using efficient two-port connections and then interconnect subcircuits using more powerful methods. It also turns out that a significant number of elements in a multiport matrix may be zero, and sparse matrix techniques [3] can be applied to reduce the computation time.

ELEMENT	Z	Y	H
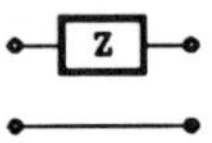	Not defined	$\begin{matrix} Y & -Y \\ -Y & Y \end{matrix}$	$\begin{matrix} Z & 1 \\ -1 & 0 \end{matrix}$
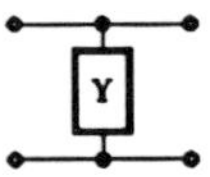	$\begin{matrix} Z & Z \\ Z & Z \end{matrix}$	Not defined	$\begin{matrix} 0 & 1 \\ -1 & Y \end{matrix}$
	Not defined	Not defined	$\begin{matrix} 0 & n \\ -n & 0 \end{matrix}$
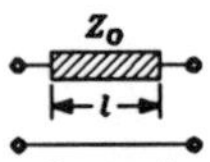 General	$\begin{matrix} Z_0\coth(\gamma l) & Z_0\operatorname{csch}(\gamma l) \\ Z_0\operatorname{csch}(\gamma l) & Z_0\coth(\gamma l) \end{matrix}$	$\begin{matrix} Y_0\coth(\gamma l) & -Y_0\operatorname{csch}(\gamma l) \\ -Y_0\operatorname{csch}(\gamma l) & Y_0\coth(\gamma l) \end{matrix}$	$\begin{matrix} Z_0\tanh(\gamma l) & \operatorname{sech}(\gamma l) \\ -\operatorname{sech}(\gamma l) & Y_0\tanh(\gamma l) \end{matrix}$
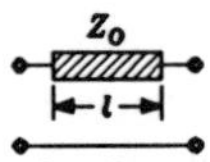 Lossless	$\begin{matrix} -jZ_0\cot(\beta l) & -jZ_0\csc(\beta l) \\ -jZ_0\csc(\beta l) & -jZ_0\cot(\beta l) \end{matrix}$	$\begin{matrix} -jY_0\cot(\beta l) & jY_0\csc(\beta l) \\ jY_0\csc(\beta l) & -jY_0\cot(\beta l) \end{matrix}$	$\begin{matrix} jZ_0\tan(\beta l) & \sec(\beta l) \\ -\sec(\beta l) & jY_0\tan(\beta l) \end{matrix}$
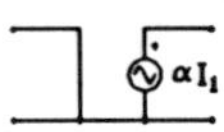	$\begin{matrix} 0 & 0 \\ \alpha & 0 \end{matrix}$	Not defined	Not defined
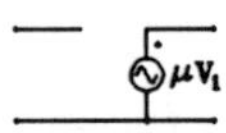	Not defined	Not defined	Not defined
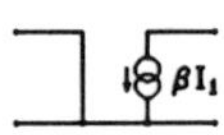	Not defined	Not defined	$\begin{matrix} 0 & 0 \\ \beta & 0 \end{matrix}$
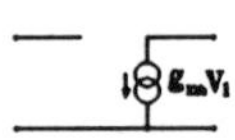	Not defined	$\begin{matrix} 0 & 0 \\ g_m & 0 \end{matrix}$	Not defined

Figure 2.8 Element *z*-, *y*- and *h*-parameters.

ELEMENT	G		ABCD		S	
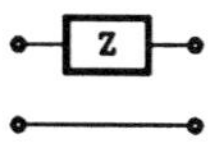	0	-1	1	Z	$\frac{Z}{Z+2Z_0}$	$\frac{2Z}{Z+2Z_0}$
	1	Z	0	1	$\frac{2Z}{Z+2Z_0}$	$\frac{Z}{Z+2Z_0}$
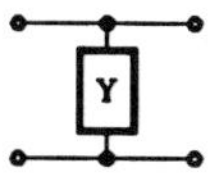	Y	-1	1	0	$\frac{-Z_0Y}{2+Z_0Y}$	$\frac{2}{2+Z_0Y}$
	1	0	Y	1	$\frac{2}{2+Z_0Y}$	$\frac{-Z_0Y}{2+Z_0Y}$
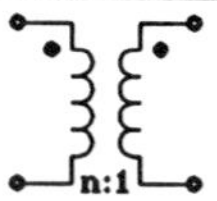	0	$-\frac{1}{n}$	n	0	$\frac{n^2-1}{n^2+1}$	$\frac{2n}{n^2+1}$
	$\frac{1}{n}$	0	0	$\frac{1}{n}$	$\frac{2n}{n^2+1}$	$-\frac{n^2-1}{n^2+1}$
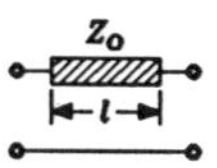	$Y_0\tanh(\gamma l)$	$-\text{sech}(\gamma l)$	$\cosh(\gamma l)$	$Z_0\sinh(\gamma l)$	$\frac{\Gamma(1-e^{-2\gamma l})}{1-\Gamma^2e^{-2\gamma l}}$	$\frac{(1-\Gamma^2)e^{-\gamma l}}{1-\Gamma^2e^{-2\gamma l}}$
General	$\text{sech}(\gamma l)$	$Z_0\tanh(\gamma l)$	$Y_0\sinh(\gamma l)$	$\cosh(\gamma l)$	$\frac{(1-\Gamma^2)e^{-\gamma l}}{1-\Gamma^2e^{-2\gamma l}}$	$\frac{\Gamma(1-e^{-2\gamma l})}{1-\Gamma^2e^{-2\gamma l}}$
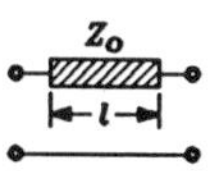	$jY_0\tan(\beta l)$	$-\sec(\beta l)$	$\cos(\beta l)$	$jZ_0\sin(\beta l)$	$\frac{\Gamma(1-e^{-j2\beta l})}{1-\Gamma^2e^{-j2\beta l}}$	$\frac{(1-\Gamma^2)e^{-j\beta l}}{1-\Gamma^2e^{-j2\beta l}}$
Lossless	$\sec(\beta l)$	$jZ_0\tan(\beta l)$	$jY_0\sin(\beta l)$	$\cos(\beta l)$	$\frac{(1-\Gamma^2)e^{-j\beta l}}{1-\Gamma^2e^{-j2\beta l}}$	$\frac{\Gamma(1-e^{-j2\beta l})}{1-\Gamma^2e^{-j2\beta l}}$
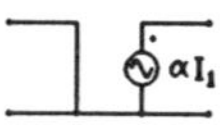	Not defined		0	0	-1	0
			$\frac{1}{\alpha}$	0	$\frac{2\alpha}{Z_0}$	-1
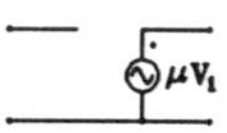	0	0	$\frac{1}{\mu}$	0	1	0
	μ	0	0	0	2μ	-1
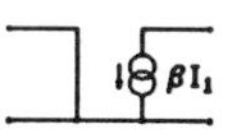	Not defined		0	0	-1	0
			0	$-\frac{1}{\beta}$	-2β	1
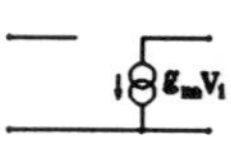	Not defined		0	$-\frac{1}{g_m}$	1	0
			0	0	$-2g_mZ_0$	1

Figure 2.9 Element *g*-, ABCD-, and *s*-parameters.

2.2.1 *S*-Parameter Analysis Using the Connection Matrix

S-parameters are a convenient representation for microwave circuits. They exist for most practical elements, are easy to measure, and provide insight into the characteristics of a network. Using (1.29) we can cascade two-ports, and by converting to other parameter sets we can connect two-ports in series, parallel, series-parallel, and parallel-series. However, many circuits are composed of multiport networks which must be interconnected as illustrated in Figure 2.10.

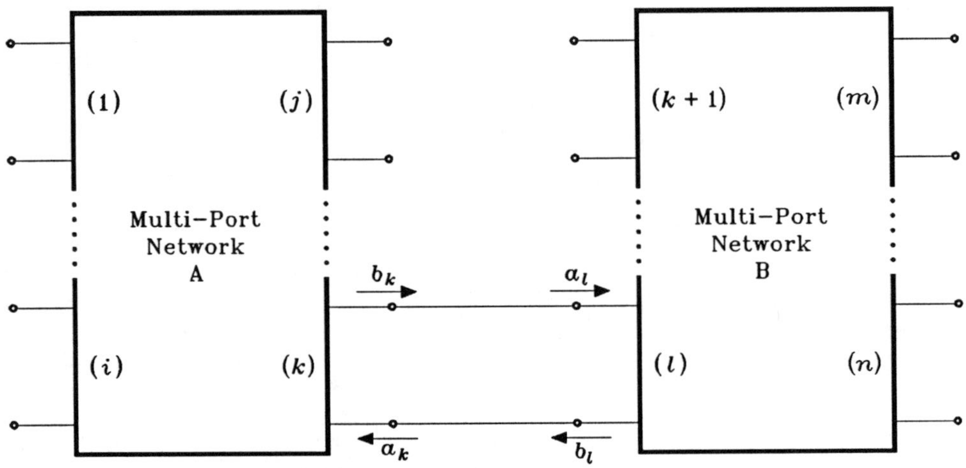

Figure 2.10 Multiport interconnection.

One approach to the multiport interconnection problem [4,5] is to partition the network into external unconnected ports and internal connected ports. External ports, which we denote with the subscript p, are those that remain after the interconnection, and connected ports c provide paths for incident and reflected waves to flow from one network to the other. We can relate the incident and reflected waves at the connected ports by

$$\boldsymbol{b}_c = \boldsymbol{\Gamma}_c \boldsymbol{a}_c \tag{2.13}$$

where $\boldsymbol{\Gamma}_c$ is called the *connection matrix*. In the case of Figure 2.10, where port k of network A is connected to port l of network B, we see that $b_k = a_l$ and $b_l = a_k$, or

$$b_k = 0 \cdot a_k + 1 \cdot a_l$$

$$b_l = 1 \cdot a_k + 0 \cdot a_l$$

and

$$\mathbf{\Gamma}_c = \begin{bmatrix} 0 & 1 \\ 1 & 0 \end{bmatrix}$$

Thus, Γ_{kl} and Γ_{lk} are one while Γ_{kk} and Γ_{ll} are zero. The connection matrix can be written by inspection, the transfer coefficients between connected ports are equal to one, and all other elements are zero.

Partitioning (1.16) into external and connected ports, we can write

$$\begin{bmatrix} \mathbf{b}_p \\ \mathbf{b}_c \end{bmatrix} = \begin{bmatrix} \mathbf{S}_{pp} & \mathbf{S}_{pc} \\ \mathbf{S}_{cp} & \mathbf{S}_{cc} \end{bmatrix} \begin{bmatrix} \mathbf{a}_p \\ \mathbf{a}_c \end{bmatrix} \tag{2.14}$$

and then expand:

$$\mathbf{b}_p = \mathbf{S}_{pp}\mathbf{a}_p + \mathbf{S}_{pc}\mathbf{a}_c \tag{2.15a}$$

$$\mathbf{b}_c = \mathbf{S}_{cp}\mathbf{a}_p + \mathbf{S}_{cc}\mathbf{a}_c \tag{2.15b}$$

Substituting (2.13) into (2.15b),

$$\mathbf{\Gamma}_c\mathbf{a}_c = \mathbf{S}_{cp}\mathbf{a}_p + \mathbf{S}_{cc}\mathbf{a}_c$$

Then collecting terms and premultiplying, we have an expression for the incident waves at the connected ports in terms of incident waves at the external ports:

$$\mathbf{a}_c = \left(\mathbf{\Gamma}_c - \mathbf{S}_{cc}\right)^{-1}\mathbf{S}_{cp}\mathbf{a}_p \tag{2.15c}$$

which can be used in (2.15a) to relate the incident and reflected waves at the external ports:

$$\mathbf{b}_p = \left[\mathbf{S}_{pp} + \mathbf{S}_{pc}\left(\mathbf{\Gamma}_c - \mathbf{S}_{cc}\right)^{-1}\mathbf{S}_{cp}\right]\mathbf{a}_p \tag{2.15d}$$

Thus, the expression for *s*-parameters of the interconnected multiports is

$$\mathbf{S}_p = \mathbf{S}_{pp} + \mathbf{S}_{pc}\left(\mathbf{\Gamma}_c - \mathbf{S}_{cc}\right)^{-1}\mathbf{S}_{cp} \tag{2.16}$$

As an example of the multiport connection, we can cascade two-port A with two-port B. Initially, we have a four-port network consisting of two disconnected two-ports:

$$S = \begin{bmatrix} A_{11} & A_{12} & 0 & 0 \\ A_{21} & A_{22} & 0 & 0 \\ 0 & 0 & B_{11} & B_{12} \\ 0 & 0 & B_{21} & B_{22} \end{bmatrix}$$

We then rearrange as shown in (2.14) to partition the external and connected ports:

$$S = \left[\begin{array}{cc|cc} A_{11} & 0 & A_{12} & 0 \\ 0 & B_{22} & 0 & B_{21} \\ \hline A_{21} & 0 & A_{22} & 0 \\ 0 & B_{12} & 0 & B_{11} \end{array}\right]$$

The connection matrix is

$$\Gamma_c = \begin{bmatrix} 0 & 1 \\ 1 & 0 \end{bmatrix}$$

and we can write (2.16) as

$$S_p = \begin{bmatrix} A_{11} & 0 \\ 0 & B_{22} \end{bmatrix} + \begin{bmatrix} A_{12} & 0 \\ 0 & B_{21} \end{bmatrix} \left\{ \begin{bmatrix} 0 & 1 \\ 1 & 0 \end{bmatrix} - \begin{bmatrix} A_{22} & 0 \\ 0 & B_{11} \end{bmatrix} \right\}^{-1} \begin{bmatrix} A_{21} & 0 \\ 0 & B_{12} \end{bmatrix}$$

$$= \begin{bmatrix} A_{11} & 0 \\ 0 & B_{22} \end{bmatrix} + \frac{1}{A_{22}B_{11} - 1} \cdot \begin{bmatrix} -A_{12}A_{21}B_{11} & -A_{12}B_{12} \\ -A_{21}B_{21} & -A_{22}B_{12}B_{21} \end{bmatrix}$$

which is identical to (1.29), derived using the signal flow graph analysis.

2.2.2 S-Parameter Analysis Using Explicit Expressions

Indications [5] are that the most efficient approach is to connect multiports one port-pair at a time and in sequence, such that the resultant network has the minimum number of ports. When we restrict the connection to a single pair of ports, a simple and efficient explicit expression can be derived[1]. We assume multiport A in Figure 2.10 has n_A ports and network B has n_B ports. Network A is assigned sequential port numbers from 1 to n_A and network B from $n_A + 1$ to n, where $n = n_A + n_B$. Writing the system of equations which describe the network in terms of incident and reflected waves,

$$\begin{aligned}
b_1 &= s_{11}a_1 + s_{12}a_2 + \cdots + s_{1k}a_k + \cdots + s_{1l}a_l + \cdots + s_{1n}a_n \\
b_2 &= s_{21}a_1 + s_{22}a_2 + \cdots + s_{2k}a_k + \cdots + s_{2l}a_l + \cdots + s_{2n}a_n \\
&\vdots \\
b_k &= s_{k1}a_1 + s_{k2}a_2 + \cdots + s_{kk}a_k + \cdots + s_{kl}a_l + \cdots + s_{kn}a_n \\
&\vdots \\
b_l &= s_{l1}a_1 + s_{l2}a_2 + \cdots + s_{lk}a_k + \cdots + s_{ll}a_l + \cdots + s_{ln}a_n \\
&\vdots \\
b_n &= s_{n1}a_1 + s_{n2}a_2 + \cdots + s_{nk}a_k + \cdots + s_{nl}a_l + \cdots + s_{nn}a_n
\end{aligned}$$

Ports k and l are to be interconnected, and from the figure we see that $a_l = b_k$ and $a_k = b_l$. Imposing these constraints on the kth and lth equations of the above matrix yields

$$b_k(1 - s_{kl}) - s_{kk}b_l = \sum_{\substack{j=1 \\ j \neq k,l}}^{n} s_{kj}a_j$$

$$b_l(1 - s_{lk}) - s_{ll}b_k = \sum_{\substack{j=1 \\ j \neq k,l}}^{n} s_{lj}a_j$$

Solving for b_k and b_l in terms of a_j yields

---

[1] This technique, jointly developed by Dr. J. Lamar Allen and the author, has not previously been published.

$$b_l = \sum_{\substack{j=1 \\ j \neq k,l}}^{n} \left[\frac{s_{kj} s_{ll} + s_{lj}(1 - s_{kl})}{(1 - s_{lk})(1 - s_{kl}) - s_{kk} s_{ll}} \right] a_j$$

$$b_k = \sum_{\substack{j=1 \\ j \neq k,l}}^{n} \left[\frac{s_{lj} s_{kk} + s_{kj}(1 - s_{lk})}{(1 - s_{lk})(1 - s_{kl}) - s_{kk} s_{ll}} \right] a_j$$

and then substituting into the original system of equations gives

$$b_i = \sum_{\substack{j=1 \\ j \neq k,l}}^{n} s_{ij} a_j + s_{ik} \sum_{\substack{j=1 \\ j \neq k,l}}^{n} \left[\frac{s_{kj} s_{ll} + s_{lj}(1 - s_{kl})}{(1 - s_{lk})(1 - s_{kl}) - s_{kk} s_{ll}} \right] a_j$$

$$+ s_{il} \sum_{\substack{j=1 \\ j \neq k,l}}^{n} \left[\frac{s_{lj} s_{kk} + s_{kj}(1 - s_{lk})}{(1 - s_{lk})(1 - s_{kl}) - s_{kk} s_{ll}} \right] a_j$$

or

$$b_i = \sum_{\substack{j=1 \\ j \neq k,l}}^{n} \left\{ s_{ij} + \frac{s_{ik}\left[s_{kj} s_{ll} + s_{lj}(1 - s_{kl})\right] + s_{il}\left[s_{lj} s_{kk} + s_{kj}(1 - s_{lk})\right]}{(1 - s_{lk})(1 - s_{kl}) - s_{kk} s_{ll}} \right\} a_j$$

Thus, the *s*-parameters of a multiport system with port *k* connected to port *l* are

$$s_{ij}' = s_{ij} + \frac{s_{ik}\left[s_{kj} s_{ll} + s_{lj}(1 - s_{kl})\right] + s_{il}\left[s_{lj} s_{kk} + s_{kj}(1 - s_{lk})\right]}{(1 - s_{lk})(1 - s_{kl}) - s_{kk} s_{ll}} \qquad (2.17)$$

Equation (2.17) is valid for any multiport that has two ports connected together. However, when we have two disconnected multiports, as in Figure 2.10, the expression can be simplified somewhat. When the networks are not connected, all transmission coefficients from one network to the other are zero; that is, $s_{lk} = s_{kl} = 0$, and (2.17) becomes

$$s_{ij}' = s_{ij} + \frac{s_{ik}\left[s_{kj} s_{ll} + s_{lj}\right] + s_{il}\left[s_{lj} s_{kk} + s_{kj}\right]}{1 - s_{kk} s_{ll}} \qquad (2.18)$$

We can further simplify (2.18) by noting that port *k* is attached to network A and port *l* to network B. As we allow *i* to range from 1 to *n*, the s_{ik} term in the numerator of (2.18) will be zero when i is greater than n_A, and s_{il} is zero when i is less than $n_A + 1$. Thus, for disconnected multiports, (2.18) reduces to

$$s_{ij}^{\prime} = s_{ij} + \frac{s_{ik}\left[s_{kj}s_{ll} + s_{lj}\right]}{1 - s_{kk}s_{ll}} \tag{2.19a}$$

for i less than or equal to the number of ports in A, and

$$s_{ij}^{\prime} = s_{ij} + \frac{s_{il}\left[s_{lj}s_{kk} + s_{kj}\right]}{1 - s_{kk}s_{ll}} \tag{2.19b}$$

for i greater than n_A.

Equation (2.19) is a surprisingly simple result for multiport interconnections of two disconnected networks. However, we must resort to (2.17) when connecting two ports of the same multiport network. In either case, the resulting network will have n - 2 ports, with port numbers in the same order as the original system and ports k and l eliminated. For example, if we have two three-port networks, we can number the ports of network A 1 through 3, and B 4 through 6. Connecting port 2 of A to port 5 of network B, we have a four-port (3 + 3 - 2) network with ports numbered as shown in Figure 2.11.

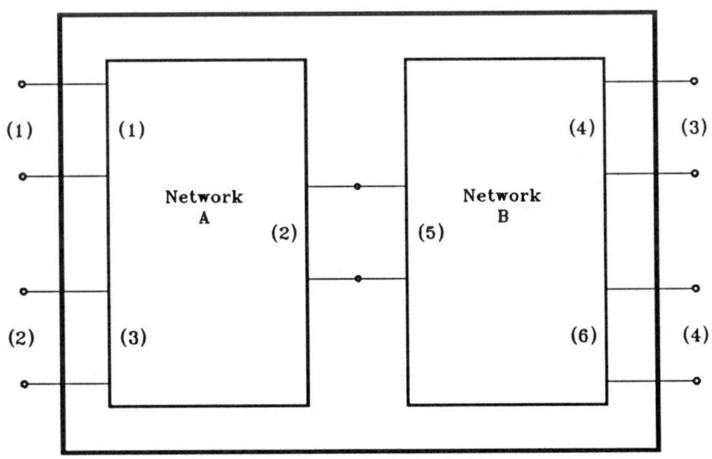

Figure 2.11 Connected three-port networks.

Another useful application of (2.17) is to terminate port k of the multiport network with a one-port that has a reflection coefficient of Γ_k. Since the networks are disconnected, we can apply (2.19), and the expression becomes

$$s'_{ij} = s_{ij} + \frac{s_{ik} s_{kj} \Gamma_k}{1 - s_{kk} \Gamma_k} \tag{2.20}$$

For example, we can create a three-port network as discussed in Section 1.3.7 and then terminate port 3 with parallel feedback. Saving $(s_{ik} s_{kj})$ and using the reciprocal of Γ_k, we can compute all four s-parameters with four additions, one subtraction, and four divisions. This approach is by far the most efficient when we wish to try different values of parallel feedback.

Turning again to the problem of cascading two-ports, we number the input and output ports of network A 1 and 2, the input and output ports of network B 3 and 4, then connect port 2 to port 3. The s-parameter matrix for this system can be written as

$$S = \begin{bmatrix} A_{11} & A_{12} & 0 & 0 \\ A_{21} & A_{22} & 0 & 0 \\ 0 & 0 & B_{11} & B_{12} \\ 0 & 0 & B_{21} & B_{22} \end{bmatrix}$$

and when $i = 1$ and $j = 1$ we can apply (2.19a):

$$s'_{11} = s_{11} + \frac{s_{12}\left[s_{21} s_{33} + s_{31}\right]}{1 - s_{22} s_{33}} = A_{11} + \frac{A_{12}\left[A_{21} B_{11} + 0\right]}{1 - A_{22} B_{11}}$$

and for $i = 1$ and $j = 4$,

$$s'_{14} = s_{14} + \frac{s_{12}\left[s_{24} s_{33} + s_{34}\right]}{1 - s_{22} s_{33}} = 0 + \frac{A_{12}\left[0 \cdot B_{11} + B_{12}\right]}{1 - A_{22} B_{11}}$$

When $i = 4$ and $j = 1$ we use (2.19b):

$$s'_{41} = s_{41} + \frac{s_{43}\left[s_{31} s_{22} + s_{21}\right]}{1 - s_{22} s_{33}} = 0 + \frac{B_{21}\left[0 \cdot A_{22} + A_{21}\right]}{1 - A_{22} B_{11}}$$

and for $i = 4$ and $j = 4$,

$$s'_{44} = s_{44} + \frac{s_{43}\left[s_{34}s_{22} + s_{24}\right]}{1 - s_{22}s_{33}} = B_{22} + \frac{B_{21}\left[B_{12}A_{22} + 0\right]}{1 - A_{22}B_{11}}$$

As this example illustrates, connecting multiports using the explicit expressions given by (2.17) through (2.20) is not at all complicated and is perhaps the most efficient means when dealing with *s*-parameters.

2.2.3 Nodal Analysis

As useful as the *s*-parameters are, they do have some drawbacks for network analysis. When we connect elements in arbitrary topologies, multiport series and parallel tees are required, which increases the number of multiport connections. For example, parallel feedback shown in Figure 1.26 requires two two-port networks (the device and feedback element) and two three-port networks (the two parallel tees), or a total of four connections. We can connect one tee to the input port of the device, forming a three-port, connect the feedback element to the tee, resulting in another three-port, and then connect the other tee to the device output port, forming a four-port network. Equation (2.19) can be used for these connections since the networks are disconnected. However, we must interconnect two ports of the four-port network, which requires (2.17). Another problem with *s*-parameter analysis is that the user is forced to add multiport tees to the circuit, figure out which ports should be connected, and then number the port after the interconnection. These tasks can be implemented as part of a computer-aided design program. Nevertheless, additional networks are required when the user is restricted to *s*-parameter analysis.

One approach that overcomes these problems is *nodal analysis*[6]. Nodal analysis is based on Kirchhoff's current law being applied to each node in the network. Figure 2.12 shows node j of an n-node network, which has an external current I_j flowing into the node. Since the sum of currents must add to zero, we can write

$$-I_j + I_{j1} + I_{j2} + \cdots + I_{jk} + \cdots + I_{jn} = 0$$

or

$$I_j = \sum_{\substack{k=1 \\ k \neq j}}^{n} I_{jk} \tag{2.21a}$$

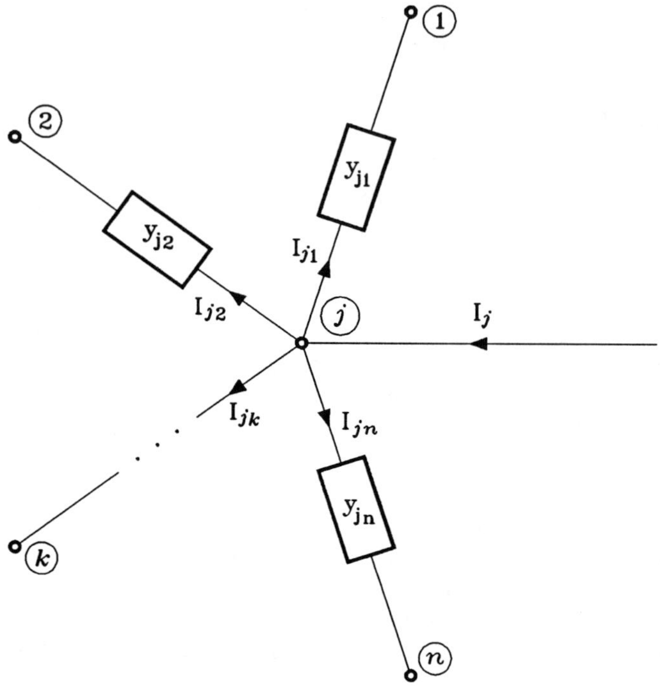

Figure 2.12 Node j of n-node network.

We can use Ohm's law to express the currents in terms of voltage and admittance:

$$I_{j1} = (V_j - V_1)\, y_{j1}$$

$$I_{j2} = (V_j - V_2)\, y_{j2}$$

$$\vdots \qquad \qquad \vdots$$

$$I_{jn} = (V_j - V_n)\, y_{jn}$$

which can be written as

$$I_{jk} = (V_j - V_k)\, y_{jk} \tag{2.21b}$$

and then substituted into (2.21a):

$$I_j = \sum_{\substack{k=1 \\ k \neq j}}^{n} (V_j - V_k)\, y_{jk} = V_j \sum_{\substack{k=1 \\ k \neq j}}^{n} y_{jk} - \sum_{\substack{k=1 \\ k \neq j}}^{n} V_k y_{jk} \qquad (2.21c)$$

Examining the first term in (2.21c), we see that the voltage at node j is multiplied by the sum of all admittances connected to node j. The total admittance at each node is known as the *self admittance*, which we denote with y_{jj}. Expanding (2.21c), we can relate the voltage and currents as

$$\begin{bmatrix} I_1 \\ I_2 \\ \vdots \\ I_k \\ \vdots \\ I_n \end{bmatrix} = \begin{bmatrix} y_{11} & -y_{12} & \cdots & -y_{1k} & \cdots & -y_{1n} \\ -y_{21} & y_{22} & \cdots & -y_{2k} & \cdots & -y_{2n} \\ \vdots & & \vdots & & & \vdots \\ -y_{k1} & -y_{k2} & \cdots & y_{kk} & \cdots & -y_{kn} \\ \vdots & & \vdots & & & \vdots \\ -y_{n1} & -y_{n2} & \cdots & -y_{nk} & \cdots & y_{nn} \end{bmatrix} \begin{bmatrix} V_1 \\ V_2 \\ \vdots \\ V_k \\ \vdots \\ V_n \end{bmatrix} \qquad (2.22)$$

The admittance matrix in (2.22) is known as the *indefinite nodal admittance matrix*, since the voltage has no definite reference similar to the indefinite scattering matrix of Section 1.3.6. The ease with which we can write the admittance matrix for a passive circuit is the most attractive feature of nodal analysis. From (2.21) and (2.22), we see that the diagonal elements y_{ii} are the sum of all admittances connected to node i, and the off-diagonal elements y_{ij} are minus the total admittance connected between nodes i and j.

We can connect a voltage source between node j and the reference node, leaving all other nodes floating. Under these conditions, no current can flow, all node voltages are equal to V_j, and we can write the ith row of (2.22) as

$$I_i = 0 = V_j \sum_{k=1}^{n} y_{ik}$$

where y_{ik} is an element of the nodal admittance matrix equal to the self admittance for $i = k$, or minus the admittance between i and k for $i \neq k$. In general, V_j is not zero; thus, the sum of all elements in any row of the indefinite nodal admittance matrix must equal zero. We can connect all nodes except j to the reference node, which results in zero node voltage for all but the jth node. All columns except j in (2.22) are zero, and we can write the total current as

$$\sum_{i=1}^{n} I_i = V_j \sum_{i=1}^{n} y_{ij}$$

where y_{ij} is an element of the nodal admittance matrix. However, the Kirchhoff current law requires the total current to be zero, and since V_j need not be zero, the sum of all elements in each column of the indefinite nodal admittance matrix is also equal to zero. The results of this analysis shows that there is a linear dependence among the rows and columns of the indefinite nodal admittance matrix; hence, the determinant is zero and (2.22) cannot be solved for the voltage vector. However, the reference node is usually ground which has zero potential and can be eliminated from (2.22), leaving a system of n - 1 equations in n - 1 unknowns. The reduced **Y** matrix is referenced to a particular node and is known as the *definite nodal admittance matrix.*

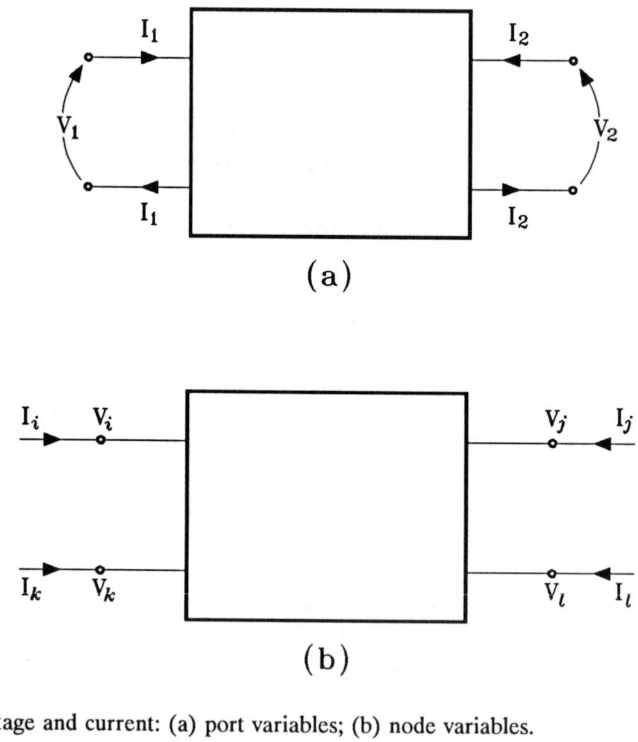

Figure 2.13 Voltage and current: (a) port variables; (b) node variables.

We can write (2.22) by inspection for a network which can be represented as admittance values connected between nodes. The current vector on the left side of (2.22) contains values of external currents imposed on each node. The diagonal elements of the admittance matrix are the total admittance connected to each node, and the off-diagonal elements y_{ij} are minus the total admittance connected between nodes i and j. Some elements cannot be represented as simple admittance values, such as the transmission line or measured transistor data. However, we can usually characterize these components using the y-parameters discussed in Section 1.1.2. Figure 2.13(a) shows the voltage and current definitions for a four-terminal, two-port network, and 2.13(b) shows the definitions in terms of nodes. We can use (1.2a) to express the port currents in terms of port voltages:

$$\begin{aligned} I_1 &= y_{11}V_1 + y_{12}V_2 \\ I_2 &= y_{21}V_1 + y_{22}V_2 \end{aligned} \tag{2.23a}$$

Referring to the figure, we can write the relationship between port and node variables:

$$\begin{aligned} V_1 &= (V_i - V_k) & \qquad V_2 &= (V_j - V_l) \\ I_i &= I_1 & I_j &= I_2 \\ I_k &= -I_1 & I_l &= -I_2 \end{aligned}$$

which can be substituted into (2.23a):

$$\begin{aligned} I_i &= I_1 = y_{11}(V_i - V_k) + y_{12}(V_j - V_l) = y_{11}V_i + y_{12}V_j - y_{11}V_k - y_{12}V_l \\ I_j &= I_2 = y_{21}(V_i - V_k) + y_{22}(V_j - V_l) = y_{21}V_i + y_{22}V_j - y_{21}V_k - y_{22}V_l \\ I_k &= -I_1 = -\left[y_{11}V_i + y_{12}V_j - y_{11}V_k - y_{12}V_l\right] \\ I_l &= -I_2 = -\left[y_{21}V_i + y_{22}V_j - y_{21}V_k - y_{22}V_l\right] \end{aligned}$$

and then written in matrix form:

$$\begin{bmatrix} I_i \\ I_j \\ I_k \\ I_l \end{bmatrix} = \begin{bmatrix} y_{11} & y_{12} & -y_{11} & -y_{12} \\ y_{21} & y_{22} & -y_{21} & -y_{22} \\ -y_{11} & -y_{12} & y_{11} & y_{12} \\ -y_{21} & -y_{22} & y_{21} & y_{22} \end{bmatrix} \begin{bmatrix} V_i \\ V_j \\ V_k \\ V_l \end{bmatrix} \tag{2.23b}$$

When the two-port in Figure 2.13(a) represents a three terminal device, we have only three nodes:

$$V_k = V_l$$

and (2.23b) reduces to

$$\begin{bmatrix} I_i \\ I_j \\ I_k \end{bmatrix} = \begin{bmatrix} y_{11} & y_{12} & -(y_{11} + y_{12}) \\ y_{21} & y_{22} & -(y_{21} + y_{22}) \\ -(y_{11} + y_{21}) & -(y_{12} + y_{22}) & (y_{11} + y_{12} + y_{21} + y_{22}) \end{bmatrix} \begin{bmatrix} V_i \\ V_j \\ V_k \end{bmatrix} \tag{2.23c}$$

In both (2.23b) and (2.23c), we simply place the four y-parameters in their normal position in the y-parameter matrix and then fill the remaining elements using the constraint that the sum of rows and columns is equal to zero.

We can write the total current at node i as

$$I_i = y_{i1}V_1 + \cdots + y_{ii}V_i + y_{ij}V_j + y_{ik}V_k + y_{il}V_l + \cdots + y_{in}V_n$$

and then when the two-port is connected to nodes i, j, k, and l, an additional current equal to

$$y_{11}V_i + y_{12}V_j - y_{11}V_k - y_{12}V_l$$

must be added to node i, and the total current becomes

$$\begin{aligned} I_i' = {} & y_{i1}V_1 + \cdots \\ & + (y_{ii} + y_{11})V_i + (y_{ij} + y_{12})V_j + (y_{ik} - y_{11})V_k + (y_{il} - y_{12})V_l \\ & + \cdots + y_{in}V_n \end{aligned}$$

where the numeric subscripts refer to the two-port y-parameters. Completing the analysis for nodes j, k, and l, we find that two-port y-parameters are included in the indefinite nodal admittance matrix by adding the appropriate parameters to the existing elements. We can extend this result to n-port networks with similar results. However, the constraint of equal but opposite currents flowing in the positive and negative terminals of a port must be observed, otherwise the transformation from port to node variables is not valid.

One of the disadvantages of nodal analysis is that some practical elements cannot be represented using y-parameters. As shown in Figure 2.8, the ideal transformer, both voltage sources and one of the current sources do not have finite y-parameters. Thus, nodal analysis in the form of (2.22) is only marginally useful, and we must modify our approach to include all elements before we have a general purpose analysis technique.

Figure 2.14(a) shows the ideal transformer in terms of port variables and 2.14(b) in terms of node voltages and currents. From Section 2.1.3, the voltage and current relationship between the primary and secondary of the ideal transformer is

$$V_1 = nV_2 \tag{2.24a}$$

$$I_2 = -nI_1 \tag{2.24b}$$

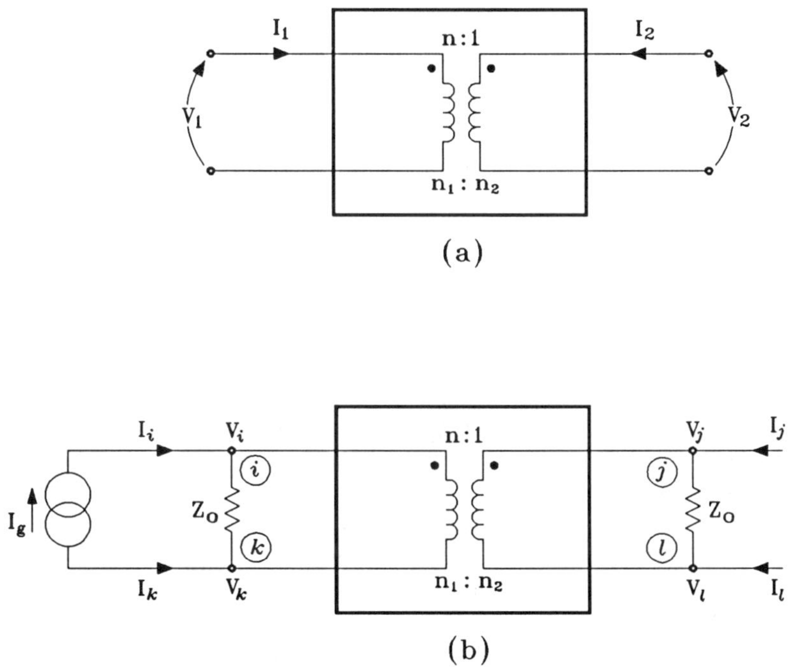

Figure 2.14 Ideal transformer: (a) port; (b) node variables.

Using (2.24a), we can relate the node voltages as

$$0 = V_1 - nV_2 = (V_i - V_k) - n(V_j - V_l) = V_i - nV_j - V_k + nV_l \tag{2.24c}$$

and from (2.24b), the node currents as

$$I_i = I_1 \qquad I_j = I_2 = -nI_1$$
$$I_k = -I_1 \qquad I_l = -I_2 = nI_1 \tag{2.24d}$$

Equation (2.24c) relates the four node voltages; however, (2.24d) expresses the four node currents in terms of the current I_1. We can augment the system of equations given in (2.22) by one row and column to include the conditions given by (2.24c) and (2.24d):

$$\begin{bmatrix} I_i \\ I_j \\ I_k \\ I_l \\ \hline 0 \end{bmatrix} = \begin{array}{c} \\ i \\ j \\ k \\ l \\ \\ \end{array} \begin{array}{c} \begin{array}{cccc} i & j & k & l \end{array} \\ \left[\begin{array}{cccc|c} 0 & 0 & 0 & 0 & 1 \\ 0 & 0 & 0 & 0 & -n \\ 0 & 0 & 0 & 0 & -1 \\ 0 & 0 & 0 & 0 & n \\ \hline 1 & -n & -1 & n & 0 \end{array}\right] \end{array} \begin{bmatrix} V_i \\ V_j \\ V_k \\ V_l \\ \hline I_1 \end{bmatrix} \tag{2.25}$$

The modification to the nodal admittance matrix shown in (2.25) is known as *modified nodal analysis* [3], and provides us with a means to solve for node voltages in a network with components which do not have **Y** matrices.

Figure 2.14(b) shows a current generator and $Z_o\Omega$ impedance connected between nodes *i* and *k*, and another $Z_o\Omega$ impedance connected from node *j* to node *l*. We assign a value of 1A to the current generator; then $I_i = 1$, $I_k = -1$, and $I_j = I_l = 0$. The admittance of the $Z_o\Omega$ impedance is Y_o and the total admittance connected to each of the *i*, *j*, *k*, and *l* nodes is Y_o, or $y_{ii} = y_{jj} = y_{kk} = y_{ll} = Y_o$. The admittance connected between the input and output nodes is also Y_o, and the off-diagonal elements are $y_{ik} = y_{ki} = -Y_o$ and $y_{jl} = y_{lj} = -Y_o$. Adding these values to (2.25), we have

$$\begin{bmatrix} 1 \\ 0 \\ -1 \\ 0 \\ 0 \end{bmatrix} = \begin{bmatrix} Y_o & 0 & -Y_o & 0 & 1 \\ 0 & Y_o & 0 & -Y_o & -n \\ -Y_o & 0 & Y_o & 0 & -1 \\ 0 & -Y_o & 0 & Y_o & n \\ 1 & -n & -1 & n & 0 \end{bmatrix} \begin{bmatrix} V_i \\ V_j \\ V_k \\ V_l \\ I_1 \end{bmatrix} \tag{2.26a}$$

Then, assuming nodes k and l are connected to ground, we can eliminate the kth and lth rows and columns, resulting in three equations and three unknowns:

$$\begin{bmatrix} 1 \\ 0 \\ 0 \end{bmatrix} = \begin{bmatrix} Y_o & 0 & 1 \\ 0 & Y_o & -n \\ 1 & -n & 0 \end{bmatrix} \begin{bmatrix} V_i \\ V_j \\ I_1 \end{bmatrix} \tag{2.26b}$$

which has the solution

$$\begin{aligned} V_i &= \frac{n^2}{Y_o(n^2 + 1)} \\ V_j &= \frac{n}{Y_o(n^2 + 1)} \\ I_1 &= \frac{1}{(n^2 + 1)} \end{aligned} \tag{2.26c}$$

We can develop the modified nodal formation for other elements using the same approach: write the voltage and current definitions and then augment the admittance matrix with auxiliary rows and columns as necessary to include these conditions. For example, the voltage source shown in Figure 2.15(a) can be described by

$$V_s = V_i - V_j$$

$$I_i = I$$

$$I_j = -I$$

and the modified nodal matrix as

$$\begin{bmatrix} I_i \\ I_j \\ \hline V_s \end{bmatrix} = \begin{array}{c} \\ i \\ j \\ \\ \end{array} \begin{bmatrix} \begin{array}{cc|c} \overset{i}{0} & \overset{j}{0} & 1 \\ 0 & 0 & -1 \\ \hline 1 & -1 & 0 \end{array} \end{bmatrix} \begin{bmatrix} V_i \\ V_j \\ \hline I \end{bmatrix} \tag{2.27}$$

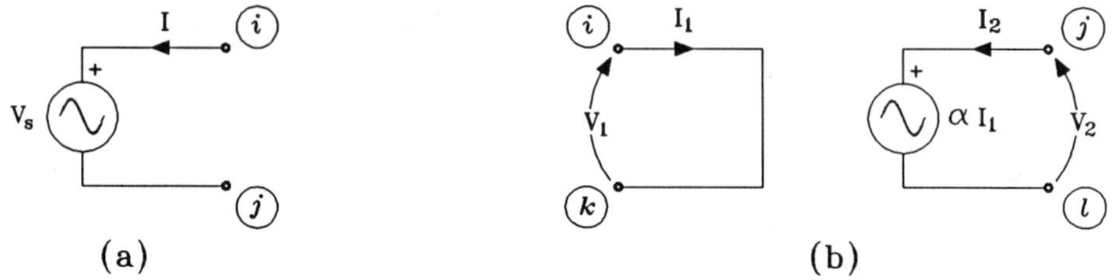

Figure 2.15 Sources: (a) voltage; (b) current-dependent.

The current-dependent voltage source shown in Figure 2.15(b) can be characterized by

$$0 = V_i - V_k \qquad\qquad 0 = V_2 - \alpha I_1 = V_j - V_l - \alpha I_1$$

$$I_i = I_1 \qquad\qquad I_j = I_2$$

$$I_k = -I_1 \qquad\qquad I_l = -I_2$$

and the system of equations becomes

$$\begin{bmatrix} I_i \\ I_j \\ I_k \\ I_l \\ \hline 0 \\ 0 \end{bmatrix} = \begin{array}{c} i \\ j \\ k \\ l \\ \\ \\ \end{array} \overset{\begin{array}{cccc} i & j & k & l \end{array}}{\left[\begin{array}{cccc|cc} 0 & 0 & 0 & 0 & 1 & 0 \\ 0 & 0 & 0 & 0 & 0 & 1 \\ 0 & 0 & 0 & 0 & -1 & 0 \\ 0 & 0 & 0 & 0 & 0 & -1 \\ \hline 1 & 0 & -1 & 0 & 0 & 0 \\ 0 & 1 & 0 & -1 & -\alpha & 0 \end{array}\right]} \begin{bmatrix} V_i \\ V_j \\ V_k \\ V_l \\ \hline I_1 \\ I_2 \end{bmatrix} \tag{2.28}$$

In this case, it was necessary to add two auxiliary rows and columns, although most elements require only one.

We can summarize the steps necessary to apply the modified nodal analysis technique for generalized circuit analysis as follows:

- Form the indefinite nodal admittance matrix by computing the total admittance connected to each node and then enter these values on the main diagonal.

- Enter off-diagonal elements y_{ij} as minus the sum of the admittance connected between nodes *i* and *j*.

- Convert *y*-parameter data to equivalent nodal representation and then add to the corresponding elements of the indefinite nodal admittance matrix.

- Include elements that cannot be represented using *y*-parameters by adding auxiliary rows and columns.

Once the system of equations has been defined for a particular network, we must solve for the parameters that characterize the network. There are a number of techniques that are applicable to solving a system of equations. However, one of the more efficient is *Gaussian elimination* [3]. Gaussian elimination consists of triangularizing the set of equations such that all elements below the main diagonal are zero, and then applying back substitution to solve for the unknowns. This approach also has the advantage that multiple source vectors can be defined, and an array of solution vectors can be generated from one application of the Gaussian elimination algorithm.

We also have a choice of several approaches that can be used to characterize the network. When the definite admittance matrix does not include auxiliary rows and columns, we can use *pivotal condensation* [3] to suppress internal nodes resulting in multiport *y*-parameters. For example, given the equations for a three-node system,

$$\begin{bmatrix} I_1 \\ I_2 \\ I_3 \end{bmatrix} = \begin{bmatrix} y_{11} & y_{12} & y_{13} \\ y_{21} & y_{22} & y_{23} \\ y_{31} & y_{32} & y_{33} \end{bmatrix} \begin{bmatrix} V_1 \\ V_2 \\ V_3 \end{bmatrix}$$

we can eliminate node 3, resulting in two-port *y*-parameters. From the bottom row,

$$V_3 = \frac{1}{y_{33}} \left[I_3 - y_{31} V_1 - y_{31} V_2 \right]$$

which can be substituted into the other two equations:

$$\begin{bmatrix} I_1 - \dfrac{y_{13}}{y_{33}} I_3 \\ \\ I_2 - \dfrac{y_{23}}{y_{33}} I_3 \end{bmatrix} = \begin{bmatrix} \left(y_{11} - \dfrac{y_{13} y_{31}}{y_{33}} \right) & \left(y_{12} - \dfrac{y_{13} y_{32}}{y_{33}} \right) \\ \\ \left(y_{21} - \dfrac{y_{23} y_{31}}{y_{33}} \right) & \left(y_{22} - \dfrac{y_{23} y_{32}}{y_{33}} \right) \end{bmatrix} \begin{bmatrix} V_1 \\ \\ V_2 \end{bmatrix}$$

In general, we can suppress node k by selecting y_{kk} as the pivot element and then computing the modified elements of the reduced **Y** matrix from

$$y'_{ij} = y_{ij} - \frac{y_{ik} y_{kj}}{y_{kk}} \tag{2.29}$$

When the eliminated node is internal, it has no external current source, $I_k = 0$, the current vector remains unchanged, and

$$\begin{bmatrix} I_1 \\ I_2 \end{bmatrix} = \begin{bmatrix} y'_{11} & y'_{12} \\ y'_{21} & y'_{22} \end{bmatrix} \begin{bmatrix} V_1 \\ V_2 \end{bmatrix}$$

We can also follow an approach similar to that derived in Section 2.2.1, where the system is partitioned into external nodes denoted by the subscript p and internal nodes denoted by subscript c:

$$\begin{bmatrix} \boldsymbol{I}_p \\ \boldsymbol{I}_c \end{bmatrix} = \begin{bmatrix} \boldsymbol{Y}_{pp} & \boldsymbol{Y}_{pc} \\ \boldsymbol{Y}_{cp} & \boldsymbol{Y}_{cc} \end{bmatrix} \begin{bmatrix} \boldsymbol{V}_p \\ \boldsymbol{V}_c \end{bmatrix} \tag{2.30a}$$

Then,

$$\boldsymbol{I}_p = \left[\boldsymbol{Y}_{pp} - \boldsymbol{Y}_{pc} \boldsymbol{Y}_{cc}^{-1} \boldsymbol{Y}_{cp}\right] \boldsymbol{V}_p = \boldsymbol{Y}'_{pp} \boldsymbol{V}_p \tag{2.30b}$$

This approach is less efficient than reducing one node at a time. However, we can compute **Y′** once and then use it for the analysis at all frequencies. Once the system of equations has been reduced to external voltage and current using either method, we can apply (1.18c) to convert the multiport y-parameters into s-parameters.

In some instances pivotal condensation cannot be applied (e.g., the system of equations given in (2.26b)). When (2.29) or (2.30) is applied to eliminate the third row and column, we find that the pivotal element is zero. However, we can use Gaussian elimination to solve for V_i, V_j, and I_1. In fact, once we have the node voltages, the branch currents or s-parameters can readily be computed. Figure 2.16(a) shows an external current source and $Z_i \Omega$ impedance connected between two nodes that forms port i of the network. The port voltage V_i and current I_i are defined as shown in the figure. Current flowing into port i is $(I_g - I_z)$, and, using (1.14), we can write the incident wave as

$$a_i = \frac{V_i + Z_i(I_g - I_z)}{2\cdot\sqrt{|Re\,Z_i|}} = \frac{V_i + Z_i(I_g - V_i/Z_i)}{2\cdot\sqrt{|Re\,Z_i|}} = \frac{Z_i I_g}{2\cdot\sqrt{|Re\,Z_i|}}$$

and the reflected wave at port i as

$$b_i = \frac{V_i - Z_i^*(I_g - I_z)}{2\cdot\sqrt{|Re\,Z_i|}} = \frac{V_i - Z_i^*(I_g - V_i/Z_i)}{2\cdot\sqrt{|Re\,Z_i|}} = \frac{V_i\left(1 + Z_i^*/Z_i\right) - Z_i^* I_g}{2\cdot\sqrt{|Re\,Z_i|}}$$

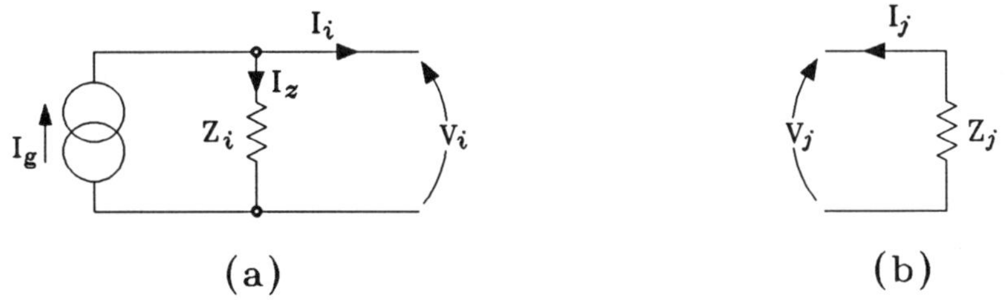

Figure 2.16 (a) Current source; (b) termination.

Figure 2.16(b) shows port j, which is terminated by a $Z_j\Omega$ impedance connected between the pair of nodes. The reflected wave at this port is

$$b_j = \frac{V_j - Z_j^* I_j}{2\cdot\sqrt{|Re\,Z_j|}} = \frac{V_j - Z_j^*(-V_j/Z_j)}{2\cdot\sqrt{|Re\,Z_j|}} = \frac{V_j\left(1 + Z_j^*/Z_j\right)}{2\cdot\sqrt{|Re\,Z_j|}}$$

We can use a_i and b_i to compute s_{ii} as

$$s_{ii} = \frac{b_i}{a_i} = \frac{V_i\left(1 + Z_i^*/Z_i\right) - Z_i^* I_g}{Z_i I_g} = \frac{V_i}{Z_i I_g}\left(1 + \frac{Z_i^*}{Z_i}\right) - \frac{Z_i^*}{Z_i} \tag{2.31a}$$

and then a_i and b_j to compute s_{ji} as

$$s_{ji} = \frac{b_j}{a_i} = \frac{V_j\left(1 + Z_j^*/Z_j\right)}{2\cdot\sqrt{|\mathrm{Re}\,Z_j|}} \Bigg/ \frac{Z_i I_g}{2\cdot\sqrt{|\mathrm{Re}\,Z_i|}} = \frac{V_j}{Z_i I_g}\cdot\frac{\sqrt{|\mathrm{Re}\,Z_i|}}{\sqrt{|\mathrm{Re}\,Z_j|}}\cdot\left(1 + \frac{Z_j^*}{Z_j}\right) \tag{2.31b}$$

A convenient value for the source current is 1A, which eliminates I_g from (2.31), and when the reference impedance is positive real,

$$s_{ii} = \frac{2V_i}{Z_i} - 1$$
$$s_{ji} = \frac{2V_j}{\sqrt{Z_i \cdot Z_j}} \qquad (2.32)$$

A better choice for source current might be 2A, which eliminates the factor of 2 in the numerator of (2.32); however, a factor of 2 is introduced into the denominator of (2.31a) and (2.31b). The best approach is to use 2A when all reference values are positive real, and 1A in other cases.

With minimal effort, modified nodal analysis can be used to provide the network *s*-parameters:

- Form the modified nodal matrix.
- Augment the matrix by connecting a reference impedance between each pair of nodes, which forms a port.
- Apply an external current to the positive node of port *i*.
- Solve the system of equations for node voltages.
- Compute port voltage as the difference between node voltages and then apply (2.31) or (2.32) to compute s_{ii} and s_{ji}, where $j = 1$ to n, and $j \neq i$.
- Repeat the previous three steps for each port in the network.

When using Gaussian elimination, we can create n source vectors, each with a single current source at a different port, and then apply the algorithm to solve for n node voltage vectors simultaneously.

The ideal transformer shown in Figure 2.14(b) has been augmented with a $Z_o\Omega$ reference impedance at the input and output ports. Equation (2.26c) shows the node voltages resulting from a 1A current source applied to the positive node of the input port when nodes k and l are grounded. Assuming Z_o is positive real, we can apply (2.32):

$$s_{11} = \frac{2V_1}{Z_o} - 1 = \frac{2 \cdot \left[n^2 Z_o / (n^2 + 1)\right]}{Z_o} - 1 = \frac{n^2 - 1}{n^2 + 1}$$

look at maximum gain, minimum noise figure, and maximum output power for a number of devices, select those that best match the design specifications, then make a final decision based on cost, availability, reliability, or a combination of factors.

Once the device is selected, we must determine the source and load impedance necessary to provide the desired performance. We can design input and output matching networks based on the source and load impedance, combine these networks with the active device, and then analyze the network to confirm that the design meets all specifications. Most modern active devices are potentially unstable, which means that the device can oscillate under certain conditions. We must therefore consider stability in the device selection process as well as in the design of input and output matching networks. In some cases it is necessary to add stabilization networks which reduce the gain and output power while increasing the noise figure. A device that is acceptable before stabilization may not be acceptable after it is stabilized.

Given the network parameters, it is imperative that we develop the means to completely characterize the network in a form that provides useful information for circuit design. Otherwise, we might as well use trial and error design techniques.

2.3.1 Active and Passive Networks

Conservation of energy requires that the total power incident at all ports of a passive network equal the power absorbed by the network, plus power emerging from the network. The power dissipated in the network is the difference between incident and reflected power, or $|a_i|^2 - |b_i|^2$ from the results of Section 1.3. Total dissipated power is the sum of the dissipated power at each port:

$$P_d = \sum_{i=1}^{n} \left(|a_i|^2 - |b_i|^2\right) = \sum_{i=1}^{n} a_i a_i^* - \sum_{i=1}^{n} b_i b_i^*$$

or

$$P_d = [a^*]^T a - [b^*]^T b \tag{2.33a}$$

where $[\mathbf{a}^*]^T$ and $[\mathbf{b}^*]^T$ are created by replacing each element of **a** and **b** with its complex conjugate and then interchanging rows and columns to form the transpose. From (1.16),

$$b = S a$$

$$[b^*]^T = [S^*]^T [a^*]^T$$

$$s_{21} = \frac{2V_2}{\sqrt{Z_o \cdot Z_o}} = \frac{2 \cdot [nZ_o/(n^2 + 1)]}{\sqrt{Z_o^2}} = \frac{2n}{n^2 + 1}$$

which are the same expressions derived in Section 2.1.3.

Nodal analysis has the advantage that the system of equations is easil generated directly from the circuit topology. There is no need for additiona connection elements, although some components do require auxiliary rows an columns. The majority of components in a circuit can be represented as a admittance, and since off-diagonal elements are equal to minus the value o admittance connected between nodes, the nodal admittance matrix has a high percentage of zero entries. Sparse matrix techniques [3] can be applied to reduce the computational effort required to solve the system of equations.

Another attractive feature of nodal analysis is realized in conjunction with optimization. The more powerful optimization techniques, which are described in Chapter 3, use a *gradient*, the derivative of the error function with respect to element values. The gradient can be determined numerically by computing the error function, perturbing the element value slightly, computing a new error function, and then dividing the change in error function by the change in element value. This approach requires an initial analysis, plus an additional analysis for each element. Each analysis may contain many frequency points, and considerable effort is required to compute a single array of gradients. In addition, the gradient is defined for an infinitesimal change in element value, and numerical problems occur when the change is made too small. Increased accuracy results when the forward and backward difference is used, wherein the element is increased and decreased a small amount. However, this requires two analyses per element. The *adjoint network method* [8,9] is based on Tellegen's theorem [10] and allows us to compute all gradients from only two analyses, an analysis of the original network and the adjoint network, which can be formed directly from the original network. Since the gradient is based on the voltage and currents in the two networks rather than small changes in element values, the accuracy is improved, as well as the computational efficiency.

2.3 NETWORK CHARACTERIZATION

We can use the techniques of Chapter 1 and the first two sections of this chapte to analyze most networks. *S*-parameters provide valuable information describin the gain versus frequency, as well as the input and output return loss and voltag standing wave ratio (VSWR). However, this information alone is usually n sufficient to commit a design to production or provide insight into the behavior an active device. One of the more common tasks is amplifier design, where t first, and perhaps most important, step is selecting an active device. Typically, which we can substitute into (2.33a):

$$P_d = [a^*]^T a - [S^*]^T [a^*]^T S\, a$$

and then rewrite as

$$P_d = [a^*]^T \{U - [S^*]^T S\} a \tag{2.33b}$$

where **U** is the unit matrix. In general, the incident power is positive and the term within braces in (2.33b) determines whether the dissipated power is positive or negative. We can define this term as

$$Q = U - [S^*]^T S \tag{2.34}$$

which is known as the *dissipation matrix* [11]. When **Q** is nonnegative, the network is passive, or the dissipated power is greater than or equal to zero.

The necessary and sufficient condition for positive real **Q** is nonnegative principle minors. Thus, for a passive two-port we have

$$|s_{11}|^2 + |s_{21}|^2 \le 1 \tag{2.35a}$$

and

$$|s_{22}|^2 + |s_{12}|^2 \le 1 \tag{2.35b}$$

When the two-port is lossless, the dissipated power is zero and we can write (2.34) as

$$[S^*]^T S = U$$

or

$$\begin{bmatrix} s_{11}^* & s_{21}^* \\ s_{12}^* & s_{22}^* \end{bmatrix} \begin{bmatrix} s_{11} & s_{12} \\ s_{21} & s_{22} \end{bmatrix} = \begin{bmatrix} 1 & 0 \\ 0 & 1 \end{bmatrix}$$

which can be expanded as

$$s_{11}^* s_{11} + s_{21}^* s_{21} = 1 \tag{2.36a}$$

$$s_{11}^* s_{12} + s_{21}^* s_{22} = 0 \tag{2.36b}$$

$$s_{12}^* s_{11} + s_{22}^* s_{21} = 0 \tag{2.36c}$$

$$s_{12}^* s_{12} + s_{22}^* s_{22} = 1 \tag{2.36d}$$

Solving for s_{22} in (2.36b), we have

$$s_{22} = -\frac{s_{11}^* s_{12}}{s_{21}^*}$$

Then substituting into (2.36d), we have

$$s_{12}^* s_{12} + \frac{s_{11} s_{11}^* s_{12}^* s_{12}}{s_{21} s_{21}^*} = 1$$

Finally, we can solve (2.36a) for $s^*_{11}s_{11}$ and then substitute:

$$s_{12}^* s_{12} + \frac{\left(1 - s_{21}^* s_{21}\right) s_{12}^* s_{12}}{s_{21} s_{21}^*} = 1$$

or

$$s_{12}^* s_{12} = s_{21} s_{21}^* \tag{2.37a}$$

Similarly, we can show that

$$s_{11}^* s_{11} = s_{22}^* s_{22} \tag{2.37b}$$

Therefore, for a lossless two-port network we have

$$|s_{11}|^2 + |s_{21}|^2 = 1 \tag{2.38a}$$

$$|s_{12}|^2 + |s_{22}|^2 = 1 \tag{2.38b}$$

$$|s_{12}|^2 = |s_{21}|^2 \tag{2.38c}$$

and

$$|s_{11}|^2 = |s_{22}|^2 \tag{2.38d}$$

2.3.2 Stability

We can investigate stability by considering the interaction of a network with its external terminations. Figure 2.17(a) shows an external load Z_k connected to port k of an n-port network and Figure 2.17(b) shows the equivalent signal flow graph. An incident wave a travels through the path s_{kj} into the termination, where it is modified by Γ_k and emerges as a reflected wave. At this node, the wave can travel through the path s_{jk} to another port, or through s_{kk} into the termination, where it is again modified by Γ_k and emerges at the same node a second time. The $s_{kk}\Gamma_k$ path forms a loop where the traveling wave can circulate an infinite number of times. The effect of this loop is to modify the main signal path $s_{kj}s_{jk}\Gamma_k$ by

$$1 + s_{kk}\Gamma_k + (s_{kk}\Gamma_k)^2 + (s_{kk}\Gamma_k)^3 + (s_{kk}\Gamma_k)^4 + \cdots \tag{2.39a}$$

If this series is convergent, its value is

$$\frac{1}{1 - s_{kk}\Gamma_k} \tag{2.39b}$$

and the overall relationship between incident and reflected waves is

$$\frac{b}{a} = \frac{s_{jk}s_{kj}\Gamma_k}{1 - s_{kk}\Gamma_k} \tag{2.39c}$$

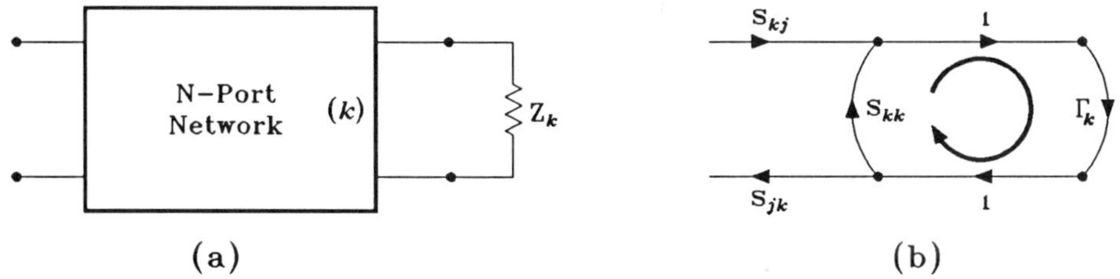

Figure 2.17 (a) Port k of n-port network; (b) flow graph.

which we recognize as the second term in the expression for input reflection coefficient resulting from an arbitrary termination given in (1.23).

The series in (2.39a) is convergent when $s_{kk}\Gamma_k < 1$, which is easily shown by considering the higher order terms. As the exponent increases, we have a number less than one raised to a large power, and the higher order terms approach zero as the exponent approaches infinity, resulting in the finite value given by (2.39b). When the product is equal to one we have an infinite series of ones, and when the product is greater than one, higher order terms become larger as the exponent approaches infinity. The nonconvergent series, in fact, provides the conditions for oscillation. A traveling wave introduced into the loop shown in Figure 2.17(b), where $s_{kk}\Gamma_k \geq 1$ would build up to an infinite value following the series in (2.39a) were it not for the physical limitations of the circuit. Circuit loss or saturation characteristics is usually the mechanism that limits the final value of the oscillating wave. The external stimulus causing oscillations to begin can be in the form of thermal noise or the transient caused by applying power to the circuit.

Thus, to ensure stability we require $s_{kk}\Gamma_k$ to be strictly less than one. When we restrict the termination to passive components, Z_k is positive real or $|\Gamma_k| \leq 1$, and $|s_{kk}|$ must be strictly less than one. We can express the condition for stability as the requirement that the magnitude of the reflection coefficient at all ports of a network be strictly less than one for any combination of passive terminations. Referring to Figure 1.9, we can represent the input or source termination as Γ_S, and the output or load termination as Γ_L. Then, using (1.25a), we let $\Gamma_S = 0$, and the two-port input reflection coefficient for an arbitrary termination becomes

$$s'_{11} = s_{11} + \frac{s_{12}s_{21}\Gamma_L}{1 - s_{22}\Gamma_L} = \frac{s_{11} - \Delta\Gamma_L}{1 - s_{22}\Gamma_L} \tag{2.40a}$$

Similarly, when $\Gamma_L = 0$, we can use (1.25d) to write the output reflection coefficient resulting from an input reflection coefficient equal to Γ_S:

$$s'_{22} = s_{22} + \frac{s_{12}s_{21}\Gamma_S}{1 - s_{11}\Gamma_S} = \frac{s_{22} - \Delta\Gamma_S}{1 - s_{11}\Gamma_S} \tag{2.40b}$$

where

$$\Delta = s_{11}s_{22} - s_{12}s_{21} \tag{2.40c}$$

If we let $\Gamma_S = \Gamma_1$ and $\Gamma_L = \Gamma_2$, then we can write the magnitude of (2.40) as

$$|s'_{11}| = \frac{|s_{11} - \Delta\Gamma_2|}{|1 - s_{22}\Gamma_2|} \tag{2.41a}$$

and

$$|s'_{22}| = \frac{|s_{22} - \Delta\,\Gamma_1|}{|1 - s_{11}\Gamma_1|} \tag{2.41b}$$

Both expressions have exactly the same form, so it is only necessary to consider one and then interchange subscripts to obtain the other. For $|s'_{11}| < 1$,

$$|1 - s_{22}\Gamma_2|^2 > |s_{11} - \Delta\,\Gamma_2|^2 \tag{2.42a}$$

where the magnitude squared is used to allow convenient manipulation of the complex expressions. The boundary separating $|s'_{11}| > 1$ and $|s'_{11}| < 1$ is

$$\left(1 - s_{22}\Gamma_2\right)\left(1 - s_{22}^*\Gamma_2^*\right) = \left(s_{11} - \Delta\,\Gamma_2\right)\left(s_{11}^* - \Delta^*\Gamma_2^*\right) \tag{2.42b}$$

Then, defining the real and imaginary components as

$$s_{11} = r_1 + jx_1$$

$$s_{22} = r_2 + jx_2$$

$$\Gamma_2 = r + jx$$

and

$$\Delta = r_\Delta + jx_\Delta$$

we can expand (2.42b) and collect terms:

$$\begin{aligned} &r^2\left[\left(r_2^2 + x_2^2\right) - \left(r_\Delta^2 + x_\Delta^2\right)\right] + r\left[2r_1r_\Delta + 2x_1x_\Delta - 2r_2\right] \\ &+ x^2\left[\left(r_2^2 + x_2^2\right) - \left(r_\Delta^2 + x_\Delta^2\right)\right] + x\left[2x_2 - 2r_1x_\Delta + 2r_\Delta x_1\right] \\ &= \left(r_1^2 + x_1^2\right) - 1 \end{aligned} \tag{2.42c}$$

Dividing (2.42c) by

$$\left[\left(r_2^2 + x_2^2\right) - \left(r_\Delta^2 + x_\Delta^2\right)\right]$$

Then, forming squares in r and x, we have

$$\left[r - \frac{\mathrm{Re}\left(s_{22} - \Delta s_{11}^*\right)^*}{|s_{22}|^2 - |\Delta|^2} \right]^2 + \left[x - \frac{\mathrm{Im}\left(s_{22} - \Delta s_{11}^*\right)^*}{|s_{22}|^2 - |\Delta|^2} \right]^2 = R^2 \tag{2.42d}$$

Equation (2.42d) defines the boundary between values of Γ_2, which cause $|s'_{11}|$ to be greater than or less than one, and is in the form of a circle, with

$$Center = \frac{\left(s_{22} - \Delta s_{11}^*\right)^*}{|s_{22}|^2 - |\Delta|^2} \tag{2.43a}$$

Expanding the right side of (2.42d), canceling terms, then factoring to form a square, we have

$$Radius = \left| \frac{s_{12}\, s_{21}}{|s_{22}|^2 - |\Delta|^2} \right| \tag{2.43b}$$

and we find that the boundary separating the stable and unstable regions of input and output reflection coefficient is a circle in the s-parameter plane. The particular circle in (2.43) is known as the *output plane stability circle*, since it defines the stable and unstable regions of the output plane. The input plane stability circle is obtained by interchanging subscripts in (2.43).

Numerous expressions for two-port characterization contain common terms often defined in the literature [12-17] as

$$\Delta = s_{11}\, s_{22} - s_{12}\, s_{21} \tag{2.44a}$$

$$C_1 = s_{11} - \Delta s_{22}^* \tag{2.44b}$$

$$C_2 = s_{22} - \Delta s_{11}^* \tag{2.44c}$$

$$D_1 = |s_{11}|^2 - |\Delta|^2 \tag{2.44d}$$

$$D_2 = |s_{22}|^2 - |\Delta|^2 \tag{2.44e}$$

$$B_1 = 1 - |s_{22}|^2 + D_1 \tag{2.44f}$$

and

$$B_2 = 1 - |s_{11}|^2 + D_2 \tag{2.44g}$$

Using this notation, the input plane stability circle can be written as

$$Center = \frac{C_1^*}{D_1} \qquad Radius = \left|\frac{s_{12}\,s_{21}}{D_1}\right| \tag{2.45a}$$

and the output plane stability circle as

$$Center = \frac{C_2^*}{D_2} \qquad Radius = \left|\frac{s_{12}\,s_{21}}{D_2}\right| \tag{2.45b}$$

We interpret the stability circles as the locus of points in one plane, which results in a magnitude equal to one for reflection coefficients in the opposite plane. Since the stability circles represent the boundary between reflection coefficients less than one and greater than one, the stable region is either entirely inside or outside the circles. The Smith chart is a mapping of all positive real impedances within a unity radius circle (see Sec. 3.1.1), and when $|s_{11}|$ and $|s_{22}|$ are less than one, the input and output impedances are positive real and stable. Unconditional stability requires that the impedance at one port be positive real for all passive terminations at the opposite port. Therefore, if we assume the stable region is inside, the stability circle must completely enclose the unity radius Smith chart. However, if we assume a stable region outside, the stability circle must exclude the unity Smith chart. Thus, no positive real impedance can cross the boundary defined by the stability circles.

We can express unconditional stability as the requirement that

$$\big||Center| - |Radius|\big| > 1$$

The stability circle either completely encloses the unity radius Smith chart or falls entirely outside. Expanding (2.45), we can write

$$\left|\frac{|s_{11} - \Delta\, s_{22}^*| - |s_{12}\,s_{21}|}{|s_{11}|^2 - |\Delta|^2}\right| > 1 \tag{2.46a}$$

for the input plane, and

$$\left| \frac{|s_{22} - \Delta s_{11}^*| - |s_{12} s_{21}|}{|s_{22}|^2 - |\Delta|^2} \right| > 1 \tag{2.46b}$$

for output plane stability circles. After considerable algebraic manipulation, both (2.46a) and (2.46b) reduce to

$$2|s_{12} s_{21}| < 1 - |s_{11}|^2 - |s_{22}|^2 + |\Delta|^2 \tag{2.47}$$

We can define a *stability factor* K as

$$K = \frac{1 - |s_{11}|^2 - |s_{22}|^2 + |\Delta|^2}{2|s_{12} s_{21}|} \tag{2.48a}$$

and then express (2.47) as

$$K > 1 \tag{2.48b}$$

K is equal to the inverse of the Linville [18-20] or the Rollet stability factor [21]. Equation (2.47) results in two additional conditions for stability:

$$|s_{12} s_{21}| < 1 - |s_{11}|^2 \tag{2.48c}$$

and

$$|s_{12} s_{21}| < 1 - |s_{22}|^2 \tag{2.48d}$$

We can summarize the necessary and sufficient conditions for unconditional stability as

$$K > 1$$

$$|s_{11}| < 1$$

$$|s_{22}| < 1$$

$$|s_{12} s_{21}| < 1 - |s_{11}|^2$$

$$|s_{12} s_{21}| < 1 - |s_{22}|^2$$

Some references cite the necessary and sufficient conditions for unconditional stability as $K > 1$ and B_1 positive. Positive B_1 is a consequence of (2.48d). However, Carson [17] shows that when (2.48d) is true, B_1 is positive, but the converse is not necessarily true. In fact, the reference includes a numerical example where B_1 is positive and (2.48d) is false.

Although we can use the numeric expressions to determine stability, the results are somewhat limited. Once we determine a device is potentially unstable, additional information is required before the situation can be corrected. Stability circles are a very convenient means for the presentation of this information, and they are easily generated using modern computer-aided design tools. Figure 2.18 shows six possible locations for the stability circles in relation to a Smith chart. When the Smith chart is referenced to the same positive real impedance as the *s*-parameters, we can easily determine the stable regions. The chart origin represents the termination impedance for measurement or calculation, and when the magnitude of the reflection coefficient in the opposite plane is less than one, Z_o must be a stable impedance. Therefore, when the stability circle does not enclose the origin, the stable region is outside, and when the circle encloses the origin, the stable region is inside. Assuming Figure 2.18 shows the input plane and $|s_{22}| < 1$, the stable region for (a) through (c) is outside the stability circle, and inside the circle for (d) through (f). Figures 2.18(a) and 2.18(d) show the condition for unconditional stability, since the entire Smith chart is outside the stability circle for 2.18(a) and inside for 2.18(d). Figures 2.18(b), 2.18(c), 2.18(e), and 2.18(f) illustrate the conditionally stable case where positive real impedance is found on both sides of the stability circle boundary. When the impedance is selected from the same region as the origin $|s_{22}| < 1$, $|s_{22}| = 1$ on the stability circle, and $|s_{22}| > 1$ when selected from the opposite region. Stable positive real impedances are shown as the shaded areas in the figure.

When the magnitude of either reflection coefficient is greater than one and Z_o is positive real, we immediately know that the network is only conditionally stable. There exists at least one positive real termination (Z_o) that causes a negative real impedance at the opposite port. The origin represents an unstable point in this case. Hence, the unstable region is outside the stability circle when the origin is enclosed, and inside if the origin is not enclosed. The reverse of the case where both reflection coefficients are less than one.

Stability Analysis Example

Given the *s*-parameters of a gallium arsenide (GaAs) FET transistor referenced to 50Ω,

$$S = \begin{bmatrix} 0.97\angle{-24^\circ} & 0.04\angle{77^\circ} \\ 3.02\angle{159^\circ} & 0.71\angle{-14^\circ} \end{bmatrix} \tag{2.49}$$

We can apply (2.48) to determine the stability characteristics of the device. Both $|s_{11}|$ and $|s_{22}|$ are less than one, which indicates the possibility of unconditional stability. Computing

$$\Delta = s_{11}s_{22} - s_{12}s_{21}$$

$$= (0.97\angle -24^\circ) \cdot (0.71\angle -14^\circ) - (0.04\angle 77^\circ)\cdot(3.02\angle 159^\circ) = 0.6909\angle -27.9547^\circ$$

then

$$K = \frac{1 - |s_{11}|^2 - |s_{22}|^2 + |\Delta|^2}{2|s_{12}s_{21}|} = \frac{1 - (0.97)^2 - (0.71)^2 + (0.6909)^2}{(2)(0.04)(3.02)} = 0.1339$$

we see that the device is potentially unstable since the stability factor is less than one. If K had turned out to be greater than one, it would still have been necessary to compute (2.48c) and (2.48d) before confirming unconditional stability.

We usually have some flexibility in the design of the input and output matching networks. However, in the case of a potentially unstable device, care must be exercised in the selection of the source and load impedances. We can define the input plane stability circle for the device given in (2.49) by first computing

$$C_1 = s_{11} - \Delta s_{22}^* = 0.97\angle -24^\circ - (0.6909\angle -27.9547)(0.71\angle 14^\circ) = 0.4944\angle -33.9652^\circ$$

and

$$D_1 = |s_{11}|^2 - |\Delta|^2 = (0.97)^2 - (0.6909)^2 = 0.4636$$

then

$$Center = \frac{C_1^*}{D_1} = \frac{(0.4944\angle -33.9652^\circ)^*}{0.4636} = 1.0664\angle 33.9652^\circ$$

and

$$Radius = \left|\frac{s_{12}s_{21}}{D_1}\right| = \left|\frac{(0.04)(3.02)}{0.4636}\right| = 0.2606$$

We construct the stability circle by first locating the center with respect to the origin of the Smith chart, $1.07\underline{/34°}$ in this case, and then drawing a circle with the specified radius of 0.2606 about this point.

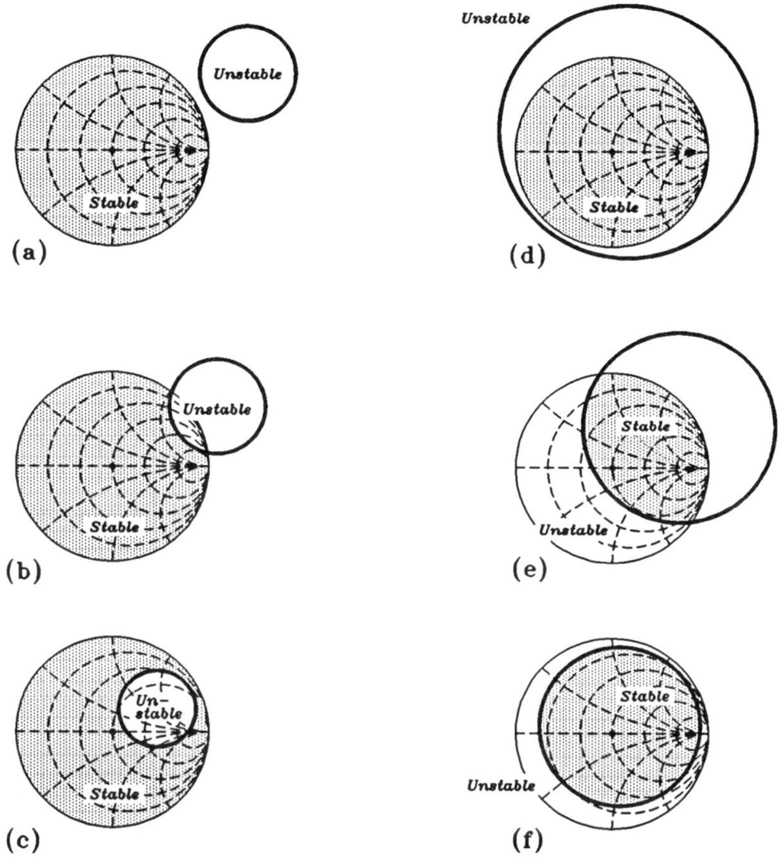

Figure 2.18 Six possible stability circle locations.

Figure 2.19 shows the stability circle plotted on the input plane. We can verify the calculations by computing the output reflection coefficient resulting from an input impedance anywhere on the stability circle. One convenient point is 75Ω + j200Ω, which translates into an input reflection coefficient of

$$\Gamma_s = \frac{(75 + j200) - 50}{(75 + j200) + 50} = 0.855\underline{/24.880°}$$

Then, using (2.40b),

$$s_{22}' = 0.71\underline{/-14^\circ} + \frac{(0.04\underline{/77^\circ})(3.02\underline{/159^\circ})(0.855\underline{/24.880^\circ})}{1 - (0.97\underline{/-24^\circ})(0.855\underline{/24.880^\circ})} = 1.001\underline{/-50.402^\circ}$$

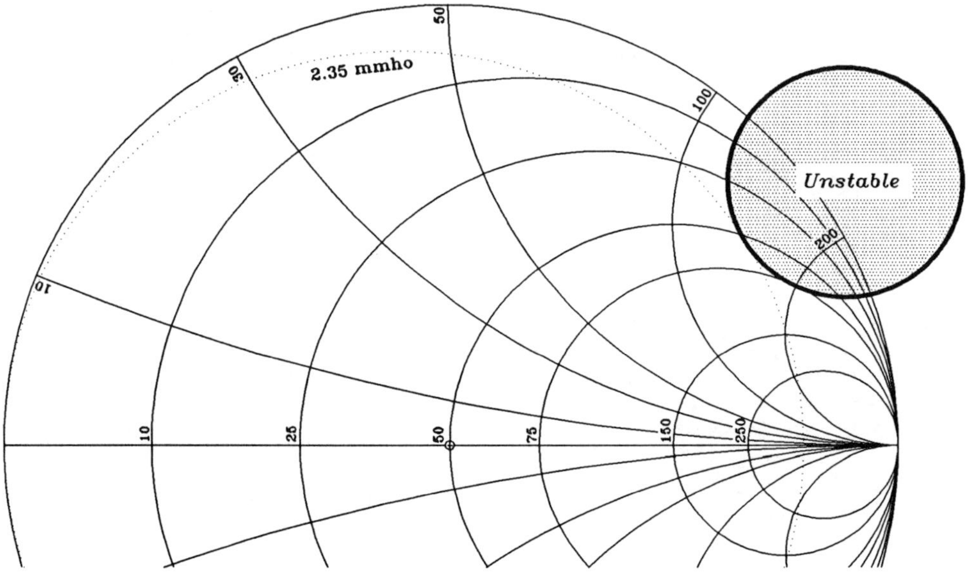

Figure 2.19 Input plane stability circle.

which confirms that the stability circle is correct. Since the magnitude of the output reflection coefficient s_{22} is less than one, 50Ω is a stable input impedance and the unstable region is inside the circle. Thus, the impedance presented to the FET gate must exclude all values within the stability circle if the output impedance is to remain positive real. However, even values of input impedance that cause negative real output impedance may not result in oscillations. The series in (2.39a) is convergent when $s_{kk}\Gamma_k < 1$, or, in terms of impedance,

$$\left(\frac{Z - Z_o}{Z + Z_o}\right)\left(\frac{Z_k - Z_o}{Z_k + Z_o}\right) < 1$$

where Z is the impedance of the device at port k and Z_k is the impedance of the termination. Expanding this expression, we have

$$\frac{ZZ_k - ZZ_o - Z_kZ_o + Z_o^2}{ZZ_k + ZZ_o + Z_kZ_o + Z_o^2} < 1$$

or

$$Z + Z_k > 0 \tag{2.50}$$

which requires the combined impedance to be less than or equal to zero before oscillations are possible.

Marginally stable designs should be avoided. *S*-parameter variations between devices and uncontrolled input and output impedances require consideration of the worst-case conditions for stability. The most severe stability problems often occur outside the operating frequency range, typically at low frequencies where the device has considerable gain and the input and output reflection coefficients may be greater than one. Input and output impedances also are usually worst-case outside the operating range. Antennas exhibit wildly varying impedance values outside the design range as do isolators and circulators used to isolate amplifiers from high VSWR loads. Stability circles provide valuable information when designing with potentially unstable devices.

Device Stabilization Techniques

Examining Figure 2.19, we find that the stability circle is tangent to the 75Ω constant resistance circle, and resistance values less than 75Ω fall within the unstable region. Similarly, reactance values between approximately +j125Ω and +j300Ω may result in negative real output impedance. When we add a 75Ω resistor in series with the gate, no passive input termination can cause an impedance less than 75Ω. All input reflection coefficients must be within the 75Ω constant resistance circle; consequently the combination is unconditionally stable. We can also add reactive elements in series with the gate; however, the input termination includes all values of reactance, so the combination does not exclude the region between +j125Ω and +j300Ω. Based on these considerations, *the input or output matching network of an unstable device must contain lossy elements if the device is to be stabilized.*

We have a second choice for stabilizing the device at the input port. As shown, the stability circle is tangent to the 75Ω constant resistance circle, as well as the 2.35-mmho constant conductance contour. We can constrain the input impedance within the 2.35-mmho conductance circle, and outside the unstable region, by adding a shunt resistor equal to 1/2.35 mmho, or approximately 425Ω. The device can be stabilized at either port, and a similar analysis of the output plane reveals that we have a total of four possibilities for stabilizing the device: a series or shunt resistor in either the input or output circuit. The best choice is dependant on the device and the particular application. Lossy elements in the input matching network degrade the noise figure, and output losses decrease the output power. Low-noise amplifiers are usually stabilized on the output plane, and power amplifiers on the input plane. When noise and power are not design considerations, we can look at gain and the matching requirements. Each of the four cases constrains impedance values to a particular region of the Smith chart, within the 75Ω constant resistance or 2.35-mmho constant conductance circle in this example. When we compute the maximum available gain for the stabilized device, the values are usually different for each of the four networks. The corresponding input and output impedances will be different as well, and one choice will probably offer an advantage over the others.

Stability must be investigated over as wide a frequency range as possible using accurate and complete circuit models [22,23]. Stability circles should be constructed as a function of frequency using worst-case device parameters, if possible. Worst-case conditions include statistical variations for a large number of devices, as well as changes due to temperature and drive level. The unstable region resulting from this exercise may be quite different from that at a single frequency, and one of the four stabilization methods often offers an advantage for broadband stabilization. Parasitic resonances in lumped elements can cause very high or low impedance values at discrete frequencies, which sometimes results in stability problems. At lower frequencies, bypass and coupling capacitors exhibit high impedance values, and at high frequencies, bonding wires and transistor leads approach open circuit values. Fortunately, physical circuit losses improves the stability.

Stability calculations of this section are based on the requirement that input and output impedance should be positive real for passive terminations. However, when designing multistage circuits, it is possible to have loops similar to the one shown in Figure 2.17(b) embedded within the network, particularly between amplifier stages. The loop gain may be greater than one, indicating the potential for oscillation although the input and output terminals exhibit positive real impedances. One analysis approach is to break the circuit into sections, compute the output impedance of one section, and compare it with the input plane stability circle of the next adjoining section to determine if a negative real output impedance is possible when the stages are cascaded. We must determine the output impedance of the first section for all passive loads at the input of the network, which can either be done by plotting numerous discrete points or using the mapping techniques discussed in Chapter 3. The combination of unconditionally stable cascaded stages is also unconditionally stable, and further analysis is not required. Each stage is constrained to positive real impedance for passive loads; thus, positive real impedance is always present between stages.

2.3.3 Conjugately Matched Two-Ports

One measure of an active device is its maximum available gain G_{MAX} as a function of frequency. Very often this is the first parameter we look at when selecting a device for a particular application. We can achieve the maximum gain by simultaneously conjugately matching the input and output ports to realize maximum power transfer into the device and its load. We can express the conjugate match in terms of two reflection coefficients, Γ_{ms} and Γ_{ml}, which represent the matched source and load, respectively. When matched, the conjugate of Γ_{ms} must equal the input reflection coefficient with the output terminated by Γ_{ml}, and the conjugate of Γ_{ml} must simultaneously equal the output reflection coefficient with an input termination of Γ_{ms}. Using (2.40), we can express these conditions as

$$\Gamma^*_{ms} = s'_{11} = s_{11} + \frac{s_{12}s_{21}\Gamma_{ml}}{1 - s_{22}\Gamma_{ml}} \tag{2.51a}$$

and

$$\Gamma^*_{ml} = s'_{22} = s_{22} + \frac{s_{12}s_{21}\Gamma_{ms}}{1 - s_{11}\Gamma_{ms}} \tag{2.51b}$$

Expanding (2.51a), we have

$$\left(1 - s_{22}\Gamma_{ml}\right)\left(s_{11} - \Gamma^*_{ms}\right) + s_{12}s_{21}\Gamma_{ml} = 0 \tag{2.51c}$$

and

$$\left(1 - s_{11}\Gamma_{ms}\right)\left(s_{22} - \Gamma^*_{ml}\right) + s_{12}s_{21}\Gamma_{ms} = 0 \tag{2.51d}$$

when (2.51b) is expanded.

We can also express the conditions for simultaneous match in terms of generalized scattering parameters [12]. When we define the reference at port 1 as the matched input impedance resulting from a conjugate output match, and the reference at port 2 as the matched output impedance resulting from an input match, s'_{11} and s'_{22} must both be zero. Substituting Γ_{ms} for Γ_S and Γ_{ml} for Γ_L, we find that (1.25a) is equal to (2.51c) for $s'_{11} = 0$, and (1.25d) is equal to (2.51d) for $s'_{22} = 0$. Solving simultaneous equations (2.51c) and (2.51d) for Γ_{ms} and Γ_{ml} yields

$$\Gamma^2_{ms} - \left(\frac{B_1}{C_1}\right)\Gamma_{ms} + \left(\frac{C^*_1}{C_1}\right) = 0$$

and

$$\Gamma_{ml}^2 - \left(\frac{B_2}{C_2}\right)\Gamma_{ml} + \left(\frac{C_2^*}{C_2}\right) = 0$$

where C_1, C_2, B_1, and B_2 are defined in (2.44). Both equations can be written in the form

$$\Gamma_{mi}^2 - \left(\frac{B_i}{C_i}\right)\Gamma_{mi} + \left(\frac{C_i^*}{C_i}\right) = 0$$

and then solved using the quadratic formula

$$\Gamma_{mi} = \frac{B_i \pm \sqrt{B_i^2 - 4\,C_i\,C_i^*}}{2\,C_i}$$

or

$$\Gamma_{mi} = \frac{C_i^*}{2|C_i|^2}\cdot\left\{B_i \pm \sqrt{B_i^2 - 4|C_i|^2}\right\} \tag{2.52}$$

Equation (2.52) has two solutions for Γ_{mi}, one using the positive sign and the other using the negative sign in front of the radical. When $|B_i/2C_i|$ is greater than one, the square root is real, and one solution has a magnitude greater than one, or negative real, and another with a magnitude less than one or positive real. The desired positive real solution is obtained by selecting the sign opposite that of B_i: a plus sign when B_i is negative and a negative sign when B_i is positive. However, when $|B_i/2C_i|$ is less than one, the square root is imaginary, and both solutions have magnitudes equal to one.

The expression $|B_i/2C_i| > 1$ can be written as $|K| > 1$, where K is the stability factor defined in (2.48a). The solution can be summarized in terms of the stability factor as follows:

$\lvert K\rvert < 1$	$\lvert\Gamma_{ms}\rvert > 1$ and $\lvert\Gamma_{ml}\rvert > 1$ for both solutions.
$\lvert K\rvert > 1$ and $K > 1$	$\lvert\Gamma_{ms}\rvert < 1$ and $\lvert\Gamma_{ml}\rvert < 1$ for one solution, $\lvert\Gamma_{ms}\rvert > 1$ and $\lvert\Gamma_{ml}\rvert > 1$ for the other solution.
$\lvert K\rvert > 1$ and $K < 1$	$\lvert\Gamma_{ms}\rvert < 1$ and $\lvert\Gamma_{ml}\rvert > 1$ for one solution, $\lvert\Gamma_{ms}\rvert > 1$ and $\lvert\Gamma_{ml}\rvert < 1$ for the other solution.

We can stabilize the device described by (2.49) and then compute the input and output reflection coefficients for a simultaneous conjugate match. The actual value of series resistance required to stabilize (2.49) is 75.2Ω. Using Figure 2.9 to compute the s-parameters for the resistor, and equation (1.29) to cascade the two matrices, we have

$$S = \begin{bmatrix} 0.8678\underline{/-21.0347^\circ} & 0.0355\underline{/61.7149^\circ} \\ 2.6834\underline{/143.7149^\circ} & 0.6666\underline{/-19.6722^\circ} \end{bmatrix} \tag{2.53}$$

From (2.48a), $K = 1.0018$, and both (2.48c) and (2.48d) are true, which confirms that the combination is unconditionally stable. We first determine the terms defined in (2.44),

$$\Delta = (0.8678\underline{/-21.0347^\circ})(0.6666\underline{/-19.6722^\circ}) - (0.0355\underline{/61.7149^\circ})(2.6834\underline{/143.7149^\circ})$$

$$= 0.6231\underline{/-32.6703^\circ}$$

$$C_1 = (0.8678\underline{/-21.0347^\circ}) - (0.6231\underline{/-32.6703^\circ})(0.6666\underline{/19.6722^\circ})$$

$$= 0.4602\underline{/-28.2845^\circ}$$

$$C_2 = (0.6666\underline{/-19.6722^\circ}) - (0.6231\underline{/-32.6703^\circ})(0.8678\underline{/21.0347^\circ})$$

$$= 0.1514\underline{/-49.6321^\circ}$$

$$D_1 = (0.8678)^2 - (0.6231)^2 = 0.3648$$

$$D_2 = (0.6666)^2 - (0.6231)^2 = 0.0561$$

$$B_1 = 1 - (0.6666)^2 + 0.3648 = 0.9204$$

and

$$B_2 = 1 - (0.8678)^2 + 0.0561 = 0.3030$$

and then let $i = 1$ in (2.52) to compute the input match. B_1 is positive, and

$$\Gamma_{ms} = \frac{0.4602\angle 28.2845^\circ}{2\,(0.4602)^2}\cdot\left\{0.9204 - \sqrt{(0.9204)^2 - 4\,(0.4602)^2}\,\right\} \tag{2.54a}$$

$$= 0.9877\angle 28.2845^\circ$$

For the output plane, $i = 2$, B_2 is positive, and

$$\Gamma_{ml} = \frac{0.1514\angle 49.6231^\circ}{2\,(0.1514)^2}\cdot\left\{0.3030 - \sqrt{(0.3030)^2 - 4\,(0.1514)^2}\,\right\} \tag{2.54b}$$

$$= 0.9630\angle 49.6231^\circ$$

2.3.4 Transducer Power Gain

Power gain is often measured in the laboratory using only a signal generator and power meter. Power output from the generator is measured. Then the network is connected to the signal source, and the power is measured at the output of the network. Power gain is defined as the ratio of output power to input power, or, in the case of decibel measurements referenced to a standard power level, the difference between output and input readings. Gain measured in this manner is known as the *transducer gain*, which is formally defined as

$$G_t = \textit{Transducer gain} = \frac{\textit{Power delivered to the load}}{\textit{Power available from the source}} = \frac{P_L}{P_{avs}} \tag{2.55}$$

G_t is the most useful gain measurement because it accounts for mismatch loss at both the input and output ports of the network. We determine the power available from the source by connecting a complex conjugate load which results in maximum power transfer. When the network is connected, any deviation from the complex conjugate impedance results in a portion of the input power reflected back into the generator. Similarly, any mismatch between the network output and load is reflected back into the network, and the measurement reflects the true performance of the network when inserted between the generator and load. The impedances of the generator and power meter are usually assumed to be 50Ω, which provides the response of the network in a 50Ω system. However, when either instrument deviates from 50Ω, errors are introduced in the measurement of P_{avs}, as well as P_L.

Figure 2.20(a) shows a signal flow graph of the generator connected to a load that equals the complex conjugate of its internal impedance. We can write

$$P_{avs} = |b|^2 - |a|^2 \tag{2.56a}$$

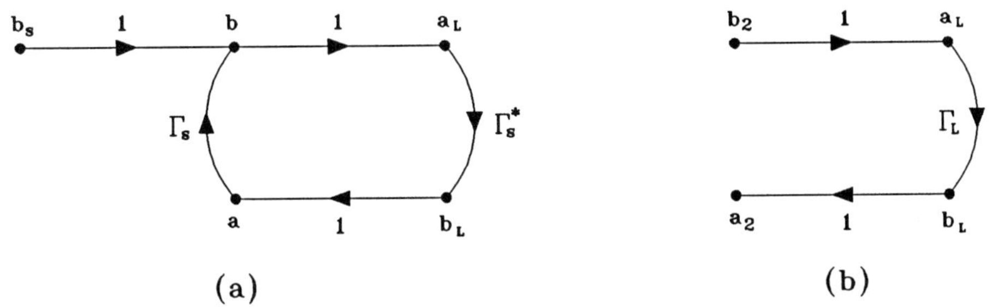

Figure 2.20 Flow graph for (a) P_{avs}; (b) P_L.

Then, using the nontouching loop rule,

$$\frac{b}{b_s} = \frac{1}{1 - \Gamma_s \Gamma_s^*} \tag{2.56b}$$

and

$$\frac{a}{b_s} = \frac{\Gamma_s^*}{1 - \Gamma_s \Gamma_s^*} \tag{2.56c}$$

Using (2.56a), (2.56b), and (2.56c), the power available from the source is

$$P_{avs} = \frac{|b_s|^2}{\left(1 - |\Gamma_s|^2\right)^2} - \frac{|b_s|^2 |\Gamma_s^*|^2}{\left(1 - |\Gamma_s|^2\right)^2} = \frac{|b_s|^2}{\left(1 - |\Gamma_s|^2\right)} \tag{2.56d}$$

Power delivered to the load shown in Figure 2.20(b) is

$$P_L = |a_L|^2 - |b_L|^2 \tag{2.57a}$$

where

$$a_L = b_2 \tag{2.57b}$$

and

$$b_L = a_L \Gamma_L = b_2 \Gamma_L \tag{2.57c}$$

Thus,

$$P_L = |b_2|^2 - |b_2|^2 |\Gamma_L|^2 = |b_2|^2 \left(1 - |\Gamma_L|^2\right) \tag{2.57d}$$

and we can write the transducer gain as

$$G_t = \frac{P_L}{P_{avs}} = \frac{|b_2|^2 \left(1 - |\Gamma_L|^2\right)\left(1 - |\Gamma_s|^2\right)}{|b_s|^2} \tag{2.58a}$$

Using Figure 1.16, which shows a signal flow graph of the generator, two-port, and load,

$$\frac{b_2}{b_s} = \frac{s_{21}}{1 - \left(s_{11}\Gamma_s + s_{22}\Gamma_L + s_{12}s_{21}\Gamma_s\Gamma_L\right) + \left(s_{11}s_{22}\Gamma_s\Gamma_L\right)} \tag{2.58b}$$

Then, substituting (2.58b) into (2.58a),

$$G_t = \frac{|s_{21}|^2 \left(1 - |\Gamma_s|^2\right)\left(1 - |\Gamma_L|^2\right)}{\left|\left(1 - s_{11}\Gamma_s\right)\left(1 - s_{22}\Gamma_L\right) - s_{12}s_{21}\Gamma_s\Gamma_L\right|^2} \tag{2.59}$$

We can also reference the input port to the generator impedance and the output port to the load impedance. Then G_t can be written in terms of the generalized scattering parameters as $|s'_{21}|^2$.

Assuming an input termination equal to the matched source given in (2.54a) and an output termination given by the matched load in (2.54b), we can determine the transducer gain for the device defined in (2.53) by first computing

$$1 - s_{11}\Gamma_s = 1 - \left(0.8678\angle{-21.0347^\circ}\right)\left(0.9877\angle{28.2845^\circ}\right) = 0.1847\angle{-35.8451^\circ}$$

$$1 - s_{22}\Gamma_L = 1 - \left(0.6666\angle{-19.6722^\circ}\right)\left(0.9630\angle{49.6321^\circ}\right) = 0.5475\angle{-35.8398^\circ}$$

and

$$\begin{aligned} s_{12}s_{21}\Gamma_s\Gamma_L &= \left(0.0355\angle{61.7149^\circ}\right)\left(2.6834\angle{143.7149^\circ}\right)\left(0.9877\angle{28.2845^\circ}\right)\left(0.9630\angle{49.6321^\circ}\right) \\ &= 0.0906\angle{-76.6536^\circ} \end{aligned}$$

Then

$$G_t = \frac{(2.6834)^2\left[1 - (0.9877)^2\right]\left[1 - (0.9630)^2\right]}{\left|(0.1847\angle -35.8451^\circ)(0.5475\angle -35.8398^\circ) - 0.0906\angle -76.6536^\circ\right|^2}$$

$$= \frac{0.0128}{\left|0.0134\angle -35.8408^\circ\right|^2} = 71.1958 = 18.5245\ dB \tag{2.60}$$

2.3.5 Maximum Available Gain

When the stability factor K is greater than one, the maximum gain is the transducer gain for the conjugately matched two-port discussed in Section 2.3.3, or

$$G_{\max} = \frac{|s_{21}|^2\left(1 - |\Gamma_{ms}|^2\right)\left(1 - |\Gamma_{ml}|^2\right)}{\left|(1 - s_{11}\Gamma_{ms})(1 - s_{22}\Gamma_{ml}) - s_{12}s_{21}\Gamma_{ms}\Gamma_{ml}\right|^2} \tag{2.61a}$$

which is known as the *maximum available gain* and defines the highest achievable gain from the stable device. We can also use the expressions for Γ_{ms} and Γ_{ml} when $|s_{12}s_{21}| \neq 0$ to write (2.61a) as

$$G_{\max} = |s'_{21}|^2 = \left|\frac{s_{21}}{s_{12}}\right| \cdot \left(K \pm \sqrt{K^2 - 1}\right) \tag{2.61b}$$

where the proper sign to be used for the radical is opposite the sign of B_1. However, when $|s_{12}s_{21}| = 0$, the stability factor is infinite and we see from (2.40) that the input and output ports are isolated. Thus, for both $|s_{11}|$ and $|s_{22}|$ less than one, $\Gamma_{ms} = s^*_{11}$, $\Gamma_{ml} = s^*_{22}$, and we can write the maximum available gain as

$$G_{\max} = |s'_{21}|^2 = \frac{1}{1 - |s_{11}|^2} \cdot |s_{21}|^2 \cdot \frac{1}{1 - |s_{22}|^2} \tag{2.61c}$$

Inspecting (2.61c), we see that the expression consists of three terms. The first term represents the input mismatch due to s_{11}, the second term is the insertion gain of the device, and the third term is the output plane mismatch due to s_{22}. When $|s_{12}|$ is equal to zero, (2.61c) is known as the *unilateral gain* [24], since power travels in one direction only. The unilateral assumption was once useful for slide rule calculations; however, there is no justification for simplifications using modern computing tools, and this subject will not be addressed further.

The calculations in (2.60) give the maximum available gain of the stable device using (2.61a). Since $|s_{12}s_{21}|$ does not equal zero, we can also use the simpler expression provided by (2.61b):

$$G_{max} = \frac{2.6834}{0.0355} \cdot \left(1.0018 - \sqrt{(1.0018)^2 - 1}\right) = 71.1958 = 18.5245\ \text{dB}$$

The matched source and load impedances are not defined for a potentially unstable device. Consequentially, the maximum available gain cannot be calculated from (2.61). However, if we assume the stability factor is equal to one, (2.61b) reduces to

$$G_{MSG} = \left|\frac{s_{21}}{s_{12}}\right| \tag{2.62}$$

which is known as the *maximum stable gain* or *MSG*. The MSG is an approximation which will be exact when $K = 1$. As K becomes smaller, the difference between the MSG and the actual stable gain increases. The stability factor for the original unstable device in (2.49) is 0.1339, and the maximum stable gain is

$$G_{MSG} = \frac{3.02}{0.04} = 75.5000 = 18.7795\ \text{dB}$$

After stabilizing the device, we computed a maximum available gain of 18.5245 dB, or an error of only 0.2550 dB between the two calculations.

2.3.6 Available Power Gain

When designing low-noise amplifiers, the noise figure is primarily determined by the input impedance. We usually design the input matching network for minimum noise figure and the output matching network for maximum gain. The minimum noise impedance rarely coincides with the complex conjugate input impedance, causing a mismatch on the input plane. We can define the gain which results when the input is mismatched and the output matched as

$$\textit{Available power gain} = \frac{\textit{Power available from the network}}{\textit{Power available from the source}} = \frac{P_{avo}}{P_{avs}}$$

The *available power gain*, G_A, is a special case of the transducer gain. When the output is matched, $\Gamma_L = (s'_{22})^*$ and (2.59) reduces to

$$G_A = \frac{|s_{21}|^2 (1 - |\Gamma_s|^2)}{(1 - |s_{22}|^2) + |\Gamma_s|^2 D_1 - 2\,\mathrm{Re}(\Gamma_s C_1)} \tag{2.63a}$$

where C_1 and D_1 are defined in (2.44b) and (2.44d), respectively. The available power gain can also be written in terms of the generalized s-parameters as

$$G_A = \frac{|s'_{21}|^2}{1 - |s'_{22}|^2} \tag{2.63b}$$

We can define a source impedance for the device in (2.49) by selecting a point just outside the stability circle shown in Figure 2.19. Subtracting the radius from the center of the stability circle yields $0.8058\underline{/33.9652°}$. Then, selecting a point slightly less, we let $\Gamma_S = 0.8050\underline{/33.9652°}$, and the available power gain is

$$G_A = \frac{(3.02)^2[1 - (0.8050)^2]}{[1 - (0.71)^2] + (0.8050)^2(0.4636) - 2\,\mathrm{Re}\{(0.8050\underline{/33.9652°})(0.4944\underline{/-33.9652°})\}}$$

$$= \frac{3.2102}{0.4959 + 0.3004 - 2(0.39801)} = 9430.8082 = 39.7455\ \mathrm{dB}$$

Although Γ_S is inside the stable region, the gain is greater than the maximum stable gain. Using (2.40b) to compute the output reflection coefficient, $s'_{22} = 0.9983\underline{/-29.1034°}$ and $\Gamma_L = 0.9983\underline{/29.1034°}$, which is positive real. However, when (2.40a) is used to compute the input reflection coefficient for Γ_L, we find that $s'_{11} = 1.2392\underline{/-33.9652°}$ and the design is unstable.

2.3.7 Operating Power Gain

The design of some networks requires a mismatch at the output port and a match at the input port. For example, we usually design the output matching network of a power amplifier to provide maximum output power, or perhaps minimum harmonic distortion at high drive levels, and the input network for maximum gain. We can follow a procedure similar to that for available gain and define

$$\textit{Operating power gain} = \frac{\textit{Power delivered to the load}}{\textit{Power into the network}} = \frac{P_L}{P_{in}}$$

The *operating power gain* is also a special case of the transducer gain with the input matched or $\Gamma_S = (s'_{11})^*$, and (2.59) reduces to

$$G_p = \frac{|s_{21}|^2 (1 - |\Gamma_L|^2)}{(1 - |s_{11}|^2) + |\Gamma_L|^2 D_2 - 2\,\mathrm{Re}(\Gamma_L C_2)} \tag{2.64a}$$

where C_2 is defined in (2.44c) and D_2 in (2.44e). When we renormalize to the respective input and output impedance, (2.64a) becomes

$$G_p = \frac{|s'_{21}|^2}{1 - |s'_{11}|^2} \tag{2.64b}$$

in terms of the generalized scattering parameters.

2.3.8 Constant Gain Contours

Available power gain given by (2.63) describes the gain as a function of source impedance with conjugately matched output, and the operating power gain in (2.64) describes gain as a function of load impedance when the input is matched. Both expressions can be written as

$$G_i = |s_{21}|^2 \cdot g_i$$

where $i = 1$ for the input plane and 2 for the output plane. We can set the gain equal to a constant; that is,

$$g_i = \frac{G_i}{|s_{21}|^2}$$

Then, following the example of stability circles in Section 2.3.2, we can expand g_i into real and imaginary components, cancel terms, factor, and eventually manipulate the expression into the form

$$(x - x_o)^2 + (y - y_o)^2 = r^2$$

which is the equation for a circle in either the input or output plane. The constant gain circles are defined by

$$Center = \left\{ \frac{G_i}{1 + D_i G_i} \right\} \cdot C_i^* \tag{2.65a}$$

and

$$Radius = \frac{\sqrt{1 - 2K|s_{12}s_{21}|G_i + |s_{12}s_{21}|^2 G_i^2}}{1 + D_i G_i} \tag{2.65b}$$

where

$$G_i = \frac{10^{(P/10)}}{|s_{21}|^2} \tag{2.65c}$$

C_i and D_i are defined in (2.44), K in (2.48a), and P is the desired power gain expressed in decibels.

Using Constant Gain Circles for Amplifier Design

The computations in Section 2.3.2 show that $K = 0.1339$, $C_1 = 0.4944\underline{/-33.9652^\circ}$, and $D_1 = 0.4636$ for the device described by (2.49), and we computed a maximum stable gain of 18.7795 dB in Section 2.3.5. We can construct the 18.7795-dB constant gain circle on the input plane by first calculating

$$G_1 = \frac{10^{(18.7795/10)}}{(3.02)^2} = 8.2782$$

then

$$Center = \left\{ \frac{8.2782}{1 + (0.4636)(8.2782)} \right\} \cdot \left(0.4944\underline{/33.9652^\circ}\right) = 0.8460\underline{/33.9652^\circ}$$

and

$$Radius = \frac{\sqrt{1 - (2)(0.1339)(0.04)(3.02)(8.2782) + [(0.04)(3.02)(8.2782)]^2}}{1 + (0.4636)(8.2782)}$$

$$= \frac{\sqrt{1.7322}}{4.8378} = 0.2721$$

Constant gain circles are drawn in the same manner as the stability circles. First locate the center as measured from the origin of the Smith chart. Then draw a circle with the specified radius. The 18.7795 dB constant gain circle is shown in Figure 2.21, along with the input plane stability circle and seven other constant gain circles, which are constructed in 1-dB increments between 11.7795 dB and the maximum stable gain.

We can select an input reflection coefficient anywhere along a constant gain circle and then compute s'_{22} using (2.40b). The output reflection coefficient is $\Gamma_L = (s'_{22})^*$ for a conjugate match on the output plane. The constant gain contours shown in Figure 2.21 exhibit maximum separation along a line connecting the origin with the centers of the stability and gain circles. As we follow a path along any gain contour toward the outside of the Smith chart, the circles become closer together until they cross at a single point. When selecting Γ_S, the best choice is usually in the region where the circles are spread apart. Small changes in the input impedance have little effect on the gain. However, when operating in the compressed region, small changes in input impedance can result in significant gain deviation.

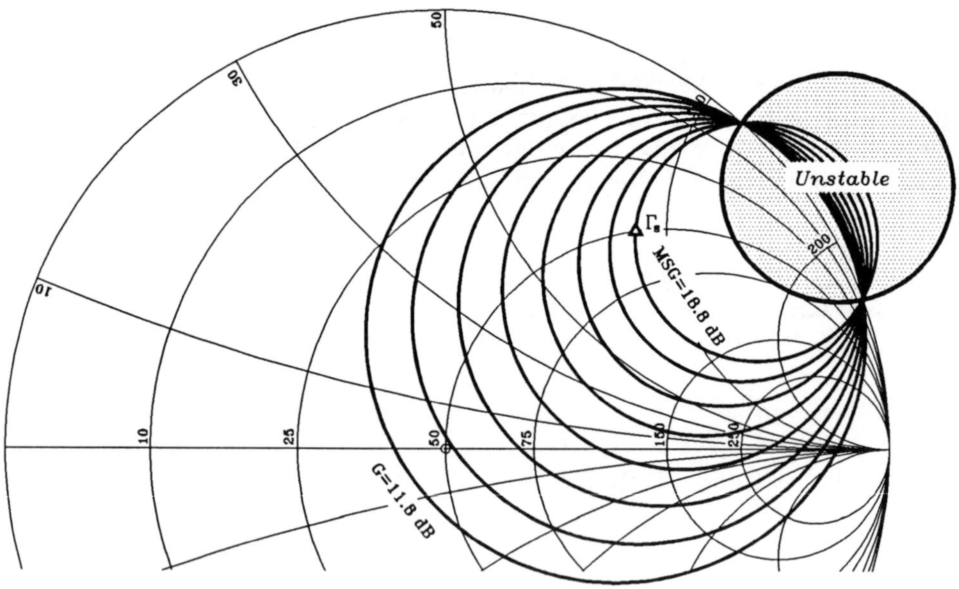

Figure 2.21 Input plane stability and constant gain circles.

The graphic presentation of gain, stability, and impedance shown in Figure 2.21 provides design information in a form immediately useful and difficult to obtain using other means. For example, we see that, with no input matching network, the device is capable of just less than 12.8 dB gain when the output is matched, and with the addition of a single element, we can achieve the full 18.8 dB maximum stable gain. Starting from the 50Ω source impedance, we can add series inductive reactance by moving along the 50Ω constant resistance circle in a clockwise direction. The two circles intersect at approximately 50Ω + j87Ω, or $\Gamma_S = 0.6564\angle 48.9767°$, as shown in the figure. Using (2.40b) to compute the output reflection coefficient, we find $\Gamma_L = 0.8523\angle 19.0958°$. The circuit is shown in Figure 2.22(a) and we can easily verify the input and output impedance by computing the transducer gain using (2.59). The transducer gain is 18.7845 dB, which is very close to the computed MSG and confirms that the impedance values are correct. However, since the device is potentially unstable, we must confirm that the input reflection coefficient is positive real before proceeding with the design. Using (2.40a) with $\Gamma_L = 0.8523\angle 19.0958°$, $s'_{11} = 1.0728\angle -37.2491°$, which is negative real and not acceptable. When the device is potentially unstable, we must look at the input and output planes simultaneously to ensure both terminations are within the stable region.

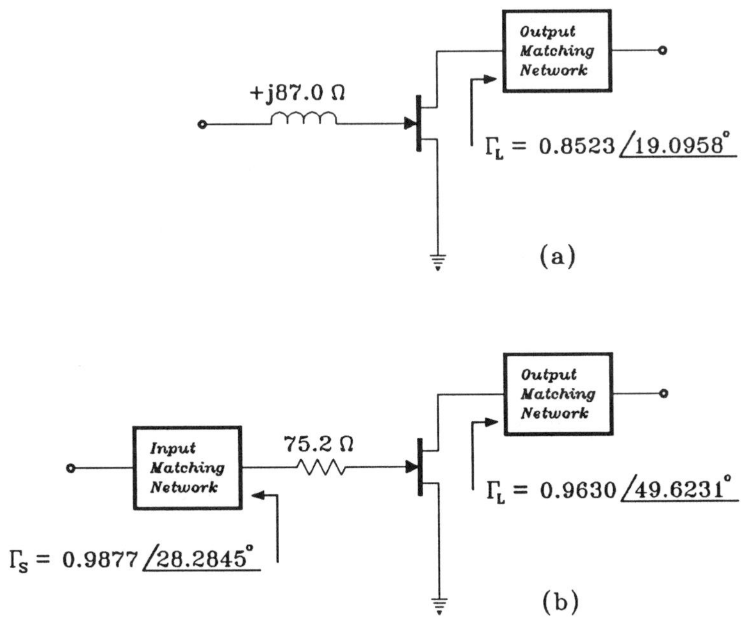

Figure 2.22 (a) Potentially unstable; (b) unconditionally stable amplifier design.

In this case, we can provide similar performance from an unconditionally stable design. Calculations in Section 2.3.2 showed that the device can be stabilized by adding a 75.2Ω resistor in series with the gate, as shown in Figure 2.22(b). The source and load reflection coefficients for the conjugately matched two-port were computed in Section 2.3.3 and given by (2.54a) and (2.54b). Using Γ_{MS} and Γ_{ML} with the expression for transducer gain in (2.59), we find that the gain is 18.5245 dB. The source and load impedances are positive real, since the device is unconditionally stable. However, we no longer have flexibility in selecting the source impedance. The input and output matching networks must be designed to present Γ_{MS} and Γ_{ML} to the gate and drain, respectively.

Amplifier Gain in Excess of the MSG

The output plane stability circle is determined by first computing C_2 and D_2 from (2.44), and then the center and radius from (2.45),

$$Center = 4.7447\angle 80.7853^\circ \qquad Radius = 4.5064$$

Figure 2.23 shows a portion of the stability circle intersecting the Smith chart. Since the origin is a stable point ($|s_{11}| < 1$) and outside the circle, the unstable region is inside. Our goal is to find a stable output impedance resulting from an input impedance on the maximum stable gain circle. We can choose a source impedance from the input plane MSG circle, compute s'_{22}, and then plot the conjugate on the output plane. Figure 2.23 shows a plot of load impedance corresponding to most of the stable region of the input plane MSG circle. As shown, the entire plot is in the unstable region, and it appears that we cannot achieve maximum stable gain with simultaneous positive real input and output device impedance.

However, we can take a different approach to realizing the maximum stable gain. We can design for the highest possible gain under the constraint that Γ_S and Γ_L are both selected from the stable regions of the input and output plane. The closest we can come to the input plane stability circle is

$$\Gamma_S = Center - Radius = (1.0664\angle 33.9652^\circ) - 0.2606 = 0.8058\angle 33.9652^\circ$$

Using (2.40b), $s'_{22} = 0.9997\angle -29.1036^\circ$, and the load impedance is the conjugate shown in Figure 2.23. The conjugately matched load is in the unstable region. However, we can first construct a line between the origin and the impedance and then select Γ_L at a point where the line intersects the stability circle. The figure shows the stable load impedance, which is $\Gamma_L = 0.3926\angle 29.1036^\circ$ and labeled as point Γ_L. The input reflection coefficient corresponding to Γ_L is $0.9988\angle -27.3650^\circ$, which confirms that both ports of the device are positive real. Calculating the transducer gain from (2.59), we find that G_t = 20.1218 dB, significantly higher than the calculated maximum stable gain. This design has a mismatch at both ports which may require two isolators or circulators for acceptable input and output VSWR. However, the cost of these components may be offset by the additional gain. MSG is based on the assumption that K = 1, and provides an estimate of the gain achieved by stabilizing the device. As this example demonstrates, it is possible to realize higher gains, although an additional analysis is required to ensure that changes in device parameters or source and load impedance do not result in an oscillator instead of the desired amplifier.

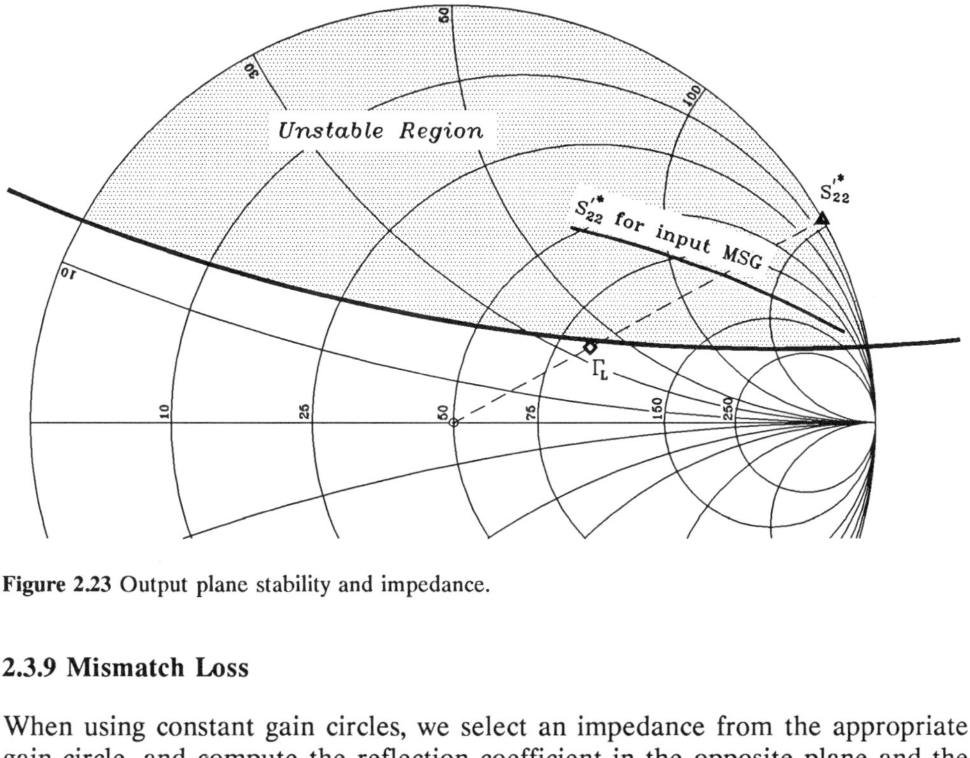

Figure 2.23 Output plane stability and impedance.

2.3.9 Mismatch Loss

When using constant gain circles, we select an impedance from the appropriate gain circle, and compute the reflection coefficient in the opposite plane and the termination impedance as its complex conjugate. This approach offers flexibility in the plane of the gain circles, and we usually take advantage by selecting an impedance from the less sensitive region, which results in a simple matching network. However, we are constrained to a single point in the opposite plane. We can overcome this limitation by selecting a circle with higher gain than necessary and then mismatching the opposite plane.

We could derive the mismatch expressions from the signal flow graph for a source and load; however, a simpler approach is to define an ideal network and use existing equations. The ideal network is perfectly matched and has no transmission loss, or $s_{11} = s_{22} = 0$ and $s_{12} = s_{21} = 1$. Substituting these s-parameters into the transducer gain equation,

$$G_{mm} = \frac{\left(1 - |\Gamma_S|^2\right)\left(1 - |\Gamma_L|^2\right)}{|1 - \Gamma_S \Gamma_L|^2} \tag{2.66}$$

Contours of constant mismatch can also be derived by substituting into either set of equations for constant gain circles, which yields

$$Center = \left\{\frac{G}{1 - |\Gamma|^2(1 - G)}\right\} \cdot \Gamma^* \tag{2.67a}$$

$$Radius = \frac{(1 - |\Gamma|^2)\sqrt{1 - G}}{1 - |\Gamma|^2(1 - G)} \tag{2.67b}$$

and

$$G = 10^{(MML/10)} \tag{2.67c}$$

where MML is the desired mismatch loss expressed in decibels, and Γ is the reflection coefficient of the impedance to be mismatched.

Amplifier Design Using Mismatch Circles

We are often required to design an amplifier with a specified gain, usually somewhat less than the maximum available or maximum stable values. Constant gain circles used in conjunction with mismatch circles make this an easy task. As an example, we can use the device described by (2.49) to design an unconditionally stable amplifier with 25 dB gain. The maximum stable gain is less than 25 dB; hence, the design will require two stages. For convenience, we elect to cascade two identical 12.5 dB gain stages, which will provide the specified 25 dB. We stabilized the device in Section 2.3.2 by connecting a 75.2Ω resistor in series with the gate, resulting in the s-parameters given by (2.53). The first step is to construct constant gain circles on either the input or output plane and then select a reflection coefficient which provides the desired gain. Figure 2.24 shows the input plane constant gain circles in steps of 1 dB from 18 to 13 dB. We could select an input impedance from the 12.5-dB gain circle; however, the output must be matched, which limits the output impedance to a single point. Instead, we choose to select Γ_S from the less sensitive region of the 15-dB gain circle, as shown in the figure. $\Gamma_S = 0.5792\angle 19.4112°$, and the next step is to design an input matching network.

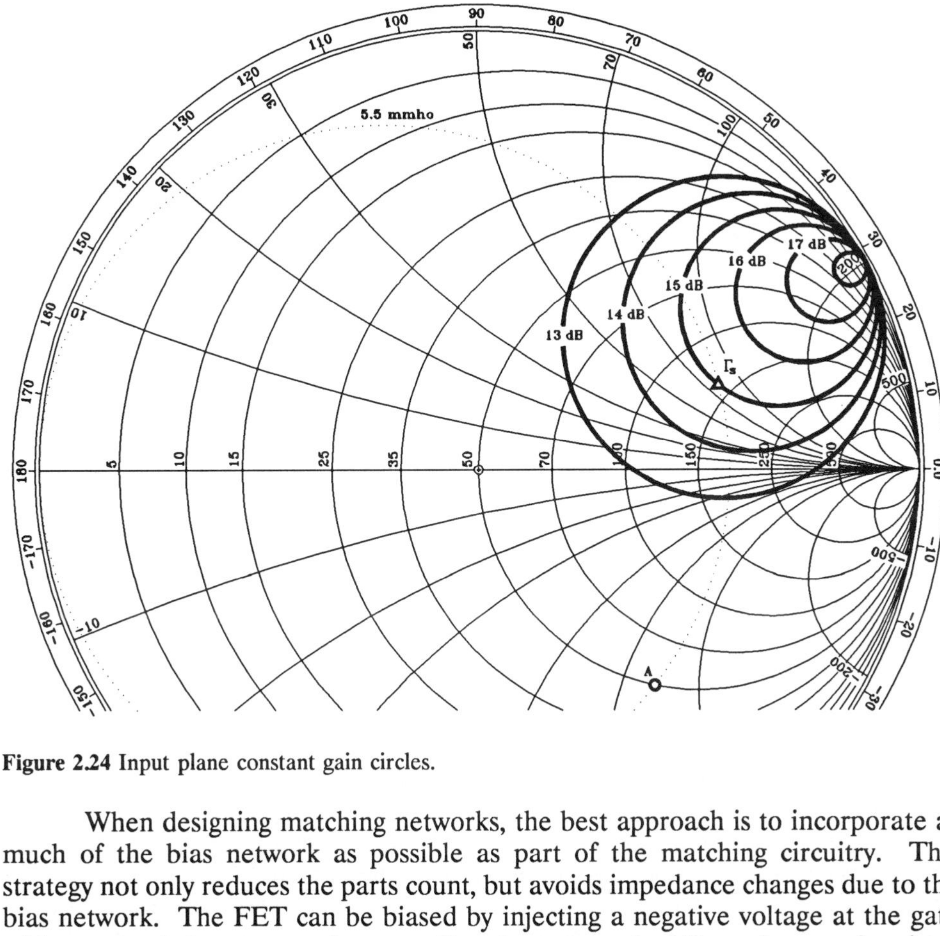

Figure 2.24 Input plane constant gain circles.

When designing matching networks, the best approach is to incorporate as much of the bias network as possible as part of the matching circuitry. This strategy not only reduces the parts count, but avoids impedance changes due to the bias network. The FET can be biased by injecting a negative voltage at the gate with a grounded source, or by grounding the gate and adding a bypassed resistor in the source, which provides the necessary negative gate-to-source voltage. In either case, the input should be isolated from external voltages and a DC path provided from the gate to ground or bias voltage. One configuration meeting these requirements is a series capacitor followed by a shunt inductor, as shown in Figure 2.26. The shunt inductor can be grounded to provide zero gate voltage, or connected to ground through a capacitor, which provides RF ground and a low impedance point to inject bias. Inspecting Figure 2.24, we see that the 5.5-mmho constant conductance contour passes through Γ_S. Starting from the 50Ω input termination, we add series capacitive reactance by following the 50Ω constant resistance circle in a counter-clockwise direction. A capacitive reactance of -j81.4Ω takes us to the intersection with the 5.5-mmho constant conductance contour, which is labeled point A in the figure. The susceptance at this point is +j8.9173 mmho, and the susceptance of Γ_S is -j3.1712 mmho. We can move along the constant conductance contour by adding shunt susceptance. The value required to move from point A to Γ_S is the difference between the two points (-j3.1712 - +j8.9173), or approximately -j12.0885 mmho. Inverting the susceptance, we find that a reactance of +j82.7Ω is required.

The output reflection coefficient resulting from Γ_S is $s'_{22} = 0.6239\angle{-28.7698°}$. If we were to design for an output match, the gain would be 15 dB, since Γ_S was selected from the 15-dB gain circle. However, the required gain is 12.5 dB, and we must design for 2.5-dB mismatch on the output plane. $\Gamma = 0.6239\angle{-28.7698°}$, and using (2.67) to compute the location of the 2.5-dB mismatch circle,

$$G = 10^{(-2.5/10)} = 0.5623$$

$$Center = \left\{\frac{0.5623}{1 - (0.6239)^2(1 - 0.5623)}\right\} \cdot \left(0.6239\angle{28.7698°}\right) = 0.4229\angle{28.7698°}$$

and

$$Radius = \frac{\left[1 - (0.6239)^2\right]\sqrt{1 - 0.5623}}{1 - (0.6239)^2(1 - 0.5623)} = 0.4870$$

Figure 2.25 shows this output plane mismatch circle, along with contours of constant 1.5- and 2.0-dB mismatch.

Inspecting the mismatch circles shown in the figure, we see that they are similar to constant gain circles, and the criteria for selecting an impedance is the same. One convenient point is the intersection of the 20-mmho constant conductance contour with the 2.5-dB mismatch circle. $\Gamma_L = 0.3244\angle{108.9287°}$, and the output matching network requires a single element. The susceptance of Γ_L is -j13.7178 mmho, or a reactance of +j72.9Ω. Figure 2.26 shows the complete 12.5-dB gain stage, where a coupling capacitor has been added to isolate the output. Substituting Γ_S and Γ_L in the expression for transducer gain, G_t = 12.4996 dB, which confirms the source and load impedance.

Although this design does provide 12.5-dB gain and two stages can be cascaded for the required 25 dB, this is not usually the best approach. Both input and output are mismatched, which may not be acceptable. However, we can match the input of the first stage and the output of the second stage, and then mismatch between stages. This design technique will be investigated in a later chapter.

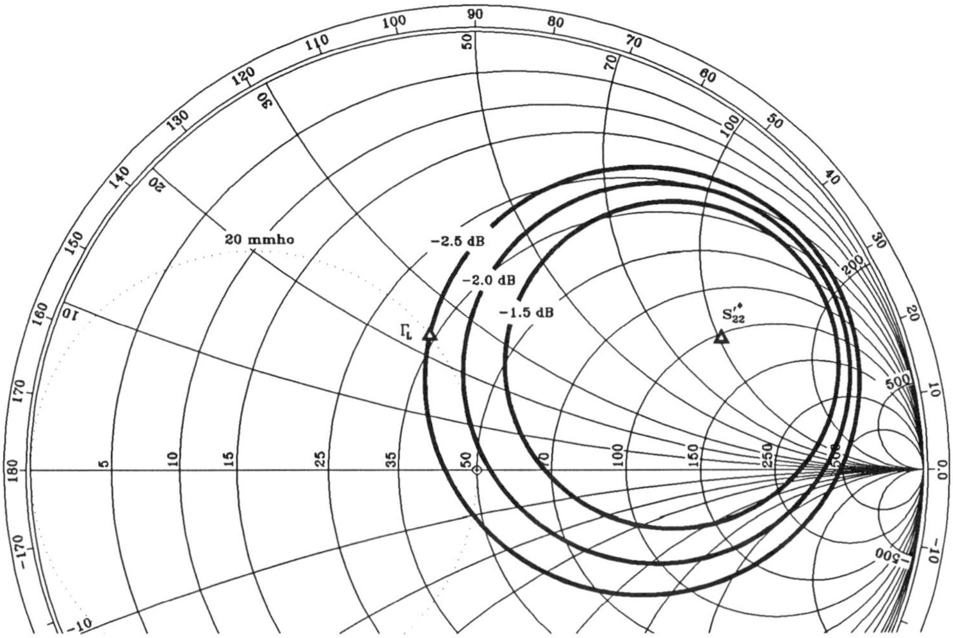

Figure 2.25 Output plane mismatch circles.

Example of Mismatch Used to Achieve MSG

The approach used in Section 2.3.8 to achieve greater than the maximum stable gain with positive real input and output impedance is based on selecting Γ_S and Γ_L from stable regions. Γ_S was selected from a point as close as possible to the stability circle, which happens to coincide with the intersection of the stability circle and locus of centers of constant gain circles. Intuitively, it seems that this would provide maximum gain and, in fact, it turns out that the stability circle is the infinite constant gain circle. We computed the load for matched output as the conjugate of s'_{22}. Since this was an unstable impedance, Γ_L was selected from the stable region, as shown in Figure 2.23. The Γ_L shown is not necessarily the best choice. We may be able to find another value from the stable region that provides more gain. The output reflection coefficient is $0.9997\underline{/-29.1036°}$, and we selected Γ_L as $0.3926\underline{/29.1036°}$. Using these values in (2.66),

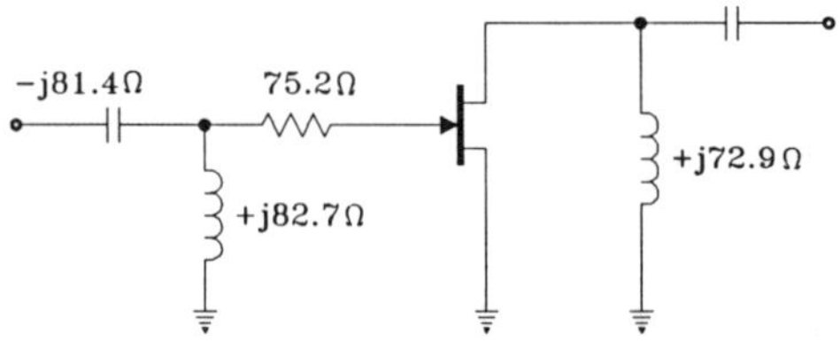

Figure 2.26 12.5-dB gain amplifier.

$$G_{mm} = \frac{\left[1 - (0.9997)^2\right]\left[1 - (0.3926)^2\right]}{\left|1 - \left(0.9997\underline{/-29.1036^\circ}\right)\left(0.3926\underline{/29.1036^\circ}\right)\right|^2} = 1.3749 \cdot 10^{-3} = -28.6173 \text{ dB}$$

and the corresponding constant mismatch circle is shown in Figure 2.27.

As the figure shows, we can achieve the same gain with reflection coefficients well into the stable region. Constructing mismatch circles from 26.0 to 27.5 dB in 0.5-dB increments, we find that the 26.5-dB circle just enters the stable region at $\Gamma_L = 0.7099\underline{/14.0558^\circ}$, which should increase the gain by more than 2 dB. The transducer gain for $\Gamma_S = 0.8058\underline{/33.9652^\circ}$ and $\Gamma_L = 0.7099\underline{/14.0558^\circ}$ is 22.2380 dB. Computing s'_{11} for this load confirms that the device input and output impedances are positive real and that the gain is almost 3.5 dB higher than the MSG.

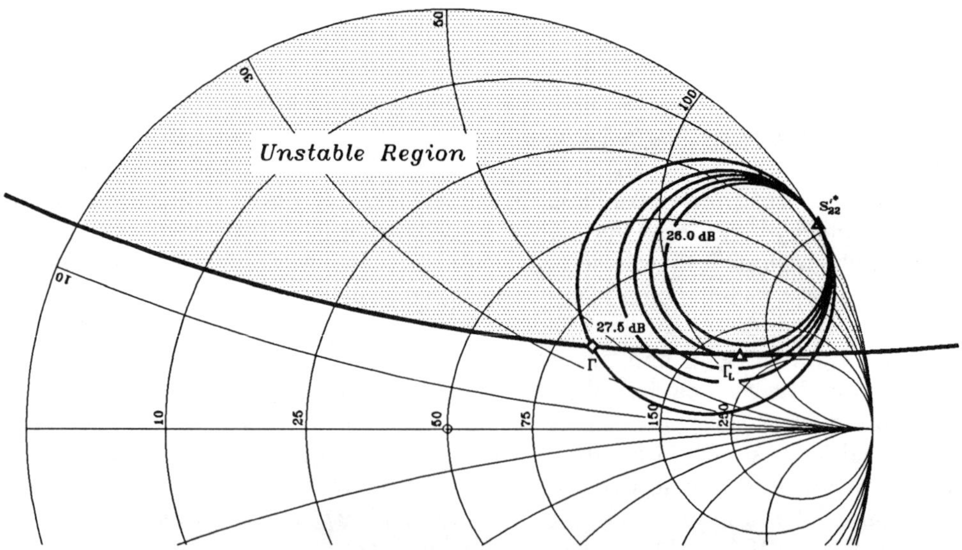

Figure 2.27 Output plane stability and mismatch.

2.3.10 Voltage Gain

We usually work with power gain; however, we can define current and voltage gains as well. The *input to output voltage gain* is defined as the ratio of port voltages, or

$$A_V = \frac{V_2}{V_1} \tag{2.68a}$$

Using the results of Section 1.3.1, where equation (1.17f) describes the voltage in terms of incident and reflected waves for real Z_o,

$$V_i = \sqrt{|Z_o|} \cdot (a_i + b_i)$$

and we can write (2.68a) as

$$A_V = \frac{a_2 + b_2}{a_1 + b_1} \tag{2.68b}$$

when the reference is real and equal at both ports. Substituting (a_i/b_s) and (b_i/b_s) from the signal flow graph analysis in Section 1.3.3,

$$A_V = \frac{(a_2/b_s) + (b_2/b_s)}{(a_1/b_s) + (b_1/b_s)} = \frac{(s_{21}\Gamma_L)/\Delta + (s_{21})/\Delta}{(1 - s_{22}\Gamma_L)/\Delta + [s_{11}(1 - s_{22}\Gamma_L) + s_{12}s_{21}\Gamma_L]/\Delta}$$

or

$$A_V = \frac{s_{21}(1 + \Gamma_L)}{(1 + s_{11})(1 - s_{22}\Gamma_L) + s_{12}s_{s1}\Gamma_L} \tag{2.68c}$$

We can also define the *source to load voltage gain* as

$$A_{VSL} = \frac{V_2}{V_S}$$

and the *insertion voltage gain* as

$$A_{VI} = \frac{V_2}{V_S} \cdot \left(1 + \frac{Z_S}{Z_L}\right)$$

Voltage gain can be derived for an arbitrary reference impedance at each port using generalized scattering parameters. Additionally, we can define current gain in a similar manner.

2.3.11 Noise Figure

Noise figure [25,26] is an important consideration in the design of amplifiers for certain applications. Manufacturers usually supply the *minimum noise figure* and *associated gain* for low-noise devices as a function of frequency. The associated gain is the maximum gain that can be achieved when the device is matched for minimum noise figure. These parameters aid in the selection of candidate devices. However, detailed analysis is usually required to select the best device.

The noise figure specification of a receiving system is based on an analysis of power levels at the input of the system. The signal arrives at a power level that is usually quite small because of path loss. Unfortunately, a certain amount of noise power is also present, and we can define a *signal-to-noise ratio* (SNR) as

$$SNR = \frac{S}{N}$$

where S is the signal power and N the noise power at the device input. When we amplify the signal, noise is also amplified, and, unless noise free, additional noise power is produced in the amplifier. H.T. Friis [27] introduced the concept of noise figure to describe the degradation in SNR due to the additional noise. The noise figure NF or F is defined as

$$F = 10\text{Log}\left\{\frac{S_i/N_i}{S_o/N_o}\right\} \text{dB} \tag{2.69}$$

where (S_i/N_i) is the SNR at the input, and (S_o/N_o) is the SNR at the output of the network. The ratio of input to output SNR in (2.69) is known as the *noise factor*. SNR is equal at the input and output of a noise-free network, and the noise figure is 0 dB. However, practical devices add noise, and the noise figure is always greater than zero, although some modern cryogenically cooled devices approach the ideal case.

The typical receiving system consists of more than a single input amplifier. There may be several stages of RF amplifiers, followed by a mixer, filter, and IF amplifiers. Each stage contributes its own noise power, and the noise figure of the complete system is the input SNR divided by the output SNR. Friis [27] provides an equation that describes the system noise factor for a cascade of n stages:

$$f_s = f_1 + \frac{f_2 - 1}{g_1} + \frac{f_3 - 1}{g_1 g_2} + \cdots + \frac{f_n - 1}{g_1 g_2 \cdots g_{n-1}} \tag{2.70}$$

where

f_s is the overall system noise factor,

f_i is the noise factor of stage i,

g_i is the associated gain of stage i,

and lower-case notation is used to indicate that all parameters are expressed as power ratios rather than in decibels. Examining (2.70), we see that the system noise figure cannot be less than the noise figure of the first stage, and higher gains reduce the noise contribution from following stages. In fact, if the gain of the first stage is high enough, the system noise figure is determined almost entirely by the first-stage noise figure.

Noise figures of low-noise systems are often expressed in terms of an *effective noise temperature* T_e, which is expressed in kelvin temperature. Noise figure and noise temperature are interchangeable; that is,

$$T_e = (f - 1) \cdot 290$$

$$f = \frac{T_e}{290} + 1$$

where 290K (16.85°C or 62.33°F) is the standard temperature for relating T_e and noise figure. The effective noise temperature converted from noise figure is referenced to the system input. The noise temperature can also be referred to the output; in fact, noise calculations are often simplified when using effective noise temperatures.

The minimum noise figure and associated gain found on typical data sheets do not directly provide useful information as to the performance of the device in a cascaded system. However, we can define a figure of merit called the *noise measure* [28-30] for an infinite cascade of identical stages:

$$M = \lim_{n \to \infty} f_s = 1 + \frac{f_{\min} - 1}{1 - 1/g_a} \tag{2.71}$$

The noise measure combines both parameters and provides some indication of the performance of a device in a cascaded system. However, minimum device noise figure does not always result in the lowest system noise figure, and each device must be evaluated to determine the true performance in a given system.

The operating noise figure of an active device is primarily determined by the source reflection coefficient. Using appropriate test equipment, we can measure three parameters [31-34] that completely describe the noise behavior of a device:

F_o - the minimum noise figure in dB.

Γ_o - the source reflection coefficient corresponding to minimum noise figure.

r_n - the normalized equivalent input noise resistance.

The actual noise resistance in ohms is sometimes given as R_n, and it is necessary to divide by the reference impedance Z_o to obtain r_n. Noise measurement is an extremely difficult task, usually requiring precise data and statistical methods to extract meaningful parameters. Once these parameters have been determined, noise figure can be related to the source reflection coefficient by

$$f = f_o + 4\,r_n \cdot \frac{|\Gamma_S - \Gamma_o|^2}{\left(1 - |\Gamma_S|^2\right)|1 + \Gamma_o|^2} \tag{2.72}$$

Solving (2.72) for r_n,

$$r_n = \frac{(f - f_o)\left[\left(1 - |\Gamma_S|^2\right)|1 + \Gamma_o|^2\right]}{4\,|\Gamma_S - \Gamma_o|^2}$$

and it is apparent that once F_o and Γ_o have been determined we only need the noise figure at any other value of Γ_S to calculate r_n.

Following the examples for stability and constant gain circles, we can set (2.72) equal to a constant which also results in source plane circles [35] with

$$Center = \frac{\Gamma_o}{1 + N} \tag{2.73a}$$

and

$$Radius = \frac{1}{1 + N} \cdot \sqrt{N^2 + N\left(1 - |\Gamma_o|^2\right)} \tag{2.73b}$$

where

$$N = \frac{f - f_o}{4 r_n} \cdot |1 + \Gamma_o|^2 \qquad (2.73c)$$

Examining (2.73), we see that the equivalent noise resistance determines the location and radius of all constant noise figure circles other than the single point at Γ_o. r_n is a measure of the sensitivity of noise figure to the source reflection coefficient.

Example of Low-Noise Amplifier Design

The s-parameters of a GaAs FET are

$$S = \begin{bmatrix} 0.71\angle{-109^\circ} & 0.12\angle{36^\circ} \\ 1.76\angle{85^\circ} & 0.54\angle{-59^\circ} \end{bmatrix}$$

At 12 GHz, the corresponding noise parameters are

$$F_o = 1.9 \text{ dB}$$

$$\Gamma_o = 0.58\angle{151^\circ}$$

$$r_n = 0.15$$

and both the s-parameters and noise parameters are referenced to 50Ω. We are given the task of designing an amplifier with 20-dB gain and the lowest possible noise figure. The first step is to characterize the device and investigate stability. Analysis reveals that $K = 0.8124$ and the MSG is 11.6633 dB, which indicates the design will require two stages. The associated gain is the maximum gain corresponding to an input reflection coefficient of Γ_o, or the available power gain G_A with $\Gamma_S = \Gamma_o$. Using (2.63) and (2.44),

$$G_A = \frac{|s_{21}|^2 \left(1 - |\Gamma_o|^2\right)}{\left(1 - |s_{22}|^2\right) + |\Gamma_o|^2 D_1 - 2\,\mathrm{Re}(\Gamma_o C_1)} = 7.2899 = 8.6272 \text{ dB}$$

The noise measure provides some indication of cascaded noise figure:

$$M = 1 + \frac{10^{(1.9/10)} - 1}{1 - 10^{(-8.6272/10)}} = 1.6361 = 2.1380\ \text{dB}$$

However, the exact value will depend on the relationship between the gain and the noise figure of both stages. Since the device is potentially unstable, we construct the input plane stability circle and constant gain contours between 8.0 and 12.0 dB in 0.5-dB increments, as shown in Figure 2.28.

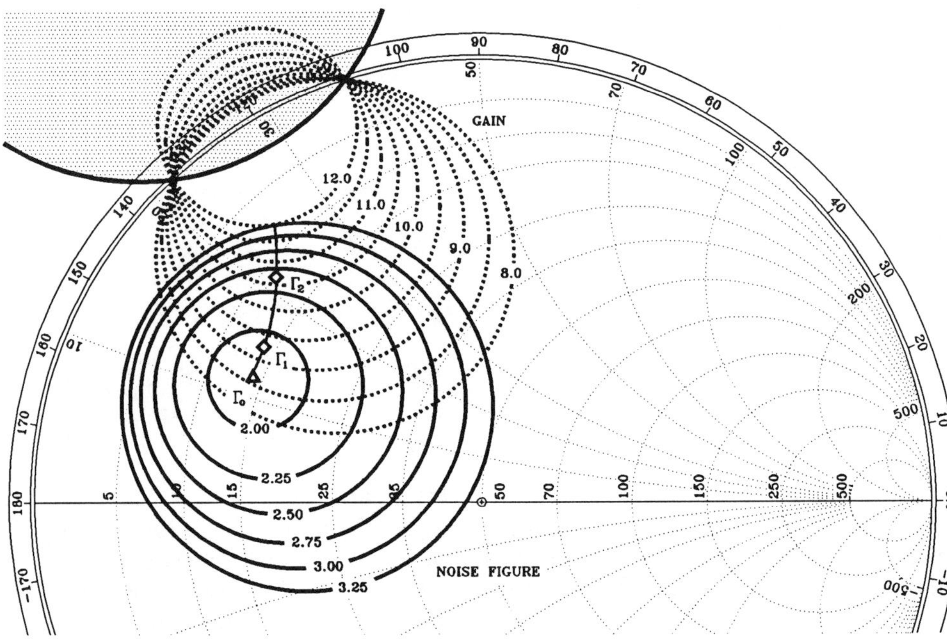

Figure 2.28 Input plane constant gain and noise circles.

We now need to look at the relationship between noise figure and gain and then select the source impedance for each stage. The easiest method is to plot contours of constant noise figure on the same plane as the constant gain circles. For the constant 2-dB noise figure circle, we can compute

$$N = \frac{10^{(2/10)} - 10^{(1.9/10)}}{(4)(0.15)} \cdot |1 + (0.58\angle 151^\circ)|^2 = 0.0194$$

$$Center = \frac{0.58\angle 151°}{1 + 0.0194} = 0.5690\angle 151°$$

and

$$Radius = \frac{1}{1 + 0.0194} \cdot \sqrt{(0.0194)^2 + (0.0194)\left[1 - (0.58)^2\right]} = 0.1128$$

which is shown in Figure 2.28, along with other circles plotted from 2.25- to 3.25-dB noise figure in increments of 0.25 dB.

Each gain circle attains the lowest noise figure at a single point closest to Γ_o. The locus of points is shown in the figure as a curve connecting Γ_o with the 12-dB gain circle. We can select the source reflection coefficients from this curve, which will result in the lowest noise figure for the corresponding gain. The curve is not in the region of least sensitivity for either gain or noise figure, and we could sacrifice noise figure in an effort to reduce sensitivity. However, we will design for minimum noise figure and use the curve as shown. Since this amplifier consists of two stages, we must determine the optimum combination of gain and noise figure for each stage, which results in the minimum amplifier noise figure. Using Figure 2.28, we see that the minimum noise figure for 9.5-dB gain is approximately 2 dB. The amplifier gain is specified as 20 dB, which means the second stage must have 10.5-dB gain. Minimum noise figure for 10.5-dB gain is 2.3 dB, and we can use (2.70) to determine the amplifier noise figure when both stages are cascaded:

$$f_s = 10^{(2/10)} + \frac{10^{(2.3/10)} - 1}{10^{(9.5/10)}} = 1.6632 = 2.2095\ \text{dB}$$

We can select another first stage gain and then repeat the process to determine the cascaded noise figure. Figure 2.29 shows a plot of cascaded noise figure versus first-stage gain for the amplifier alone, and the system noise figure when the amplifier is operating into a 10-dB noise figure, $F_3 = 10$ dB.

The figure shows that the optimum noise figure of 2.19 dB occurs at a first-stage gain of 9.25 dB. The reflection coefficient corresponding to 9.25-dB gain is $\Gamma_1 = 0.5975\angle 143.9849°$, and we can use (2.72) to confirm the noise figure:

$$f = 10^{(1.9/10)} + (4)(0.15) \cdot \frac{\left|0.5975\angle 143.9849° - 0.58\angle 151°\right|^2}{\left[1 - (0.5975)^2\right] \cdot \left|1 + 0.58\angle 151°\right|^2} = 1.5647 = 1.9444\ \text{dB}$$

The output reflection coefficient is the conjugate of s'_{22} or $0.7162\angle 62.8942°$, and the transducer gain is 9.2611 dB. The input reflection coefficient corresponding to

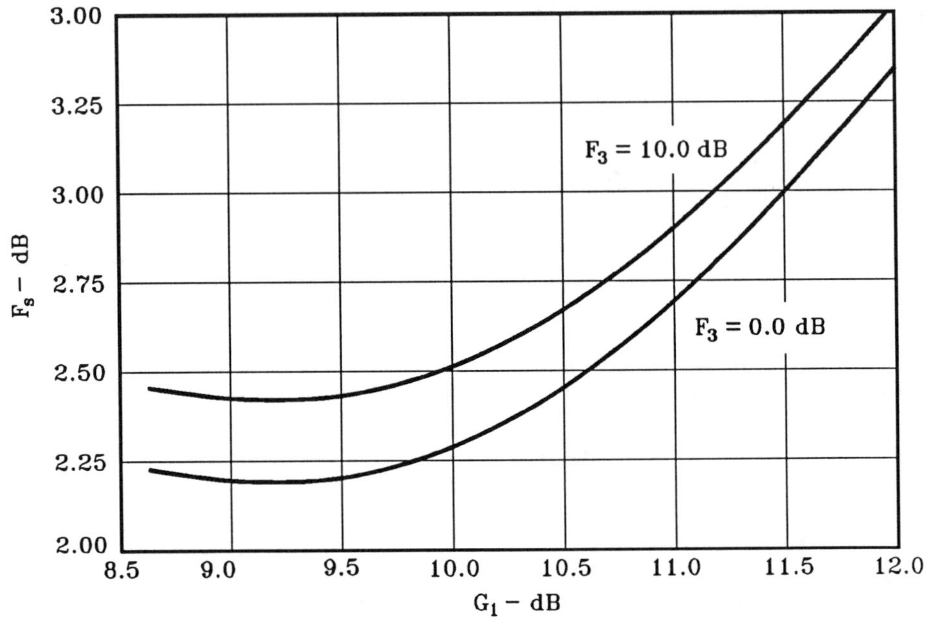

Figure 2.29 Two-stage noise figure.

Γ_L is 0.8451/-124.2569°, and the device is positive real at both ports. Second-stage gain is 10.75 dB, and the corresponding reflection coefficient is Γ_2 = 0.6821 /-131.8592°. The computed noise figure is 2.3997 dB, and the associated gain is 10.6612 dB. Both ports are positive real, and the two-stage amplifier has a noise figure of 2.19 dB at 19.92-dB gain.

The amplifier has a high-input VSWR because of the noise match, and may require additional components to reduce the VSWR, or provide unconditional stability. Any loss in additional components or matching network elements increases the noise figure and must be considered as part of the design. Most modern computer-aided design programs are capable of analyzing the effects of lossy elements, and the amplifier should be modeled in as much detail as possible to predict the hardware performance.

The noise figure and optimum source reflection coefficient are affected by series and shunt feedback. In some cases, it is possible to find certain combinations of series and shunt feedback that stabilize the device and cause the matched source impedance to coincide with the optimum noise impedance [23]. Usually the noise figure will be somewhat higher and the gain lower than that possible without feedback. However, when loss from additional components is considered, feedback can result in a superior design.

2.3.12 Power Amplifiers

Traditional power amplifier design usually requires a somewhat empirical approach. We connect a tuner at the input and output of the device, apply an input signal, and then tune the output for maximum power and the input for maximum gain. There will be a single impedance on the output plane where the power is maximum, and we can find numerous values of impedance where the output power is equal but less than maximum. We can create contours of constant output power by either fitting circles to a number of equal power points or taking enough data to form a closed contour. This technique is known as *load pull* [36-41] and results in a design approach similar to that for noise figure, except that it uses the output plane. Maximum gain corresponding to a power match is the operating power gain given by (2.64), where the output is arbitrarily terminated and the input is conjugately matched. Although output power is primarily determined by the output impedance, we can also create contours of constant power on the source plane using a similar technique known as *source pull*.

Power is usually measured at some predetermined level of gain compression, and the effects of cascading stages must be considered [42,43]. For example, if the first and second stages are compressed by 1 dB, the combination is 2 dB into compression, assuming the power level from the first stage causes 1-dB gain reduction in the second stage. We must also consider power levels at each stage of the chain. When any stage reaches maximum output power, the entire amplifier is saturated. The output intercept point [44,45], which is a measure of the ratio of fundamental to harmonic power levels, is also affected by cascading stages, and careful analysis is required when this parameter is specified.

The device may be operating at or near the nonlinear region and, strictly speaking, *s*-parameter analysis is not valid. Nevertheless, power amplifier design using these methods does produce hardware that agrees reasonably well with the computer predictions. The most recent generation of computer-aided design products includes the capability of nonlinear analysis, which provides us with tools capable of accurately analyzing fundamental and harmonic output power [45].

2.4 ACTIVE DEVICE CONVERSIONS

Most manufacturers furnish data for FET or bipolar transistors in the common source or common emitter configuration. Other configurations are useful for certain applications, and it is necessary to transform the parameters before the techniques of Section 2.3 can be applied to characterize the device. The discussion of indefinite scattering parameters in Section 1.3.6 revealed that for a three-port network formed from a three terminal device only four of the nine parameters are independent. Given the two-port parameters, we can compute the three-port indefinite matrix and then short-circuit the common port to obtain s-parameters for the desired configuration.

Example of Common Source to Common Gate Conversion

The 50Ω s-parameters given by (2.49) are those for the common source configuration. We can easily convert and then evaluate the characteristics of the device when used in the common gate configuration. Figure 1.24(a) shows that Z_o is connected from each terminal to the common node, forming three-port indefinite parameters. The common source s-parameters are measured with the source grounded. However, we can connect a shunt 50Ω resistor in series, as illustrated in Figure 1.5, which provides the necessary source reference and four of the nine s-parameters:

$$S = \begin{bmatrix} 0.9914\angle -11.6445^\circ & s_{12} & 0.0617\angle 86.6228^\circ \\ s_{21} & s_{22} & s_{23} \\ 1.1584\angle 162.8052^\circ & s_{32} & 0.8962\angle -7.9433^\circ \end{bmatrix}$$

The three-port matrix has been filled following the numbering convention shown in Figure 1.24(a). The gate is port 1, source is port 2, and the drain is port 3. The sum of any row or column must equal one, and we can compute the remaining five parameters:

$$S = \begin{bmatrix} 0.9914\angle -11.6445^\circ & 0.1408\angle 79.6207^\circ & 0.0617\angle 86.6228^\circ \\ 1.1445\angle -7.1444^\circ & 0.2572\angle 161.8536^\circ & 0.1253\angle 29.7865^\circ \\ 1.1584\angle 162.8052^\circ & 1.2385\angle -10.1664^\circ & 0.8962\angle -7.9433^\circ \end{bmatrix}$$

The desired configuration is formed by connecting a short to the appropriate port. Equation (2.20) describes the s-parameters of a multiport, with a one-port termination at port k. $\Gamma_k = -1$ for the short circuit termination, and

$$s'_{ij} = s_{ij} - \frac{s_{ik}s_{kj}}{1 + s_{kk}} \tag{2.74}$$

When we connect a termination, the row and column corresponding to the port is removed from the three-port matrix, and (2.74) is used to compute the remaining four s-parameters. The numbering convention provides a simple means to keep track of the input and output ports. Once the three-port is reduced, the lowest port number remaining is the input, and the highest is the output port.

We can form the common gate parameters by grounding the gate. The short circuit termination is connected at port 1, eliminating the first row and column. Using (2.74), the input reflection coefficient of the common gate configuration in terms of the three-port matrix indices is

$$s_{22}' = s_{22} - \frac{s_{21}s_{12}}{1 + s_{11}}$$

$$= 0.2572\angle 161.8536^\circ - \frac{(1.1445\angle -7.1444^\circ)(0.1408\angle 79.6207^\circ)}{1 + 0.9914\angle -11.6445^\circ}$$

$$= 0.2610\angle 179.8987^\circ$$

and the common gate s-parameters are

$$S = \begin{bmatrix} 0.2610\angle 179.8987^\circ & 0.1092\angle 14.1765^\circ \\ 1.2577\angle -6.4893^\circ & 0.9012\angle -5.6656^\circ \end{bmatrix}$$

The stability factor is 0.9349, which is much better than the corresponding value for the common source configuration. However, the MSG is only 10.6137 dB, approximately 8 dB lower. Figure 2.30 shows a comparison of all three configurations for the device.

Common source provides the highest gain, common gate the lowest input reflection coefficient, and common drain the lowest output reflection coefficient. Once we have the s-parameters for any configuration, we can easily convert to another configuration and investigate advantages and disadvantages for a particular application.

When noise figure is a design consideration, the four noise parameters, as well as the scattering parameters, must be transformed from one configuration to another. The conversion is not difficult [46-49], most computer-aided design programs offer this capability, which is transparent to the user. We simply specify

CONFIGURATION	$\lvert S_{11} \rvert$	$\lvert S_{22} \rvert$	*K*	*MSC*
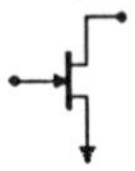 Common Source	0.9700	0.7100	0.1337	18.7795 dB
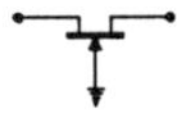 Common Gate	0.2610	0.9012	0.9349	10.6137 dB
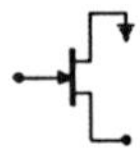 Common Drain	0.9955	0.3233	0.0848	10.8280 dB

Figure 2.30 FET configuration comparison.

the configuration for the measured data and then connect the device terminals to any three nodes. However, it often turns out that a configuration that has desirable gain, stability, or matching characteristics, has less than desirable noise performance.

REFERENCES

[1] C. Holmes and N. Kuhn, *Microwave Solid State Circuit Design*, Hewlett-Packard Company Seminar, Palo Alto, California, 1973.

[2] S. Ramo, J. Whinnery, and T. Van Duzer, *Fields and Waves in Communication Electronics*, John Wiley & Sons, Inc., New York, 1965.

[3] J. Vlach and K. Singhal, *Computer Methods for Circuit Analysis and Design*, Van Nostrand Reinhold Company, Inc., New York, 1983.

[4] V.A. Monaco and P. Tiberio, "Automatic Scattering Matrix Computation of Microwave circuits," *Alta Freq.*, Vol. 39, February 1970, pp. 59-64.

[5] V.A. Monaco and P. Tiberio, "Computer-Aided Analysis of Microwave Circuits," *IEEE Trans. Microwave Theory Tech.*, Vol. MTT-22, March 1975, pp. 249-263.

[6] J.K. Fidler and C. Nightingale, *Computer Aided Circuit Design*, Halsted Press, a Division of John Wiley & Sons, Inc., New York, 1978.

[7] S. Pissanetsky, *Sparse Matrix Technology*, Academic Press, New York, 1984.

[8] J.W. Bandler and R.E. Seviora, "Wave Sensitivities of Networks," *IEEE Trans. Microwave Theory Tech.*, Vol. MTT-20, February 1972, pp. 138-147.

[9] S.W. Director and R.A. Rohrer, "The Generalized Adjoint Network and Network Sensitivities," *IEEE Trans. on Circuit Theory*, Vol. CT-16, August 1969, pp. 318-323.

[10] B.D.H. Tellegen, "A General Network Theorem, with Applications," *Proc. Inst. Radio Engrs.*, Australia, 14 November 1953, pp. 265-270.

[11] J. Helszajn, *Passive and Active Microwave Circuits*, John Wiley & Sons, Inc., New York, 1978.

[12] George E. Bodway, "Two-Port Power Flow Analysis Using Generalized Scattering Parameters," *Microwave Journal*, Vol. 10, No. 6, May 1967.

[13] *S-Parameters...Circuit Analysis and Design*, Hewlett-Packard Company, Application Note 95, September 1968.

[14] *S-Parameter Design*, Hewlett-Packard Company, Application Note 154, April 1972.

[15] G. Gonzalez, *Microwave Transistor Amplifiers Analysis and Design*, Prentice-Hall, Inc., Englewood Cliffs, New Jersey, 1984.

[16] T. Ha, *Solid-State Microwave Amplifier Design*, John Wiley & Sons, New York, 1981.

[17] R.S. Carson, *High-Frequency Amplifiers*, John Wiley & Sons, Inc., New York, 1975.

[18] D. Woods, "Reappraisal of the Unconditional Stability Criteria for Active 2-Port Networks In Terms of S-Parameters," IEEE *Trans. on Circuits and Systems*, Vol. CAS-23, No. 2, February 1976, pp. 73-81.

[19] W.H. Ku, "A Simple Derivation for the Stability Criterion of Linear Active Two Ports," *Proc. IEEE*, Vol. 53, No. 3, March 1965, pp. 310-311.

[20] E.R. Hauri, "Overall Stability Factor of Linear Two-Ports in Terms of Scattering Parameters," *IEEE Journal of Solid-State Circuits*, Vol. ISC, No. 4, December 1971, pp. 413-415.

[21] J.M. Rollett, "Stability and Power-Gain Invariants of Linear Two Ports," *IEEE Trans. on Circuit Theory*, Vol. CT-9, No. 1, March 1962, pp. 29-32, and corrections, Vol. CT-10, No. 1, March 1963, p. 107.

[22] L. Besser and L. Olsen, "Transistor Stability CAD Helps to Find Stabilizing Elements," *Microwave System News and System Technology*, Vol. 17, No. 4, April 1987, pp. 54-64.

[23] L. Besser, "Stability Considerations of Low Noise Transistor Amplifiers with Simultaneous Noise and Power Match," *1975 IEEE International Microwave Symposium Digest*, May 1975, pp. 327-329.

[24] W.H. Ku, "Unilateral Gain and Stability Criterion of Active Two-Ports in Terms of Scattering Parameters," *Proc. IEEE*, Vol. 54, No. 11, November 1966, pp. 1617-1618.

[25] H. Fukui, "The Noise Performance of Microwave Transistors," *IEEE Trans. on Electron Devices*, Vol. ED-13, No. 3, March 1966, pp. 329-341.

[26] H.A. Haus and R.B. Adler, "Optimum Noise Performance of Linear Amplifiers," *Proc. IRE*, Vol. 46, No. 8, August 1958, pp. 1517-1533.

[27] H.T. Friis, "Noise Figures of Radio Receivers," *Proc. IRE*, Vol. 32, July 1944, pp. 419-422.

[28] H. Fukui, "Available Power Gain, Noise Figure, and Noise Measure of Two Ports and Their Graphical Representation," *IEEE Trans. on Circuit Theory*, Vol. CT-13, No. 2, June 1966, pp. 137-142.

[29] H. Fukui, "Noise Measure of Microwave Transistors," *IEEE Proc.*, Vol. 54, No. 9, September 1966, pp. 1204-1205.

[30] *Derivation, Definition and Application of Noise Measure*, Hewlett-Packard Company, Application Bulletin 9, August 1976.

[31] A. van der Ziel, *Noise in Measurements*, John Wiley & Sons, Inc., New York, 1976.

[32] M. Mitama and H. Katoh, "An Improved Computational Method for Noise Parameter Measurements," *IEEE Trans. on Microwave Theory and Techniques*, Vol. MTT-27, No. 6, June 1979, pp. 612-615.

[33] *Fundamentals of RF and Microwave Noise Figure Measurements*, Hewlett-Packard Company, Application Note 57-1, July 1983.

[34] R.Q. Lane, "The Determination of Device Noise Parameters," *Proc. IEEE*, Vol. 57, August 1969, pp. 1461-1462.

[35] C.R. Poole and D.K. Paul, "Optimum Noise Measure Terminations for Microwave Transistor Amplifiers," *IEEE Trans. on Microwave Theory and Techniques*, Vol. MTT-33, No. 11, November 1985, pp. 1254-1257.

[36] M. Pierpoint et al., "An Automated Measurement Technique for Measuring Amplifier Load-Pull and Verifying Large-Signal Device Models," *1986 IEEE MTT-S International Microwave Symposium Digest*, June 1986, pp. 625-628.

[37] D. Poulin, "Load-Pull Measurements Help You Meet Your Match," *Microwaves*, Vol. 19, No. 11, November 1980, pp. 61-65.

[38] J.M. Cusack et al., "Automatic Load Contour Mapping for Microwave Power Transistors," *IEEE Trans. on Microwave Theory and Techniques*, Vol. MTT-22, No. 12, December 1974, pp. 1146-1152.

[39] F. Sechi et al., "A Computer Controlled Microwave Tuner for Automated Load-Pull," *RCA Review*, Vol. 44, No. 4, December 1983, pp. 566-583.

[40] S.M. Perlow, "New Algorithms for the Automated Microwave Tuner Test System," *RCA Review*, Vol. 46, No. 3, September 1985, pp. 341-355.

[41] Y. Takayama, "A New Load-Pull Characterization Method for Microwave Power Transistors," *NEC Technical Note TN78401*, March 1978.

[42] G. Sorger, "The 1 dB Gain Compression Point for Cascaded Two-Port Networks," *Microwave Journal*, Vol. 31, No. 7, July 1988, pp. 136-142.

[43] S.M. Perlow, "Basic Facts about Distortion and Gain Saturation," *Applied Microwave*, May 1989, pp. 107-117.

[44] J-M. Mourant, "Predicting Amplifier 1dB Compression Point," *Applied Microwave*, Winter 1990, pp. 91-99.

[45] S.A. Maas, *Nonlinear Microwave Circuits*, Artech House, Inc., Norwood, Massachusetts, 1988.

[46] K.B. Niclas, "Active Matching with Common-Gate MESFET's," *IEEE Trans. on Microwave Theory and Techniques*, Vol. MTT-33, No. 6, June 1985, pp. 492-498.

[47] K. Hartmann and M.J.O. Strutt, "Changes in the Four Noise Parameters Due to General Changes of Linear Two-Port Circuits," *IEEE Trans. on Electron Devices*, Vol. ED-20, No. 10, October 1973, pp. 874-877.

[48] W.A. Suter, "Feedback and Parasitic Effects on Noise," *Microwave Journal*, Vol. 26, No. 2, February 1983, pp. 123.

[49] G.D. Vendelin, "Feedback Effects on the Noise Performance of GaAs MESFETs," *1975 IEEE MTT-S International Microwave Symposium Digest*, June 1975, pp. 324-326.

[50] J.D. Sifri, "Matching Technique Yields Optimum LNA Performance," *Microwaves and RF*, Vol. 25, No. 2, February 1986, pp. 87-90.

[51] G.D. Vendelin, *Design of Amplifiers and Oscillators by the S-Parameter Method*, John Wiley & Sons, Inc., New York, 1982.

[52] J.A. Eisenberg, "Designing Amplifiers for Optimum Noise Figure," *Microwaves*, Vol. 13, No. 4, April 1974, pp. 36-44.

[53] L. Besser and S. Swenson, "Take the Hassle Out of FET Amp Design," *Microwave System News*, Vol. 7, No. 9, September 1977, pp. 97-101.

[54] M. Staloff, "Computer-Aided Design and Analysis of Microwave Transistor Amplifiers,"34-*Microwave Journal*, Vol. 30, No. 5, May 1987, pp. 253-268.

[55] C.D. Motchenbacher and F.C. Fitchen, *Low-Noise Electronic Design*, John Wiley & Sons, Inc., New York, 1973.

[56] H. Fukui, *Low-Noise Microwave Transistors and Amplifiers*, IEEE Press, New York, 1981.

Chapter 3

Mapping and Optimization

The techniques discussed in Chapters 1 and 2 provide the means for analyzing complex networks. Although there is no distinct division, two-port, multiport, and nodal analyses are generally used to analyze a circuit, given its topology and element values, while other analysis methods are useful for synthesizing the design. The topics introduced in this chapter can also be characterized as design techniques, and are perhaps as powerful as any we have looked at thus far.

3.1 MAPPING

The designer is frequently required to determine what effect an impedance connected at one port has on the other network parameters. For instance, contours of constant gain can be constructed on the source plane and a source impedance can be selected that provides the desired gain when the output is conjugately matched. Is there a better set of corresponding source and load impedances? Perhaps a combination can be found that requires simple input and output matching networks. Another combination may yield impedance values far removed from the regions of instability, and still another set may be located that is easier to match over a broad frequency range.

There are many other instances where the network characteristics are required as a function of an impedance connected between two terminals. The oscillator designer looks for an impedance that can be used in a series or shunt feedback configuration to provide negative resistance at one port, an impedance that can easily be matched to the characteristic impedance of the system at another port, and, it is hoped, isolation between the input and output ports. Feedback also offers significant advantage in certain types of amplifier designs. The proper selection of feedback impedance can be used to design amplifiers with specified gain and simultaneous matched input and output ports, improved stability, and high reverse isolation. Designing with four-terminal devices such as the dual gate field-effect transistor (FET) is especially difficult and requires even more information about the network behavior as a function of the impedance at one or more ports.

Accurate and complete characterization can often be the difference between a marginal design, which just meets specifications after considerable tuning, and a design that is easy to produce. How can the designer obtain this information? The stability, gain, mismatch, noise, and power contours discussed in Chapter 2 are one method. Another technique consists of calculating the parameters for many discrete values of an impedance connected at the appropriate port. This approach requires the resources of a computer because the equations are complex and must be evaluated for each impedance value, perhaps at many frequencies. A computer listing does not provide much insight into the variation of the network parameters. However, each data point can be plotted on either a Smith or polar chart and, with enough points, a map of the desired parameter as a function of frequency can be obtained. This technique is not used by many designers because it is time consuming. Fortunately, there is a better method.

3.1.1 Mapping Theory and the Bilinear Transform

In order to determine the effect of an impedance connected at one port or across two terminals of a network, we must first find the relationship between the desired parameters as a function of impedance. Furthermore, as it turns out, some of the more useful mappings are for networks with more than two ports, and the analysis must be valid for multiport networks.

The *bilinear* or *linear fractional transformation*,

$$w = \frac{a \cdot z + b}{c \cdot z + d} \tag{3.1}$$

is the basis for the mappings we will develop. a, b, c, and d are arbitrary complex constants, and for $(ad - bc) \neq 0$, Equation (3.1) maps the totality of circles and straight lines in the z-plane into the totality of circles and straight lines in the w-plane in a conformal manner. Note that for $(ad - bc) = 0$, w is equal to a constant. The linear transformation depends on three essential constants, the ratios of any three of the constants a, b, c, d to the fourth. Therefore, we can rewrite (3.1) as

$$w = \frac{(a/c)\cdot z + (b/c)}{z + (d/c)} \tag{3.2}$$

and then use a theorem from linear transformation theory which states that *three given distinct points* z_1, z_2, z_3 *can always be mapped onto three prescribed distinct points* w_1, w_2, w_3 *by one and only one linear mapping,* $w = f(z)$, to solve for the three constants.

Since we are primarily interested in the behavior of the network parameters as a function of impedance, the particular mapping which will be investigated is the one that relates the impedance plane to the scattering parameter plane through the bilinear transform. The *s*-parameters are referenced to an impedance Z_o, and it will be convenient to normalize the impedance plane Z and then map the normalized z-plane onto the s-parameter plane. The reflection coefficient developed in Section 1.3 is given by

$$\Gamma = \frac{Z - Z_o^*}{Z + Z_o} \tag{3.3a}$$

where Z is the equivalent port impedance and Z_o is the reference impedance. Substituting $(R + jX)$ for Z and $(R_o + jX_o)$ for Z_o, we can write

$$\frac{Z - Z_o^*}{Z + Z_o} = \frac{(R + jX) - (R_o - jX_o)}{(R + jX) + (R_o + jX_o)} = \frac{R + j(X + X_o) - R_o}{R + j(X + X_o) + R_o} \tag{3.3b}$$

The normalized reflection coefficient is

$$s = u + jv = \frac{z - 1}{z + 1} = \frac{r + jx - 1}{r + jx + 1} \tag{3.3c}$$

where z is the normalized impedance, r is the real component, and x is the imaginary component. Comparing the two, we see that the normalized impedance is

$$z = r + jx = \frac{R}{R_o} + j\frac{(X + X_o)}{R_o} \tag{3.4}$$

Thus, to normalize an impedance, we divide the real part by the real part of the normalizing impedance and then divide the sum of both imaginary components by the real part of the normalizing impedance.

Comparing the bilinear transform in (3.2) with the normalized reflection coefficient in (3.3c), we see that the coefficients are $(a/c) = 1$, $(b/c) = -1$, and $(d/c) = 1$. Selecting three points on the normalized impedance plane, we can use (3.3c) and (3.4) to compute the corresponding points on the impedance and *s*-planes. Table 3.1 shows tabulated data for the three points, and Figure 3.1 shows both planes graphically.

Table 3.1 Relationship Between Three Points in the *Z*-, *z*-, and *s*-Planes

Point	*Z*	*z*	*s*
A	$0 - jX_o$	$0 + j0$	$-1 + j0$
B	$R_o - jX_o$	$1 + j0$	$0 + j0$
C	$0 + j(R_o - X_o)$	$0 + j1$	$0 + j1$

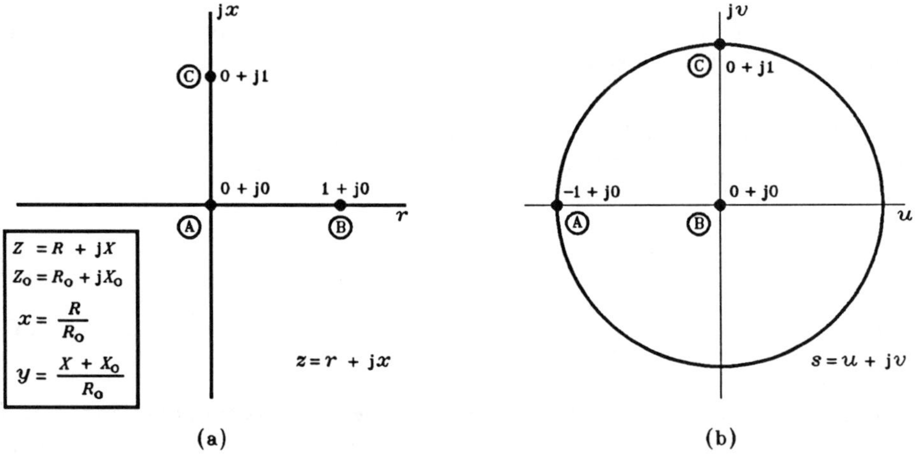

Figure 3.1 Bilinear transform mapping: (a) z-plane; (b) s-plane.

We observe from (3.3c) that for *r* greater than 0, $|s|$ is less than 1. When *r* is less than 0, $|s|$ is greater than 1, and $|s|$ equals 1 when $r = 0$. Therefore, the entire positive real portion of the *z*-plane is mapped into the unit circle on the *s*-plane. The normalization equation (3.4) shows that *r* is positive when the sign of *R* and R_o is the same, and we conclude that the real part of impedance values corresponding to reflection coefficients within the unit circle has the same sign as the real component of the reference impedance.

We can write (3.3b) as

$$s = u + jv = \frac{(R - R_o) + j(X + X_o)}{(R + R_o) + j(X + X_o)} \tag{3.5a}$$

expand and then equate real terms:

$$R(u - 1) + X(-v) + R_o(u + 1) + X_o(-v) = 0 \tag{3.5b}$$

and imaginary terms:

$$R(v) + X(u - 1) + R_o(v) + X_o(u - 1) = 0 \tag{3.5c}$$

We can multiply (3.5b) by (u - 1), (3.5c) by (-v), and then subtract to eliminate X:

$$u^2(R + R_o) + u(-2R) + v^2(R + R_o) + (R - R_o) = 0 \tag{3.6a}$$

Dividing by ($R + R_o$) and then forming a square term in u,

$$\left(u - \frac{R}{R + R_o}\right)^2 + v^2 = \left(\frac{R_o}{R + R_o}\right)^2 \tag{3.6b}$$

which is the equation for a circle in the s-plane with

$$Center = \left(\frac{R}{R + R_o}\right) + j0 = \left(\frac{r}{r + 1}\right) + j0 \tag{3.6c}$$

and

$$Radius = \left|\frac{R_o}{R + R_o}\right| = \left|\frac{1}{r + 1}\right| \tag{3.6d}$$

Thus, straight lines of constant resistance in the impedance plane are mapped into circles in the s-plane.

Similarly, we can eliminate R from (3.5b) and (3.5c):

$$u^2(X + X_o) + u(-2[X + X_o]) + v^2(X + X_o) + v(-2R_o) + (X + X_o) = 0 \tag{3.7a}$$

divide by $(X + X_o)$, factor u and v into squares:

$$(u - 1)^2 + \left(v - \frac{R_o}{X + X_o}\right)^2 = \left(\frac{R_o}{X + X_o}\right)^2 \tag{3.7b}$$

and we see that constant reactance also maps into s-plane circles with

$$Center = 1 + j\frac{R_o}{X + X_o} = 1 + j\frac{1}{x} \tag{3.7c}$$

and

$$Radius = \left|\frac{R_o}{X + X_o}\right| = \left|\frac{1}{x}\right| \tag{3.7d}$$

The admittance of a point in the normalized impedance plane is formed by inverting the impedance, $y = (1/z)$. Substituting into (3.3c),

$$s = u + jv = \frac{1 - y}{1 + y} = \frac{1 - (g + jb)}{1 + (g + jb)} \tag{3.8a}$$

where g is the normalized conductance and b is the normalized susceptance. We can expand and then equate real terms:

$$g(u + 1) + b(-v) + (u - 1) = 0 \tag{3.8b}$$

and imaginary terms:

$$g(v) + b(u + 1) + v = 0 \tag{3.8c}$$

Eliminating the susceptance b from (3.8b) and (3.8c), we have

$$u^2(g + 1) + u(2g) + v^2(g + 1) + (g - 1) = 0 \tag{3.9a}$$

Then, dividing by $(g + 1)$ and forming squares in u and v,

$$\left(u + \frac{g}{g+1}\right)^2 + v^2 = \left(\frac{1}{g+1}\right)^2 \tag{3.9b}$$

Thus, constant conductance in the normalized admittance plane maps into circles in the s-plane with

$$Center = -\frac{g}{g+1} + j0 \tag{3.9c}$$

and

$$Radius = \left|\frac{1}{g+1}\right| \tag{3.9d}$$

We derive contours of constant susceptance by eliminating g from (3.8b) and (3.8c):

$$u^2(b) + u(2b) + v^2(b) + v(2) + b = 0 \tag{3.10a}$$

and then dividing by b and forming squares in u and v:

$$(u+1)^2 + \left(v + \frac{1}{b}\right)^2 = \left(\frac{1}{b}\right)^2 \tag{3.10b}$$

which also results in s-plane circles with

$$Center = -1 - j\frac{1}{b} \tag{3.10c}$$

and

$$Radius = \left|\frac{1}{b}\right| \tag{3.10d}$$

We can replace the load impedance with an admittance and write the input reflection coefficient in terms of admittance:

$$\Gamma = \frac{b_i}{a_i} = \frac{V_i - Z_o^*(V_i Y_i)}{V_i + Z_o(V_i Y_i)} = \frac{1 - Z_o^* Y}{1 + Z_o Y} \tag{3.11a}$$

and then expand into real and imaginary components:

$$\frac{1 - (R_o - jX_o)(G + jB)}{1 + (R_o + jX_o)(G + jB)} = \frac{(1 - X_oB + jX_oG) - (R_oG + jR_oB)}{(1 - X_oB + jX_oG) + (R_oG + jR_oB)} \tag{3.11b}$$

Comparing (3.11b) with (3.8a), we see that the normalized admittance is

$$y = g + jb = \frac{R_o(G + jB)}{1 - X_o(B - jG)} \tag{3.11c}$$

which, unfortunately, does not turn out to be as simple as the impedance normalization given in (3.4). However, we can use (3.4) and (3.11c) to confirm that the admittance is the reciprocal of impedance in both planes, and it is perhaps easier to use impedance and then invert as necessary. We can also equate (3.11b) with $s = u + jv$, eliminate susceptance, and then solve for contours of constant conductance in the s-plane

$$Center = \frac{G(X_o^2 - R_o^2)}{G(X_o^2 + R_o^2) + R_o} + j\frac{2GR_oX_o}{G(X_o^2 + R_o^2) + R_o} \tag{3.12a}$$

and

$$Radius = \left|\frac{R_o}{G(X_o^2 + R_o^2) + R_o}\right| \tag{3.12b}$$

When conductance is eliminated from the two equations, we find that contours of constant susceptance also map into circles given by

$$Center = \frac{B(X_o^2 - R_o^2) - X_o}{B(X_o^2 + R_o^2) - X_o} + j\frac{2BR_oX_o - R_o}{B(X_o^2 + R_o^2) - X_o} \tag{3.13a}$$

and

$$Radius = \left|\frac{R_o}{B(X_o^2 + R_o^2) - X_o}\right| \tag{3.13b}$$

Given a reflection coefficient in the s-plane, we can solve (3.3c) for normalized impedance:

$$z = \frac{1 + s}{1 - s} \tag{3.14a}$$

and (3.8a) for normalized admittance:

$$y = \frac{1 - s}{1 + s} \tag{3.14b}$$

Comparing the two expressions, we see that we can convert between impedance and admittance by changing the sign of s, which corresponds to rotating the original point through 180°. Once we have the normalized impedance, (3.4) can be used to compute the actual real and imaginary components:

$$\begin{aligned} R &= r \cdot R_o \\ X &= x \cdot R_o - X_o \end{aligned} \tag{3.14c}$$

and (3.11c) can be used to calculate Y from the normalized admittance:

$$G = \frac{gR_o}{(R_o + bX_o)^2 + (gX_o)^2} \tag{3.14d}$$

$$B = \frac{bR_o + X_o(g^2 + b^2)}{(R_o + bX_o)^2 + (gX_o)^2} \tag{3.14e}$$

3.1.2 The Smith Chart

The bilinear transform described by (3.2) was introduced by Philip Smith [1,2] and is perhaps the most important tool we have for visually relating impedance, admittance, and reflection coefficient. The mapping is known as a *Smith chart* and turns out to be a very convenient means for designing matching networks.

The normalized Smith chart mapping is easily generated using equations (3.6), (3.7), (3.9), and (3.10). The contour for $r = 0$ can be drawn immediately, since it is represented by the unity circle; any other constant resistance circle requires (3.6). For example, the location of the $r = 1$ circle is given by

$$Center = \left(\frac{r}{r+1}\right) + j0 = \left(\frac{1}{1+1}\right) + j0 = 0.5 + j0$$

and

$$Radius = \left|\frac{1}{r+1}\right| = \left|\frac{1}{1+1}\right| = 0.5$$

This contour is shown in Figure 3.2 as the solid circle that passes through the origin. Similarly, the location of constant reactance circles is computed using equation (3.7). The location of constant -j1 is

$$Center = 1 + j\frac{1}{x} = 1 + j\frac{1}{(-1)} = 1 - j1$$

and

$$Radius = \left|\frac{1}{x}\right| = \left|\frac{1}{(-1)}\right| = 1.0$$

A portion of this contour is also shown in Figure 3.2 as a solid curve that intersects the unity circle at -90°. Although contours of constant reactance and susceptance are full circles, normally only the portion within the unity circle is shown. However, if we are interested in reflection coefficients greater than 1, we can construct negative constant resistance circles and extend the reactance and susceptance circles.

The figure shows positive constant resistance circles from 0.1 to 10.0, as well as several positive and negative reactance circles. The straight line along the u, or horizontal, axis is the infinite radius circle corresponding to $x = 0$. Real and imaginary components of impedance are plotted as solid lines in the figure. Commercially available Smith charts usually include a scale that can be used to measure the magnitude of reflection coefficient, standing wave ratio, transmission loss, and reflection loss. The periphery of the unit circle may include a scale graduated in wavelengths, as well as in degrees. We can immediately convert between impedance and reflection coefficient using the appropriate scales. For example, we can plot the impedance point corresponding to $z = 1.0 + j1.0$ by locating the intersection of the 1.0 constant resistance and the +1.0 constant reactance contour, which is labeled point A in the figure. Measuring the magnitude and angle from the origin, we have $\Gamma = 0.447\angle 63.43°$. Conversely, if the point is plotted from a reflection coefficient, we read r from the constant resistance circle and x from the constant reactance circle passing through the point.

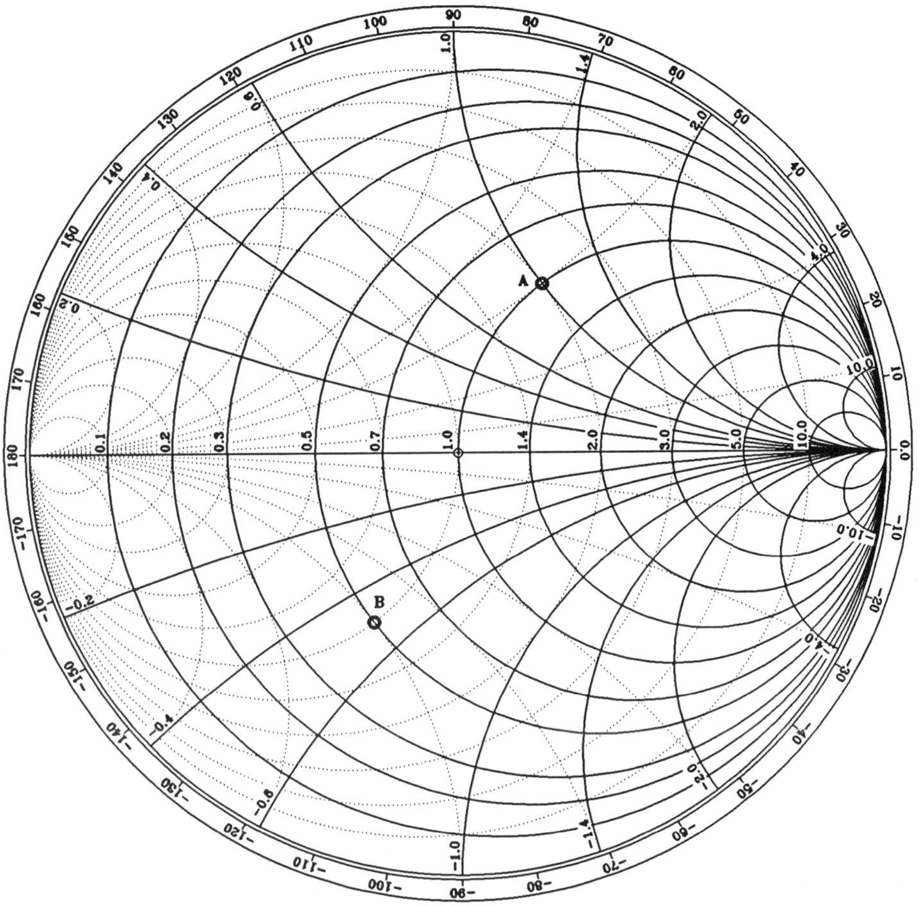

Figure 3.2 Normalized Smith chart.

When the mapping is normalized, we must use (3.4) to transform from Z to z, or (3.14c) to transform from z to Z. The most common reference is 50Ω, and in the case of point A we compute

$$R = r \cdot R_o = (1.0) \cdot (50.0) = 50.0\Omega$$

and

$$X = x \cdot R_o - X_o = (1.0) \cdot (50.0) - 0.0 = 50.0\Omega$$

or $Z = 50.0\Omega + j50.0\Omega$. When the reference is real, we can multiply the normalized values by R_o, the origin becomes $R_o + j0\Omega$, and all other values are scaled by R_o. Smith charts are available that are referenced to 50Ω and labeled with the actual resistance and reactance rather than normalized values. However, a complex reference changes the interpretation somewhat [3,4]. The origin of the normalized chart is $z = 1 + j0$. Using (3.14c),

$$R = r \cdot R_o = (1.0) \cdot (R_o) = R_o$$

$$X = x \cdot R_o - X_o = (0.0) \cdot (R_o) - X_o = -X_o$$

and we see that the origin corresponds to the complex conjugate of Z_o. The normalized infinite radius reactance circle or horizontal axis is j0, which becomes $-jX_o$ in the Z-plane, and zero reactance in the Z-plane corresponds to a normalized value of $0 = (xR_o - X_o)$ or $+j(X_o/R_o)$ in the z-plane. Figure 3.3 shows the Smith chart referenced to $25\Omega + j25\Omega$. Constant resistance and reactance contours are labeled with the corresponding Z-plane values. Comparing Figure 3.3 with Figure 3.2, we see that the mappings are identical and that only the assigned values differ. Figure 3.3 also shows that the entire region of positive reactance has been compressed into one quadrant of the chart, and negative reactance is expanded. Reading directly from Figure 3.3, the impedance of point A is $25\Omega + j0\Omega$. Using (3.3a),

$$\Gamma = \frac{Z - Z_o^*}{Z + Z_o} = \frac{(25 + j0) - (25 - j25)}{(25 + j0) + (25 + j25)} = \frac{+j25}{50 + j25} = 0.447\underline{/63.43°}$$

which is the same answer we obtain by measuring the polar coordinates of the point. Thus, we can obtain the reflection coefficient from either normalized or Z-plane values.

We can convert between impedance and admittance simply by rotating the original point by 180° [5,6]. Assuming point A shown in Figure 3.2 is an impedance, we can plot the corresponding admittance located at $0.447\underline{/(63.43°-180°)}$ or $0.447\underline{/-116.57°}$, labeled as point B in the figure. After converting to admittance, the constant resistance contours become constant conductance contours, and constant reactance becomes constant susceptance. Point B falls on the z-plane $r = 0.5$ and $x = -0.5$ circles, which translates into $y = 0.5 - j0.5$. We can use (3.8a) to compute the reflection coefficient for an admittance at point B:

$$\Gamma = \frac{1 - y}{1 + y} = \frac{1 - (0.5 - j0.5)}{1 + (0.5 - j0.5)} = \frac{0.5 + j0.5}{1.5 - j0.5} = 0.447\underline{/63.43°}$$

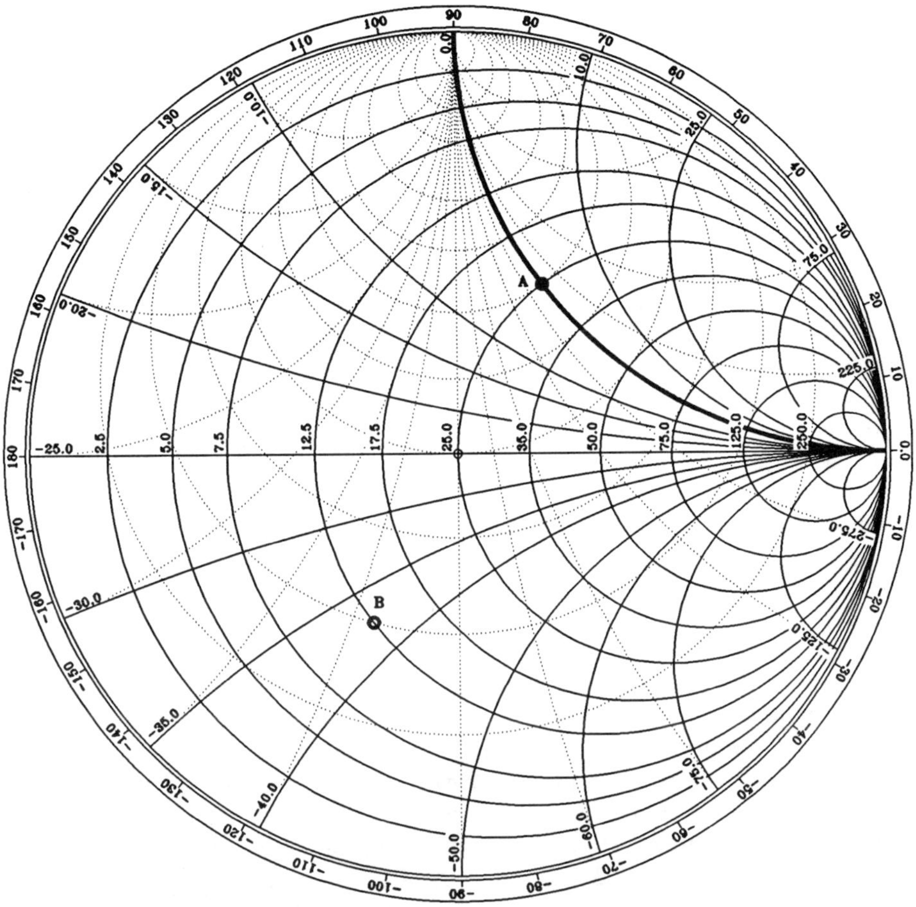

Figure 3.3 Smith chart referenced to $25\Omega + j25\Omega$.

which is identical to the value measured at point A, *not point B*. The reflection coefficient must be measured from a point that represents impedance. When the reference impedance is real, $X_o = 0$, equation (3.11c) reduces to

$$y = g + jb = R_o(G + jB) = R_oY$$

and we can calculate Y as y/R_o. Assuming a reference impedance of 50Ω, the admittance at point B is $Y = (0.5 - j0.5)(1/50)$, or $Y = (10 - j10)$ mmho. Inverting Y we have $Z = 50\Omega + j50\Omega$, which agrees with the value computed above. When the chart is referenced to a real impedance, we divide by R_o to normalize; thus, admittance can be determined directly from Z-plane values divided by R_o^2. Point B is located at (25Ω - j25Ω) on the 50Ω Smith chart, and Y = (25Ω - j25Ω)(1/50Ω)(1/50Ω) or Y = (10 - j10) mmho.

Figure 3.3 shows that the impedance of point A is 25Ω + j0Ω, with the chart referenced to 25Ω + j25Ω. Rotating by 180° to point B, we see that the admittance is located at the intersection of the 12.5Ω constant resistance and -37.5Ω constant reactance circle. Applying (3.4), we calculate the normalized values as r = 0.5 and x = -0.5, which are the same results obtained in the previous example, y = 0.5 - j0.5. Substituting g = 0.5, b = -0.5, R_o = 25Ω, and X_o = 25Ω into (3.14d) and (3.14e) reveals that the admittance Y is equal to (40.0 + j0) mmho. Admittance formed by rotating an impedance on a complex referenced chart provides little information as to the composition of the admittance, and, in fact, requires considerable calculations before actual values can be determined. It is easier to simply invert the impedance. The impedance at point A is real and we expect the admittance to be real as well. However, the imaginary component at B is not zero, nor is it a simple function of Z_o.

A better approach when dealing with admittance is to construct contours of constant conductance and susceptance, along with the contours of constant resistance and reactance. The combined mapping is sometimes called an *immittance chart*, a combination of *impedance* and *admittance*, and can be used with either a real or complex reference. Figures 3.2 and 3.3 show the admittance contours as dotted lines. The value of each admittance contour is obtained by inverting the corresponding impedance, although labels are sometimes printed on the chart, making this step unnecessary. The real axis is convenient for determining conductance values, since $g = 1/r$ and susceptance can easily be obtained from the r = 0 contour where $b = -(1/x)$. For example, the constant conductance circle passing through point A in Figure 3.2 crosses the real axis at r = 2 and g = (1/2) or 0.5. Following the constant susceptance contour from point A to the chart boundary where r = 0, x = 2 and b = -0.5. The impedance of point B shown in Figure 3.3 is 12.5Ω - j37.5Ω. Following the constant conductance contour to the intersection with the real axis, we find that G = (1/125), or G = 8 mmho. The constant susceptance contour passing through point B intersects the R = 0 circle at approximately -42Ω and B = 23.8 mmho, or Y = (8 + j23.8) mmho. Thus, we can read the reflection coefficient, impedance, or admittance directly from the immittance chart.

The impedance and admittance contours shown in Figure 3.2 are rotated 180° with respect to each other. This relationship allows us to use the constant resistance and reactance contours as conductance and susceptance after rotating an impedance point by 180°. In effect, we rotate an impedance rather than the contours. However, when the reference is complex, the relationship between impedance and admittance is not 180°, as illustrated in Figure 3.3. *Constant resistance and reactance circles are not translated into constant conductance and susceptance contours when the reference is complex.* In the preceding, we were able to compute the correct admittance only because we normalized the impedance, inverted, and then converted back to the complex reference.

3.1.3 Impedance Matching Using the Smith Chart

Smith and immittance charts provide a very effective means for synthesizing simple matching networks [5,6]. We are provided with a visual display that is easy to interpret and can be combined with other information, such as stability, gain, or noise figure. Numerous topologies can be synthesized, and one can be selected based on convenient element values or bias considerations.

The impedance corresponding to any reflection coefficient is a series combination of resistance and reactance determined directly from the contours of constants R and X. When we add a series reactive element X_1, the new impedance is $R + j(X + X_1)$, which is a point on the same constant resistance circle. Thus, series reactance follows contours of constant R. Inductive reactance moves the impedance clockwise in the direction of increasing reactance values, and capacitive reactance moves the point counterclockwise. The reactance necessary to move from one point to another is simply the difference between the final and initial points. For example, when we move from jX_1 to jX_2, the required value is $jX_2 - jX_1$ or $j(X_2 - X_1)$ Ω.

Admittance represents a parallel equivalent network, and the addition of a shunt reactive element jB_1 results in $G + j(B + B_1)$. Shunt reactive elements follow contours of constant conductance, capacitive reactance clockwise and inductive reactance counterclockwise. The susceptance required to move between two points is determined in the same fashion as series reactance, the difference between the final and initial susceptance. We can determine the equivalent parallel resistance and reactance by inverting the real and imaginary admittance components separately. For example, the admittance of point A in Figure 3.2 was determined to be (10 - j10) mmho when referenced to 50Ω. The real part of the parallel equivalent is (1/0.01) or 100Ω, and the imaginary equivalent (1/-j0.01) or +j100Ω and $Z_p = (100 + j100)$ Ω.

The simplest network that will transform 50Ω into the impedance at point A is a series inductive reactance. Figure 3.4 shows that we follow the constant resistance circle from the origin in a clockwise direction. The value is (j50 - j0) or $X_L = 50$. However, we may require a specific topology for bias or other reasons, which should include a series input capacitor for DC blocking and a shunt inductor at point A which can be used for bias. Starting from 50Ω, we add series capacitance by moving counterclockwise along the $R = 50$Ω circle. Shunt inductance also moves counterclockwise along a constant conductance circle, and $-jX_C$ should be selected to move from the origin to the conductance circle passing through point A, 10 mmho in this case. Reading from the impedance scale, $X_C = (-50 - 0)$ or 50Ω capacitive reactance. At the intersection of the $G = 10$ mmho and $R = 50$Ω circle, we follow the constant susceptance contour to the edge of the chart, invert the reactance, and find that B_1 = +10 mmho. Similarly, we find B_2 at point A is -10 mmho, and B_L = (B_2 - B_1) or B_L = -20 mmho. The matching network consists of a series -j50Ω capacitor followed by a shunt +j50Ω inductor, as shown in Figure 3.5(a).

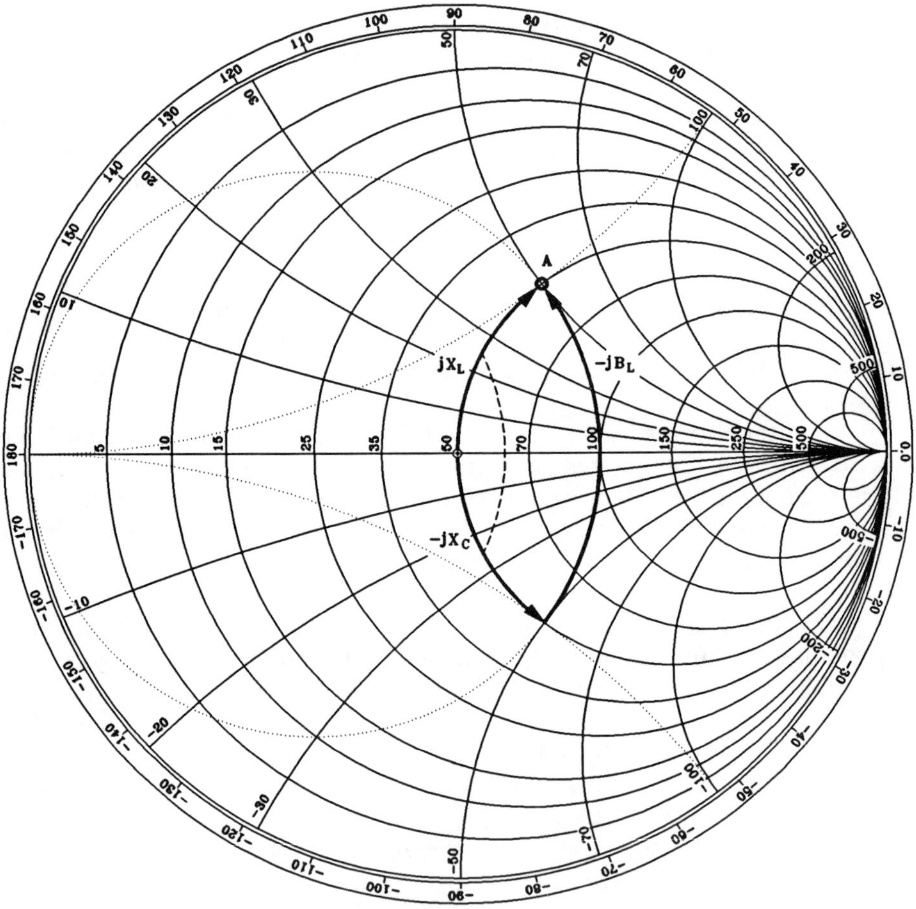

Figure 3.4 Matching networks from 50Ω to point A.

As we move higher in frequency, the lead inductance of packaged devices and bonding wire inductance of chip devices become significant, and we cannot realize a shunt element exactly at the input or output plane; the matching network must be modified accordingly. Again starting from the origin, we add a series capacitor, but only 25Ω reactance, which brings us to the 16-mmho constant conductance contour shown as the dashed line in Figure 3.4. The susceptance at this point is 8 mmho, and, following the G = 16 mmho contour in a counterclockwise direction until it intersects the R = 50Ω circle at +j25, B = -8 mmho. The total susceptance of this element is -16 mmho, or +62.5Ω reactance. Finally we add 25Ω inductive reactance, which brings us to point A, and the matching network consists of a series -j25Ω capacitor, a +j62.5Ω shunt inductor, and a +j25Ω series inductor, as shown in Figure 3.5(b). Inductive reactance due to the transistor leads or bonding wires is subtracted from the +j25Ω series reactance, which results in the desired impedance at the device reference plane.

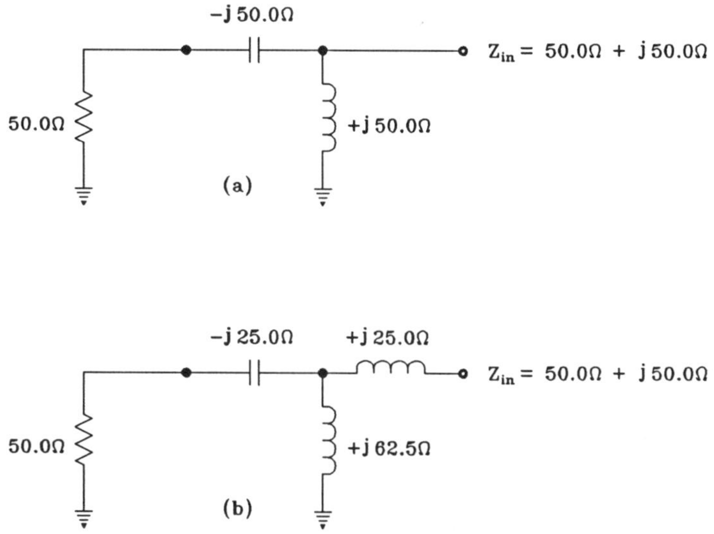

Figure 3.5 Matching networks for point A in Figure 3.4.

When we restrict the matching network topology, the range of realizable impedance values is also limited. Using our series C/shunt L network as an example, the series capacitor can transform the 50Ω load into any impedance on the R = 50Ω circle within the capacitive reactance region of the Smith chart. Shunt inductance then transforms this impedance in a counterclockwise direction following a constant conductance contour. All possible impedance values resulting from this network are shown within the shaded region of Figure 3.6(c). We also see from the figure that it is not possible to provide an impedance at point A using either the topology in Figure 3.6(b) or 3.6(f). The networks shown in Figure 3.6(e) and 3.6(f) have the same matching regions when the two elements are interchanged.

In addition to parasitics caused by the physical layout of the circuit, we must also consider the behavior of lumped elements at high frequencies. Inductors usually have some capacitive coupling between coils, and capacitors show a small inductive reactance due to their internal geometry. The net effect is that both elements exhibit series and parallel resonances. The inductor may actually look capacitive and the capacitor may appear inductive at certain frequencies. Even without the effects of parasitics, the lumped element assumption upon which the analysis is based may not be valid. At 18 GHz, a quarter-wavelength is approximately 164 mils, or 0.4 cm in air. A relative dielectric constant of 40 reduces the physical length to 26 mils, or about 0.07 cm. The voltage and current in a lumped element is not a function of distance which limits the physical size to a fraction of a wavelength. Component manufacturers are continually improving miniature lumped elements that can be used at high frequencies. However, these elements are costly and require skilled assemblers. High-frequency matching networks are usually designed using distributed elements, and the Smith chart proves to be equally useful for this task [5,6].

The s-parameters for the transmission line element were derived in Chapter 2 and given by equation (2.8). When the characteristic impedance of the line is the same as the chart reference, $\Gamma = 0$, and (2.8) reduces to

$$s_{11} = s_{22} = 0$$

$$s_{12} = s_{21} = e^{-j\theta}$$

Equation (1.23) can be used to calculate the input reflection coefficient with the transmission line terminated in Γ_L:

$$s_{11}' = s_{11} + \frac{s_{12}s_{21}\Gamma_L}{1 - s_{22}\Gamma_L} = 0 + \frac{(e^{-j\theta})(e^{-j\theta})\Gamma_L}{1 - 0\cdot\Gamma_L} = (e^{-j2\theta})\Gamma_L$$

or

$$s_{11}' = \rho_L \underline{/\theta_L - 2\theta} \tag{3.15}$$

where ρ_L is the magnitude and θ_L is the angle of Γ_L. θ is the electrical length of the transmission line. Thus, the original reflection coefficient is translated 2θ degrees in a clockwise direction about a constant radius circle equal to $|\Gamma_L|$.

When we open-circuit the transmission line, $\Gamma_L = +1$, and (3.15) becomes

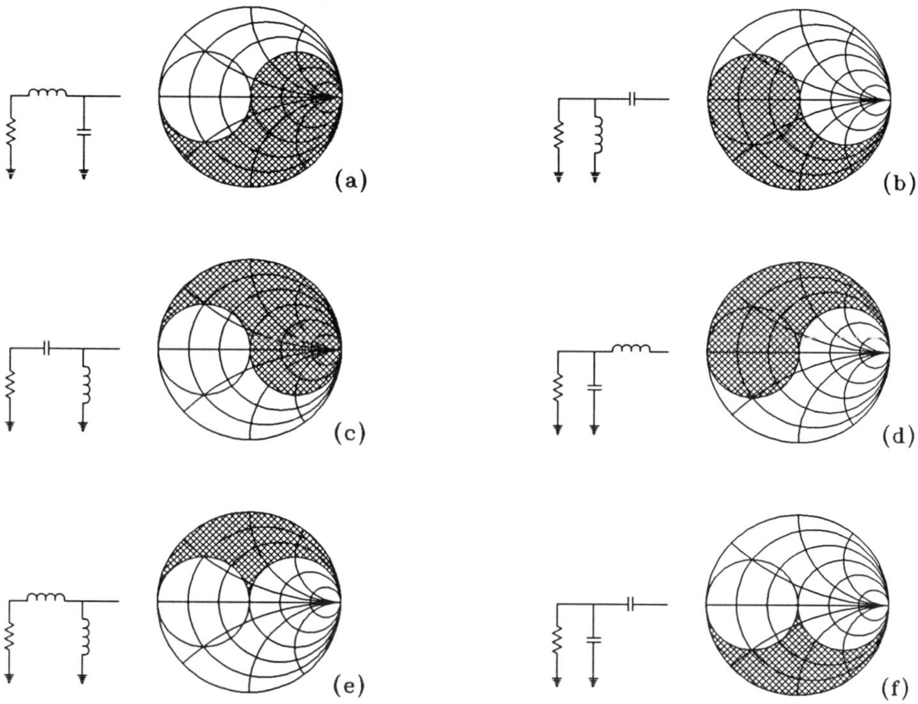

Figure 3.6 Matching regions for two-element networks.

$$s'_{11} = e^{-j2\theta} = 1\underline{/-2\theta} \tag{3.16a}$$

Then, substituting into (3.14a) gives us

$$z = \frac{1+s}{1-s} = \frac{1+e^{-j2\theta}}{1-e^{-j2\theta}} = \frac{e^{j\theta}+e^{-j\theta}}{e^{j\theta}-e^{-j\theta}} = -j\frac{1}{\tan\theta} = -j\mathrm{Cot}\,\theta \tag{3.16b}$$

and we see that the open-circuited stub provides capacitive reactance. When the transmission line is short-circuited, $\Gamma_L = -1$ and

$$s'_{11} = -e^{-j2\theta} = 1\underline{/180-2\theta} \tag{3.17a}$$

Then, using (3.14a),

$$z = \frac{1+s}{1-s} = \frac{1-e^{-j2\theta}}{1+e^{-j2\theta}} = \frac{e^{j\theta}-e^{-j\theta}}{e^{j\theta}+e^{-j\theta}} = j\tan\theta \tag{3.17b}$$

which corresponds to inductive reactance. The electrical length necessary to realize a specified reactance can be determined by solving (3.16b) and (3.17b):

$$\begin{aligned} \theta_{OC} &= \tan^{-1}(R_o/X_C) \\ \theta_{SC} &= \tan^{-1}(X_L/R_o) \end{aligned} \tag{3.18}$$

where R_o is the chart reference, equal to one if the Smith chart is normalized. We can also use the reflection coefficient and impedance scales to determine electrical length. The open circuit stub starts at 0° and rotates clockwise by an angle equal to twice its electrical length. Therefore, we can locate the reactance, determine the degrees of rotation from 0°, and then divide by two. Similarly, the short circuit stub starts at 180° and also rotates clockwise. The electrical length is one half the total rotation from 180°.

Although we usually consider the short circuit stub to be inductive and open circuit stubs to be capacitive, the reactance changes signs for 2θ between 180° and 360° or electrical lengths of 90° to 180°. Thus, between one-quarter and one-half wavelength ($\lambda/4$ and $\lambda/2$), the roles are reversed, and the response is repeated for every additional half wavelength. This behavior can be very useful. A short circuit is usually difficult to achieve at microwave frequencies. However, we can use a $\lambda/4$ open stub, which is equivalent to a zero length shorted stub. We can also insert $\lambda/2$ sections to physically separate components at frequencies where the physical length is very short. However, the longer we make the distributed elements, the more selective they become. For example, if we have a shorted stub with an electrical length of 45° at 1.0 GHz, we can use an open stub with an electrical length of (45° + 90°), or 135°, to avoid the problems associated with a short circuit. Using (3.17b), the shorted stub has a normalized impedance of jTan(45°), or j1, and from (3.16b) the normalized impedance of the open stub is -jCot(135°), which is also j1. At 1.2 GHz the shorted stub has an electrical length of (1.2/1.0) • 45°, or 54.0°, and the open stub (1.2/1.0)(135°), or 162.0°. The impedance of the shorted stub is j1.38, and the impedance of the open stub is j3.08, almost a 3 to 1 difference for 20% change in frequency! Broadband designs usually require the shortest practical lengths.

We have a number of choices for the topology of distributed element matching networks. However, the series-connected shorted and open stubs are difficult to realize using microstrip or stripline construction and are usually avoided. One topology that provides the required grounding for bias and series element used to correct for transistor lead or bonding wire inductance is shown in Figure 3.7(a) and in the corresponding Smith chart construction in Figure 3.8. The network consists of a shunt shorted stub labeled TL_S, followed by a transmission line labeled TL in Figure 3.8. We first construct a circle about the origin which passes through point A. The transmission line translates any reflection coefficient on the circle into any other point rotating in a clockwise direction. We can move from the origin to the circle by following the 20-mmho constant conductance contour in a counterclockwise direction. The susceptance of the shunt element is (-20 - 0), or -20 mmho, which corresponds to 50Ω inductive reactance. The angle of the reflection coefficient at this point is approximately 116.6°, labeled θ_1 in the figure. Rotating clockwise to point A, θ_2 is 63.4°, and the length of the transmission line -(63.4° - 116.6°)/2, or 26.6°. The length of the shorted stub is measured from 180° to the desired reactance of 50Ω, and is (180° - 90°)/2, or 45°. We complete the design by adding a coupling capacitor if necessary. The capacitor should be selected to operate below resonance if possible, or, alternatively, a value can be selected that has a series resonance in the vicinity of the operating frequency and has minimal effect on the matching network. We may also need to bypass the shorted stub to provide a low impedance point for bias injection. When good bypass capacitors are not available, we can add a λ/4 open stub, which does provide a good RF ground at a single frequency.

The characteristic impedance Z_c of all transmission lines is assumed equal to the Smith chart reference, which results in simple design techniques. When Z_c does not equal the reference, the open and shorted stubs can be calculated from the expressions for input impedance:

$$Z_s = jZ_c\tan(\theta_s)$$

$$Z_o = -jZ_c\cot(\theta_o)$$

However, the input reflection coefficient of the terminated transmission line is no longer a simple expression, and the constant radius circle cannot be used. One approach is to renormalize the chart to Z_c, rotate about a circle, and then change the reference back. The special case of a quarter-wavelength line where Z_c is not equal to Z_o is particularly useful and does not require renormalization. The input impedance of a transmission line terminated with Z_L can be written as

$$Z_{in} = Z_c \cdot \left\{ \frac{Z_L + jZ_c\tan\theta}{Z_c + jZ_L\tan\theta} \right\}$$

and at θ = 90°, where the tangent function is infinity, the input impedance becomes

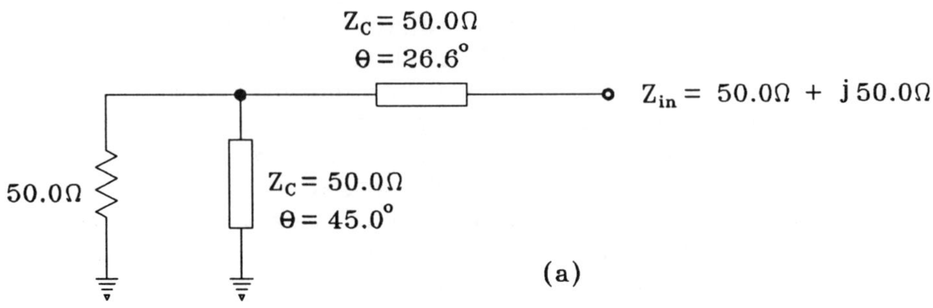

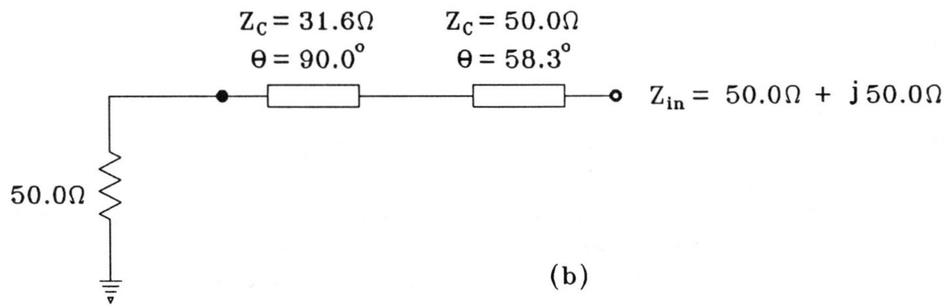

Figure 3.7 Distributed element matching networks.

$$Z_{in} = \frac{Z_c^2}{Z_L}$$

or

$$Z_c = \sqrt{Z_{in} Z_L}$$

When the product of Z_{in} and Z_L is real, Z_c is also real and we can translate Z_L into Z_{in} using a single quarter-wavelength transmission line. For example, we can transform 50Ω into 20Ω using a $\lambda/4$ line with a characteristic impedance equal to the square root of (50)(20), or 31.62Ω. Examining Figure 3.8, we see that 20Ω falls on the circle passing through point A, and a rotation of -(63.4° - 180°), or 116.6°, is required to shift 20Ω into point A. Thus, the matching network consists of a 31.62Ω, 90° line cascaded with a 50Ω line, which has an electrical length of 58.3°, as shown in Figure 3.7(b).

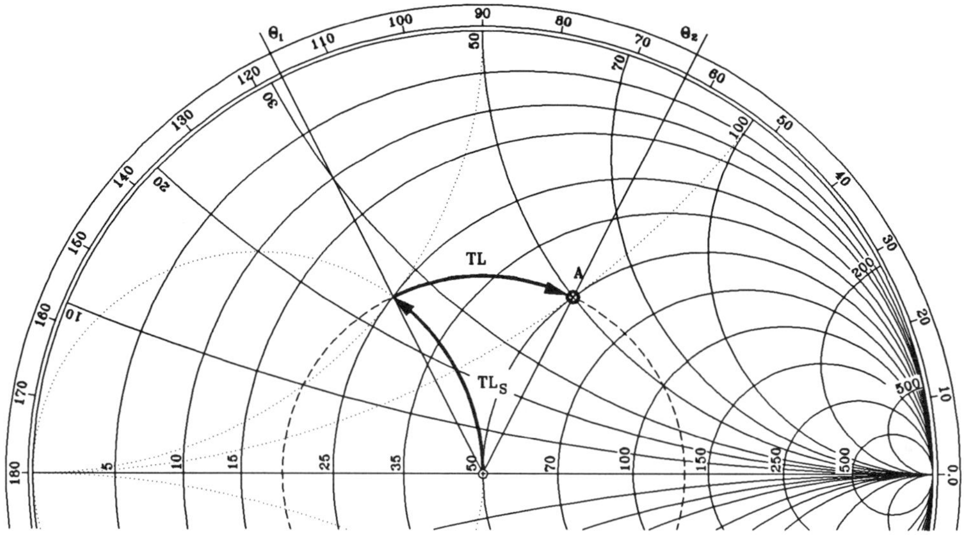

Figure 3.8 Distributed element matching network.

The matching examples in this section started at the origin and followed constant resistance, conductance, or constant radius contours to the desired reflection coefficient. In effect, we are working from the network termination towards a specified impedance. The first element is connected to the termination, and the last element to the load which may be the input or output of a transistor or another network. The reflection coefficient is usually determined using the methods of Chapter 2 and represents an impedance from a gain or mismatch circle, Γ_{MS}, Γ_{ML}, or optimum noise Γ_O. However, we can also design the matching network starting from the impedance. Generally, we compute s'_{11} or s'_{22} and then define Γ as the complex conjugate. This step can be avoided by starting at the impedance and following the appropriate contours to the origin. The design is exactly the same; the only difference is the sequence of the elements.

3.1.4 Source and Load Mappings

When using constant gain circles, we select an impedance from a point on the desired circle and then use (2.40) to compute the reflection coefficient in the opposite plane. We can select a number of points from the gain circle, evaluate (2.40), and then plot the corresponding reflection coefficient, which eventually provides some idea of the behavior of impedance in both planes. However, more useful results are obtained when we apply (2.40) to transform the impedance from one plane to the other and then map the transformed impedance.

The source and load reflection coefficients in (2.40) can be written as

$$\Gamma_L = \frac{z_L - 1}{z_L + 1} \tag{3.19a}$$

and

$$\Gamma_S = \frac{z_S - 1}{z_S + 1} \tag{3.19b}$$

Γ_S and Γ_L are referenced to the same impedance as the *s*-parameters; hence, z_S and z_L are also normalized. Substituting (3.19a) into (2.40a),

$$s'_{11} = s_{11} + \frac{s_{12}s_{21}\left\{\dfrac{z_L - 1}{z_L + 1}\right\}}{1 - s_{22}\left\{\dfrac{z_L - 1}{z_L + 1}\right\}} \tag{3.20a}$$

and (3.19b) into (2.40b),

$$s'_{22} = s_{22} + \frac{s_{12}s_{21}\left\{\dfrac{z_S - 1}{z_S + 1}\right\}}{1 - s_{11}\left\{\dfrac{z_S - 1}{z_S + 1}\right\}} \tag{3.20b}$$

After considerable algebraic manipulation, we can separate out z_L in the numerator and denominator of the second term in (3.20a):

$$s'_{11} = s_{11} + \frac{s_{12}s_{21}s_{22}^*}{1 - |s_{22}|^2} + \frac{s_{12}s_{21}\left(1 - s_{22}^*\right)}{\left(1 - |s_{22}|^2\right)\left(1 - s_{22}\right)} \cdot \left\{\frac{z_L - \dfrac{1 + s_{22}^*}{1 - s_{22}^*}}{z_L + \dfrac{1 + s_{22}}{1 - s_{22}}}\right\} \tag{3.21a}$$

and z_S in (3.20b):

$$s_{22}' = s_{22} + \frac{s_{12}s_{21}s_{11}^*}{1 - |s_{11}|^2} + \frac{s_{12}s_{21}\left(1 - s_{11}^*\right)}{\left(1 - |s_{11}|^2\right)\left(1 - s_{11}\right)} \cdot \left\{ \frac{z_S - \dfrac{1 + s_{11}^*}{1 - s_{11}^*}}{z_S + \dfrac{1 + s_{11}}{1 - s_{11}}} \right\} \tag{3.21b}$$

Unfortunately, (3.21) is not a simple expression in the form of (3.3a). However, we note that the terms added to and subtracted from z_S and z_L are similar to equation (3.14a) and represent normalized impedance. Changing variables, we can rewrite (3.21) in the form

$$s' = C + R \cdot \left\{ \frac{Z - N^*}{Z + N} \right\} \tag{3.22}$$

and it is apparent that we can obtain mappings of constant resistance and reactance from the term within braces in (3.22). The centers and radii of the contours must be multiplied by R and then added to C, which results in a mapping in one plane as a function of impedance values in the opposite plane.

R, C, and N each affect the mapping in an independent manner. Comparing equation (3.22) with (3.3a) shows that N is the normalization impedance or the conjugate of the impedance represented by the origin of the mapping. The bilinear transform results in a mapping where the real portion of the z-plane, which has the same sign as the real component of the normalization impedance, is mapped into the unit circle. Multiplication by R results in the unity radius circle becoming $|R|$ and the entire mapping rotated by the angle of R. We can also consider R as the vector that points from the origin to the infinite impedance point. The original vector is $1\angle 0°$, which becomes R after multiplication. The addition of C results in a translation of the entire mapping, or C corresponds to the center with respect to the original coordinate system.

To summarize, the mapping constants for the input or source plane mapping of z_L are

$$C_1 = s_{11} + \frac{s_{12}s_{21}s_{22}^*}{1 - |s_{22}|^2} \tag{3.23a}$$

$$R_1 = \frac{s_{12}s_{21}\left(1 - s_{22}^*\right)}{\left(1 - |s_{22}|^2\right)\left(1 - s_{22}\right)} \tag{3.23b}$$

and

$$N_1 = \frac{1 + s_{22}}{1 - s_{22}} \tag{3.23c}$$

The constants for the output or load plane z_S mappings are

$$C_2 = s_{22} + \frac{s_{12}s_{21}s_{11}^*}{1 - |s_{11}|^2} \tag{3.24a}$$

$$R_2 = \frac{s_{12}s_{21}\left(1 - s_{11}^*\right)}{\left(1 - |s_{11}|^2\right)\left(1 - s_{11}\right)} \tag{3.24b}$$

and

$$N_2 = \frac{1 + s_{11}}{1 - s_{11}} \tag{3.24c}$$

Source and Load Mapping Example

We can apply the source and load mappings to the amplifier design example in Section 2.3.7, which is based on the device defined by (2.49). Figure 2.21 shows the input plane stability and gain circles used to design an amplifier with maximum stable gain. Although we selected a source impedance from the maximum stable gain (MSG) circle, the design turned out to have a negative real input impedance when conjugately matched at the output. We could select another point on the MSG circle, compute s'_{22}, and then calculate s'_{11} with the output conjugately matched. After enough iterations, we may be able to find a point on the gain contour that provides the desired gain with positive real input and output impedance. However, the mapping technique proves to be a far more effective approach to this problem.

We can map the source impedance onto the s_{22}-plane using equations (3.22) and (3.24). The three mapping constants are

$$C_2 = 0.71\angle -14^\circ + \frac{(0.04\angle 77^\circ)(3.02\angle 159^\circ)(0.97\angle 24^\circ)}{1 - |0.97|^2} = 2.1521\angle -80.7853^\circ$$

from equation (3.24a),

$$R_2 = \frac{(0.04\angle 77^\circ)(3.02\angle 159^\circ)(1 - 0.97\angle 24)}{(1 - |0.97|^2)(1 - 0.97\angle -24^\circ)} = 2.0440\angle 88.1958^\circ$$

from equation (3.24b), and

$$N_2 = \frac{1 + 0.97\angle -24^\circ}{1 - 0.97\angle -24^\circ} = 4.6926\angle -85.7166^\circ = 0.3505 - j4.6795$$

from equation (3.24c). The reference impedance for the mapping is N_2 or R_o = 0.3505 and X_o = -4.6795. The three mapping constants are computed from the normalized s-parameters; hence, the constants are also normalized to Z_o, or 50Ω in this case. We first compute the constant R = 0 contour, which defines the boundary between positive and negative real impedance. Using (3.6c),

$$Center = \left(\frac{0}{0 + 0.3505}\right) + j0 = 0 + j0$$

and (3.6d),

$$Radius = \left|\frac{0.3505}{0 + 0.3505}\right| = 1$$

Substituting into (3.22), we have

$$Center = 2.1521\angle -80.7853^\circ + (2.0440\angle 88.1958^\circ)(0\angle 0^\circ) = 2.1521\angle -80.7853^\circ$$

and

$$Radius = (2.0440)(1) = 2.0440$$

The mapping is drawn on the s_{22}-plane and uses the same coordinate system. We construct the R = 0 circle by locating a point at $2.1521\angle -80.7853^\circ$ as measured from the origin of the Smith chart, and then drawing a circle with a radius of 2.0440, as shown in Figure 3.9.

The normalizing impedance N_2 is positive real; hence, the entire positive real portion of the Z_S-plane is contained within the R = 0 contour. The figure shows the positive real portion of the Z_S mapping, along with a Smith chart which encloses the positive real portion of the s_{22}-plane and the output plane stability circle. We can determine the output reflection coefficient for any input impedance very easily. Locate the input impedance on the Z_S mapping and then read the output reflection coefficient or impedance directly from the s_{22} coordinates. For example, we locate the intersection of the R = 50 and X = 0 circles on the Z_S mapping and then read the reflection coefficient as $0.71\angle -14°$. This is the output reflection coefficient with the input terminated in 50Ω or s_{22} of the device.

The figure shows that a large portion of the mapping falls outside the positive real Smith chart. In fact, the mapping provides the same information as

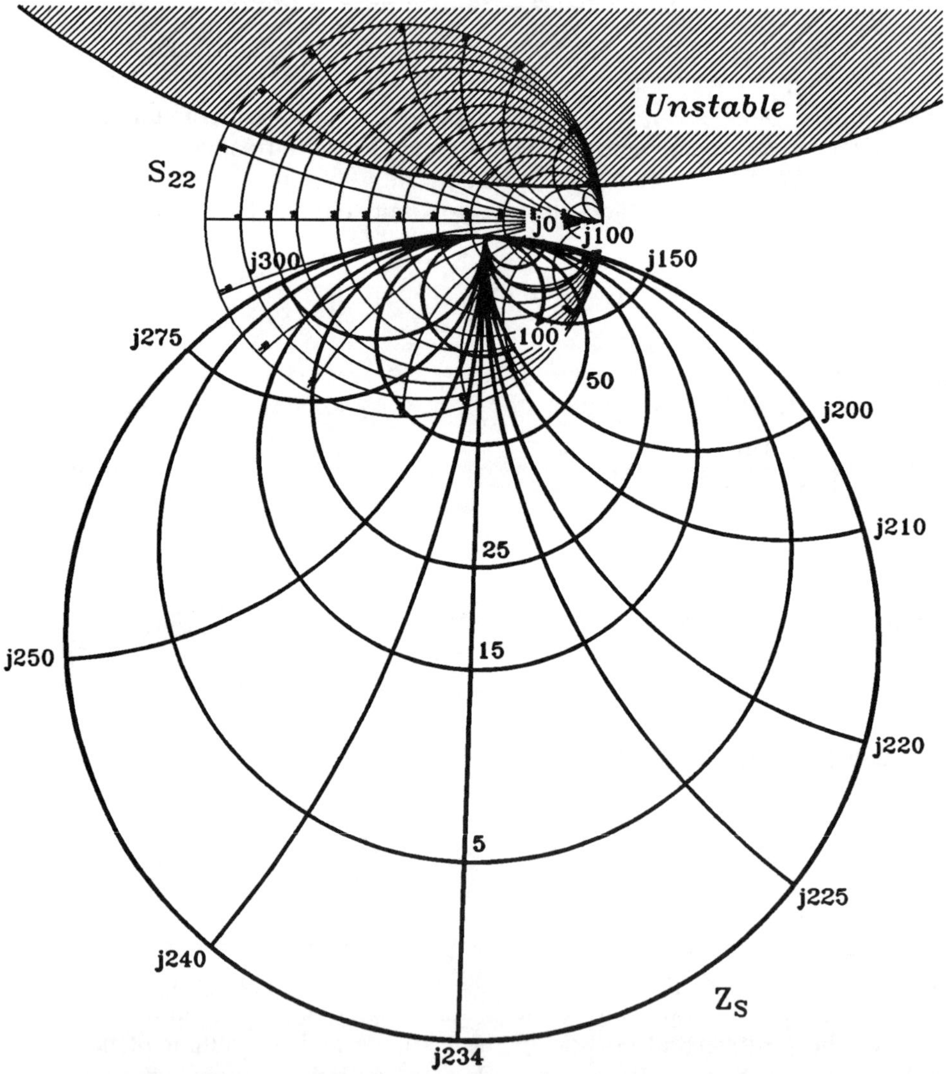

Figure 3.9 Z_S mapping onto S_{22}-plane.

the input plane stability circle. Values of source impedance resulting in negative real output impedance fall outside the unity Smith chart; hence, the s_{22}-plane Smith chart represents the stable region for Z_S. Examining the mapping we find that resistive values less than 75Ω are potentially unstable, as are reactive values between approximately 110Ω and 280Ω. These are the same results obtained using the input plane stability circle, but presented in a slightly different form. We are not usually interested in the unstable region for amplifier design, and we can expand the positive real portion of the mapping which provides improved resolution, as shown in Figure 3.10.

Once Z_S is mapped onto the s_{22}-plane, we can refer to the input plane constant gain circles shown in Figure 2.21, locate a point on the desired gain circle, and then plot the impedance on the mapping and compare s'_{22} conjugate with the output plane stability circle to determine if Z_S is acceptable. However, we can also construct input plane gain circles on the Z_S mapping, which eliminates the necessity of transferring an impedance from the input plane to the mapping. Equation (3.22) shows that the Z_S mapping is identical to the source plane mapping except that it is referenced to a different impedance, scaled and rotated by R, and translated by C. Therefore, circles on the input plane are translated into circles on the mapping. Expressions for gain circles on the Z_S or Z_L mapping are not simple, and it is perhaps more efficient to locate three points on the desired circle, use (2.40) to compute the corresponding three points on the mapping, and then use these to determine the circle. The load impedance is the complex conjugate of s'_{22}. We can plot the conjugate of the output plane stability circle which defines the stable region for s_{22}.

Figure 3.10 shows the conjugate output plane stability circle plotted as a dashed line, along with constant gain circles of 18.8, 16.0, 14.0, and 12.0 dB. It is immediately apparent from the figure that all gain circles greater than 13 dB fall completely within the unstable region for positive real Z_S, and it is not possible to realize more than 13 dB gain with a conjugately matched output. We can, however, design for positive real input and output impedance with 12-dB gain. The source impedance can be selected from any point on the 12-dB gain contour. One convenient point is the intersection with R = 50Ω, which is marked with a triangle in the figure. The reactance is approximately +j407Ω; thus the input matching network consists of a series inductor. Reading from the s_{22}-plane Smith chart, the corresponding output impedance is Z = 50 - j21.6Ω. The conjugate is +j21.6Ω, and we find that the output matching network is also a series inductor.

The information contained in Figure 3.10 is more difficult to obtain using other methods. For example, just by examining the mapping we see that for positive real input and output impedance and conjugately matched output, the maximum achievable gain is 13 dB. Using other methods, it is necessary to evaluate Z_S, s'_{22}, and s'_{11} for a number of points on several gain circles. In addition, we can immediately see the relationship between Z_S and Z_L, which can be a definite advantage for matching network design, as this example illustrates.

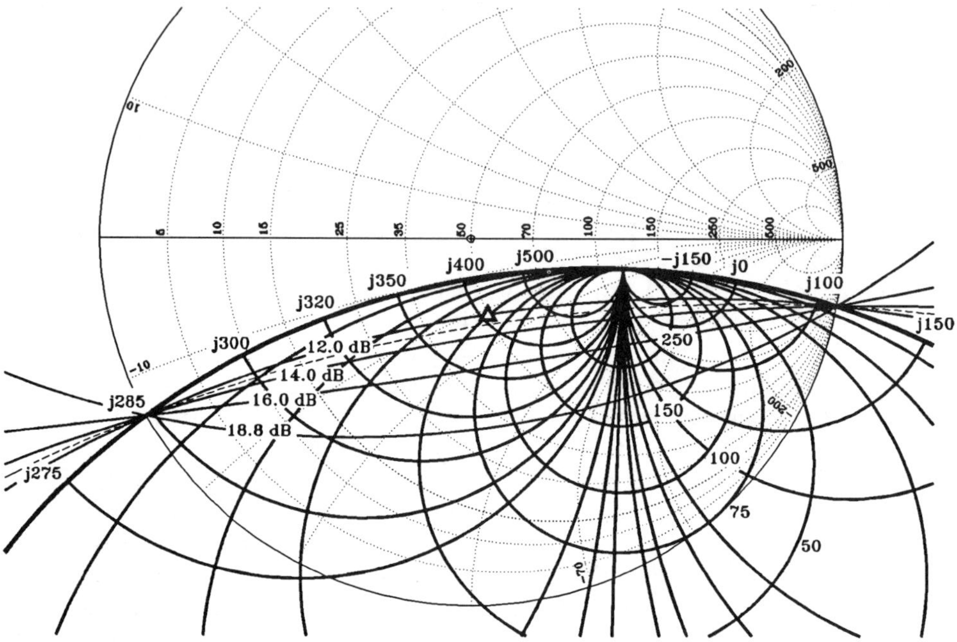

Figure 3.10 S_{22}-plane Z_S and input gain mapping.

The load impedance can also be mapped onto the s_{11}-plane, as shown in Figure 3.11. However, this mapping is typically much smaller than Z_S for most microwave devices and does not offer as much flexibility. In fact, if $s_{12} = 0$, the mapping reduces to a single point; that is, the input impedance is independent of the load.

3.1.5 Multiport Mappings

The two-port mapping of source and load impedance developed in Section 3.1.4 provides the designer with valuable information describing the behavior of the network as a function of terminating impedance. However, the full impact of this design technique is only realized when extended to allow mapping the network parameters as a function of a variable impedance placed anywhere in the circuit.

As the number of ports or complexity of a network is increased, it becomes increasingly more difficult to interpret the results of numeric calculations, and the designer must rely on computer-aided optimization or other time-consuming analysis to achieve the desired results. The multiport mapping provides an attractive alternative because it enables the designer to simultaneously evaluate the effects of any impedance on several network parameters without additional

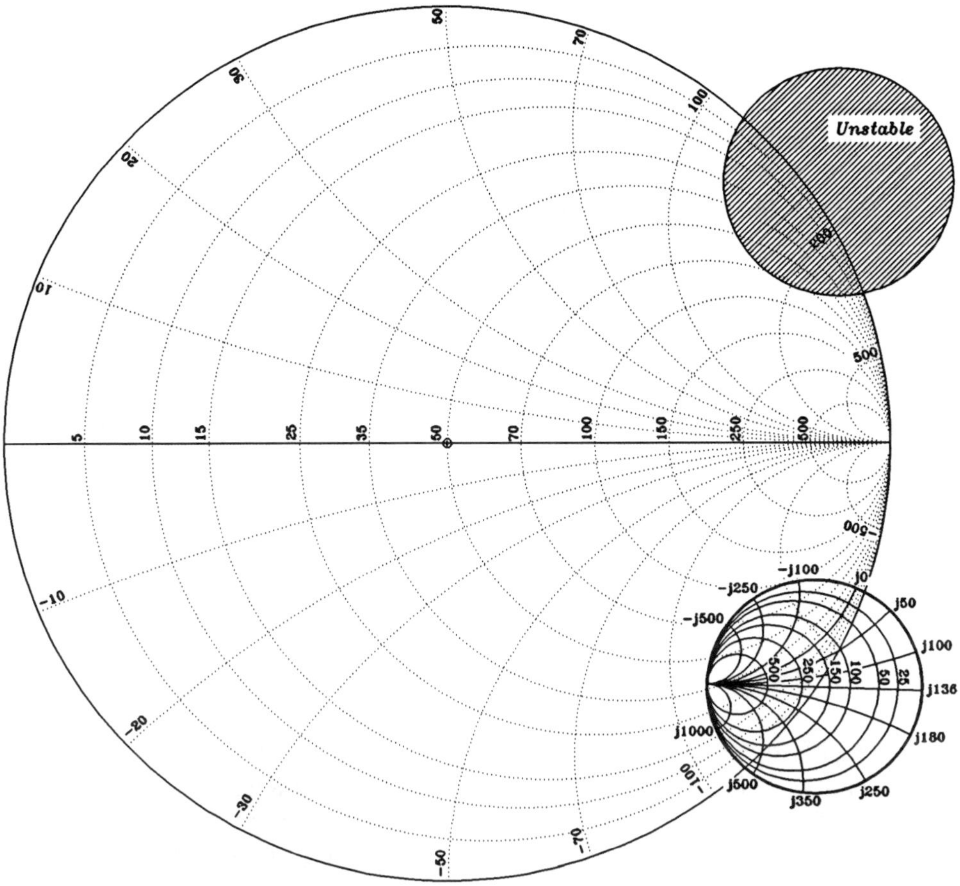

Figure 3.11 Z_L mapping onto S_{11}-plane.

calculations. Although there is no substitute for computer-aided analysis and optimization when dealing with complex circuits or wide bandwidths, the designer is nevertheless required to supply the topology and initial element values as inputs. Both the topology and initial element values have a significant effect on the optimized circuit performance and physical realization. As illustrated in Figure 3.6, some topologies cannot be used to match certain impedance values, and no amount of optimization will be successful if the matching network is not feasible. The multiport mapping provides the circuit designer with the means to conveniently interpret the complex calculations and make engineering judgments necessary to achieve the design objective in an efficient manner.

Figure 3.12 shows three examples of networks used in the design of amplifiers, where the circuit parameters must be calculated as a function of a variable impedance. Figure 3.12(a) shows an impedance connected between the gate and drain of an FET, forming a parallel feedback path. Figure 3.12(b) shows the series feedback configuration, and Figure 3.12(c) shows the impedance connected at one port of a three-port circulator, forming a reflection amplifier. Each circuit is actually a three-port network with one port terminated by an arbitrary impedance, resulting in a two-port amplifier which has characteristics dependent on the impedance at the third port.

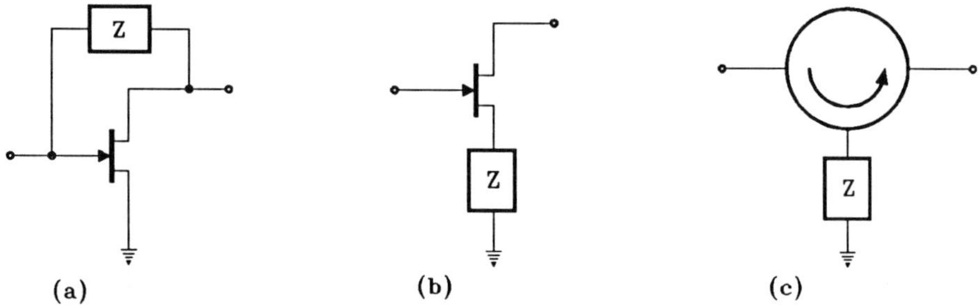

Figure 3.12 Examples of networks with variable impedance.

The derivation of multiport mapping parallels that of the two-port mapping in Section 3.1.4. However, we use an arbitrary n-port parameter s'_{ij} instead of s'_{11} or s'_{22}. The expression that relates any n-port s-parameter to a one-port termination at port k was derived in Section 2.2.2 and given by equation (2.20):

$$s'_{ij} = s_{ij} + \frac{s_{ik} s_{kj} \Gamma_k}{1 - s_{kk} \Gamma_k}$$

Writing Γ_k in terms of impedance,

$$\Gamma_k = \frac{z_k - 1}{z_k + 1} \tag{3.25a}$$

substituting into (2.20),

$$s'_{ij} = s_{ij} + \frac{s_{ik}s_{kj}\left\{\dfrac{z_k - 1}{z_k + 1}\right\}}{1 - s_{kk}\left\{\dfrac{z_k - 1}{z_k + 1}\right\}} \tag{3.25b}$$

and then simplifying in terms of z_k in the numerator and denominator results in

$$s'_{ij} = s_{ij} + \frac{s_{ik}s_{kj}s^*_{kk}}{1 - |s_{kk}|^2} + \frac{s_{ik}s_{kj}\left(1 - s^*_{kk}\right)}{\left(1 - |s_{kk}|^2\right)\left(1 - s_{kk}\right)} \cdot \left\{\frac{z_k - \dfrac{1 + s^*_{kk}}{1 - s^*_{kk}}}{z_k + \dfrac{1 + s_{kk}}{1 - s_{kk}}}\right\} \tag{3.25c}$$

Comparing (3.25c) with (3.22), the three mapping constants are

$$C = s_{ij} + \frac{s_{ik}s_{kj}s^*_{kk}}{1 - |s_{kk}|^2} \tag{3.26a}$$

$$R = \frac{s_{ik}s_{kj}\left(1 - s^*_{kk}\right)}{\left(1 - |s_{kk}|^2\right)\left(1 - s_{kk}\right)} \tag{3.26b}$$

and

$$N = \frac{1 + s_{kk}}{1 - s_{kk}} \tag{3.26c}$$

The physical meaning of these constants is exactly the same as the two-port constants discussed in Section 3.1.4 because the form of the bilinear transform is the same. The interpretation of the mappings is also identical. However, it is usually more convenient to plot forward and reverse transmission coefficients using a polar chart instead of the Smith chart. Generally, the magnitude of these coefficients is of more interest than the impedance.

The mapping techniques considered to this point assume that one port of the network is available for the mapping impedance which, in some cases, requires an n-port network to be transformed into an $n + 1$ port network. For example, the series and parallel feedback shown in Figure 3.12(a) and 3.12(b) requires the basic two-port transistor to be transformed into a three-port. Although the transformation in this case is trivial (see Secs. 1.3.6 and 1.3.7), larger, more complex networks require considerable resources for this conversion. However, a different approach to the solution of the linear fractional transformation given in (3.1) provides a means of solving for the mapping constants without requiring a network transformation. The theorem from linear fractional transformation theory stated in Section 3.1.1 is given implicitly in equation (3.27):

$$\frac{w - w_1}{w - w_3} \cdot \frac{w_2 - w_3}{w_2 - w_1} = \frac{z - z_1}{z - z_3} \cdot \frac{z_2 - z_3}{z_2 - z_1} \tag{3.27}$$

where w_1, w_2, w_3, and z_1, z_2, z_3 are the three distinct points in the two planes related by $w = f(z)$. When one of the three points is the point ∞, the quotient of the two differences containing this point is to be replaced by one. We can solve (3.27) for w, or s in our notation, with

$$\frac{s - s_1}{s - s_3} = \frac{z - z_1}{z - z_3} \cdot \frac{z_2 - z_3}{z_2 - z_1} \cdot \frac{s_2 - s_1}{s_2 - s_3} = \frac{z - z_1}{z - z_3} \cdot A \tag{3.28a}$$

and

$$s = \frac{z(s_1 - As_3) - s_1 z_3 + As_3 z_1}{z(1 - A) - z_3 + Az_1} \tag{3.28b}$$

We can compare (3.28) with equation (3.22) to solve for the three mapping constants R, C, and N. Expanding (3.22), we have

$$s' = C + R \cdot \frac{z - N^*}{z + N} = \frac{z(C + R) - RN^* + CN}{z + N}$$

Then, dividing the numerator and denominator of (3.28b) by $(1 - A)$, and comparing the denominators of both equations gives us

$$N = \frac{Az_1 - z_3}{1 - A} \tag{3.29a}$$

Comparing the numerators, we can write

$$C + R = \frac{s_1 - As_3}{1 - A}$$

and

$$CN - RN^* = \frac{As_3 z_1 - s_1 z_3}{1 - A}$$

Then, solving for the other constants,

$$C = \frac{s_1(N^* - z_3) + As_3(z_1 - N^*)}{(N + N^*)(1 - A)} \tag{3.29b}$$

$$R = \frac{s_1(N + z_3) - As_3(N + z_1)}{(N + N^*)(1 - A)} \tag{3.29c}$$

and

$$A = \frac{(z_2 - z_3)(s_2 - s_1)}{(z_2 - z_1)(s_2 - s_3)} \tag{3.29d}$$

Thus, equation (3.29) only requires that we know three corresponding values of the desired s-parameter for three distinct values of the mapping impedance. This approach is known as the *three-point mapping method* and is sometimes convenient for complex networks.

As an example of the three-point method, we can compute the constants for the Z_S mapping of Section 3.1.4. We are interested in s'_{22} as a function of a source impedance, and can use (2.40b) to calculate the output reflection coefficient. The source impedance can be any value; however, one convenient point is Z_S = 50Ω, where the output is s_{22}. The three points are given in Table 3.2.

Table 3.2 Three Points in the Z-, z-, and s_{22}-Planes

Point	Z	z	s
1	50Ω + j0Ω	1 + j0	$0.71\angle -14°$
2	25Ω + j25Ω	0.5 + j0.5	$0.75279\angle -12.16642°$
3	25Ω - j25Ω	0.5 - j0.5	$0.68996\angle -11.20106°$

First we compute

$A = 1.07776\underline{/-6.76983°}$

using (3.29d), then

$N = 4.69351\underline{/-85.75268°} = 0.34761 - j4.68062$

from equation (3.29a),

$C = 2.16824\underline{/-80.94007°}$

from equation (3.29b), and

$R = 2.06116\underline{/88.12078°}$

using (3.29c). The numeric values agree with those in Section 3.1.4 to a few decimal places. Had we carried out the arithmetic using higher precision, the mapping constants would be identical using either method.

Multiport Mapping Example

As an example of multiport mapping, we can examine the effects of parallel feedback on s_{21} of the device defined by (2.49). We have a choice of computing the mapping constants using the explicit expression in (3.26) or the three-point method given by (3.29). The three-port conversion described in Section 1.3.7 is not difficult, so we elect to use (3.26). Port numbers are assigned as shown in Figure 1.26; the gate is port 1, the drain is port 2, and port 3 is referenced from the gate to the drain. A 50Ω resistor connected in parallel with the FET provides four of the nine s-parameters, and then (1.35) is applied to compute the remaining five parameters. The three-port s-parameters are

$$S = \begin{bmatrix} 0.16694\underline{/-168.95520°} & 0.38635\underline{/-2.84912°} & 0.45047\underline{/-1.62539°} \\ 0.32323\underline{/163.93076°} & 0.21956\underline{/177.39345°} & 1.09416\underline{/175.83414°} \\ 1.15317\underline{/-6.04553°} & 0.39588\underline{/-175.77163°} & 0.54935\underline{/-9.66835°} \end{bmatrix}$$

We intend to map s_{21} as a function of an impedance connected at port 3; hence, $i = 2$, $j = 1$, and $k = 3$. Using (3.26a), we compute

$$C = s_{21} + \frac{s_{23}s_{31}s_{33}^*}{1 - |s_{33}|^2} = 0.32323\angle 163.93076$$

$$+ \frac{(1.09416\angle 175.83414)(1.15317\angle -6.04553)(0.54935\angle 9.66835)}{1 - (0.54935)^2}$$

$$= 1.30704\angle 175.66138°$$

and then

$$R = \frac{s_{23}s_{31}(1 - s_{33}^*)}{(1 - |s_{33}|^2)(1 - s_{33})}$$

$$= \frac{(1.09416\angle 175.83414)(1.15317\angle -6.04553)(1 - 0.54935\angle 9.66835)}{(1 - [0.54935]^2)(1 - 0.54935\angle -9.66835)}$$

$$= 1.80711\angle 147.03180°$$

from (3.26b), and

$$N = \frac{1 + s_{33}}{1 - s_{33}} = \frac{1 + 0.54935\angle -9.66835}{1 - 0.54935\angle -9.66835}$$

$$= 3.30231\angle -14.80343° = 3.19270 - j0.84375$$

from (3.26c). The mapping reference is N and the $R = 0$ circle is located at

$$Center = \left(\frac{0}{0 + 3.19270}\right) + j0 = 0 + j0$$

and has

$$Radius = \left|\frac{3.19270}{0 + 3.19270}\right| = 1.0$$

Applying (3.22), which transforms coordinates to the s_{21}-plane, we have

$$Center = 1.30704\underline{/175.66138} + (1.80711\underline{/147.03180})(0.0\underline{/0.0}) = 1.30704\underline{/175.6613}$$

and

$$Radius = (1.80711)(1.0) = 1.80711$$

Measuring the center of the circle from the origin of the s_{21}-plane and the radius from the scale of the plane, we can plot the zero resistance contour, as shown in Figure 3.13. The figure shows a mapping of the feedback impedance plotted on a polar chart, and we can immediately determine s_{21} for any value of feedback. For example, the infinite impedance point corresponds to no feedback and can be used to check the mappings. Measuring from the origin, we have $3.02\underline{/159°}$ at $Z = \infty$, which is the original two-port s_{21}.

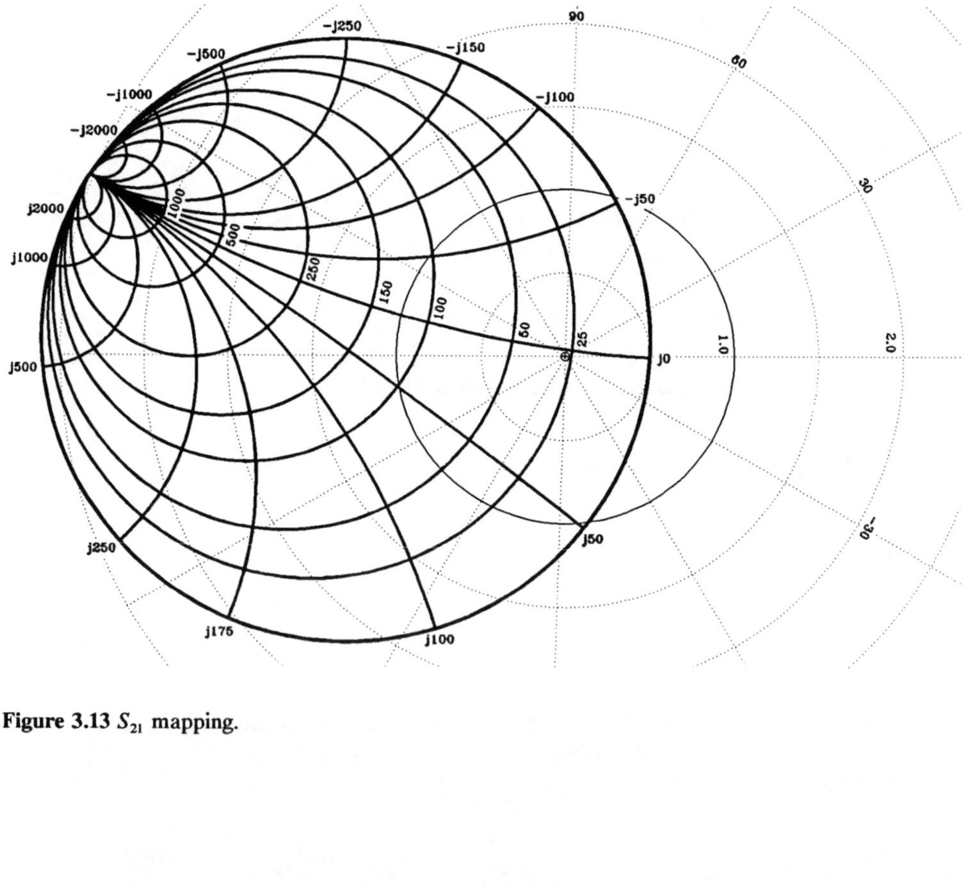

Figure 3.13 S_{21} mapping.

Examining Figure 3.13, we see that the magnitude of s_{21} can be increased to its maximum value by connecting a +j750Ω feedback element from drain to gate. The s-parameters for this configuration are

$$S = \begin{bmatrix} 0.98923\underline{/-4.79518^\circ} & 0.07268\underline{/-94.39030^\circ} \\ 3.11297\underline{/173.79816^\circ} & 0.74261\underline{/7.51134^\circ} \end{bmatrix}$$

which is not significantly different from the original parameters in (2.49), with the possible exception of $|s_{12}|$, which has almost doubled. The stability factor is -0.0348 and the maximum stable gain is 16.3179 dB. Compared with a stability factor of 0.1339 and MSG of 18.7795 dB computed in Chapter 2, maximizing s_{21} does not improve the characteristics of this device.

3.2 MAPPING APPLICATIONS

Mapping is a powerful design tool which allows us to visually interpret the effects of an impedance embedded anywhere in a circuit. We can create multiple mappings and then simultaneously observe the effects of the impedance on several parameters. Modern computer-aided design (CAD) programs can create these mappings almost instantly, and high-resolution graphics coupled with color and the ability to zoom in on regions of interest make the interpretation a simple matter. Nevertheless, these techniques are often overlooked by many circuit designers. This section contains several mapping applications that will serve to illustrate the usefulness of this design technique.

3.2.1 Feedback Amplifier Mapping

We found from the example in Section 3.1.5 that simply adding parallel feedback to maximize s_{21} resulted in reduced stability factor and maximum stable gain compared with the original device. Once we determine the reason for this degradation, we can apply the knowledge to improve the characteristics of the network. The one parameter that changed significantly was the magnitude of s_{12}, which almost doubled compared with only a few percent change for the other parameters. Looking at the equation for stability factor given by (2.48a) and maximum stable gain in (2.62), we see that s_{12} appears in the denominator of both expressions. Therefore, if we can reduce the magnitude of s_{12} without significantly affecting s_{21}, both the stability factor and maximum stable gain will be improved. We can investigate the effects of parallel feedback on s_{12} by creating a mapping.

We follow the same procedure that was used to create the s_{21} mapping in Section 3.1.5, except that, in this case, $i = 1, j = 2$, and $k = 3$. The mapping constants are

$$C = 0.25355\angle -11.14862°$$

$$R = 0.25541\angle 159.84581°$$

and

$$N = 3.30231\angle -14.80343° = 3.19270 - j0.84375$$

The mapping is shown in Figure 3.14 plotted on a polar chart.

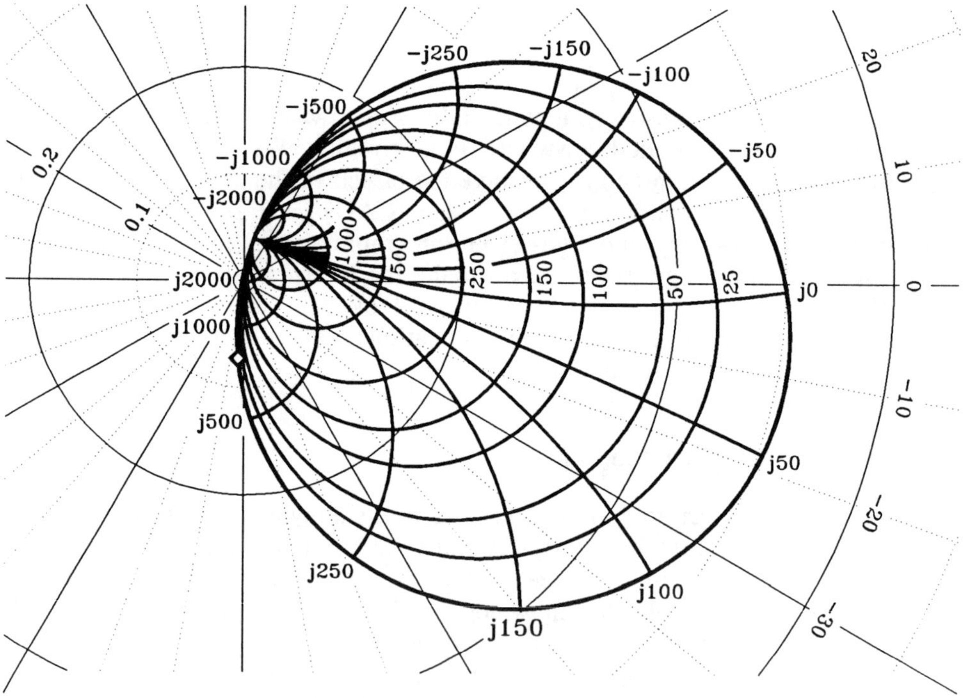

Figure 3.14 S_{12} mapping.

Figure 3.14 shows that the original choice of +j750Ω for the feedback element does in fact increase the magnitude of s_{12}. However, we see that a feedback element with +j2000Ω will reduce $|s_{12}|$ by at least an order of magnitude.

We must investigate the effect on s_{21} before proceeding with the design. Figure 3.13 shows that the magnitude of s_{21} is reduced slightly but is still greater than that of the device without feedback. The parallel combination of the device and +j2000Ω feedback element has an **S** matrix given by

$$S = \begin{bmatrix} 0.98120\angle{-16.92408^\circ} & 0.00233\angle{-153.77158^\circ} \\ 3.07121\angle{164.44332^\circ} & 0.72411\angle{-6.01542^\circ} \end{bmatrix}$$

and a corresponding stability factor of 0.6493 and an MSG of 31.1981 dB. We have succeeded in increasing the maximum stable gain of this device by more than 12 dB, and in simultaneously improving the stability simply by adding one element. Actually, a blocking capacitor will also be required to avoid a DC short from the gate to the drain. However, had we relied on the manufacturer's data sheet or characterization of the device alone, we would not have realized the full potential of this FET, resulting in more complex and costly designs than necessary.

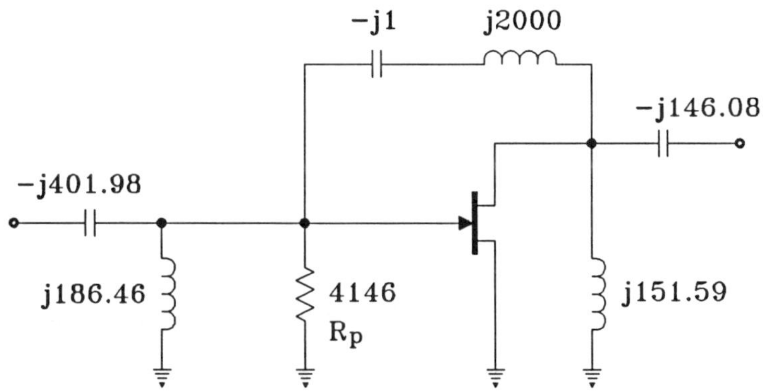

Figure 3.15 25-dB gain amplifier.

We used this device in Section 2.3.9 to design a 25-dB gain amplifier. The circuit required two stages since the MSG was calculated as 18.8 dB. In addition, the input and output impedances were not 50Ω, which requires an isolator or circulator at both ports or a design that provides the mismatch between stages. We can redesign our amplifier using a single stage and provide matched input and output as well. The input plane stability circle is located at $1.010\angle{16.575^\circ}$ and has a radius of 0.015. We can use either a small series resistor or large shunt resistor to stabilize the device. Expanding the Smith chart around the stability circle, we find that a shunt resistor equal to 18,350Ω is enough to provide stability and G_{MAX} =

31.070 dB. However, analyzing the network for other resistor values shows that when the resistor is reduced to 4146Ω, $K = 2.003$, $G_{MAX} = 25.0$ dB, $\Gamma_{MS} = 0.97060\angle 16.57934°$, and $\Gamma_{ML} = 0.81026\angle 2.71605°$. Using techniques discussed in Section 3.1.3, we can design the input and output matching networks and then analyze the complete amplifier shown in Figure 3.15. The design is unconditionally stable, and has 25-dB gain and an input and output voltage standing wave ratio (VSWR) less than 1.001:1. However, before we complete the design, stability must be investigated at all frequencies. At high frequencies the feedback network exhibits high inductive reactance caused by the inductor, and at low frequencies the capacitor has a large capacitive reactance. The result is an increased magnitude

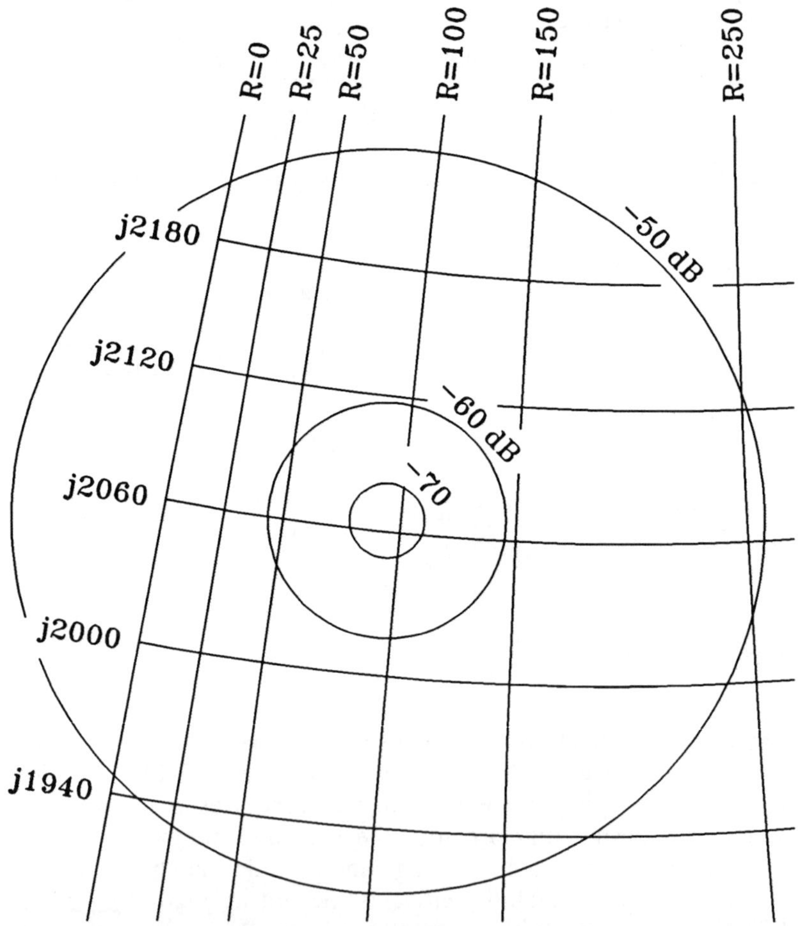

Figure 3.16 Expanded s_{12} mapping.

of s_{12} as the feedback impedance approaches infinity and the device becomes unstable outside the operating range. We might consider a shunt stabilization network consisting of a parallel *L-C* in series with a resistor. At resonance the network has infinite impedance; however, either above or below the resistive component tends to stabilize the device. In any case, the feedback design offers superior performance with a lower parts count, and has the added benefit that the shunt resistor can be trimmed in production to set the gain exactly on 25 dB.

3.2.2 Active Isolator

Some designs require isolation or buffers between one circuit and another. For example, an oscillator is usually followed by a buffer amplifier, which isolates the circuit from changes in load impedance which can affect the oscillating frequency. We reduced the magnitude of s_{12} from 0.04 to 0.00233 in Section 3.2.1 by adding a j2000Ω feedback element, and it is possible to further improve the reverse isolation using the s_{12} mapping. Figure 3.16 shows the mapping of Figure 3.14 expanded about the origin or $|s_{12}| = 0$ point. The three circles represent s_{12} magnitudes of -50, -60, and -70 dB.

The figure shows that the j2000Ω feedback element provides approximately 53 dB of isolation. However, we can achieve better results with an impedance of approximately 95Ω + j2065Ω. The *s*-parameters for this combination are

$$S = \begin{bmatrix} 0.97546\underline{/-17.14080^\circ} & 0.00001\underline{/-1.29318^\circ} \\ & \\ 3.05669\underline{/164.29355^\circ} & 0.71916\underline{/-6.23902^\circ} \end{bmatrix}$$

which indicates a reverse isolation exceeding 96 dB. The stability factor is 261.22, and the maximum available gain is 26.00 dB. As with most designs that use potentially unstable devices, we must investigate stability over as much of the frequency spectrum as practical, and modify the design as necessary to ensure stability.

As a practical matter, we would not expect to achieve anywhere near 96 dB of reverse isolation from this design. Variations in the device parameters and component tolerance combine to reduce the isolation. Radiation from components as well as higher-order modes also becomes a problem at high frequencies. Although mode suppressors and careful design of the cavity housing the amplifier can help, two or more well shielded stages may be required to achieve isolation on the order of 90 to 100 dB.

3.2.3 Reflection Amplifier

The state-of-the-art in microwave transistor performance is continually advancing. Modern devices are commercially available that have more than 10 dB gain above 18 GHz, and maximum frequencies of oscillation in excess of 100 GHz. However, the need for high-frequency components that operate into the millimeter wavelengths often results in design requirements that cannot be met using conventional amplifier design methods. Microwave transistors may have only a few decibels of gain at the design frequency, and four or five stages may be necessary to achieve the desired performance. Circuit loss may actually be greater than the maximum available gain, and the design is not feasible. Traditionally, designers have used the negative resistance characteristics of IMPATT, Gunn, and tunnel diodes [7-11] to provide amplification at frequencies where transistors were not available. We can also use the characteristics of microwave transistors to design similar circuits and realize far more gain than possible from normal two-port design techniques.

We examined the effects of an arbitrary termination at port k in Section 2.3.2. and derived equation (2.39), which relates the incident and reflected waves:

$$\frac{b}{a} = \frac{s_{jk}s_{kj}\Gamma_k}{1 - s_{kk}\Gamma_k}$$

When the ratio of reflected to incident waves is greater than one, the wave emerging from the interface has greater amplitude than the incident wave. The s-parameters are a function of the network or device and usually cannot be altered. However, we do have control over Γ_k and, under certain conditions, a specified reflection gain can be achieved.

Amplifiers are two-port components with one port associated with incident power and the other port with amplified output power. Therefore, we must separate the incident and reflected waves if reflection gain is to prove useful. One such device is the three-port circulator shown in Figure 3.17. The circulator has properties such that power can flow in only one direction, as shown by the arrow. Thus, power is coupled from port 1 to port 2, port 2 to port 3, and port 3 to port 1, but not in the reverse direction. The ideal circulator has no loss $s_{21} = s_{32} = s_{13} = 1$, infinite reverse isolation $s_{12} = s_{23} = s_{31} = 0$, and perfect match at all ports $s_{11} = s_{22} = s_{33} = 0$. The s-parameter matrix for this device is

$$S_C = \begin{bmatrix} 0 & 0 & 1 \\ 1 & 0 & 0 \\ 0 & 1 & 0 \end{bmatrix}$$

and when we terminate port 2 with Γ_k, the two-port parameters are

$$S = \begin{bmatrix} 0 & 1 \\ \Gamma_k & 0 \end{bmatrix} \tag{3.30a}$$

The input and output ports are matched, the gain from the input to output port is Γ_k, and the reverse isolation from output to input is 0 dB. Physically, a wave incident at port 1 is coupled to port 2, enters the termination and is modified by Γ_k, and then reenters port 2 where it is coupled to port 3.

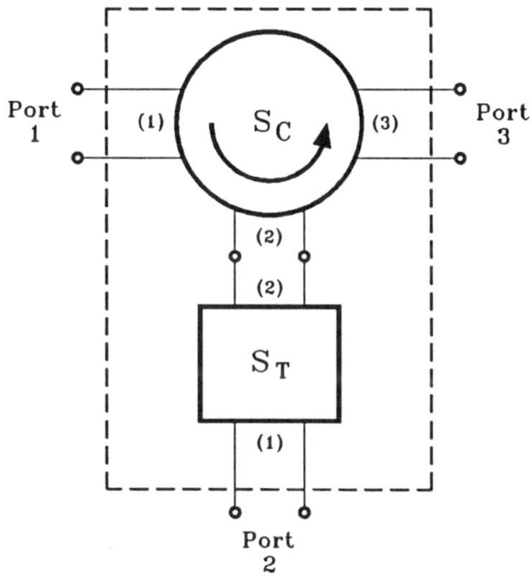

Figure 3.17 Reflection amplifier.

When the reference is positive real, Γ is greater than one for negative real impedance, and we can design for specified gain by terminating the circulator with the appropriate negative real load. Certain types of diodes exhibit negative real impedance when biased properly. However, we can also use the characteristics of the microwave transistor, which offers several advantages over diode designs. Transistor designs are easier to stabilize, and a third port is available for gain control or signal monitoring. Figure 3.17 shows port 2 of a two-port network with

$$S_T = \begin{bmatrix} s_{11} & s_{12} \\ s_{21} & s_{22} \end{bmatrix}$$

connected to port 2 of the circulator. The three-port s-parameters for this network are

$$S = \begin{bmatrix} 0 & 0 & 1 \\ s_{12} & s_{11} & 0 \\ s_{22} & s_{21} & 0 \end{bmatrix} \tag{3.30b}$$

which reduces to (3.30a), with $\Gamma_k = s'_{22}$ when we terminate port 1 of the two-port network. The gain from port 1 to port 3 is the output reflection coefficient of the two-port network, and we need only design for a specified magnitude of s_{22} to realize the required gain.

Mapping techniques provide the designer with valuable information pertaining to the behavior of networks and can be applied to the design of reflection amplifiers. This information is especially critical when designing with negative resistance elements, since the possibility of oscillation always exists. As an example of this application, we will design a reflection amplifier with 20-dB gain using an FET that has 50Ω common source s-parameters given by

$$S = \begin{bmatrix} 0.604\angle{-114^\circ} & 0.076\angle{121^\circ} \\ 1.297\angle{89^\circ} & 0.738\angle{-83^\circ} \end{bmatrix} \tag{3.31a}$$

The stability factor is 1.212 and the maximum available gain is 9.538 dB. The design would require three stages using a conventual design approach; however, we may be able to realize 20-dB gain from a single device using the reflection amplifier configuration. The required magnitude of Γ_k is $10^{(20/20)}$ or 10.0. Since this device is unconditionally stable, there is no passive source or load impedance that results in a reflection coefficient greater than one. Therefore, we must use feedback or a different device orientation. Using the techniques discussed in Section 2.4, we compute the common gate s-parameters as

$$S_g = \begin{bmatrix} 0.3642\angle{-165.93^\circ} & 0.3811\angle{21.59^\circ} \\ 0.9654\angle{-41.29^\circ} & 1.0168\angle{-47.72^\circ} \end{bmatrix} \tag{3.31b}$$

and common drain *s*-parameters as

$$S_d = \begin{bmatrix} 0.8231\angle{-53.80^\circ} & 0.5616\angle{12.76^\circ} \\ 1.0400\angle{-32.71^\circ} & 0.2094\angle{-158.14^\circ} \end{bmatrix} \tag{3.31c}$$

The common gate orientation is a good candidate for this application; $K = 0.50$, MSG = 4.04 dB, s_{22} is greater than one, and s_{11} is less than one. However, s_{22} is only slightly greater than one, and any circuit loss would reduce the magnitude below one. Therefore, we must increase the magnitude of s_{22}.

The maximum reflection coefficient at the drain terminal due to an impedance at the source or gate can easily be determined from the mapping constants. The original impedance mapping referenced to *N* is scaled and rotated by *R* and then translated by *C*. Assuming a positive real *Z* and *N*, the maximum magnitude of Γ is equal to one on the constant zero resistance circle, and the maximum magnitude of the mapping in the *s*-parameter plane is in the direction of *C* and given by the sum of both radii:

$$|S_{max}| = |C| + |R| \tag{3.32a}$$

The vector from the center of the mapping to the impedance which results in S_{MAX} has a magnitude of one and an angle such that when rotated by *R* coincides with *C*, or

$$|R|\angle{\theta_R} \cdot \left\{ \frac{z_{max} - N^*}{z_{max} + N} \right\} = |R|\angle{\theta_C} \tag{3.32b}$$

where θ_R and θ_C are the angles of *R* and *C*, respectively. The impedance that satisfies this constraint is

$$z_{max} = \frac{N^* + N\left(1\angle{\theta_C - \theta_R}\right)}{1 - 1\angle{\theta_C - \theta_R}} \tag{3.32c}$$

When the real component of N is negative, (3.32) provides the impedance and reflection coefficient for zero resistance, but not S_{max}. The best approach is to generate a mapping on the appropriate parameter plane.

The common gate s_{22} mapping constants for an impedance connected at the source terminal is

$$C = 0.86765\underline{/-50.17961^\circ}$$

$$R = 0.42419\underline{/-27.19200^\circ}$$

and

$$N = 0.48134\underline{/-11.54164^\circ}$$

Thus, the maximum output reflection coefficient is

$$|s_{22}|_{max} = |C| + |R| = 0.86765 + 0.42419 = 1.29184$$

The techniques of Section 1.3.6 can be used to transform the two-port common gate s-parameters into three-port parameters, with each port referenced from one terminal to ground. An impedance connected at port 3 forms series feedback from gate to ground. The mapping constants for the s_{22}-plane are

$$C = 1.01040\underline{/-91.35840^\circ}$$

$$R = 0.75347\underline{/67.57072^\circ}$$

$$N = 1.36813\underline{/-66.21926^\circ}$$

and the maximum output reflection coefficient due to series feedback is

$$|s_{22}|_{max} = 1.01040 + 0.75347 = 1.76387$$

Thus, series feedback is a better choice, and since N is positive real, we can use (3.32c) to compute z_{MAX}:

$$z_{max} = \frac{1.36813\underline{/66.21926^\circ} + (1.36813\underline{/-66.21926^\circ})(1\underline{/-158.92912^\circ})}{1 - 1\underline{/-158.92912^\circ}}$$

$$= 1.14937\underline{/90.00000^\circ}$$

or Z_{MAX} = j57.46850Ω. Connecting 57.46850Ω inductive reactance in series with the common gate parameters in (3.31b) provides the *s*-parameters of the two-port:

$$S_T = \begin{bmatrix} 1.24471\angle 96.04186° & 0.85594\angle 99.27492° \\ 1.53835\angle -140.86902° & 1.76387\angle -91.35868° \end{bmatrix} \tag{3.33}$$

and we see that the magnitude of s_{22} is the predicted value. Unfortunately, s_{11} is also greater than one, which is undesirable because of the possibility of oscillations at this port. We must consider the effects of the series feedback on s_{11} and s_{22} simultaneously, and then select an impedance that provides negative real output impedance and an associated positive real input impedance, if possible. Since the magnitude of s_{22} cannot be increased to the desired value of 10 (20-dB reflection gain) using feedback, a matching network will be necessary between the transistor output and port 2 of the circulator. The matching network will have an effect on s'_{11} and may very well cause a magnitude greater than one, even though the feedback impedance is selected to ensure a positive real input impedance. However, minimizing s_{12} improves the isolation between the input and output ports and should also be considered when selecting a feedback impedance.

Using the common gate *s*-parameters shown in (3.31b), we can compute the constants for the s_{11} mapping as a function of series feedback:

$$C = 0.54138\angle 67.11625°$$

$$R = 0.81413\angle -86.26679°$$

and

$$N = 1.36813\angle -66.21926°$$

and then map the impedance on the s_{11}-plane, as shown in Figure 3.18. The constants for the s_{22}-plane were computed above, and the mapping is shown in Figure 3.19. The s_{12} mapping is shown in Figure 3.20 for mapping constants of

$$C = 0.74687\angle 62.17279°$$

$$R = 0.52031\angle -41.80915°$$

and

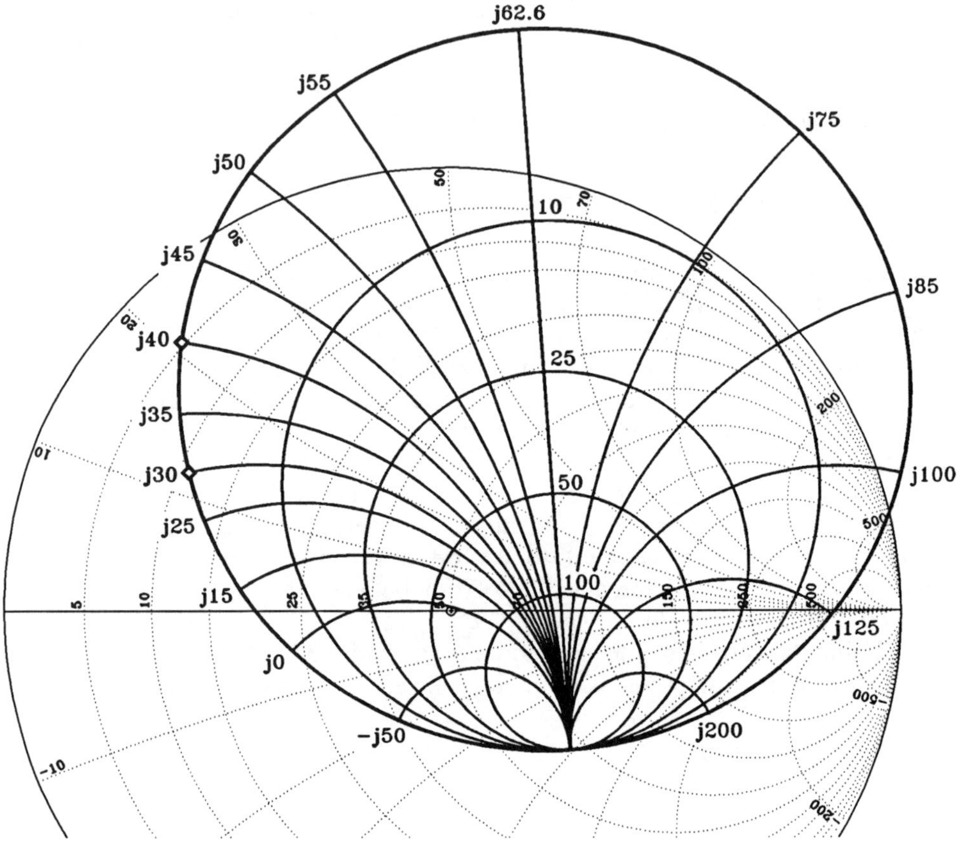

Figure 3.18 S_{11} mapping of series feedback.

$N = 1.36813\underline{/-66.21926^\circ}$

Examining the s_{22} mapping in Figure 3.19, we see that inductive feedback less than approximately j90Ω results in a magnitude greater than one. The s_{11} mapping in Figure 3.18 shows that the input reflection coefficient is less than one for values below j46Ω. Based on this information, we can select a feedback impedance of j40Ω which provides $|s_{11}| = 0.85$ and $|s_{22}| = 1.55$. However, when we also consider the s_{12} mapping in Figure 3.20, a better choice is j30, which reduces the magnitude of s_{12} from 0.36 to 0.23, an improvement of almost 4 dB.

The s-parameters for the common gate transistor with j30Ω series feedback are

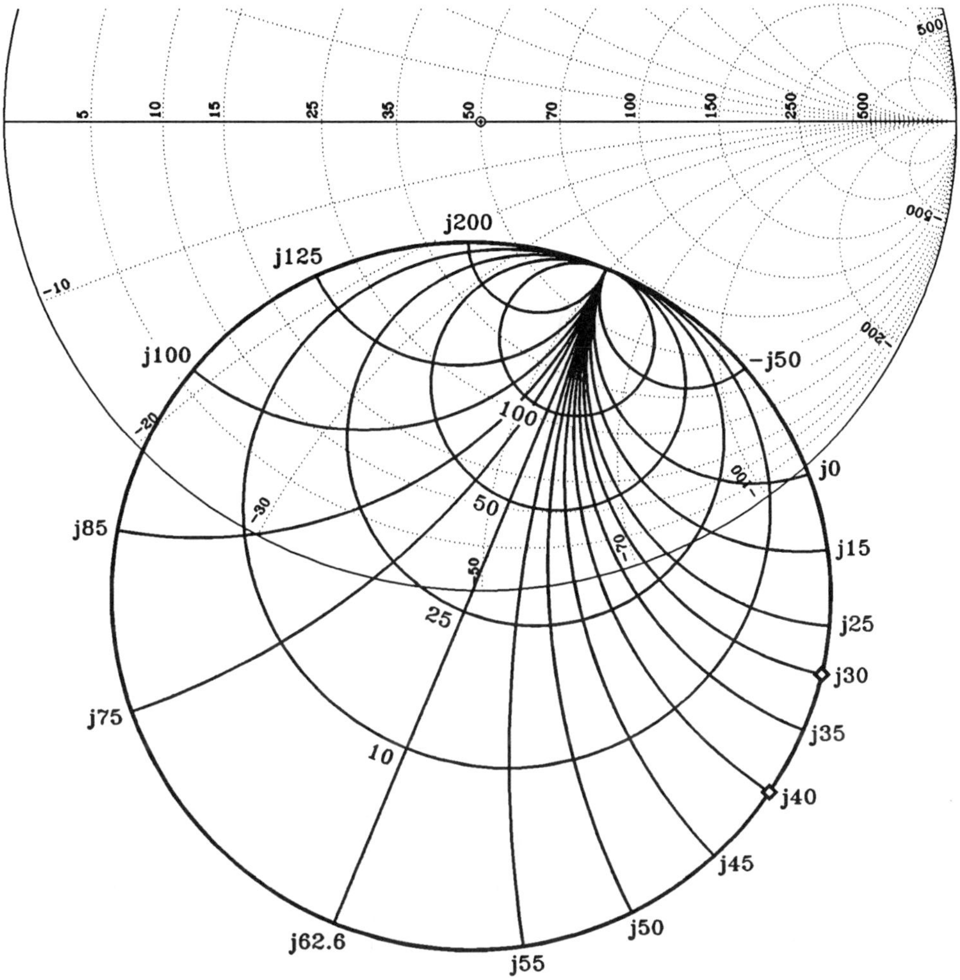

Figure 3.19 S_{22} mapping of series feedback.

$$S_T = \begin{bmatrix} 0.66030\underline{/151.79774^\circ} & 0.23169\underline{/72.22188^\circ} \\ & \\ 1.17771\underline{/-75.84495^\circ} & 1.37599\underline{/-58.91228^\circ} \end{bmatrix} \qquad (3.34)$$

which has the required characteristics. The magnitude of the input reflection coefficient is less than one; the magnitude of the output reflection coefficient is greater than one; and $|s_{12}|$ is minimized.

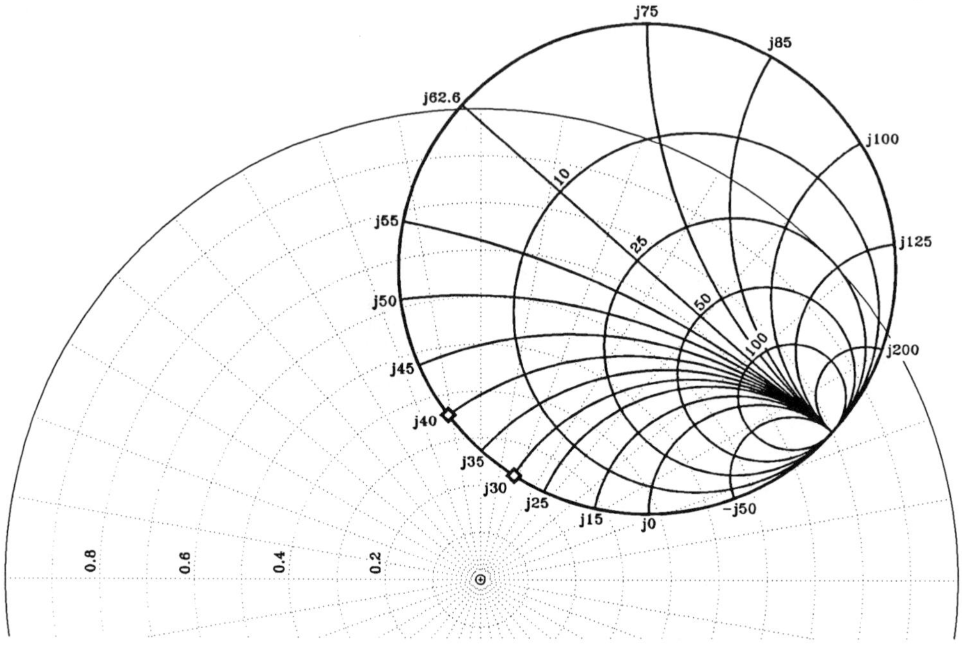

Figure 3.20 S_{12} mapping of series feedback.

Before continuing with the design, we should examine the mappings in more detail. The figures contain additional information that can be useful for the practical aspects of the design. We have assumed an ideal feedback element which has zero resistive component. However, at higher microwave and millimeter wave frequencies, this is not a good assumption. Skin effect, radiated energy, and other factors combine to create loss or a nonzero resistive component. If the Q is known either through measurements or calculations, we can account for the loss by using the appropriate constant resistance circle on the mappings. Worst-case conditions for changes in inductance, as well minimum and maximum Q, can also be investigated and the components specified accordingly.

The next step is to design a matching network that transforms s_{22} into a reflection coefficient with a magnitude equal to ten at port 2 of the circulator to provide the required 20-dB gain. However, we are dealing with negative resistance and must carefully analyze the matching requirement to ensure that the amplifier does not turn into an oscillator. Figure 3.21 shows a Smith chart referenced to 50Ω, which is expanded to encompass a reflection coefficient of 10, illustrated in the figure as the broken circle. The normal positive real portion of the Smith chart is the small shaded circle in the center. Examining the figure, we see that only resistive components between approximately -41Ω and -61Ω provide the required reflection coefficient magnitude. Equation (3.3a) can be used to determine the impedance as a function of reflection coefficient. For example, when we substitute $\Gamma = 10\angle 0°$ into (3.3a), $Z = -61.11111\Omega + j0\Omega$, and $Z = -40.90909\Omega + j0\Omega$ for $\Gamma = 10\angle 180°$.

We analyzed the incident and reflected waves circulating around a loop formed by one port of a network and its associated termination in Section 2.3.2. The condition where the circulating wave attains a finite value is given by (2.50):

$$Z + Z_k > 0$$

in terms of the impedance Z at one port of the network and the termination impedance Z_k. When the sum is less than or equal to zero, the circulating wave builds up following the series in (2.39a) until limited by physical properties of the network or termination. We can design a type of oscillator known as a *one-port negative resistance oscillator* by choosing appropriate values for Z and Z_k. Detailed analysis of the negative resistance oscillator [12,13] reveals that two conditions must be satisfied to sustain steady-state oscillations:

$$R + R_k = 0 \tag{3.35a}$$

and

$$X + X_k = 0 \tag{3.35b}$$

where $Z = R + jX$ and $Z_k = R_k + jX_k$. Considering the reflection amplifier shown in Figure 3.17, Z is the impedance presented to port 2 of the circulator by the two-port network, and Z_k is the impedance looking into port 2, $50\Omega + j0\Omega$ for the ideal device. We can ensure that the design will not oscillate by requiring R to be more positive than -50Ω, that is, $R + R_k > 0$. Referring to Figure 3.21, we see that the largest value of R that provides a reflection coefficient of 10 is -40.90909Ω, and the associated reactive component is zero. Therefore, the matching network should be designed to transform s_{22} into $\Gamma = 10\angle 180°$.

Although we can use an expanded Smith chart as shown in Figure 3.21 for the matching network design, the enlarged scale makes it difficult to accurately read impedance values. However, we can invert the magnitude of the reflection coefficients, which provides a convenient means for dealing with negative real impedance using a standard Smith chart. This operation has the effect of turning the Smith chart inside out, positive real impedance outside the unity radius circle and negative real inside. For example $s_{22} = 1.37599\angle -58.91228°$ or $Z = -30.33721\Omega - j80.03232\Omega$. When the magnitude of s_{22} is inverted, the reflection coefficient is $0.72675\angle -58.91228°$ and the impedance is $30.33721\Omega - j80.03232\Omega$, which can be plotted on a standard Smith chart. Most commercial network analyzers can in fact measure $1/\Gamma$ by either reversing the reference and test channels for manual systems or defining a new parameter for automated systems. This has the effect of inverting the ratio of incident and reflected waves a/b. However, when this technique is used, the sign of the reactance is changed as well and must be taken into account. When the reference is pure resistive, as is usually the case for test equipment, inverting Γ results in measurements referenced to $-R_o$, which is easily confirmed by substituting $-R_o$ into (3.3a) and (3.4).

Figure 3.22 shows a portion of the normal Smith chart used for the design of the matching network. The magnitude of s_{22} and Γ are inverted and labeled

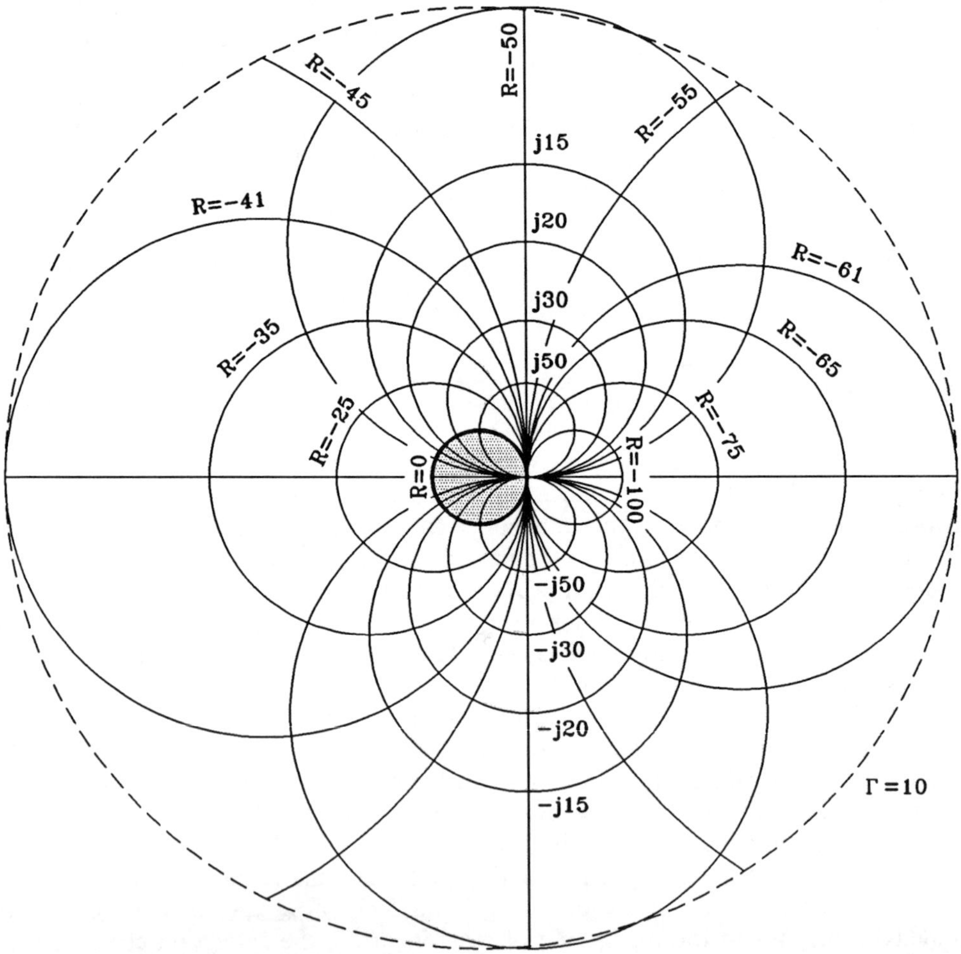

Figure 3.21 Smith chart with magnitude = 10.

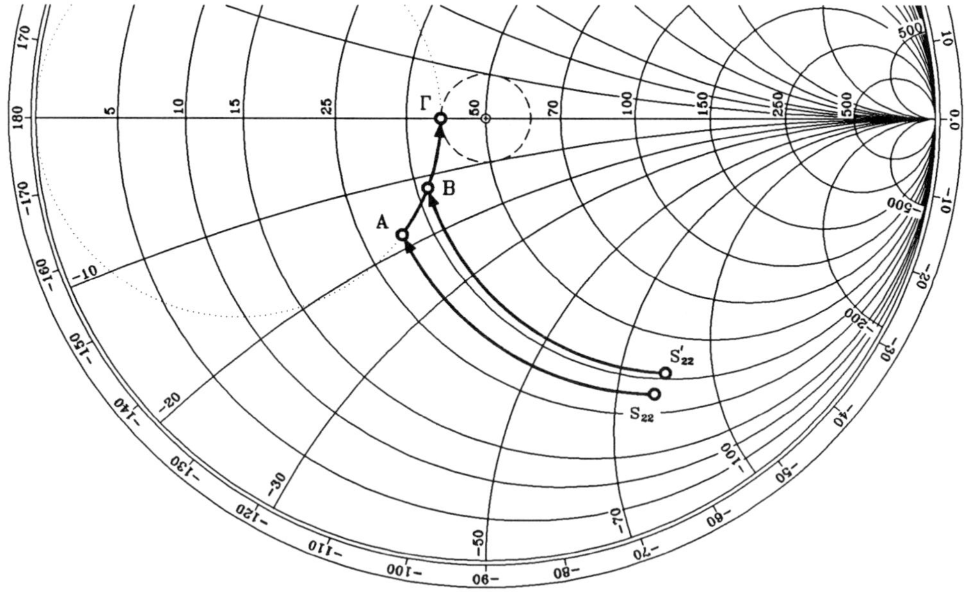

Figure 3.22 Reflection amplifier matching network design.

accordingly. At high microwave and millimeter-wave frequencies, chip devices are normally used to avoid package parasitics. The inductive reactance of the wires used to bond the source, gate, and drain pads to the circuit can be significant at these frequencies, and the first element of the matching network should be a series inductor. Within limits, the series inductive reactance can be realized by adjusting the lengths of the bond wires. The second requirement is to provide drain bias without upsetting the impedance of the matching network. Referring to the figure, we see that we can add series reactance to transform s_{22} to point A, which falls on the constant conductance circle passing through Γ. Either a shunt inductor or shunt shorted transmission line can be used to move from point A to Γ. Selecting the shorted transmission line and then using the techniques of Section 3.1.3 to determine the element values, we have the two-port portion of the reflection amplifier shown in Figure 3.23.

The *s*-parameters for the network shown in Figure 3.23 are

$$S_T = \begin{bmatrix} 2.47546\underline{/-49.30943^\circ} & 2.43896\underline{/-34.53202^\circ} \\ 12.39769\underline{/177.40116^\circ} & 9.99998\underline{/179.99710^\circ} \end{bmatrix}$$

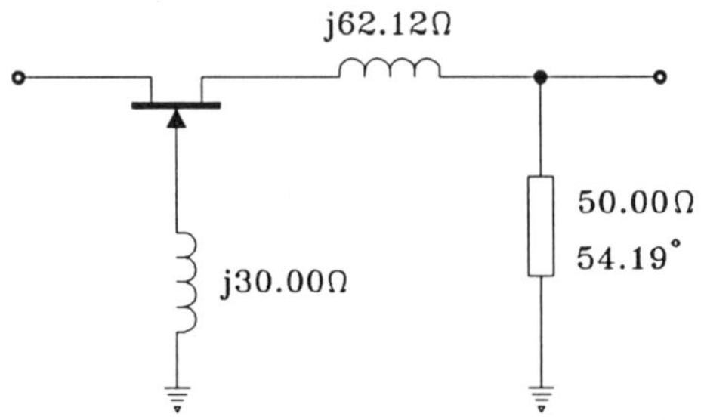

Figure 3.23 Matching network.

which does have the required output reflection coefficient of 10. However, s_{11} is also much greater than one which is not desirable because we must now be concerned with the possibility of oscillations at both ports. What happened to cause the magnitude of s_{11} to become so large? Especially since we minimized s_{12} in an effort to provide some isolation between the input and output ports. The s_{11}-plane mapping of load impedance for the common gate device with j30Ω series feedback shown in Figure 3.24 provides the answer. The mapping is turned inside out compared with the previous mappings. That is, the region of positive real load impedance is *outside* the R = 0 contour. As the constant resistance values are increased from 0 to 30Ω, the radii of the contours increases to infinity. Then, as constant resistance is increased further, the radii decrease again. The R = 50Ω load impedance contour falls completely within the positive real region of the s_{11}-plane, which explains why the input impedance shown in Figure 3.18 appears to be acceptable. The mapping constant N computed from the s-parameters given in (3.34) is -0.60675 - j1.60064, and the reference impedance for the mapping is -30.33750 - j80.03200Ω, which provides a warning.

The load impedance presented to the drain terminal by the matching network of Figure 3.23 is 32.88237Ω + j85.84847Ω, which is labeled as s'_{11} in Figure 3.24. Other points on the $|\Gamma|$ = 10 circle can be used in an attempt to reduce the magnitude of s_{11}. However, the resistive component will be closer to -50Ω and the possibility of oscillations enhanced. Instead, we can add a series resistor in the source lead which cancels the input negative resistance and eliminates the possibility of oscillations. The input impedance is -65.74178Ω - j48.12772Ω with the matching network shown in Figure 3.23, and a 100Ω series resistor is sufficient to provide a net positive resistance. The s-parameters of the 100Ω resistor cascaded with the common gate FET with j30Ω series feedback are

$$S_T = \begin{bmatrix} 0.38451\angle 6.89127^\circ & 0.08909\angle 79.11315^\circ \\ 0.45284\angle -68.95368^\circ & 1.42798\angle -55.18652^\circ \end{bmatrix}$$

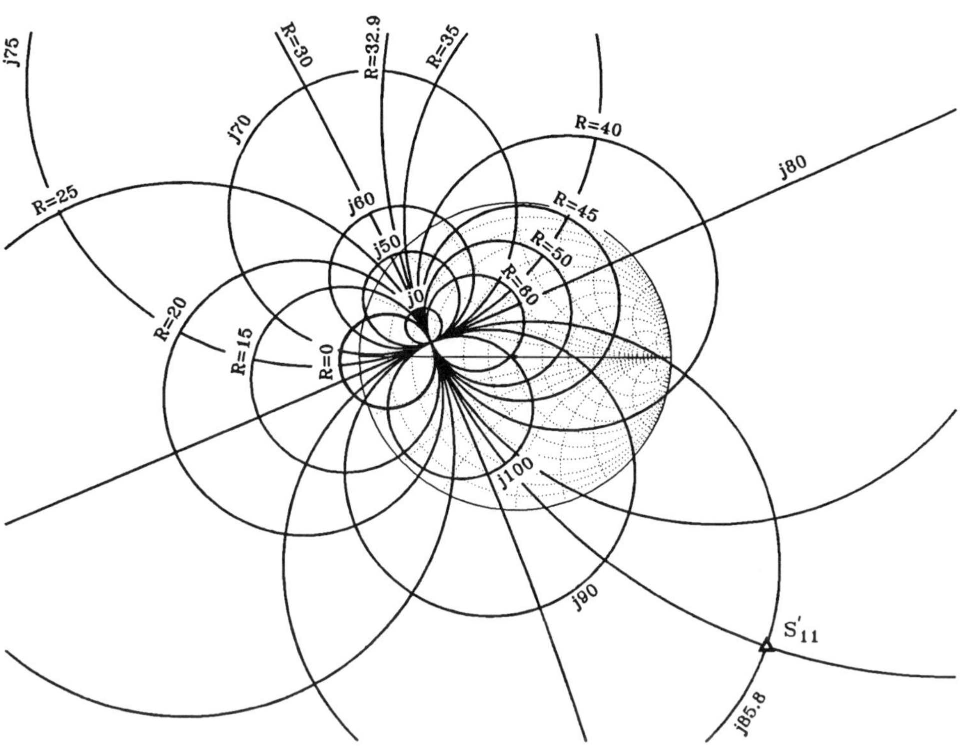

Figure 3.24 S_{11} plane mapping of load impedance.

The magnitude of the output reflection coefficient is inverted and then plotted on Figure 3.22 as s'_{22}. This impedance is not too different than the original, and we can use the same topology for the matching network, but with different element values. The series reactance is +j71.04415Ω and the shunt reactance is +j123.83343Ω. Bonding wire will be used to realize the series reactance and a shunt shorted stub used to provide the shunt reactance. We can use a shunt transmission line of any impedance and select 70Ω, which has a convenient linewidth on most microstrip and stripline material. The input impedance of a shorted transmission line is $jZ_o\mathrm{Tan}(\theta)$, and an electrical length of 60.52159° provides the required reactance. The s-parameters of the two-port network are

$$S_T = \begin{bmatrix} 0.74514\angle{-4.74035^\circ} & 0.86955\angle{-20.21243^\circ} \\ 4.42011\angle{-168.27925^\circ} & 9.99999\angle{179.99998^\circ} \end{bmatrix} \qquad (3.36)$$

which is positive real at the input port, and the magnitude of s_{22} is 10, as required for 20-dB reflection gain. The complete reflection amplifier is shown in Figure 3.25.

The amplifier is matched at the input and output ports and has 20-dB gain. However, the reverse isolation is 0 dB, and it is necessary to add an isolator if a higher value is required. The shunt stub is RF-grounded through a capacitor, and drain bias introduced at this low-impedance junction. Depending on how well the capacitor performs at the operating frequency, it may be necessary to add an open stub with an electrical length of 90° to provide a better RF ground. The gate is at zero potential, and negative gate-to-source voltage is provided by the current flowing through the two resistors. This configuration is known as *self bias* [14,15], and the drain current is determined by the value of the source resistors. The resistors are necessary to ensure stability, and we have some flexibility in selecting their value. However, should there be a conflict between the RF and bias requirements, we can always bypass a portion of the total resistance with either a large capacitor or inductor.

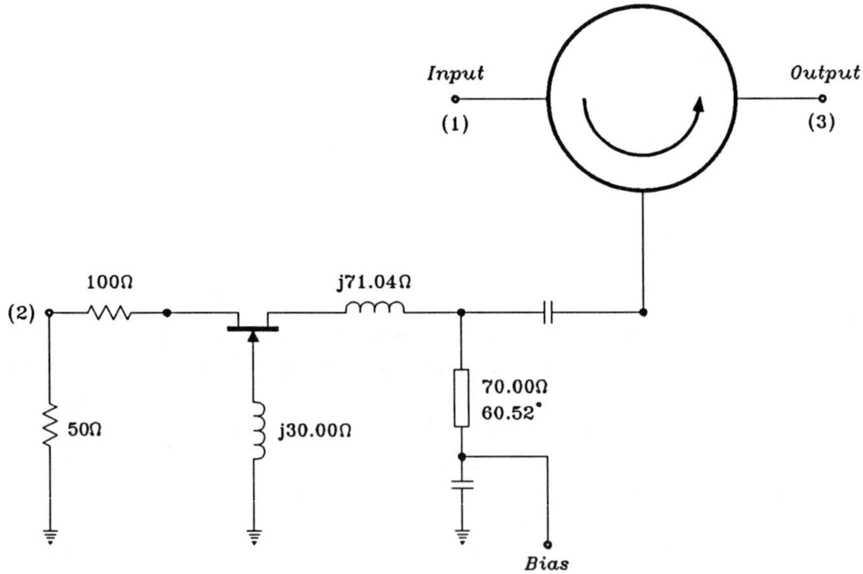

Figure 3.25 Complete reflection amplifier.

The two-port reflection amplifier does meet the specifications. However, the configuration that results when the 100Ω source resistor is connected to an external port has some interesting characteristics. The *s*-parameters for the three-port numbered as shown in Figure 3.25 are

$$S = \begin{bmatrix} 0.00000\angle 0.00000° & 0.00000\angle 0.00000° & 1.00000\angle 0.00000° \\ 0.86955\angle -20.21243° & 0.74514\angle -4.74035° & 0.00000\angle 0.00000° \\ 9.99999\angle 179.99998° & 4.42011\angle -168.27925° & 0.00000\angle 0.00000° \end{bmatrix}$$

The gain from port 1 to port 2, s_{21}, is -1.21 dB, and s_{32}, the gain from port 2 to port 3, is 12.91 dB. Many system applications require provisions to monitor an input signal or inject a test signal. Port 2 of the reflection amplifier provides a convenient means to achieve both requirements. However, there is also a problem created by the gain from port 2 to port 3. Any thermal noise generated in the resistors is amplified, added to the noise of the active device, and then combined with the signal at port 3. We can minimize the noise somewhat by careful selection of the resistive material, although it may be necessary to reduce s_{32} if noise is an important design factor.

During production we expect device and component variations, which affect the performance of the design and may require some form of adjustment to

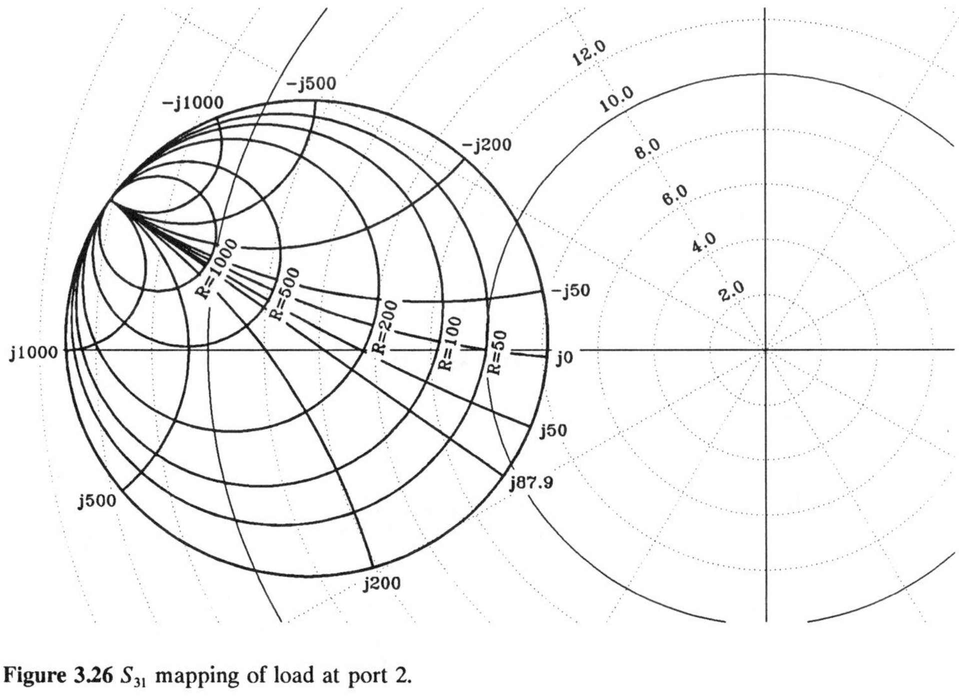

Figure 3.26 S_{31} mapping of load at port 2.

improve the yield. We can investigate the effects of the termination at port 2 by mapping s_{31} as shown in Figure 3.26. The mapping reveals that the resistive component can change by $\pm 25\Omega$ and the reactive component by $\pm j70\Omega$ for gain variations of ± 1 dB. The resistor provides a convenient means for adjusting the gain during production. We can design the resistor as an area of metal film with an initial value of 25Ω. Then, during test, the area can be trimmed, which increases the resistance and sets the gain exactly on the design value. We could also use a PIN diode to adjust the gain by as much as 8 dB. The PIN diode acts as a variable resistor at microwave and millimeter-wave frequencies. The resistance is a function of bias and varies from a few tenths of an ohm to several thousand ohms.

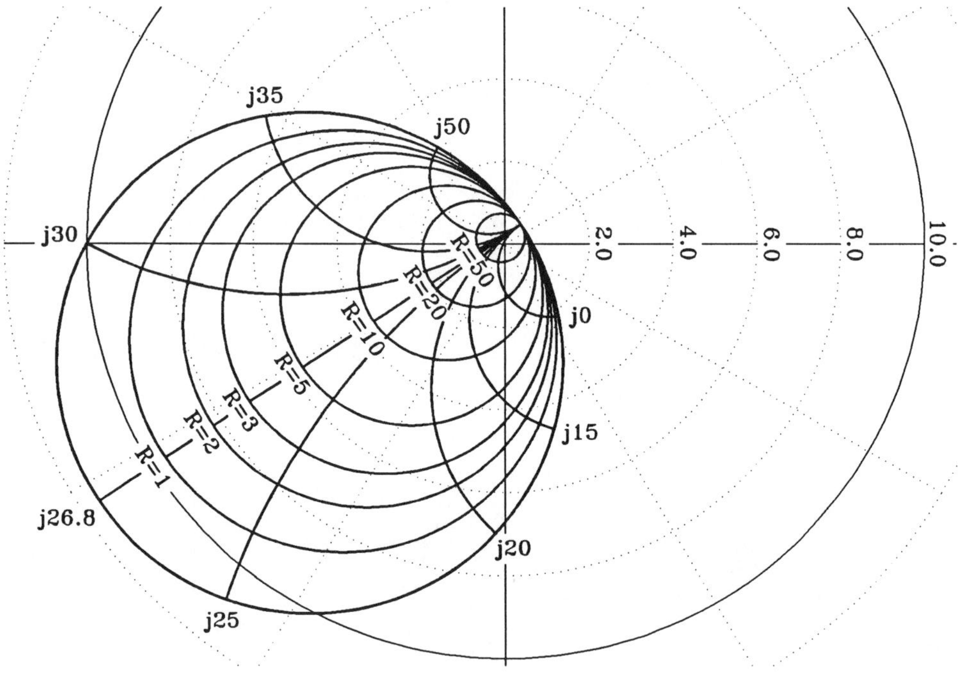

Figure 3.27 S_{31} mapping of feedback impedance.

Series feedback from gate to ground is also critical to the operation of this design and should be investigated as well. Figure 3.27 shows the mapping of gain as a function of feedback. This mapping illustrates an example where the three-point method is more convenient for computing the mapping constants. We could transform the three-port shown in Figure 3.25 into a four-port network and then use (3.26). However, we already have s_{31} for one value of feedback impedance and need only compute s_{31} for two additional values and then use (3.29). The mapping shows that gain is sensitive to the resistive component of the feedback, and we must reduce the inductance to compensate for loss. In fact, if the loss in this element is more than about 0.8Ω, we will not be able to achieve 20-dB gain, although we can adjust the termination at port 2 to increase the gain. The mapping also shows that we can use a PIN diode in the feedback path to control the gain by more than 20 dB. Gain control will be nonlinear because of the sensitivity for small resistance values, but may be useful for certain systems. We can make another interesting observation from the figure. When the reactance is slightly greater than j26.8Ω, the phase remains constant across the entire gain control range, which is a desirable characteristic for some applications.

The three examples in this section clearly illustrate the power of mapping as a circuit design tool. We were able to simultaneously increase the maximum stable gain and stability factor of the FET circuit in Section 3.2.1 simply by looking at the mappings of s_{12} and s_{21}. Then, in Section 3.2.2, we used the s_{12} mapping to create an unconditionally stable device and achieve greater than 90-dB reverse isolation. Design of the reflection amplifier in Section 3.2.3 presents the greatest challenge, since we must design for a specific impedance at all three terminals of the active device. We used the mappings to determine the circuit topology and element values during the design, and then investigated the characteristics of the complete circuit. This design used common gate series feedback; however, we can add parallel feedback in an effort to improve the input-to-output isolation, or try the common drain configuration. The mapping approach to circuit design provides information in a format easy to comprehend and difficult to obtain using other techniques.

3.3 OPTIMIZATION

We approach most design problems in stages, progressing from the simplest circuit model to the most complex. For example, an amplifier design usually begins by characterizing one or more devices to determine maximum performance. At this stage, we assume that we can design any matching networks necessary to achieve the maximum performance, and then select a device and define the circuit specifications: device types, number of stages, gain per stage, noise figure, and output power of each stage. The next step is to define a source and load impedance for each device and then design the required matching networks. The networks may be as simple as one or two elements that can be designed using graphical techniques, or as complex as a synthesized network of six to ten elements. We can model each stage individually, confirm the performance, and then perhaps add auxiliary circuits to improve stability and include the effects of bias networks. A complete circuit model is constructed by cascading stages. Then we can analyze the effects of interactions due to nonzero s_{12} and less than perfect matching networks. Finally, the model is modified to include loss, parasitics, and discontinuities which are functions of the actual components used and the physical layout. Careful attention to the details of modeling the design usually results in computer predictions that agree very well with hardware measurements, even into millimeter wavelength frequencies.

At some point in this process, we may find that our computer model deviates from the desired response, and it becomes necessary to alter the design. Usually, changing the element values is sufficient. However, in some cases portions of the network will have to be redesigned or the entire configuration changed. How do we determine the set of element values that provides the response closest to that desired? The error is often due to differences between design assumptions and the actual circuit, which means there is little to be gained from another design iteration. However, we can identify parameters in the circuit model that can be controlled, and then vary these in a search for the best combination. This approach is the mathematical equivalent of circuit alignment using test equipment. The CAD simulator replaces the test equipment, and changing element values correspond to adjusting components in the circuit. The obvious advantages of using a computer model is speed and cost; it is not necessary to actually build and test hardware. CAD modeling also provides a means of easily adjusting elements that are difficult, if not impossible, to change once the circuit has been fabricated. The mask set for a monolithic microwave integrated circuit (MMIC) can cost in excess of $50,000 and often requires weeks or months between design and test. Transmission lines are fabricated in microstrip or stripline as areas of metalization. The width determines the impedance and the physical length controls the electrical length of the transmission line. We can easily adjust the length of open stubs by adding rectangular segments called pads which are connected using bond wires. Breaking wire bonds shortens the length and adding pads increases the length. Although the complete equivalent circuit includes the inductance of the wire bonds and coupling between pads, the dominant effect is a change in length of the open stub. We can also adjust the impedance somewhat by using the crude and time-consuming technique of scribing off metalization to narrow the line, or by adding small pieces of metal to make the line wider. However, we cannot easily change the electrical length of either the shunt shorted stub or cascaded transmission line.

The process of finding the set of element values that provides a circuit response as close as possible to the specified target response is known as *optimization*. To be more precise, the problem of maximizing or minimizing a given function,

$$E = f(x_1, x_2, \ldots, x_n) \tag{3.37}$$

subject to certain constraints, is known as the *general constrained optimization problem*, and the literature abounds with various approaches to this problem. References 16 through 23 provide an overview of some of the more common types of optimization methods and contain literally hundreds of references to other sources. The function E is known as the *objective function*, *performance measure*, *cost function*, or *error function*, and denotes that which is to be minimized or maximized. We usually create an error function that describes the difference between the actual and desired response of a circuit. Hence, we are concerned with minimizing E or reducing the error between the calculated and target responses. If necessary, we can minimize $-E$, which is mathematically the same as maximizing E. For example, the reflection coefficient of a circuit can be maximized for use in an oscillator, or the reflection amplifier by minimizing $-|s_{ii}|$.

We often use iterative techniques to solve simple design problems, the use of which, in fact, is a form of optimization. For example, in Section 3.2.1 we constructed a 25-dB gain amplifier by adjusting the value of the gate stabilization resistor, R_p in Figure 3.15, until G_{MAX} equaled 25 dB, and then designed matching networks to simultaneously match the input and output ports. The problem in this case was to determine the value of R_p, which results in a G_{MAX} of 25 dB. The mathematical solution to this problem requires writing the s-parameters, substituting into the expression for G_{MAX}, and then solving for R_p. A much easier approach is to simply model the circuit as a resistor and transistor and then analyze G_{MAX} as we manually enter various values of R_p. We may eventually find the solution by using random values; however, a specific optimization strategy is more efficient. The circuit is stable for resistor values less than or equal to 18.35kΩ, where G_{MAX}=31.196 dB. Also, we know that at R_p = 0 the signal path is shorted to ground and G_{MAX} = 0 dB. One strategy that seems reasonable is to define an interval for R_p, where G_{MAX} is greater than 25 dB at one end and less than 25 dB at the other, compute G_{MAX} at the midpoint, and then define a new interval from the midpoint to the appropriate end. This technique is known as a *half-interval search*, and reduces the uncertainty of R_p by one-half for each iteration.

The gain at R_p = 18.35 kΩ is too high and the gain at R_p = 0 too low. Therefore, the first interval is between 0 and 18.35 kΩ, and the midpoint is (18.35 kΩ - 0)/2 or 9.175 kΩ, with a corresponding G_{MAX} of 27.657 dB. The gain is still too high, and we can define a new interval between 0 and 9.175 kΩ, which reduces the uncertainty by one-half. Table 3.3 shows the first 15 iterations, and Figure 3.28 shows a plot of G_{MAX} as a function of R_p, along with the first five points of the half-interval search.

The half-interval search is not a particularly efficient means for solving this problem. Fifteen iterations were required to reduce the uncertainty from 18,350Ω to approximately 1Ω. In this case, the execution time is minimal. However, the optimization problem may involve a network with more than a hundred components, several hundred frequency points, and significant execution time. We know that the exact solution falls somewhere within the interval between where the gain is too high and where it is too low. Rather than simply divide the interval in half, we can assume a linear function between the two end points, interpolate to find the 25-dB intercept, and then define a new interval based on the interpolation and the

Table 3.3 Half-Interval Search

Iteration	R_p (Ω)	G_{MAX} *(dB)*	*Uncertainty*
1	18350.00000	31.196	18350.00000
2	9175.00000	27.657	9175.00000
3	4587.50000	25.346	4587.50000
4	2293.75000	22.893	2293.75000
5	3440.62500	24.352	1146.87500
6	4014.06250	24.889	573.43750
7	4300.78125	25.126	286.71875
8	4157.42188	25.010	143.35938
9	4085.74219	24.950	71.67969
10	4121.58203	24.980	35.83984
11	4139.50196	24.995	17.91992
12	4148.46192	25.002	8.95996
13	4143.98194	24.999	4.47998
14	4146.22193	25.001	2.23999
15	4145.10194	25.000	1.12000

two end points. Defining one end point as R_1, G_1 and the other as R_2, G_2, where R_i is the resistance and G_i is the maximum available gain at the point, we can write the equation for a straight line and then solve for the value of R_p, which corresponds to G = 25 dB:

$$R_p = \frac{(R_2 - R_1)(25 - G_1)}{(G_2 - G_1)} + R_1 \tag{3.38}$$

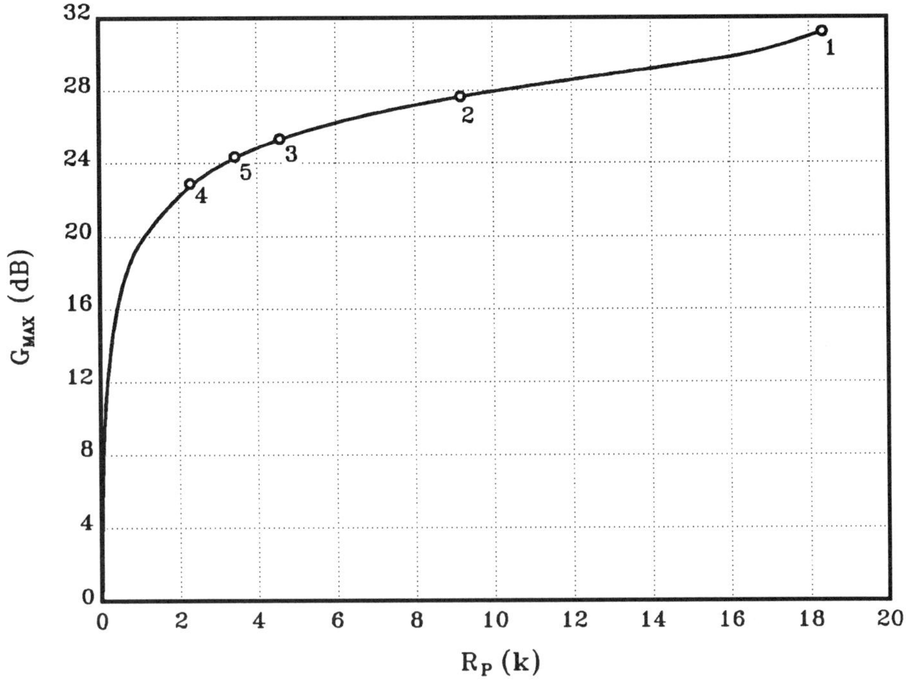

Figure 3.28 G_{MAX} as a function of R_p.

After the fourth iteration of the half-interval search shown in Figure 3.28, we have determined that R_p is located somewhere between 2293.75Ω and 4587.50Ω. Figure 3.29 shows an expanded view of these two points and the straight line assumption depicted as the broken line connecting the two points. Substituting into (3.38), we can compute

$$R_p = \frac{(4587.500 - 2293.750)(25.000 - 22.893)}{(25.346 - 22.893)} + 2293.75 = 4263.962$$

which is plotted as a diamond in the figure. G_{MAX} equals 25.096 dB, and the new interval is located between 2293.750Ω and 4263.962Ω.

Table 3.4 shows five iterations of the linear interpolation search, sufficient to determine a value of R_p such that the maximum available gain is 25.000 ± 0.001 dB. Comparing Table 3.3 with Table 3.4 we see that interpolation has improved the convergence significantly, although the uncertainty has not been similarly improved. The solution to each of the five linear interpolation iterations resulted in G_{MAX} greater than 25.0 dB. Thus, the lower limit remained fixed. However, we

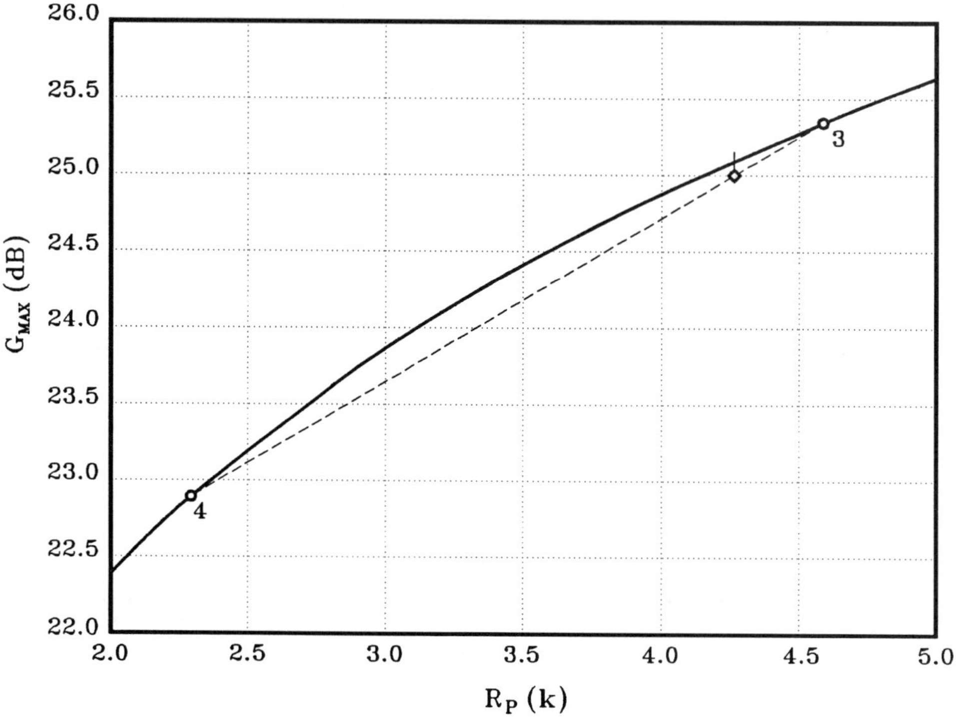

Figure 3.29 Linear interpolation.

can easily devise schemes that improve the convergence and uncertainty. For example, we can alternate the half-interval and linear interpolation search.

3.3.1 Error Functions

The purpose of optimization is to minimize a given error function that describes the difference between computed and target parameters. We did not explicitly specify an error function in Section 3.3. Instead, we merely searched for a value of R_p that provides G_{MAX} = 25 dB. However, we must formulate an error function before we can apply optimization to the general class of circuit design problems. The obvious choice is the difference between the calculated parameters, X_C, and target parameters, X_T; that is, $E = (X_C - X_T)$. Unfortunately, this error function becomes negative for X_T greater than X_C, and minimization does not have the desired effect. However, if we square the difference, $E = (X_C - X_T)^2$, then the error function is zero when the computed and target parameters are equal, and positive otherwise. Minimization using this form of the error function is known as *least*

Table 3.4 Linear Interpolation Search

Iteration	R_p	G_{MAX} *(dB)*	*Uncertainty*
5	4263.96249	25.096	1970.21249
6	4178.10666	25.027	1884.35666
7	4154.26522	25.007	1860.51522
8	4148.10458	25.002	1854.35458
9	4146.34606	25.001	1852.59606
10	4145.46722	25.000	1851.71722

squared optimization.

We can use the half-interval search, linear interpolation, or some combination to minimize this error function. However, the function contains a squared term and a reasonable assumption is that we can approximate it as a quadratic in the vicinity of the minimum:

$$y = ax^2 + bx + c \tag{3.39a}$$

The three coefficients a, b, and c can be determined from three points on the curve, and we can set the first derivative equal to zero which defines the minimum:

$$x_m = -\frac{b}{2a} \tag{3.39b}$$

Using three values of x and the corresponding function values, $<x_A, y_A>$, $<x_B, y_B>$, $<x_C, y_C>$, ordered such that $x_A < x_B < x_C$, as shown in Figure 3.30, we can solve for a and b and then substitute into (3.39b):

$$x_m = \frac{1}{2} \cdot \frac{y_A\left(x_B^2 - x_C^2\right) + y_B\left(x_C^2 - x_A^2\right) + y_C\left(x_A^2 - x_B^2\right)}{y_A\left(x_B - x_C\right) + y_B\left(x_C - x_A\right) + y_C\left(x_A - x_B\right)} \tag{3.39c}$$

Returning to the problem in Section 3.3, we can write the error function as

$$E = f(R_p) = \left(G_{MAX} - 25\right)^2$$

and then starting with the third and fourth iterations of the half-interval search in Table 3.3, the error is 0.11972 at R_p = 4587.5 and 4.43945 at R_p = 2293.75Ω. We need three points for quadratic interpolation and require that y_B be less than either y_A or y_C, which guarantees that a minimum exists somewhere in the interval between x_A and x_C. In other words, the error at the interior point (not necessarily the midpoint) is less than the error at either end point. Although we know that $(X_C - X_T)$ is positive for R_p above the optimum, and negative for R_p below the optimum, the sign may not change in the general optimization problem and cannot be used to test the interval. Dividing the interval in half, we compute the interior point as R_p = 3440.625 and E = 0.41990, which is greater than the error at the upper end. Figure 3.31 shows points 3, 4, and 5, which we have examined so far: x_A corresponds to point 4, x_B to point 5, and x_C to point 3. Since y_B is less than y_A, we eliminate point 4 and assume that the minimum is between points 5 and 3. Dividing this interval in half, we have R_p = 4014.0625 and E = 0.01232, shown as point 6 in the figure. The error at the interior point is less than that at either end, and we can now approximate the function over this interval and then compute the

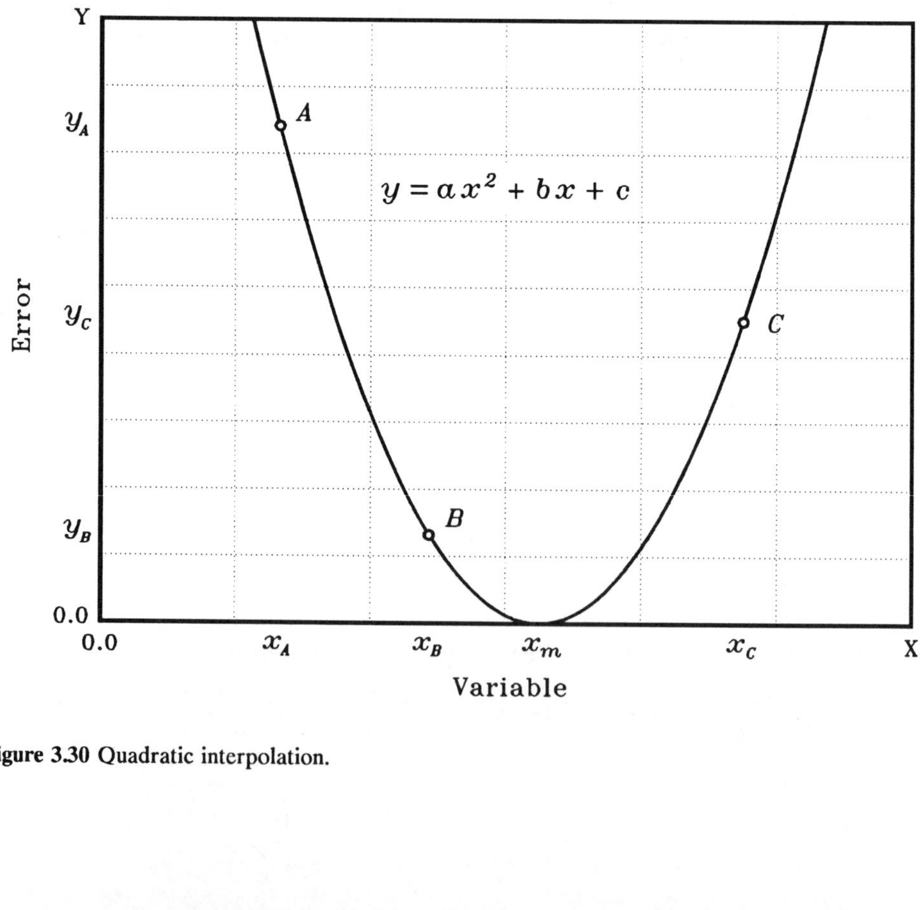

Figure 3.30 Quadratic interpolation.

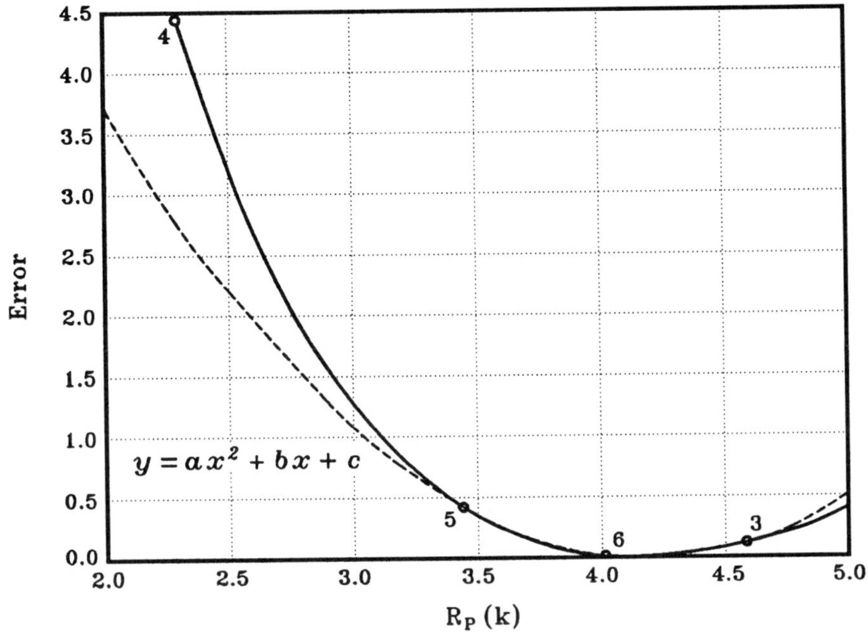

Figure 3.31 Quadratic curve fit.

minimum. Using data from points 5, 6, and 3, we can solve for the coefficients of the quadratic equation:

$$a = 7.83048 \cdot 10^{-7}$$

$$b = -6.54814 \cdot 10^{-3}$$

$$c = 1.36800 \cdot 10^{1}$$

which is plotted as the dashed line in Figure 3.31. As shown, the approximation is very good between points 5 and 3 and not too bad between 3000Ω and 5000Ω. Substituting into either (3.39b) or (3.39c), we find that x_m = 4181.18983, and the corresponding error is $8.4 \bullet 10^{-4}$, a reduction of more than an order of magnitude. Since x_m is the new minimum replacing the previous interior point, we now have four points and can eliminate one to reduce the interval for the next interpolation. We can test x_B to determine if it is less than or greater than x_m, and replace the appropriate end point with x_B. In this case the points are ordered $x_A < x_B < x_m < x_C$, and we can eliminate x_A and then define a new interval for interpolation between x_B and x_C, or 4014.0625Ω to 4587.5Ω. Table 3.5 shows the half-interval searches as iterations 3 through 6 and the quadratic interpolation as iterations 7 through 10.

Table 3.5 Quadratic Interpolation Search

Iteration	R_p	G_{MAX} *(dB)*	$(G_{MAX} - 25)^2$
3	4587.50000	25.346	$1.197 \cdot 10^{-1}$
4	2293.75000	22.893	4.439
5	3440.62500	24.352	$4.199 \cdot 10^{-1}$
6	4014.06250	24.889	$1.232 \cdot 10^{-2}$
7	4181.18983	25.029	$8.400 \cdot 10^{-4}$
8	4152.13734	25.005	$3.000 \cdot 10^{-5}$
9	4143.54768	24.998	$4.000 \cdot 10^{-6}$
10	4146.04806	25.000	0.000

The quadratic search procedure as developed by Powell [24] and modified by Zangwill [25] turns out to be a very powerful optimization tool, even for the n-dimensional case and error functions which are not quadratic. In many instances the optimizer will spend more effort locating an interval about the minimum point than the quadratic search requires to actually find the minimum. In fact, this was the case in the previous example. We required six function evaluations to find the interval, and only four to find the minimum.

The error function described above forms the basis for the performance measure used with most CAD systems, although it must be modified somewhat to cover a broad range of specifications. Figure 3.32 shows typical design specifications for an amplifier. Figure 3.32(a) depicts the gain versus frequency response and design goals, shown using dashed lines. We require flat gain of G_P dB in a passband extending from f_1 to f_2, and gain below G_L dB from zero to f_L and f_U to f_3. The design must also have a stability factor greater than 1.0 across the entire analysis range, as shown in Figure 3.32(b). Also, the input and output return loss must be below L_1 and L_2 dB, respectively, in the passband, as shown in Figures 3.32(c) and 3.32(d). In this case, we have four different specifications which must be satisfied simultaneously for a successful design. However, the design goals change as a function of the analysis frequency. For example, between zero and f_L, the gain should be less than G_L, and the stability factor greater than 1.0, but between f_1 and f_2, the gain should equal G_P, input return loss should be less than L_1, output return loss less than L_2, and the stability factor greater than 1.0. We can write the total error function by breaking the problem into frequency ranges and summing the individual error components in each range. Referring to Figure 3.32, we have

$$E = \sum_{f=0}^{f_3} (K(f) - 1.0)^2$$

$$+ \sum_{f=0}^{f_L} \left(G(f) - G_L\right)^2$$

$$+ \sum_{f=f_1}^{f_2} \left[\left(G(f) - G_P\right)^2 + \left(20\mathrm{Log}|s_{11}(f)| - L_1\right)^2 + \left(20\mathrm{Log}|s_{22}(f)| - L_2\right)^2\right]$$

$$+ \sum_{f=f_U}^{f_3} \left(G(f) - G_L\right)^2$$

where the frequency range is indicated by the summation limits. This problem requires four ranges. The full analysis range is from zero to f_3 for the stability factor, zero to f_L for the lower gain specification, f_1 to f_2 for the gain and return loss, and f_U to f_3 for the upper gain specification. We could include the stability factor in the other three ranges rather than create a separate range; however, we would miss frequencies from f_L to f_1 and f_2 to f_U. Broadband amplifiers sometimes tend to exhibit a gain peak just below the lower passband edge, and it might be necessary to specify a maximum gain of G_P from f_L to f_1 and f_2 to f_U. In this case, we could include the stability factor in each range without missing any frequencies.

Optimizing the error function as written above would set $K = 1$, $G = G_L$, $|s_{11}| = L_1$, and $|s_{22}| = L_2$ if successful. In fact, we actually want an error contribution only if the specification is exceeded. Most optimizers allow us to specify inequalities, as well as equalities, for the optimization goals, for example, gain *equal to* G_P, gain *less than* G_L, or stability factor *greater than* 1.0. In the case of an inequality, we are not concerned with the actual value of a parameter so long as it is above or below a threshold. Internally, the optimizer usually tests the inequality for a logical true or false. If true, the error component is set to zero. Otherwise, the difference is computed, squared, multiplied by the weight, and then added to the total error function. We can also use inequality goals to specify a

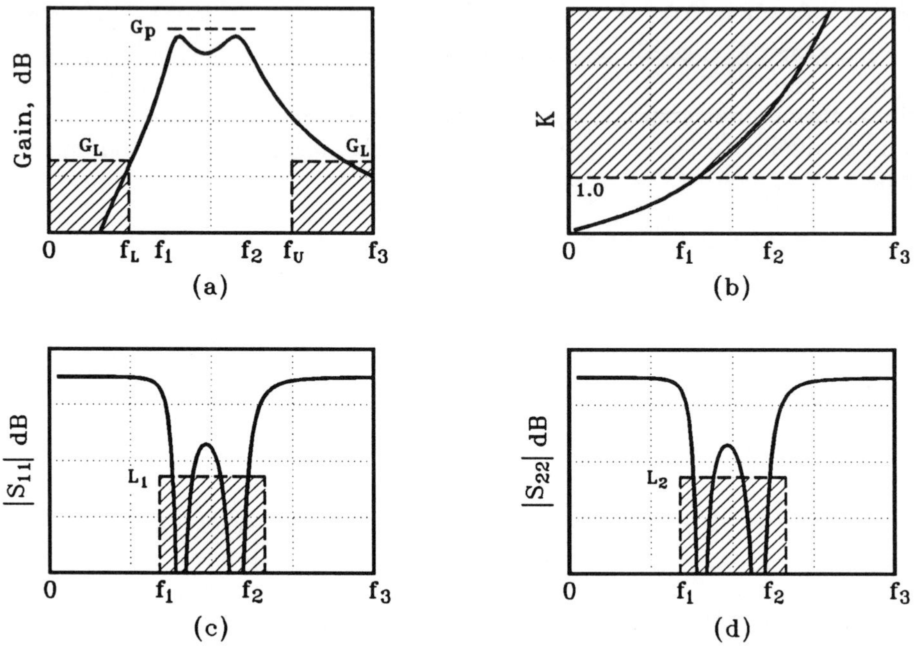

Figure 3.32 Typical amplifier design specifications.

range of acceptable values, such as ripple about some median gain. Two goals are required, one greater than the lower bound and the other less than the upper bound. When the parameter is inside the limits, both terms are zero, and when either limit is exceeded, one of the two specifications contributes to the overall error function.

The error function created by summing all applicable error components at each analysis frequency has the desired characteristics. The total error function is a positive combination of the squared individual differences between computed and specified parameters, which equals zero when all design goals are met. However, we seldom achieve zero error and must accept a compromise between certain specifications. The relative importance of each design goal can be controlled by multiplying the squared difference by a factor known as a *weight*. We can easily determine the weights by defining differences for each goal that has equal significance in the overall performance. Using the passband shown in Figure 3.32 as an example, we might decide that 0.1-dB gain error, 1.0-dB return loss error, and a stability factor error of 0.05 should contribute equally to the error function. The gain difference squared is 0.01, and a weight of 100 provides a contribution of 1. Similarly, a weight of 1 for return loss and 400 for the stability factor also results in an error component of 1. The total contribution is also affected by the number of frequencies. In this example, the stability factor is summed over the entire analysis range, but the gain and return loss only contribute when the frequency is between f_1 and f_2. We could compensate by changing the weights. However, dividing the total error in each range by the number of frequencies removes this dependance and makes the error specification somewhat simpler for the designer. The numerical value of the error function affects the convergence of the optimizer to some degree. Examining the error function shown in Figure 3.31, we see that a weight of 100 changes the error at point 4 from 4.5 to 450. Since the minimum remains at zero, the slope between the two points is steeper and small roundoff errors become significant.

The error function used for *each* frequency range by most CAD optimizers is

$$E = \frac{1}{n}\sum_{f=f_L}^{f_U}\left\{W_1\cdot(P_1(f) - G_1(f))^2 + W_2\cdot(P_2(f) - G_2(f))^2 + \cdots\right\} \tag{3.40a}$$

where

f_L is the lower frequency limit.
f_U is the upper frequency limit.
n is the number of frequencies in the range.
$P_i(f)$ is the parameter to be optimized.
$G_i(f)$ is the corresponding goal for $P_i(f)$ at each frequency.
W_i is the weight.

One problem with the error formulation in (3.40a) is the tendency to average individual error contributions over the entire analysis range. This characteristic sometimes results in unacceptable results after optimization. For example, broadband amplifiers often exhibit a tendency to roll off short of the upper end, in addition to the low-end gain peak mentioned earlier. The gain peak and rolloff may occur over only 10% to 15% of the analysis frequency range and not contribute substantially to the total error. We can add more analysis frequencies in these regions or specify three error ranges and increase the weights at each band edge. However, we can also change the error function:

$$E = \frac{1}{n}\sum_{f=f_l}^{f_u}\left\{W_1\cdot(P_1(f) - G_1(f))^p + W_2\cdot(P_2(f) - G_2(f))^p + \cdots\right\} \tag{3.40b}$$

where p is an even integer. This type of error function is known as a *least* p*th approximation* [18,19,26] and reduces to (3.40a) when $p = 2$. Assuming the difference between the computed parameter and goal is greater than one (we can always add 1), the largest error contributes more as we increase the value of p. Finally, as p approaches infinity, the maximum error almost completely dominates the total error, and we have a *minimax approximation* [16,27]. Minimax optimization minimizes the maximum error and can often be used to provide equal-ripple response for amplifiers and filters.

3.3.2 Optimization Specifications

Optimization may be the single most powerful design tool we have at our disposal, with the possible exception of the simulator itself. Success or failure often depends on how we specify the problem to the optimizer and on the strategy used to reach the final design objective. What constitutes success or failure? In the case of RF or microwave circuit design, we can define success as a network that meets all specifications, contains realizable elements, is reproducible, and is cost-effective. That is, we can produce competitive hardware useful for a specific application. The most important factor governing the success or failure is the underlying design approach. Ideally, we want to make the most efficient use of expensive active devices and design simple circuits that are relatively insensitive to fabrication tolerances and are not labor-intensive. Once we reach the point where optimization is necessary, the basic topology is well defined and the optimizer certainly cannot find a minimum if one does not exist. In other words, the design must be *feasible*; that is, a solution exists for at least one set of element values.

As circuit designers using CAD, we have control over three aspects of optimization. We can define which elements are to be optimized and their limits or *constraints*, the optimization goals and weights, and the method of optimization used to search for the minimum. The error function, which is defined by the goals, weights, and elements, is a surface in $n + 1$ dimensional space, where n is the number of elements. The error shown in Figure 3.31 is a function of one variable (R_p) and is two-dimensional. We can carefully select the optimization parameters, resulting in an error surface which is well behaved, or we can simply use the *brute force* method and optimize all elements using a number of design goals. Convergence is directly related to the error surface, and the relationship between the error function and optimization parameters can best be illustrated through an example.

We designed a 12.5-dB gain amplifier in Section 2.3.9 by defining a source impedance on the 15-dB constant gain circle and then mismatching the output by 2.5 dB, as shown in Figure 2.26. We can increase the gain to 15 dB by conjugately matching the output. Assuming the operating frequency is 2.0 GHz, the topology and element values are shown in Figure 3.33 for the 15-dB gain amplifier. Analysis shows that the gain is 15 dB and the magnitude of s_{22} must equal zero. However, the network is ideal, and it is necessary to add loss, parasitics, and a bias network. We will restrict our attention to the two output elements, L_2 and C_2, so we can depict the error function as a three-dimensional surface. We also assume both elements are lossless, which provides an error function for the ideal circuit shown in the figure.

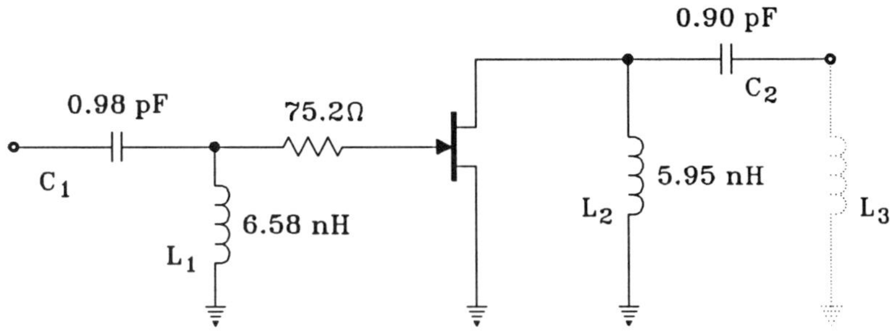

Figure 3.33 2.0-GHz, 15.0-dB gain amplifier.

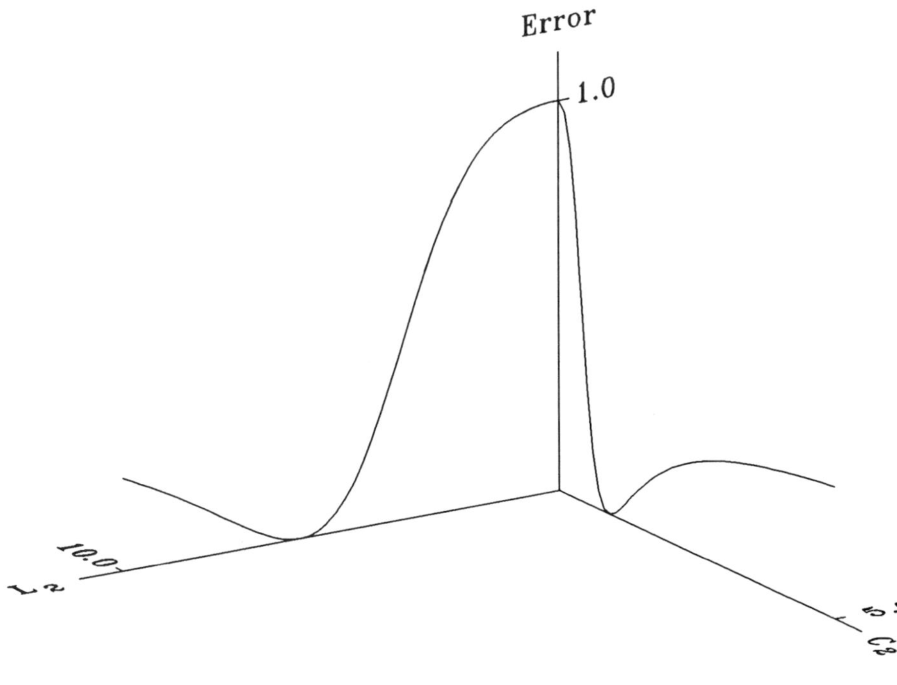

Figure 3.34 Error function versus L_2 and C_2.

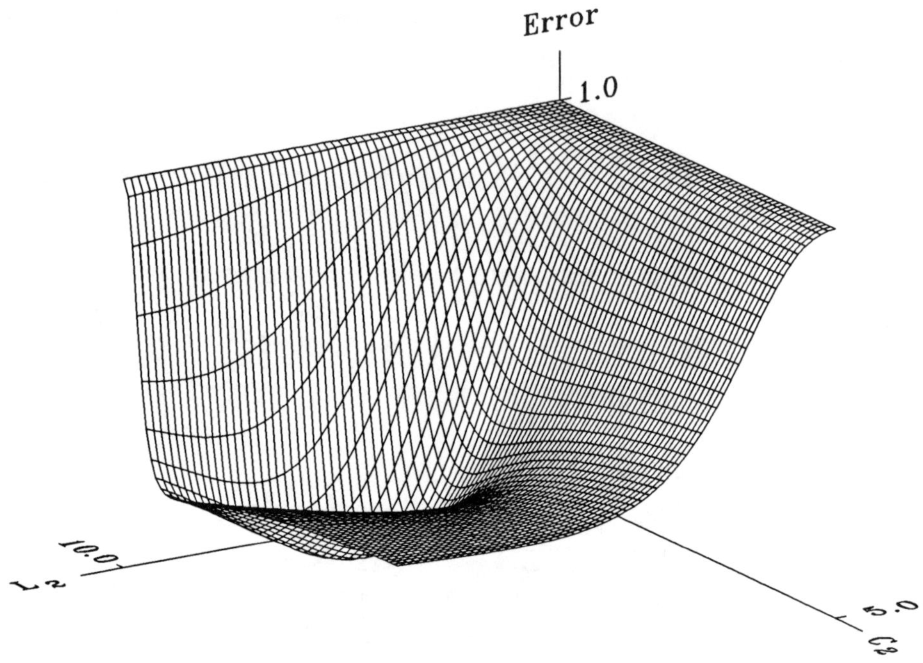

Figure 3.35 Error function for goal equal output match.

The only design requirement is that the amplifier provide 15-dB gain at 2.0 GHz; however, we can define the optimization goal in several different ways. The input impedance was selected from the 15-dB gain circle, which means that the output must be matched, or the magnitude of the output reflection coefficient must equal zero. We can express this optimization goal as

$$|S_{22}| = 0$$

and the error function as

$$E = \left(|S_{22}| - 0.0\right)^2$$

assuming a weight of one. We can examine the two-dimensional error for L_2 and C_2 independently by varying one element at a time, as shown in Figure 3.34. In the figure, the error function for each element is well behaved and exhibits one minimum at the optimum value shown in Figure 3.33. Examining the load impedance shown in Figure 2.25, we see that a matching network consisting of L_2 and C_2 has only one unique solution similar to the example shown in Figure 3.4.

However, when we specify both output elements as optimization variables, the error surface becomes three-dimensional, as shown in Figure 3.35. The drawing shows the error as a function of discrete L_2 and C_2 values. The error increases from zero at the L_2-C_2 plane to 1.0 in a vertical direction along the z-axis. L_2 increases from zero at the intersection with the z-axis to 10 nH, as shown, and C_2 ranges from 0 to 5 pF. Lines shown on the error surface represent constant element values and are equally spaced between the minimum and maximum for each element. If we look down from the positive z direction, the L_2-C_2 plane would appear to be divided into a number of squares. As shown, the error is 1.0 for small element values and decreases rapidly as we increase the element values. The error function is affected to a greater degree by capacitance, as illustrated by the separation between lines of constant capacitance.

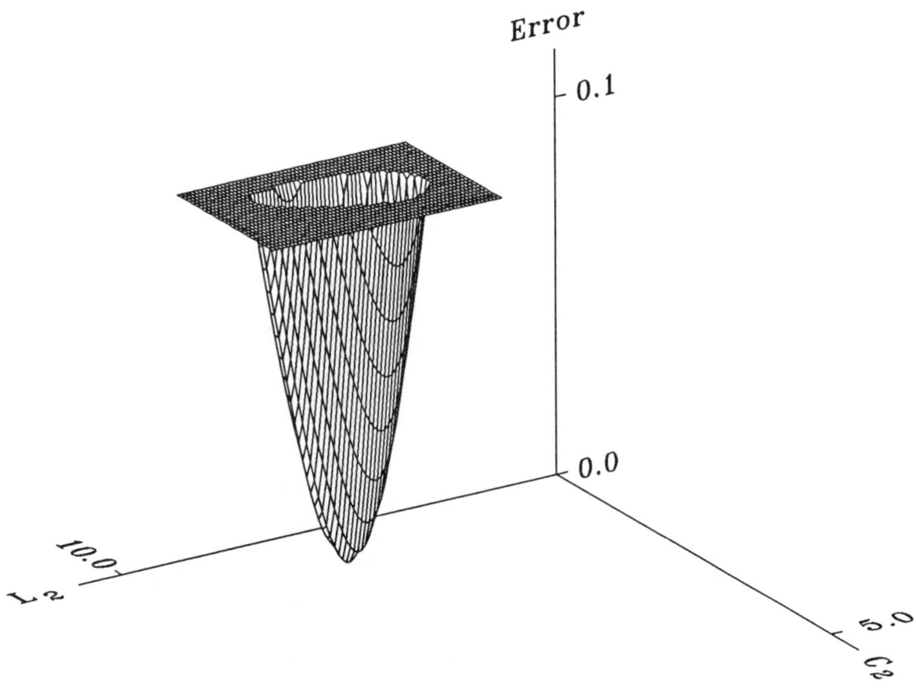

Figure 3.36 Expanded error for goal equal output match.

The minimum in Figure 3.35 appears to be broad and poorly defined. However, when we expand the scale to display errors between 0.0 and 0.1, corresponding to a magnitude between 0.0 and 0.3 for s_{22}, as illustrated in Figure 3.36, the minimum is in fact well defined. This problem should present little difficulty for the optimizer. We could start anywhere on the error surface, search for a minimum using either element, switch to the other element, and locate the minimum. Then, after a few iterations alternating between L_2 and C_2, we would rapidly arrive at the minimum shown in Figure 3.36. Examining Figure 3.35, we see that the initial element values have an effect on convergence. For example, if we start with very small values, the error function is almost flat and may require a number of iterations to find the minimum. Similarly, the error function is nearly flat for large element values and will exhibit slow convergence during the early iterations. However, if we start somewhere around midscale for both elements, the optimizer should converge very rapidly to the minimum. The initial values often determines the success or failure of an optimization. When we take time to provide reasonable starting values based on good design techniques, the chance of success is very good. On the other hand, if we simply specify random element values, the optimizer may fail to find a solution, or, if it does find one, it may not be the best. In this case, if we start with a capacitance of 1000 pF or an inductance of 1000 nH, most optimizers will fail because the error function appears to be flat due to the limited precision of the mathematical calculations.

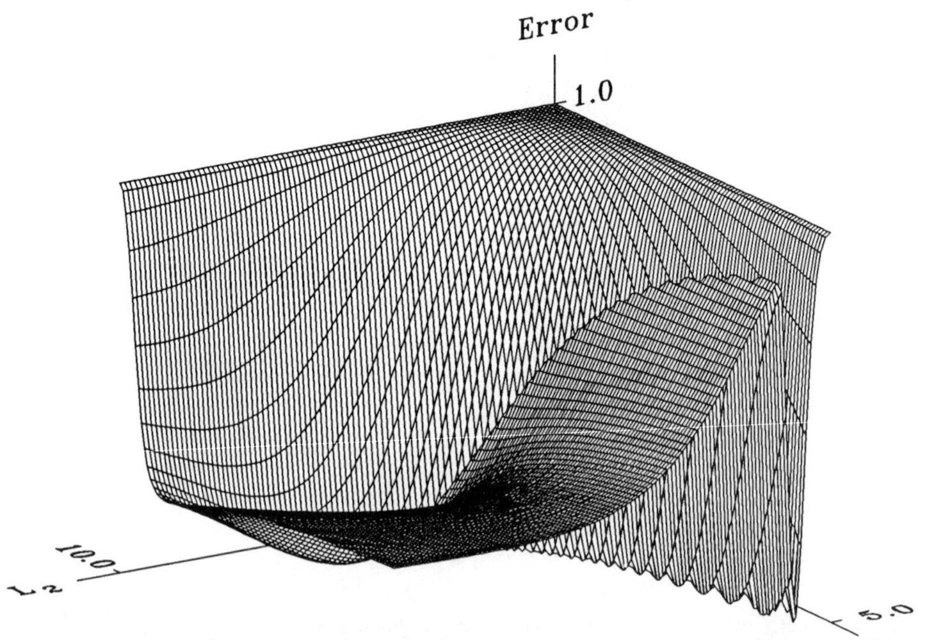

Figure 3.37 Error function for three output elements.

The error surface shown in Figures 3.35 and 3.36 is simple and well behaved. However, as we add more optimization variables, the complexity increases and the problem becomes more difficult. A minimum of two elements is required to provide the load impedance shown in Figure 2.25. However, we can also design the matching network using three elements as shown by L_2, C_2, and L_3 in Figure 3.33. The error surface is four-dimensional and cannot easily be represented by a two-dimensional drawing. Instead, we can plot L_2 and C_2 versus error for optimum values of L_3, as shown in Figure 3.37. In other words, for each set of L_2 and C_2 values, the error shown in the figure is the lowest that can be found by adjusting L_3 to any positive value. Comparing Figure 3.37 with 3.35, we see that the error surface for the three-element matching network is indeed more complex than that of the two-element network. We also observe that the original two-element minimum appears in Figure 3.37. This point corresponds to infinite inductance for L_3 or zero shunt admittance at the output. The figure shows that as L_2 decreases below 5.95 nH, C_2 increases above 0.90 pF, and we find that the minimum is a narrow valley rather than a single point. In this case, the optimizer

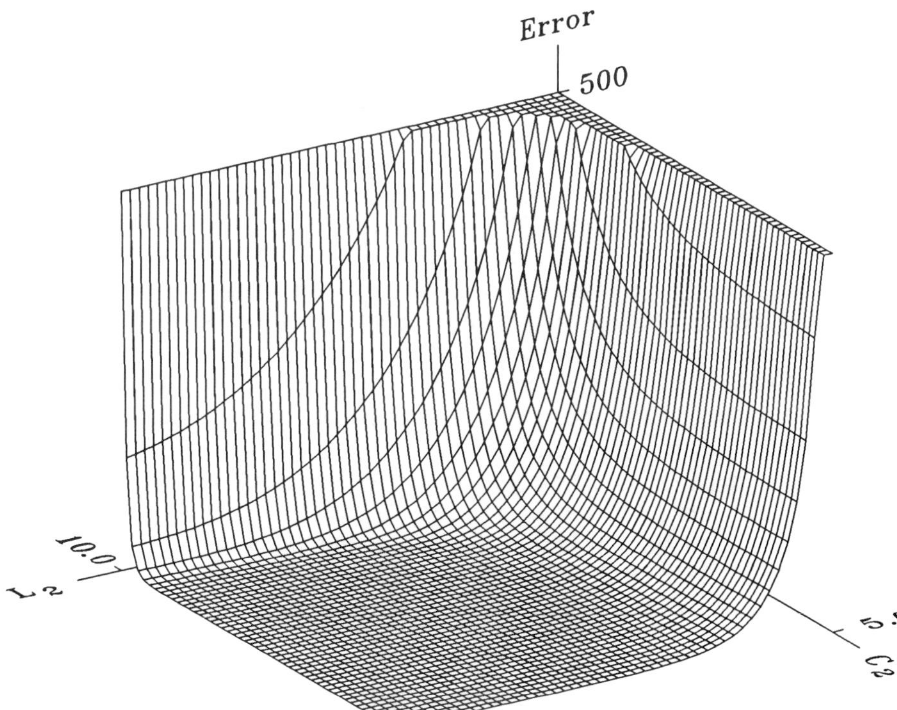

Figure 3.38 Error function for goal equal 15-dB gain.

will converge to a solution determined by the initial starting point, since there are infinite combinations of L_2, C_2, and L_3 that satisfy the specifications. As shown, the best solution is the two-element matching network, where the minimum exhibits the least sensitivity to component changes. One can only imagine what the error surface must look like when we specify matching networks using the trial and error approach and then optimize all components at once.

The obvious choice of an optimization goal for this design is gain,

$$20\text{Log}|s_{21}| = 15$$

where the error function is defined by

$$E = \left(20\text{Log}|s_{21}| - 15\right)^2$$

and plotted in Figure 3.38 for values between 0 and 500, approximately 22-dB gain error. As shown, the error surface is similar to that of Figure 3.35. In fact, the optimization goals are all equivalent; that is, 15-dB gain is only possible when the output is matched. Figure 3.39 shows the surface for errors less than or equal to 5.0, and, since the surface is smooth and monotonic, the optimizer will have no difficulty converging to the minimum shown in the figure. However, the surface appears to be somewhat flat in the vicinity of the optimum, which may slow the convergence somewhat. We have defined the error as the deviation from maximum gain as a function of element values in the output matching network, or mismatch loss (squared) as a function of load impedance. Comparing the output plane mismatch circles shown in Figure 2.25 with Figures 3.38 and 3.39, we can see the correlation. The mismatch circles are compressed toward the outside of the Smith chart and expanded in the direction of the origin. This results in an error surface with a steep slope in one region and a gradual slope in another. The mismatch circles also tend to expand as we approach the maximum gain, which is reflected in Figure 3.39 as a broad minimum.

We can also define the optimization goal in terms of output VSWR or impedance. The optimization problem can be reduced to its simplest form when we select impedance. The conjugately matched output reflection coefficient was calculated in Section 2.3.9 as $\Gamma_L = 0.6239\underline{/28.7698^\circ}$, or $z_L = 2.0670 + j2.0325$ normalized to 50Ω. We can model the output matching network of L_2 and C_2, shown in Figure 3.33, and then optimize for the required input impedance. The input matching network and active device parameters are not necessary. The optimization goal is

$$z_L = 2.0670 + j2.0325$$

and the error function is

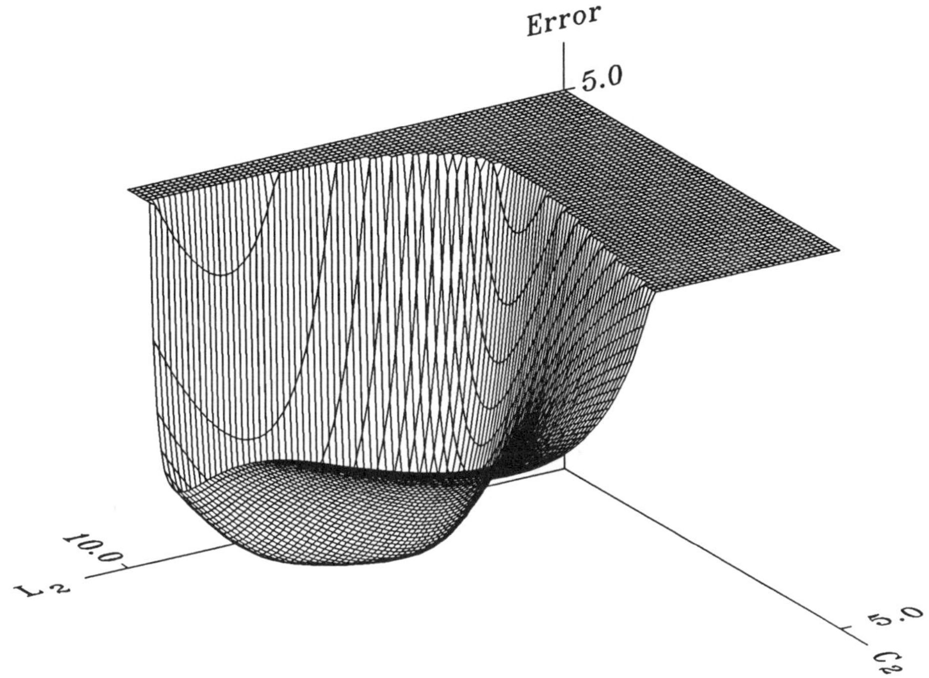

Figure 3.39 Expanded error for goal equal 15-dB gain.

$$E = \left[\left|z_{in} - (2.0670 + j2.0325)\right|\right]^2$$

Figure 3.40 shows the surface for this error function, which has a well-defined minimum. Impedance is the best choice for this problem because the circuit model is very simple and the optimizer will converge rapidly. However, there are other factors that must be considered when selecting optimization goals.

We defined four optimization goals which are all equivalent for determining the element values of the output matching network. Although we might see some minor differences in the convergence, all four will converge to the same minimum. The design meets the 15-dB gain specification. However, it does not accurately represent the physical circuit. The lumped elements exhibit loss and internal parasitics, and we must include other parasitics resulting from the layout in the complete model. For example, packaged devices have inductance associated with the leads, and bonding pads used to connect elements on a printed circuit board cause shunt capacitance. The best approach when increasing the complexity of the circuit model (and the error surface) is to change only a few elements at a time, then reoptimize. This strategy ensures that we do not stray too far from the minimum. Using the amplifier in Figure 3.33 as an example, we might add loss and parasitics to L_2 and C_2 using equivalent *R-L-C* circuits provided by the manufacturer or derived from measurements, and then optimize only the two elements for an output match. Next, we bypass L_2 to ground with a capacitor, add the equivalent circuit for the drain bias network, and then optimize L_2 and C_2 again. Finally, we model the FET drain lead, bonding pads, and perhaps a short wire from the output of C_2 to a connector, and then optimize L_2 and C_2 again. At this point we have an accurate model of the drain matching network, and the output is matched, although the gain is somewhat less than 15 dB because of component loss. We can follow the same procedure for the input matching network, optimizing only L_1 and C_1 for an optimization goal of 15-dB gain. Once the model is complete, we optimize all four elements for simultaneous goals of 15-dB gain and an output match.

We must be careful not to specify conflicting, redundant, or impossible goals, which tend to unduly complicate the error surface. Specifying gain and output match for the ideal amplifier shown in Figure 3.33 is redundant since we can only achieve 15-dB gain when the output is matched. However, once we add loss, both specifications are necessary because an output match alone results in

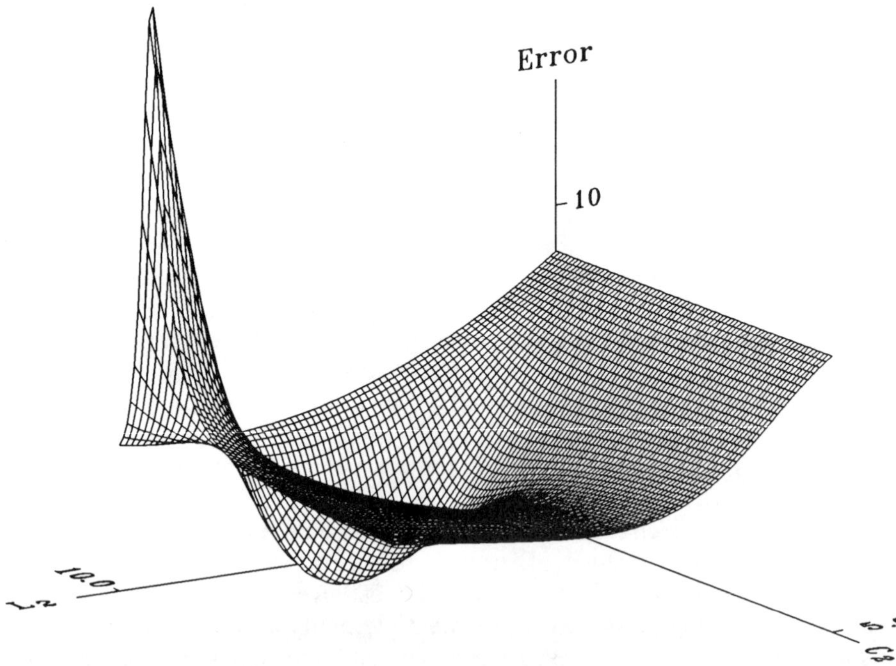

Figure 3.40 Error function for goal equal Z_L.

something less than 15-dB gain. In this case, the gain specification is impossible if we only optimize the output network. The input plane gain must be increased to compensate for loss in the output matching network. We can examine the effects of conflicting goals using the original amplifier design from Section 2.3.9. The input impedance was selected from the 15-dB gain contour and the output impedance from the 2.5-dB mismatch circle, resulting in 12.5-dB gain. Both input and output are mismatched; however, we ignore this fact and define the optimization goals as

$$20\text{Log}|s_{21}| = 12.5$$

$$|s_{11}| = 0$$

and

$$|s_{22}| = 0$$

which results in an error function of

$$E = \left(20\text{Log}|s_{21}| - 12.5\right)^2 + \left(|s_{11}| - 0.0\right)^2 + \left(|s_{22}| - 0.0\right)^2$$

where equal weights of 1.0 have been used. That portion of the error surface below 1.0 is shown in Figure 3.41 as a function of L_2 and C_2. The drawing shows two separate regions where the error decreases, one at C_2 = 10 pF and another at L_2 = 15 nH. These regions are called *local minima* to denote the fact that they are minimum points only in a restricted area of the surface. The *global minimum* is the lowest point on the entire error surface and is the optimum solution. Once the optimizer finds a local minimum, it is trapped. Usually we do not know if a minimum is local or global. However, we can start the optimizer from a number of different points and develop some confidence for a global minimum if the convergence is always to the same set of element values.

The optimization goals for the 12.5-dB gain amplifier conflict with one another. We cannot expect to optimize only the output network and simultaneously achieve the desired gain and a match at both ports. However, when we take time to design the circuit properly, the results are quite different. In this case, we optimize the gate resistor for G_{MAX} = 12.5 dB, compute Γ_{MS} and Γ_{ML}, and then design the input and output matching networks. Selection of the optimization variables should be based on the goals as well as on their effect on the circuit performance. Using the error function defined above for the 12.5-dB gain amplifier, we must simultaneously optimize L_1, C_1, L_2, C_2, and the gate resistor to ensure that all three error components can be minimized. When the circuit is based on a good preliminary design and the optimization goals and variables are

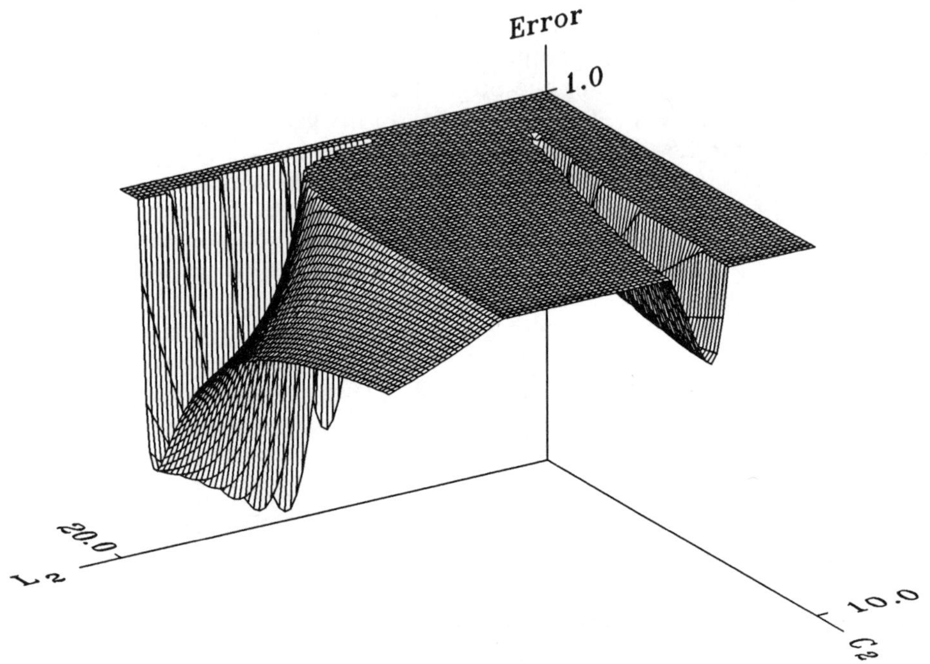

Figure 3.41 Goal equal 12.5-dB gain, input, and output match.

carefully defined, the optimizer is usually successful. However, when convergence is slow or the results are not as expected, we have the option of trying other optimization methods.

3.3.3 Direct Search Optimization Methods

Numerous optimization strategies are available that can generally be classified as either a *direct* or *indirect* method, depending on how the error function is used to locate the minimum. Direct methods use the performance measure as a function of element values to minimize the function, and indirect methods use the differential properties of the error function. Some techniques employ both approaches, and it is not always possible, or necessary, to categorize a particular method.

One of the simplest direct search techniques is the *grid search*. We simply define a minimum and maximum search limit for all elements, divide each interval into a number of segments, and then evaluate the error function for all possible combinations of discrete element values. Figures 3.35 through 3.41 were, in fact, generated using a grid search. This method can often locate the global minimum, at least within the defined boundaries, but is very inefficient. Except for the minimum, previous function evaluations are discarded. In addition, the grid search can require an astronomical number of calculations. Two elements divided into 100 segments requires 10,000 function evaluations, and five elements requires 10,000 *million* evaluations! We can always reduce the number of segments, although the probability of locating the minimum is diminished.

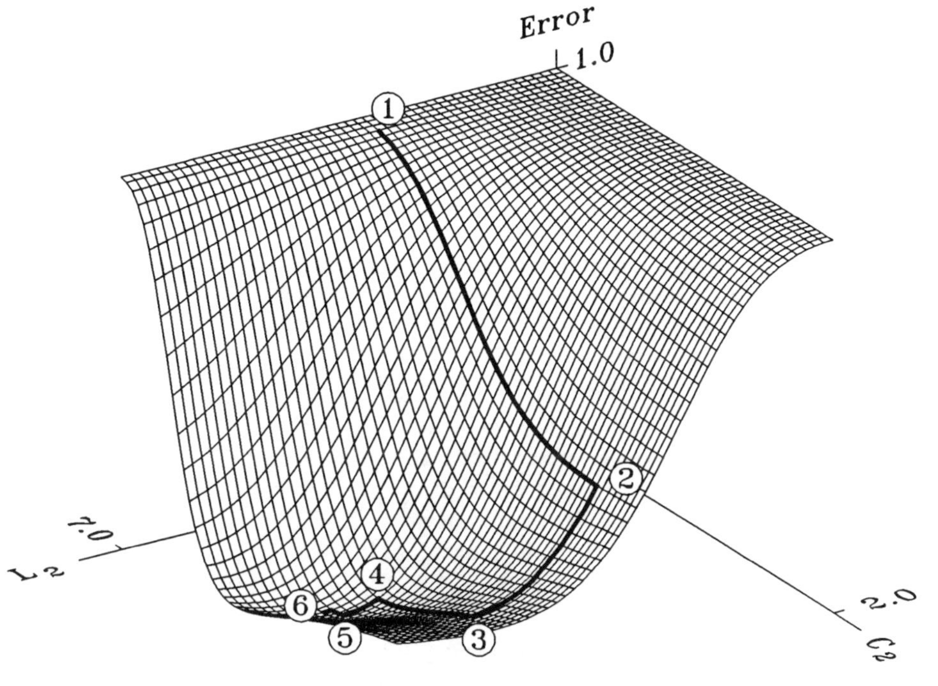

Figure 3.42 Univariate search.

We can improve the grid search by using a more efficient means of locating the minimum in each interval. For example, quadratic interpolation, discussed in Section 3.3.1, converges rapidly and is not a complicated procedure. We can adopt a strategy which is the mathematical equivalent of the technique we often use on the bench to align a circuit. That is, we adjust each element in turn for a minimum. This method is known as a *univariate search* [22] and is illustrated in Figure 3.42. Starting at point 1, we locate the minimum by changing only the value of C_2, which takes us to point 2. Then, from point 2 we minimize the error function by searching in the L_2 direction. Point 3 represents one complete cycle through the optimization variables (2 in this case), and we simply repeat the process as many times as it takes to locate the minimum. The search is usually terminated when a specified error function value is reached, the improvement from one iteration to the next is insignificant, or when a designated number of iterations are complete.

The univariate search procedure makes some use of past function evaluations to predict the location of a new minimum. The search along each element axis is accomplished by interpolation rather than by simply dividing the interval into a number of discrete points. However, we can significantly improve convergence by using previous calculations to predict a search direction in n-dimensional space. Examining points 2 through 6 in Figure 3.42, we can see a pattern emerging. A line drawn from point 2 and passing between points 3, 4, and 5 comes very close to the optimum at point 6. We can take advantage of this pattern for the n-dimensional optimization problem by conducting n univariate searches and then searching in the direction of a vector connecting the first and last points. In other words, after all variables have been optimized individually, we assume the true minimum is somewhere in the direction of a line connecting the starting point with the current minimum point. Figure 3.43 shows an example of this technique, which is known as a *pattern search*. Starting at point 1, a univariant search in C_2 then L_2 defines point 3, which is one complete cycle through all optimization variables. We define the pattern direction as the line connecting point 1 and point 3, which is shown as a broken line in the figure. Searching in the pattern direction, we arrive at point 4, where the univariant search is restarted. Point 5 is the minimum for C_2 and point 6 the minimum for L_2. The new pattern search direction is along the line connecting points 4 and 6, again shown as a broken line, and the optimum is located at point 7. In this case the pattern search was not as efficient as the univariant search, mainly due to the poor starting point. However, the pattern search is usually more efficient than the univariant method.

The pattern search method illustrated in Figure 3.43 was published by Powell in 1964 [24]. A similar method was published in 1961 by Hooke and Jeeves [28]. The Hooke and Jeeves method uses an exploratory move about each base point to determine the pattern direction, rather than a univariant search. Figure 3.44 shows the Hooke and Jeeves search for the same error surface illustrated in Figure 3.43. Starting at point 1, we evaluate the error function, increase L_2 by a predetermined increment (four grid squares or about 8% in this case), and evaluate the error function again. The error at point 2 is lower, so we retain this point as a temporary base, increase C_2, and evaluate the error. The error at point 3 is less than that at point 2, which provides another temporary base. When an exploratory move fails to find a lower error, we try the opposite direction; that is, we subtract the incremental element value from L_2 or C_2. Should both directions fail, the increment is reduced and the exploratory move repeated. Since we have completed an exploratory move for each variable, the pattern search proceeds in a direction defined by a vector from point 1 and passing through point 3.

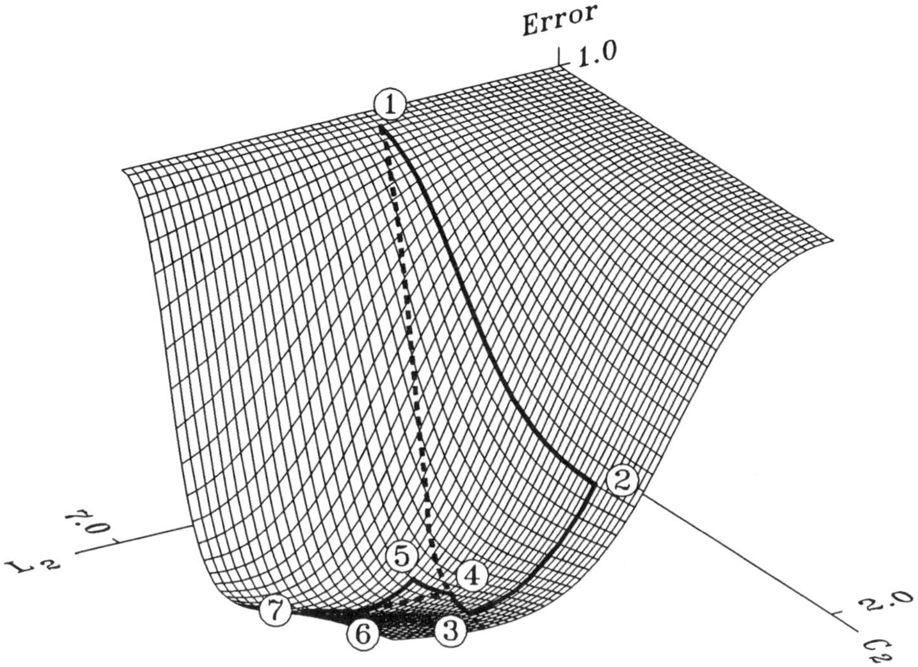

Figure 3.43 General pattern search method.

The minimum for the first pattern search in Figure 3.44 is shown as point 4, which is very close to the optimum for this error surface. The seemingly outstanding convergence properties of this method is strictly due to the starting point and incremental element values selected for this example. However, it does illustrate the point that various optimization strategies can converge quite differently on a particular error surface, even from the same starting point. In fact, if we change the incremental element values for the initial exploratory move, convergence may be altered significantly. For example, if we use 0.1 nH and 0.1 pF as incremental element values to explore about point 1, lower errors are found by increasing L_2 and C_2 as above, but the pattern direction is different, as shown by the broken line in Figure 3.44. Point 5 is the minimum in this case, which will require several more pattern searches before the optimum is located at point 4.

Central to the pattern search, and other methods as well, is the concept of searching for a minimum along a direction vector in n-dimensional space [17]. If $\mathbf{x}_a$ and $\mathbf{x}_b$ are two points in n-dimensional space, then

$$\boldsymbol{x}(\alpha) = (1 - \alpha)\boldsymbol{x}_a + \alpha \boldsymbol{x}_b \tag{3.41a}$$

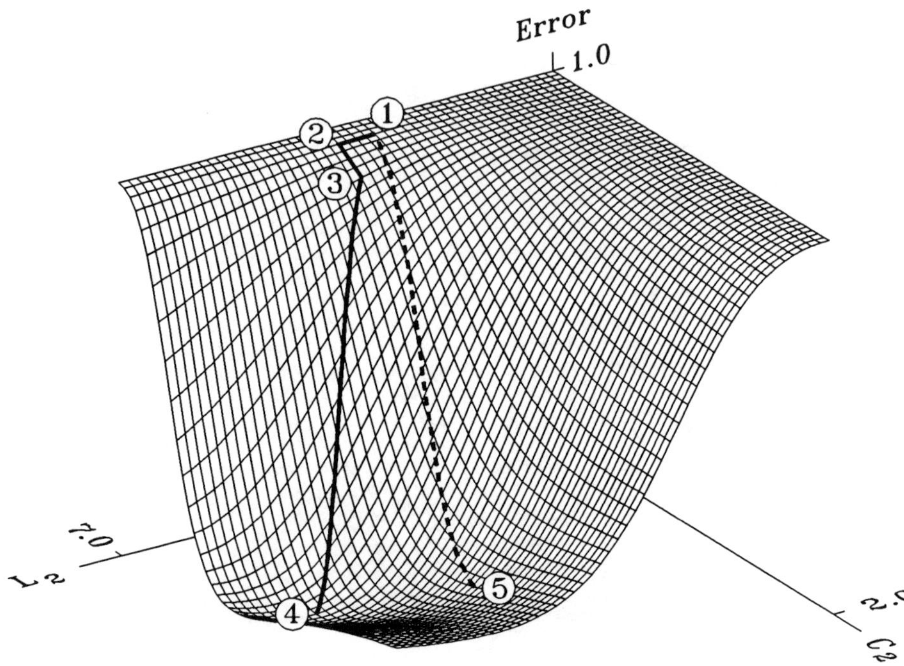

Figure 3.44 Hooke and Jeeves pattern search.

defines a straight line passing through $\mathbf{x}_a$ and $\mathbf{x}_b$. When $\alpha = 0$, $\mathbf{x}(\alpha) = \mathbf{x}_a$, and when $\alpha = 1$, $\mathbf{x}(\alpha) = \mathbf{x}_b$. If instead of two points we have one point, $\mathbf{x}_a$, and a direction vector, $\mathbf{r}$, then (3.41a) can be written as

$$x(\alpha) = x_a + \alpha r \tag{3.41b}$$

where

$$r = \begin{bmatrix} r_1 & r_2 & r_3 & \cdots & r_n \end{bmatrix}^T$$

Equation (3.41b) is identical to (3.41a) when we derive the direction vector by subtracting the two points; that is, $\mathbf{r} = \mathbf{x}_b - \mathbf{x}_a$. We can use (3.41b) to rewrite the expression for the basic error function, given in (3.37) as

$$E = f(x_a + \alpha r) \tag{3.42}$$

and we see that minimizing the error function along a direction vector reduces to a *one-dimensional search* for the optimum value of α. This type of search is known as a *line search*, and any of the methods discussed in Section 3.3.1 can be used to locate the minimum.

Table 3.6 Line Search

α	L_2	C_2	*Error*
0	3.14286	0.12245	0.99398
1	3.71429	0.28571	0.93893
2	4.28572	0.44898	0.74276
3	4.85715	0.61225	0.36223
4	5.42857	0.77551	0.06124
5	6.00000	0.93878	0.00167
6	6.57143	1.10205	0.05074

The error surface in Figure 3.44 shows 50 equally spaced values for each element. Thus, the increment for L_2 is (7/49) or 0.14286 nH, and (2/49) or 0.04082 pF for C_2. The coordinates for point 1 are L_2 = 3.14286 and C_2 = 0.12245; units are omitted for brevity. We added four increments to each element which defined point 3 and the search direction. The direction vector is the difference between the two points or four increments, $\mathbf{r} = [(4)(0.14286)\ \ (4)(0.04082)]^T$ and we can write (3.41b) as

$$x(\alpha) = x_a + \alpha r = \begin{bmatrix} 3.14286 \\ \\ 0.12245 \end{bmatrix} + \alpha \begin{bmatrix} 0.57143 \\ \\ 0.16327 \end{bmatrix}$$

The error at point 1, α = 0, is 0.99398, and at point 3, α = 1, 0.93893. Table 3.6 lists values of α in increments of 1 from 0 to 6, along with the corresponding element and error function values. The last three points meet the criterion for applying quadratic interpolation; that is, the error at the interior point is less than at either end point. Substituting into (3.39c), we compute α_m=5.04832 and 0.00301 for the error. After several iterations to reduce the uncertainty, we find α_m = 4.75000, L_2 = 5.85714, C_2 = 0.89796, and an error of 0.00044.

3.3.4 Indirect Search Optimization Methods

Numerous other direct search methods can be found in the literature [17,19,22,23]. However, most are not as efficient as indirect methods once in the vicinity of the optimum. Among the most powerful indirect methods are those based on the *gradient*:

$$\nabla f = g = \left[\frac{\partial f}{\partial x_1} \;\; \frac{\partial f}{\partial x_2} \;\; \frac{\partial f}{\partial x_3} \;\; \cdots \;\; \frac{\partial f}{\partial x_n} \right]^T \tag{3.43}$$

which is a direction vector in n-dimensional space pointing in the direction of maximum function value increase [17]. The gradient provides a measure of the sensitivity of the error function to changes in element values, and can be used for *sensitivity analysis* [23]. When the partial is zero, the error function is unaffected by incremental changes in element values; larger magnitudes indicate increasing sensitivity. Each gradient component can be normalized to the corresponding element value, which provides relative sensitivities. Intuitively, it seems that this approach would be more efficient than the methods considered in Section 3.3.3. We simply move in the negative gradient direction where the error function decreases at the fastest rate. This technique is known as the *steepest descent (ascent)* method and was introduced by Cauchy [29] in 1847 for solving simultaneous equations.

As a practical matter, the gradient is usually computed using numerical techniques [30]; that is,

$$\frac{\partial f}{\partial x_i} = \lim_{h \to 0} \frac{f(x_i + h) - f(x_i)}{h} \tag{3.44}$$

because closed form expressions are not available. Equation (3.44) can be a significant source of error, often resulting in convergence problems. The numerical gradient is only accurate when h is small; however, a small increment results in the subtraction of two almost equal numbers in the numerator divided by another small number in the denominator. Finite precision computer arithmetic limits the smallest value of h that can be used reliably, even using double precision calculations. Thus, the gradient may exhibit large errors when the error surface is not well behaved. We can improve the accuracy somewhat by using the forward and backward difference, $f(x + h)$ and $f(x - h)$. However, this requires one additional analysis for each variable and, in most cases, does not improve the convergence enough to justify its use. The adjoint method [16,31-34] discussed in Section 2.2.2 overcomes these problems. We can compute the complete gradient vector directly from the circuit voltage and current values resulting from two analyses: one for the original network and a second for the adjoint network. The gradient is not dependent on incremental values, and the number of analyses is independent of the number of variables.

We can use the method of steepest descent to solve the optimization problem shown in Figure 3.45. The starting point is the same as the previous example: L_2 = 3.14286 nH and C_2 = 0.12245 pF, which is labeled as point 1. Grid spacing is also the same: ΔL = 0.14286 nH and ΔC = 0.04082 pF. The first step is to evaluate the gradient at the starting point by applying (3.44) for each variable. Typically, h is some small percentage of the element value, which is determined by the precision of the arithmetic algorithms. In this case, we will let ΔL and ΔC equal one grid division for one set of calculations and 0.1% of the element value for another set of calculations. Table 3.7 shows the error function value at the initial point $f(x)$, positive increment $f(x + h)$, and negative increment $f(x - h)$.

Table 3.7 Error Function Evaluation

Point	*Error* Δ = *Grid*	*Error* Δ = *0.1%*
$f(x)$	0.99398	0.99398
$f(x + \Delta L)$	0.99316	0.99397
$f(x - \Delta L)$	0.99455	0.99400
$f(x + \Delta C)$	0.98869	0.99397
$f(x - \Delta C)$	0.99738	0.99399

Using one grid division as the increment and substituting values from Table 3.7 into equation (3.44), we have

$$\frac{\partial E}{\partial L_2} = \frac{0.99316 - 0.99398}{0.14286} = -0.00574$$

for the L_2 component,

$$\frac{\partial E}{\partial C_2} = \frac{0.98869 - 0.99398}{0.04082} = -0.12961$$

for the C_2 component, and the gradient is

$$g = [-0.00574 \quad -0.12961]^T$$

Table 3.8 shows the gradients resulting from both sets of incremental element values using three different numerical techniques. The first line shows the forward difference, the second line the backward difference, and the third line shows a combination of the forward and backward difference. We see that there is not much difference when we use an increment equal to 0.1% of the element value. However, there is a significant difference when we use one grid increment which turns out to be approximately 4.5% of L_2 and 33.33% of C_2 at point 1. The forward-backward combination is the most accurate in this instance, although two additional function evaluations are required.

Table 3.8 Gradient Calculations

Gradient Method	*Gradient Vector* Δ = *Grid*	*Gradient Vector* Δ = *0.1%*
$\dfrac{f(x+h) - f(x)}{h}$	$[-0.00574 \quad -0.12961]^T$	$[-0.00478 \quad -0.10482]^T$
$\dfrac{f(x) - f(x-h)}{h}$	$[-0.00399 \quad -0.08330]^T$	$[-0.00478 \quad -0.10471]^T$
$\dfrac{f(x+h) - f(x-h)}{2h}$	$[-0.00487 \quad -0.10645]^T$	$[-0.00478 \quad -0.10476]^T$

Figure 3.45 shows the gradient as a vector emanating from point 1. The magnitude of the gradient vectors shown in the figure are not drawn to scale; however, the directions are accurate. Starting from point 1, we conduct a line search in the negative gradient direction, which brings us to point 2. At point 2, we simply repeat the procedure: evaluate the gradient at the current element coordinates and then conduct a line search in the negative gradient direction. The figure shows that this procedure locates the minimum after four iterations.

Steepest descent is similar to the pattern search methods in that the one-dimensional search direction is determined by exploring about a base point. These methods are relatively inefficient and exhibit poor convergence as we approach the optimum point. Once a line search is complete, previous evaluations are discarded and the search is restarted. However, we can extend the quadratic interpolation discussed in Section 3.3.1 to an *n*-dimensional surface and then estimate the location of the minimum point. We can approximate the function with

$$f(\mathbf{x}) = \frac{1}{2}\mathbf{x}^T A \mathbf{x} + B^T \mathbf{x} + C \tag{3.45a}$$

which usually turns out to be a good estimate in the vicinity of the minimum. The gradient of the function evaluated at point $\mathbf{x}_a$ is

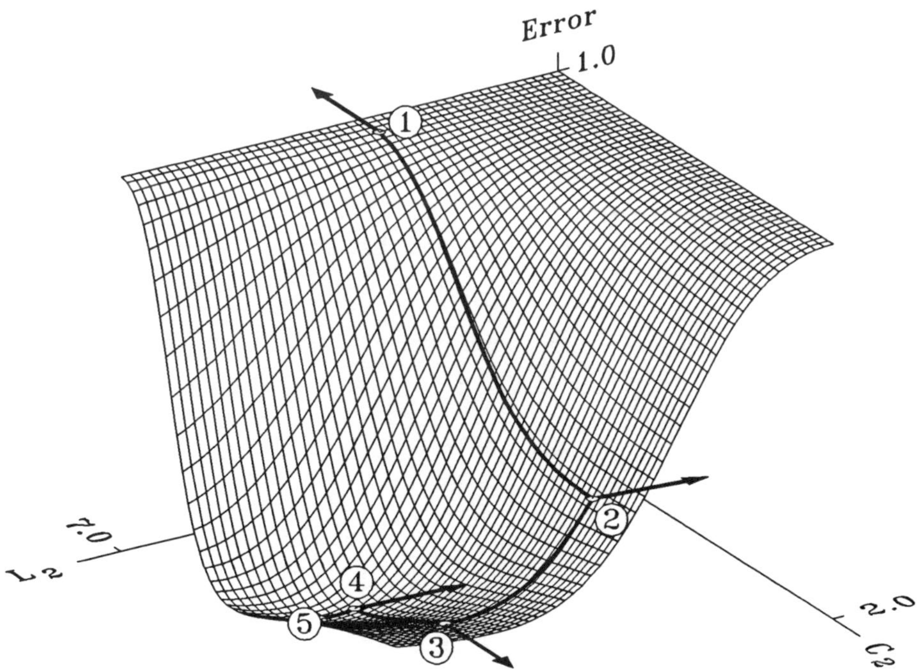

Figure 3.45 Gradient or steepest descent method.

$$\nabla f(x_a) = g_a = A x_a + B \tag{3.45b}$$

and at the minimum point $\mathbf{x}_m$ is

$$\nabla f(x_m) = g_m = 0 \tag{3.45c}$$

Thus,

$$x_m = x_a - A^{-1} g_a \tag{3.45d}$$

and, if (3.45a) is a good approximation, we can solve for the minimum in one iteration, provided we know $\mathbf{A}^{-1}$. Determining the matrix **A** from discrete function evaluations is often difficult. However, we can use certain mathematical properties to avoid this problem. When the n x n matrix **A** is positive definite, it can be shown [17,19,22] that two direction vectors, $\mathbf{r}_1$ and $\mathbf{r}_2$, are conjugate with respect to **A** when

$$r_1^T A\, r_2 = 0$$

Furthermore, when $\mathbf{r}_1$ and $\mathbf{r}_2$ are **A**-conjugate, they are linearly independent, and, at most, n one-dimensional searches are required to locate the minimum of (3.45a).

Fletcher and Reeves [35] published a simple iterative algorithm for generating **A**-conjugate search directions:

$$r_i = -g_i + \frac{g_i^T g_i}{g_{i-1}^T g_{i-1}} \cdot r_{i-1} \tag{3.46a}$$

where

$$r_0 = -g_0 \tag{3.46b}$$

Their procedure uses the gradient at the current base point, along with the previous gradient and direction vector, to determine a new search direction and is known as the *conjugate gradient method*. As an example, we can optimize the following quadratic function:

$$f(x,y) = \frac{1}{2}\begin{bmatrix} x & y \end{bmatrix}\begin{bmatrix} 3.0 & 0.01 \\ 0.01 & 1.5 \end{bmatrix}\begin{bmatrix} x \\ y \end{bmatrix} + \begin{bmatrix} -2 & -1 \end{bmatrix}\begin{bmatrix} x \\ y \end{bmatrix} + 1 \tag{3.47a}$$

which is written in the same form as (3.45a). In this case we can compute the gradient analytically. The first partial derivatives are

$$\frac{\partial f}{\partial x} = 3x + 0.01y - 2$$

$$\frac{\partial f}{\partial y} = \frac{3}{2}y + 0.01x - 1 \tag{3.47b}$$

and the gradient is

$$g(x,y) = [(3x + 0.01y - 2) \quad (3y/2 + 0.01x - 1)]^T \tag{3.47c}$$

Starting from $x = 0$ and $y = 0$, $f(0,0) = 1.000$,

$$g(0,0) = g_0 = [-2 \quad -1]^T$$

and from (3.46b),

$$r_0 = [2 \quad 1]^T$$

which is the steepest descent direction for the first iteration. Figure 3.46 shows the gradient vector at the starting point, which is labeled point 0. Searching in the negative gradient direction, $\mathbf{r}_0$, we find that $\alpha_m = 0.36928$, $f(\alpha_m) = 0.07681$, $x = 0.73855$, and $y = 0.36928$, which is shown as point 1 in the figure. Evaluating the gradient at point 1,

$$g_1 = [0.21935 \quad -0.43870]^T$$

We can use (3.46a) to compute the second search direction:

$$\mathbf{r}_1 = -\begin{bmatrix} 0.21935 \\ -0.43870 \end{bmatrix} + \frac{(0.21935)^2 + (0.43870)^2}{(2)^2 + (1)^2} \begin{bmatrix} 2 \\ 1 \end{bmatrix} = \begin{bmatrix} -0.12312 \\ 0.48681 \end{bmatrix}$$

Minimizing the function in the $\mathbf{r}_1$ direction, we arrive at the optimum, where x = 0.66446 and y = 0.66224, which is labeled point 2 in the figure.

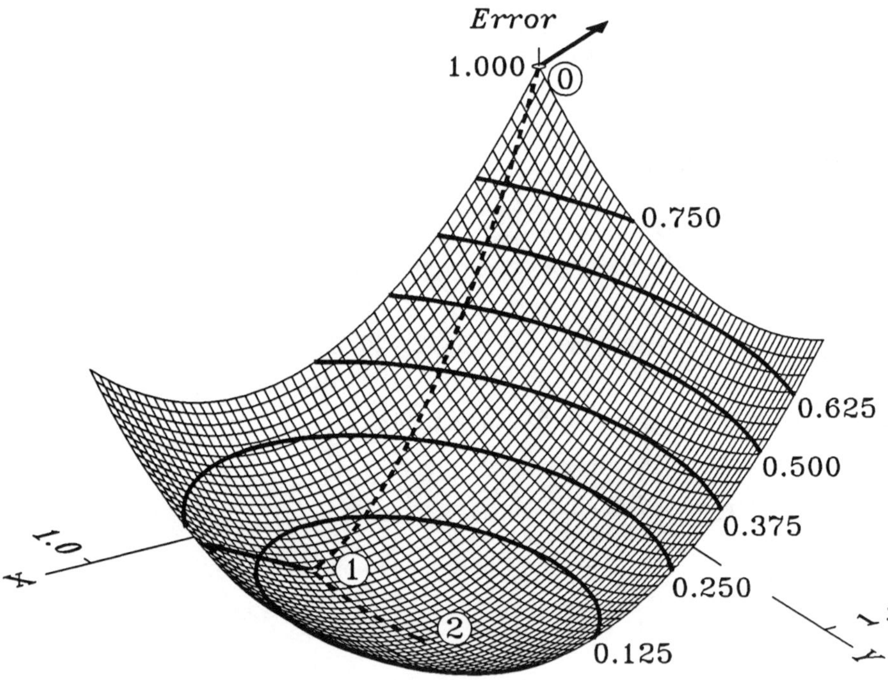

Figure 3.46 Conjugate gradient search.

Figure 3.47 shows the quadratic error surface from a viewpoint perpendicular to the *x-y* plane. Contours of constant function values illustrated and labeled in Figure 3.46 project onto the *x-y* plane as concentric ellipses, with a common center located at the optimum point. The figure shows that the gradient at point 1 is orthogonal to the search direction from point 0 to point 1. If we were to search in the negative gradient direction (steepest descent) as shown by the dashed-line projection in the figure, we would miss the optimum point, and would require additional iterations. However, by choosing the **A**-conjugate direction, the search proceeds directly from point 1 to the center, and the method converges to an exact solution in two iterations.

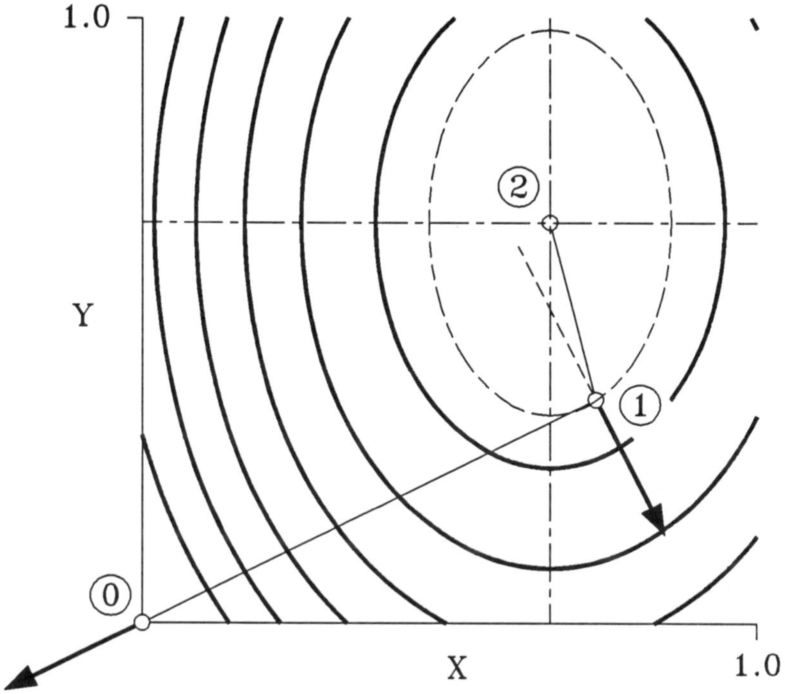

Figure 3.47 Quadratic error contours.

Comparing the function in (3.47a) with the general form of (3.45a), we see that matrix **A** is known and (3.45d) can be applied to solve for the minimum directly. Starting from point 0 where we have already computed the gradient,

$$x_m = \begin{bmatrix} 0 \\ 0 \end{bmatrix} - \begin{bmatrix} 3.0 & 0.01 \\ 0.01 & 1.5 \end{bmatrix}^{-1} \begin{bmatrix} -2 \\ -1 \end{bmatrix} = \begin{bmatrix} 0.66446 \\ 0.66224 \end{bmatrix}$$

Using a Taylor series expansion of (3.45a), it can be shown [17] that **A** is a matrix of second partial derivatives:

$$A = \begin{bmatrix} \dfrac{\partial^2 f}{\partial x_1 \partial x_1} & \dfrac{\partial^2 f}{\partial x_1 \partial x_2} & \cdots & \dfrac{\partial^2 f}{\partial x_1 \partial x_n} \\ \dfrac{\partial^2 f}{\partial x_2 \partial x_1} & \dfrac{\partial^2 f}{\partial x_2 \partial x_2} & \cdots & \dfrac{\partial^2 f}{\partial x_2 \partial x_n} \\ \vdots & \vdots & & \vdots \\ \dfrac{\partial^2 f}{\partial x_n \partial x_1} & \dfrac{\partial^2 f}{\partial x_n \partial x_2} & \cdots & \dfrac{\partial^2 f}{\partial x_n \partial x_n} \end{bmatrix} \tag{3.48}$$

which is known as a *Hessian matrix* and usually denoted by **H**. We can compute the Hessian for our quadratic function from the first partial derivatives given by (3.47b):

$$A = \begin{bmatrix} \dfrac{\partial}{\partial x}\left\{\dfrac{\partial f}{\partial x}\right\} & \dfrac{\partial}{\partial x}\left\{\dfrac{\partial f}{\partial y}\right\} \\ \dfrac{\partial}{\partial y}\left\{\dfrac{\partial f}{\partial x}\right\} & \dfrac{\partial}{\partial y}\left\{\dfrac{\partial f}{\partial y}\right\} \end{bmatrix} = \begin{bmatrix} 3.0 & 0.01 \\ 0.01 & 1.5 \end{bmatrix}$$

However, we usually only have discrete function values, and the matrix cannot be determined with sufficient accuracy. The errors associated with evaluation of the first partial derivatives are compounded when we attempt to compute second partial derivatives numerically. Davidon [36] developed a procedure for generating the $\mathbf{H}^{-1}$ matrix after n one-dimensional searches *without the use of second partial derivatives*. The method was later extended by Fletcher and Powell [37] and is known as the *DFP* or *Davidon-Fletcher-Powell optimization method*. All gradient methods are based on a common search procedure:

$$x(\alpha) = x_a - \alpha H g_a \tag{3.49}$$

where

$\mathbf{x}_a$ is the initial n element coordinates in column-vector form.
$\mathbf{g}_a$ is the gradient evaluated at $\mathbf{x}_a$, also in column-vector form.

H is an n x n matrix.

α is a real number determined such that $f[\mathbf{x}(\alpha)]$ is minimized.

The methods differ primarily in the selection of α and **H**. When **H** is the unit matrix, element values equal to 1 on the main diagonal and 0 elsewhere, (3.49) describes the steepest descent method.

The DFP algorithm belongs to a class of gradient methods known as *quasi-Newton* or *variable metric*, which seek to update **H** in (3.49) such that the inverse Hessian is determined after n iterations without using matrix inversion or second partial derivatives.

The distance between two points in n-dimensional space is defined as [22]

$$d^2 = (x_a - x_b)^T A (x_a - x_b) \tag{3.50}$$

where **A** is a positive definite matrix of order n called the *metric matrix*. Usually **A** is the unit matrix and (3.50) provides the distance between $\mathbf{x}_a$ and $\mathbf{x}_b$ in the original coordinate system. However, we can rescale the variables to more convenient values, which changes the metric matrix relative to the original system and introduces a new coordinate system. Variable metric methods have the desirable property that the variables are transformed or rescaled during the update sequence, which enhances numerical stability and convergence. The DFP method is generally considered the best gradient optimization technique and is available on most CAD systems. Compared with other gradient techniques, the DFP method is usually more robust and converges in fewer iterations.

3.3.5 Optimization Strategy

The most effective optimization strategy is to begin with a sound idealized design, formulate a realistic error function based on the specifications, and then gradually increase the complexity, as discussed in Section 3.3.2. This approach ensures that the optimizer is never very far from the minimum and will rapidly converge. However, we must also consider the physical aspects of the design, which requires that we place certain constraints on the element values. Microstrip or stripline construction limits the characteristic impedance of transmission lines to a relatively narrow range. Similarly, lumped element values are limited by physical constraints or parasitics. We can restrict the variables to specific values, which results in *constrained optimization* [16]. The optimizer is only allowed to search for an optimum within a restricted region of n-dimensional space. In fact, all the error surfaces illustrated in Figures 3.35 through 3.47 are constrained. Unfortunately, imposing constraints may force the optimizer to start some distance from the optimum, and it certainly complicates the error surface.

The best approach is to consider the physical constraints at the earliest possible stage in the design process. If the matching networks are simple enough to use a Smith chart, we can immediately determine element values and change the topology as necessary to provide realizable components. More complex designs may require synthesized matching networks, and there are usually a number of electrically equivalent networks with different topologies and element values. If not, it may be possible to change the synthesis parameters in such a manner that the element values are improved. However, when faced with a design that includes one or more nonrealizable elements, we can first try unconstrained optimization using different optimizers and starting points if necessary. Unconstrained optimization is generally more effective than constrained optimization, and it is possible that one of the methods will find a minimum where all elements are realizable. If unsuccessful, we can constrain one or two elements and then repeat the optimization. Once the initial design meets the performance and realizability specifications, we can proceed with some assurance that the circuit is feasible.

Sections 3.3.3 and 3.3.4 clearly illustrates the difference between optimization methods. Each technique converges to a minimum following a unique sequence of one-dimensional searches. When the error surface contains local minima as illustrated in Figure 3.41, individual optimizers may converge to different solutions, even from the same starting point. Therefore, unless a solution is found that is acceptable in all respects, it is usually advantageous to try several optimizers. Direct search methods are usually superior when far from the optimum, while the indirect class of optimizers converges better in the vicinity of the minimum.

REFERENCES

[1] P.H. Smith, "An Improved Transmission Calculator," *Radio Engineer Handbook*, 1st ed., McGraw-Hill, Inc., New York, 1940.

[2] P.H. Smith, "Transmission Line Calculator," *Electronics*, Vol. 12, January 1944, pp. 130-133 and 318-325.

[3] W.E. Blanchard, "New and More Accurate Inside-Out Smith Chart," *Microwaves*, Vol. 6, No. 11, November 1967, pp. 38-42.

[4] G.D. Vendelin et al., "Computer Analyzes RF Circuits with Generalized Smith Charts," *Electronics*, March 21, 1974, pp. 102-109.

[5] J.F. White, "The Smith Chart: An Endangered Species," *Microwave Journal*, Vol. 22, November 1979, pp. 49-54.

[6] A.B. Bereskin, Letter to the Editor concerning J.F. White's article, *Microwave Journal*, March 1980, pp. 65-66.

[7] H.C. Okean, "Synthesis of Negative Resistance Reflection Amplifiers Employing Band-Limited Circulators," *IEEE Trans. on Microwave Theory and Techniques*, Vol. MTT-14, No. 7, July 1966, pp. 323-337.

[8] H.C. Okean and P.J. Meier, "Microwave Integrated Tunnel Diode Amplifiers for Broad-band High Performance Receivers," *IEEE Trans. on Microwave Theory and Techniques*, Vol. MTT-20, No. 2, February 1972, pp. 165-172.

[9] J.W. Bandler, "Stability and Gain Predictions of Microwave Tunnel-Diode Reflection Amplifiers," *IEEE Trans. on Microwave Theory and Techniques*, Vol. MTT-13, No. 6, November 1965, pp. 814-819.

[10] R.W. Laton and G.I. Haddad, "Characteristics of IMPATT-Diode Reflection Amplifiers," *IEEE Trans. on Microwave Theory and Techniques*, Vol. MTT-21, No. 11, November 1973, pp. 668-680.

[11] S.O. Scanlan and J.T. Lim, "A Design Theory for Optimum Broadband Reflection Amplifiers," *IEEE Trans. on Microwave Theory and Techniques*, Vol. MTT-13, No. 6, November 1965, pp. 827-836.

[12] K. Kurokawa, "Some Basic Characteristics of Broadband Negative Resistance Oscillator Circuits," *Bell System Tech. J.*, Vol 48, No. 6, July 1969, pp. 1937-1955.

[13] D.J. Esdale and M.J. Howes, "A Reflection Coefficient Approach to the Design of One-Port Negative Impedance Oscillators," *IEEE Trans. on Microwave Theory and Techniques*, Vol. MTT-29, No. 8, August 1981, pp. 770-776.

[14] G.E. Brehm and G.D. Vendelin, "Biasing FETs for Optimum Performance," *Microwaves*, February 1974, pp. 38-44.

[15] G.D. Vendelin, "Five Basic Bias Designs For GaAs FET Amplifiers," *Microwaves*, February 1978.

[16] J.W. Bandler, "Computer-Aided Circuit Optimization," Chapter 6 in *Modern Filter Theory and Design*, G.C. Temes and S.K. Mitra, eds., John Wiley & Sons, Inc., New York, 1973, pp. 211-271.

[17] D.A. Pierre, *Optimization Theory With Applications*, John Wiley & Sons, Inc., New York, 1969.

[18] T.R. Cuthbert, Jr., *Optimization Using Personal Computers*, John Wiley & Sons, Inc., New York, 1987.

[19] G.R. Walsh, *Methods of Optimization*, John Wiley & Sons, Inc., New York, 1975.

[20] L. Hasdorff, *Gradient Optimization and Nonlinear Control*, John Wiley & Sons, Inc., New York, 1976.

[21] J.E. Dennis, Jr., and R.B. Schnabel, *Numerical Methods for Unconstrained Optimization and Nonlinear Equations*, Prentice-Hall, Inc., Englewood Cliffs, New Jersey, 1983.

[22] S.S. Rao, *Optimization Theory and Applications*, Wiley Eastern Limited, New Delhi, India, 1979.

[23] K.C. Gupta, R. Garg, and R. Chadha, *Computer-Aided Design of Microwave Circuits*, Artech House, Inc., Dedham, Massachusetts, 1981.

[24] M.J.D. Powell, "An Efficient Method for Finding the Minimum of a Function of Several Variables Without Calculating Derivatives," *The Computer Journal*, Vol. 7, July 1964, pp. 155-162.

[25] W.I. Zangwill, "Minimizing a Function Without Calculating Derivatives," *The Computer Journal*, Vol. 10, November 1967, pp. 293-296.

[26] G.C. Temes and D.Y.F. Zai, "Least p*th* Approximation," *IEEE Trans. on Circuit Theory*, Vol. CT-16, May 1969, pp. 235-237.

[27] J.W. Bandler, T.V. Srinivasan, and C. Charalambous, "Minimax Optimization of Networks by Grazor Search," *IEEE Trans. on Microwave Theory and Techniques*, Vol. MTT-20, No. 9, September 1972, pp. 596-604.

[28] R. Hooke and T.A. Jeeves, "Direct Search Solution of Numerical and Statistical Problems," *Journal of the Association for Computing Machinery*, Vol. 8, April 1961, pp. 212-229.

[29] A. Cauchy, "Méthode Générale pour la Résolution des Systémes d' Équations Simultanées," *Comptes Rendus Hebdomadaires des Séances de l' Académie des Sciences*, Paris, Vol. 25, October 1847, pp. 536-538.

[30] E. Kreyszig, *Advanced Engineering Mathematics*, 4th ed., John Wiley & Sons, Inc., New York, 1979.

[31] S.W. Director and R.A. Rohrer, "The Generalized Adjoint Network and Network Sensitivities," *IEEE Trans. on Circuit Theory*, Vol. CT-16, August 1969, pp. 318-323.

[32] S.W. Director and R.A. Rohrer, "Automated Network Design - The Frequency-Domain Case," *IEEE Trans. on Circuit Theory*, Vol. CT-16, August 1969, pp. 330-337.

[33] J.W. Bandler and R.E. Seviora, "Current Trends in Network Optimization," *IEEE Trans. on Microwave Theory and Techniques*, Vol. MTT-18, No. 12, December 1970, pp. 1159-1170.

[34] J.W. Bandler and R.E. Seviora, "Wave Sensitivities of Networks," *IEEE Trans. on Microwave Theory and Techniques*, Vol. MTT-20, No. 2, February 1972, pp. 138-147.

[35] R. Fletcher and C.M. Reeves, "Function Minimization by Conjugate Gradients," *The Computer Journal*, Vol. 7, July 1964, pp. 149-154.

[36] W.C. Davidon, "Variable Metric Method for Minimization," *A.E.C. Research and Development Report*, ANL-5990, 1959.

[37] R. Fletcher and M.J.D. Powell, "A Rapidly Convergent Descent Method for Minimization," *The Computer Journal*, Vol. 6, July 1963, pp. 163-168.

Chapter 4

Modern Network Synthesis

This chapter addresses the problem of designing a network from a given set of electrical specifications. From a practical viewpoint, this is not a trivial task. Most components contain parasitics, resulting in significant deviation from ideal characteristics, and the physical properties of the layout compound the problem further. Additionally, active and passive components have tolerances that cause unit-to-unit variations, possibly resulting in a limited number of production circuits that meet specifications.

Network synthesis proves to be the best approach to the design problem. Not only is the circuit topology automatically determined from electrical specifications, but frequently the effect of parasitics and component tolerances can be minimized resulting in a near-optimum design. Several computer-aided synthesis programs are commercially available, providing the designer with the essential software. These programs are very easy to use and often require no more effort than defining a few electrical specifications and then selecting a topology. However, a complete understanding of the underlying theory will enable the engineer to take full advantage of these powerful design tools.

4.1 INTRODUCTION TO NETWORK SYNTHESIS

Network synthesis can be characterized as the process of defining a network that provides predetermined electrical performance. Perhaps the most common application of network synthesis is filter design. Most engineers at one time or another have designed an electrical filter using explicit equations or tables of element values. Equations are often derived from closed-form solutions of a particular synthesis problem, and tables of element values are frequently generated using a digital computer to determine numerical values for a number of slightly different synthesis problems. Design equations and tables can be very useful; however, both are limited to specific topologies and electrical parameters. Network synthesis overcomes these restrictions and provides significant design flexibility as well.

Several characteristics of network synthesis combine to provide the engineer with an extremely powerful design tool useful for a number of applications:

- Networks are optimum as synthesized, and no amount of optimization can improve the performance.[1]

- There are a number of different topologies that result in exactly the same electrical performance, literally hundreds of thousands in some cases, and the designer is free to choose the network based on realizability considerations such as convenient element values, incorporation of physical parasitics such as bonding pad capacitance or transistor lead inductance into the circuit, minimum numbers of certain types of elements, and topologies that can be used for bias or other useful functions.

- Synthesis programs can be made simple to use, very fast, and, when combined with artificial intelligence, they enable the designer to rapidly evaluate several alternatives to a particular design problem.

As RF and microwave designers, we are most familiar with the characterization of one-port and two-port networks in terms of s-parameters, impedance, or admittance. Given a network, we can easily measure any of these parameters using a network analyzer. However, the objective of most design efforts is to produce a network that has one or more prescribed electrical parameters. Once we express the desired parameters in the proper form, the process of synthesizing a network is, in fact, remarkably simple. The basic constituents of synthesis are the elements that combine to form the desired network. Figure 4.1 shows four ideal lumped elements and their corresponding complex-frequency or s-plane impedance or admittance. The s-plane ($s = \sigma + j\omega$) is a particularly convenient representation of a response as a function of the neper frequency σ and radian frequency ω, which turns out to be very useful for modern network synthesis.

We can write the s-plane input impedance or admittance of a network almost by inspection. For example, using Figure 4.1, the input impedance for the series combination of an inductor and resistor is

[1] This statement applies to the ideal network; that is, lossless elements with resonance-free and nondispersive characteristics. Strictly speaking, when we consider loss and parasitics, the network may require predistortion [1,2] or optimization to realize the desired response.

$$Z(s) = sL + R \tag{4.1a}$$

and if we add a shunt capacitor,

$$Y'(s) = Y(s) + sC \tag{4.1b}$$

or

$$Y'(s) = \frac{s^2LC + sRC + 1}{sL + R} \tag{4.1c}$$

Element	*Z(s) or Y(s)*	*Pole*	*Test Z(s)*	*Value*
	$Z(s) = sL$	$Z(s)\big\|_{s\to\infty}$	$n > m$	$L = \frac{p_n}{q_m}$
	$Y(s) = \frac{1}{sL}$	$Y(s)\big\|_{s\to 0}$	$p_0 = 0$, $q_0 > 0$	$L = \frac{p_1}{q_0}$
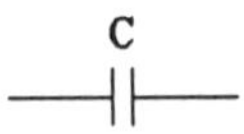	$Z(s) = \frac{1}{sC}$	$Z(s)\big\|_{s\to 0}$	$p_0 > 0$, $q_0 = 0$	$C = \frac{q_1}{p_0}$
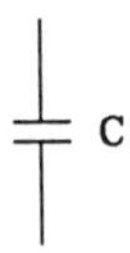	$Y(s) = sC$	$Y(s)\big\|_{s\to\infty}$	$m > n$	$C = \frac{q_m}{p_n}$

$$Z(s) = \frac{P(s)}{Q(s)} = \frac{p_n s^n + \cdots + p_2 s^2 + p_1 s + p_0}{q_m s^m + \cdots + q_2 s^2 + q_1 s + q_0}$$

Figure 4.1 Element extraction test and value.

We now turn our attention to the problem of synthesizing a network which provides an input impedance or admittance in the form of (4.1). The type and value of elements in a network having the impedance shown in (4.1a) is quite simple and can be determined by inspection; however, an impedance or admittance in the form of (4.1c) is not obvious and requires some effort to synthesize. One approach is to use partial fraction decomposition or simple polynomial division to manipulate $Z(s)$ or $Y(s)$ into a form in which an element can be identified and removed. Dividing the numerator polynomial of (4.1c) by the denominator polynomial yields

$$Y'(s) = sC + \frac{1}{sL + R} \tag{4.2}$$

which is (4.1b) in expanded form. Subtracting the shunt capacitor and inverting to form the impedance function results in (4.1a), and the last two elements are identified by inspection.

The process of identifying and removing elements from the impedance and admittance functions is known as *element extraction* and can certainly be carried out using synthetic division or partial fraction decomposition, as demonstrated above. However, another approach, based on the poles of $Z(s)$ or $Y(s)$, is often much simpler for complex functions. The frequency at which $Z(s)$ or $Y(s)$ becomes infinite is known as a *pole*, and can be very useful for element extraction. Similarly, the frequency where $Z(s)$ or $Y(s)$ becomes zero is called a *zero*. As shown in Figure 4.1, the poles of inductors and capacitors occur at $s = 0$ or $s = \infty$ and cause $Z(s)$ or $Y(s)$ to have a corresponding pole if the element is connected to the input port. For example, a network with a series capacitor as the input element has infinite impedance (pole) at zero frequency independent of any other elements in the network. Once it is known that a particular element is attached to the input port, $Z(s)$ or $Y(s)$ can be manipulated into the sum of the element impedance or admittance plus a remainder term, multiplied or divided by s to obtain the sum of the element value and remainder, and then evaluated at a value of s which eliminates the remainder and provides the element value.

The input admittance of a network having a shunt capacitor as the input element can be expressed as

$$Y(s) = \frac{P(s)}{Q(s)} = \frac{p_n s^n + \cdots + p_1 s + p_0}{q_m s^m + \cdots + q_1 s + q_0} \tag{4.3a}$$

Our goal is to design a network that can be built using real elements, and we must turn to realizability theory to ensure that this will be the case. Realizability theory [3-7] states that the driving-point immittance of a passive network must be positive real: $Z(s)$ or $Y(s)$ real for real s, and $\mathrm{Re}[Z(s)]$ or $\mathrm{Re}[Y(s)] \geq 0$ for $\mathrm{Re}[s] \geq 0$. The positive real requirement stems from the fact that for a sinusoidal input, the average power dissipated in a passive network must be non-negative. The following are two characteristics of positive real functions:

- The highest powers of the numerator and denominator polynomials may differ at most by unity; there can be no multiple poles or zeros at $s = \infty$.

- The lowest powers of the numerator and denominator polynomials may differ at most by unity; there can be no multiple poles or zeros at $s = 0$.

Since $Y(s)$ in (4.3a) has a pole at $s = \infty$ because of the shunt capacitor, the order of $P(s)$ is one degree higher than $Q(s)$; that is, $n = m + 1$. Dividing $Q(s)$ into $P(s)$ results in a form similar to (4.2) above:

$$Y(s) = sC + \frac{R(s)}{Q(s)} \tag{4.3b}$$

where $R(s)$ is the remainder from $P(s)$ after sC has been factored out. $R(s)$ is one degree lower than $P(s)$, or the same order as $Q(s)$. Dividing both sides by s to isolate the element value yields

$$\frac{Y(s)}{s} = C + \frac{R(s)}{s\,Q(s)} \tag{4.3c}$$

and since $sQ(s)$ is one degree higher than $R(s)$, we can evaluate at $s = \infty$ to eliminate the remainder:

$$\left.\frac{Y(s)}{s}\right|_{s \to \infty} = C \tag{4.3d}$$

or

$$C = \left.\frac{P(s)}{s\,Q(s)}\right|_{s \to \infty} \tag{4.3e}$$

$P(s)$ and $sQ(s)$ are the same degree and the quotient indeterminate at $s = \infty$. However, we can apply L'Hopital's Rule to find that the value of the shunt capacitor is just *the highest order coefficient of* P(s) *divided by the highest order coefficient of* Q(s). Figure 4.1 shows the test and value for the shunt capacitor in terms of the impedance function $Z(s)$. When the degree of the denominator is one higher than the numerator, the admittance function has a pole at infinity caused by the shunt capacitor. The capacitor value is the highest order coefficient of the denominator, divided by the highest order coefficient of the numerator. Returning to the admittance of (4.1c), we see that the order of the numerator is one degree higher than the denominator, so $Y(s)$ has a pole at $s = \infty$, indicating a shunt capacitor, and, by inspection, the value of the capacitor is C.

Derivations for the test and value of the other three lumped elements is similar to the preceding and is shown in Figure 4.1 for the impedance function $Z(s)$. $Z(s)$ contains a series inductor if the degree of the numerator is one higher than that of the denominator, and the value is the highest order coefficient of the numerator divided by the highest order coefficient of the denominator. A shunt capacitor can be removed from $Z(s)$ if the degree of the denominator is one higher than the numerator, and the value is the quotient of the highest order coefficient in the denominator divided by the highest order coefficient in the numerator. The shunt inductor and series capacitor have poles at zero and are determined by the lowest order nonzero coefficients in the numerator and denominator polynomials of $Z(s)$, as shown in the figure.

4.2 ELEMENT EXTRACTION

Network synthesis has many useful applications, including the design of filters, multiplexers, and impedance-matching networks. Each application has its own set of design specifications requiring slightly different approaches to the synthesis problem. However, these differences occur only in the methods used to determine $Z(s)$. Once $Z(s)$ is defined, the element extraction process is common for all types of network synthesis, which provides one of the major advantages of synthesis as a design tool.

The input impedance shown in (4.4) is typical of that encountered during the design of a filter or impedance-matching network.

$$Z(s) = \frac{20s^4 + 6s^3 + 35s^2 + 2s + 5}{20s^3 + 6s^2 + 15s} \tag{4.4}$$

We shall investigate the steps involved in synthesizing the corresponding network in some detail. The degree of the numerator of $Z(s)$ is one higher than the denominator, $n = 4$ and $m = 3$, and thus has a pole at $s = \infty$. Figure 4.1 shows that the input element is a series inductor with a value of p_4 divided by q_3, or 1. Subtracting $1 \cdot s$ from (4.4) to remove the inductor yields

$$Z'(s) = \frac{20s^2 + 2s + 5}{20s^3 + 6s^2 + 15s} \tag{4.5a}$$

and we have reduced the highest order in the numerator of $Z(s)$ by two. The impedance of (4.5a) has a shunt capacitor as the input element (pole of $Y(s)$ at s

= ∞), with a value of (q_3/p_2), or 1. Subtracting the capacitor from (4.5a) results in (4.5b):

$$Z''(s) = \frac{20s^2 + 2s + 5}{4s^2 + 10s} \tag{4.5b}$$

and the highest order of the denominator has been reduced by one. The degree of the numerator and denominator is the same; hence, there are no poles at $s = \infty$; however, there is a pole of $Z(s)$ at $s = 0$ ($q_0 = 0$), and a series capacitor is available for extraction. The capacitor has a value of (q_1/p_0), or 2, and is subtracted from (4.5b) to yield

$$Z'''(s) = \frac{10s}{2s + 5} \tag{4.5c}$$

which also has no poles at $s = \infty$. The lowest order coefficient of the numerator is zero, which corresponds to a shunt inductor with a value of (p_1/q_0), or 2. The impedance obtained after subtracting the inductor is $Z(s) = 5$, or a 5Ω resistor, and the synthesis is complete as shown in Figure 4.2(a).

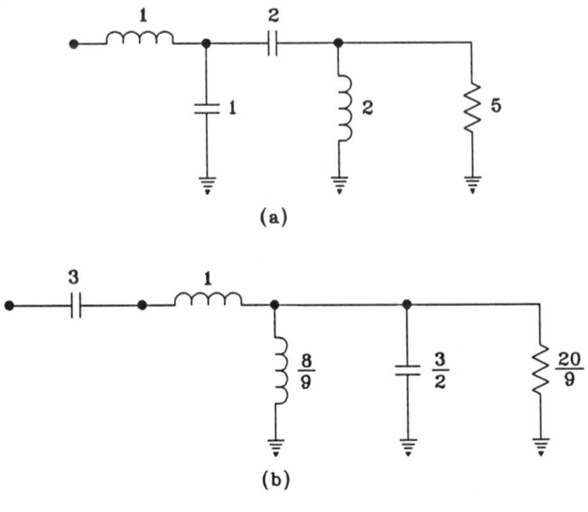

Figure 4.2 Two networks from *Z(s)* of (4.4).

The astute reader may have noticed that although we extracted a series inductor and shunt capacitor from (4.4) and (4.5a), there was a second choice in each case. Although $Z(s)$ of (4.4) has a pole at infinity due to the series inductor, there is also a pole at zero due to a series capacitor. We can synthesize another network from (4.4) by selecting different elements as follows:

- Pole of $Z(s)$ at $s = 0$, extract a series capacitor $= 3$.
- Pole of $Z(s)$ at $s = \infty$, extract a series inductor $= 1$.
- Pole of $Y(s)$ at $s = 0$, extract a shunt inductor $= 8/9$.
- Pole of $Y(s)$ at $s = \infty$, extract a shunt capacitor $= 3/2$.
- Termination resistor $= 20/9$.

The second network shown in Figure 4.2(b) has exactly the same input impedance as the network in Figure 4.2(a), which can be confirmed by writing the impedance for both circuits as shown.

Both networks have the same number and type of elements. However, the topology and range of element values are different. In addition, the termination resistance is significantly different, the first more than double the value of the second. We could continue this process and synthesize all possible networks that have the prescribed input impedance, and then select one for our design. However, in so doing we end up with discrete values of terminating resistance, and the chances are very good that none of the values are exactly what we want. Adjustment of the termination is one of the more difficult tasks in network synthesis. Fortunately, there is an additional degree of flexibility during the element extraction process which proves to be very useful for this purpose. We do not have to extract the exact element value at any one step; instead, we can extract a portion of the element in one step and the remainder several steps later.

This process is known as *partial element extraction* [4,6,7] or *partial pole removal*, and can sometimes be used to adjust the termination. For example, (4.5a) is the input impedance after a 1H series inductor has been removed from (4.4). A 1F shunt capacitor was extracted for the network in Figure 4.2(a); however, we can extract one-half of the capacitor, which yields

$$Z(s) = \frac{20s^2 + 2s + 5}{10s^3 + 5s^2 + (25/2)s} \tag{4.6a}$$

and we see that the coefficients of the denominator have changed, but not the degree. $Z(s)$ has a pole at zero due to a series 2.5F capacitor, and $Y(s)$ has a pole at infinity due to a 0.5F shunt capacitor, the other half of the element just extracted. Removing the 2.5F series capacitor results in

$$Z(s) = \frac{32s}{20s^2 + 10s + 25} \tag{4.6b}$$

which has a shunt inductor equal to 32/25 and a shunt capacitor equal to 5/8 available for extraction. After extracting the shunt capacitor we have

$$Z(s) = \frac{32\,s}{10\,s + 25} \tag{4.6c}$$

Then, removing the shunt inductor reveals that the termination is equal to 16/5. The complete network is shown in Figure 4.3.

This type of network is known as *nonminimum* because it contains more than the minimum number of elements necessary to synthesize the desired function. The termination of the network in Figure 4.3 falls between the values in Figures 4.2(a) and 4.2(b), which in fact are the limiting values for the termination. Figure 4.2(a) represents the case where we extract the full value of the first shunt capacitor, and 4.2(b) is the case where we do not extract the capacitor at all. By carefully selecting the value of the capacitor, we can adjust the termination to any value between 5 and approximately 2.2. The problem of termination adjustment will be considered in more detail in Section 4.9. However, for now we will turn our attention to the problem of defining an appropriate $Z(s)$ function.

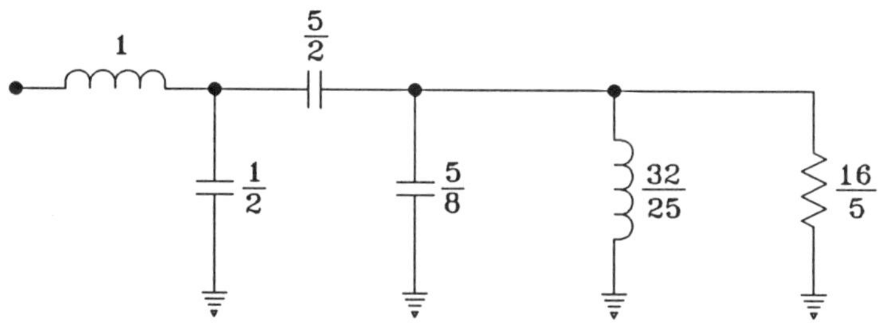

Figure 4.3 Partial element extraction from the network of Figure 4.2(a).

4.3 THE APPROXIMATION PROBLEM

Once $Z(s)$ is defined, we can apply the methods of Section 4.2 to synthesize one or more networks. Unfortunately, most design specifications are not in a form that we can directly use to synthesize a network. Insertion loss versus frequency is often specified for filters, channel frequencies for multiplexers, and load models or s-parameters for matching networks. However, we can use these parameters to define an appropriate $Z(s)$ function and then simply extract elements to determine the network.

A common design requirement is prescribed insertion loss or gain as a function of frequency, which is typical of filter design. Equation (2.59) is the expression for the two-port transducer gain as a function of arbitrary source and load terminations. When we terminate both ports with Z_o, the reference impedance, (2.59) reduces to

$$G_t = |s_{21}|^2 \tag{4.7a}$$

We are interested in the response at real frequencies and can express (4.7a) as

$$G_t(\omega^2) = |s_{21}(j\omega)|^2 \tag{4.7b}$$

Assuming an ideal network with lossless elements, we can apply the analysis from Section 2.3.1. Equation (2.38a) relates $|s_{21}|^2$ to the input reflection coefficient $|s_{11}|^2$ as

$$|s_{11}|^2 + |s_{21}|^2 = 1$$

Transforming into the s-plane ($s = j\omega$) and using (2.38a), we can write

$$G_t(\omega^2) = 1 - |s_{11}(s)|^2$$

or

$$|s_{11}(s)|^2 = 1 - G_t(\omega^2) \tag{4.8}$$

It can be shown [6,8] that $s_{11}(s)$ is a bounded real function for a passive network; that is, $|s_{11}(s)| \leq 1$ for $\mathrm{Re}\{s\} \geq 0$ and $s_{11}(s)$ real for real s; then the conjugate is $s_{11}(-s)$, and $|s_{11}(s)|^2$ can be factored into

$$s_{11}(s)\, s_{11}(-s) = 1 - G_t(\omega^2) \tag{4.9}$$

Note that $0 \leq G_t(\omega^2) \leq 1$, since $|s_{11}(s)|^2$ is bounded real. Also, the degree of the numerator of $1 - G_t(\omega^2)$ must be less than or equal to the degree of the denominator because $|s_{11}(s)|^2$ is analytic along the entire $j\omega$ axis, including infinity. Factoring $s_{11}(s) \bullet s_{11}(-s)$, we have

$$s_{11}(s) = \pm\frac{h(s)}{g(s)} \tag{4.10}$$

where $h(s)$ and $g(s)$ are real polynomials, and $g(s)$ is strictly Hurwitz. Strictly Hurwitz requires that all zeros of $g(s)$ occur only in the left half-plane and that the $j\omega$ axis is excluded.

Using the familiar relationship between reflection coefficient and impedance:

$$Z(s) = Z_o \frac{1 + s_{11}(s)}{1 - s_{11}(s)} \tag{4.11a}$$

we have

$$Z(s) = \frac{g(s) \pm h(s)}{g(s) \mp h(s)} \tag{4.11b}$$

which is the expression for the input impedance of the network referenced to 1Ω.

Belevitch [6,9] has shown that any lossless two-port lumped-element network has a scattering matrix which can be written as

$$S(s) = \frac{1}{g(s)} \begin{bmatrix} h(s) & f(s) \\ \varepsilon f(-s) & -\varepsilon h(-s) \end{bmatrix}$$

where

$g(s)$, $h(s)$, and $f(s)$ are real polynomials, $\epsilon = \pm 1$.
$g(s)$ is strictly Hurwitz.
$h(s)h(-s) + f(s)f(-s) = g(s)g(-s)$.

If the lossless network is reciprocal, that is, $s_{12} = s_{21}$,

$$S(s) = \frac{1}{g(s)} \begin{bmatrix} h(s) & f(s) \\ f(s) & -\varepsilon h(-s) \end{bmatrix}$$

Comparing s_{21} in the general and reciprocal forms, we find that $f(s) = \epsilon \bullet f(-s)$, and $f(s)$ is either an even or odd polynomial when the lossless lumped-element network is reciprocal. ϵ is +1 when $f(s)$ is an even polynomial and -1 when $f(s)$ is an odd polynomial. The input and output reflection coefficients can be written as

$$s_{11}(s) = \frac{h(s)}{g(s)} \tag{4.12a}$$

$$s_{22}(s) = \frac{-\varepsilon h(-s)}{g(s)} \tag{4.12b}$$

and we see that once $s_{11}(s)$ is known, we can form $s_{22}(s)$ by replacing $h(s)$ with $h(-s)$ when the numerator of $s_{21}(s)$ is an odd polynomial, and $-h(-s)$ when $f(s)$ is even. We can associate the sign uncertainty in (4.10) with $1/g(s)$, and we note that the sign must be applied to both $s_{11}(s)$ and $s_{22}(s)$.

Given a design specification of gain versus frequency, we need only approximate the specification as the ratio of two polynomials in ω^2, transform into the s-plane, subtract the ratio from unity to form $|s_{11}(s)|^2$, factor out $s_{11}(s)$ and form $Z(s)$, and then extract the elements to synthesize one or more networks. This procedure is not at all complicated, although some care must be exercised to ensure that all conditions for a passive network are met.

4.4 LOW-PASS FILTER SYNTHESIS

Low-pass filters have characteristics such that they pass all signals below a specified frequency and attenuate signals above. The low-pass filter often finds applications in systems where it is necessary to eliminate higher order harmonics or spurious signals (e.g., signal sources, frequency multipliers, or mixers). Ideally, the low-pass filter would pass the desired signals without distortion of any kind and completely reject all other signals. Unfortunately, we cannot achieve the characteristics of an ideal filter, even with an infinite number of elements, and must resort to approximations. There are several useful approximations that generally fall into the category of amplitude or phase. Amplitude approximations usually have poor phase response, resulting in phase or group delay distortion, and approximations based on phase usually have undesirable amplitude characteristics. Some approximations can be characterized as *transitional*; that is, they provide acceptable phase response in the passband and then make a transition to an amplitude approximation in the stopband. When faced with a particular design problem, the engineer must be familiar with the characteristics of various approximations and select the one best suited for the application.

4.4.1 The Butterworth or Maximally Flat Low-Pass Filter

The gain function shown in (4.13) has the property that all derivatives of G_t with respect to ω are zero at $\omega = 0$ and $\omega = \infty$; hence, it is known as a maximally flat or Butterworth response, commonly given by

$$G_t(\omega^2) = \frac{K_m}{1 + \varepsilon^2 \omega^{2n}} \tag{4.13}$$

The value of this function is K_m at $\omega = 0$ and monotonically decreases to 0 at $\omega = \infty$. Since G_t is constrained between 0 and 1, K_m must be less than or equal to 1, and it defines the maximum gain or *minimum insertion loss* (MIL) of the function. Normally, K_m is specified as 1, which results in a filter with a minimum insertion loss of 0 dB, however there are instances where a nonzero minimum insertion loss is required. The edge of the passband, or *cutoff frequency* f_c, is at $\omega = 1$, where G_t has a value of $K_m/(1 + \epsilon^2)$. ϵ^2 defines the loss at the cutoff frequency and is often set equal to 1, which results in an additional 3-dB loss at the cutoff frequency. However, from (4.8) we see that with $K_m = 1$ and $\epsilon^2 = 1$, $|s_{11}|^2 = 0.5$ at f_c, which corresponds to a voltage standing wave ratio (VSWR) of 5.8:1 and would not be acceptable for most filter designs. Figure 4.4 shows the response of (4.13) with $K_m = 1$ and $\epsilon^2 = 1$ from $\omega = -4$ to $\omega = 4$ for n equal to 2, 3, and 4.

The figure shows that all three curves have exactly the same value (-3.01 dB) at $\omega = \pm 1$; but the slope of the response above f_c becomes steeper as n is increased and tends toward a slope of $6n$ dB per octave or $20n$ dB per decade. The passband response from 0 to f_c flattens out and remains near zero over a larger percentage of the band as n is increased. For example, the -0.5-dB point for $n = 2$ is at $\omega = 0.59$, at $\omega = 0.77$ for $n = 4$, and at $\omega = 0.90$ for $n = 10$. Figure 4.4 also shows that the response is symmetrical about zero. Negative frequencies are shown in the shaded area.

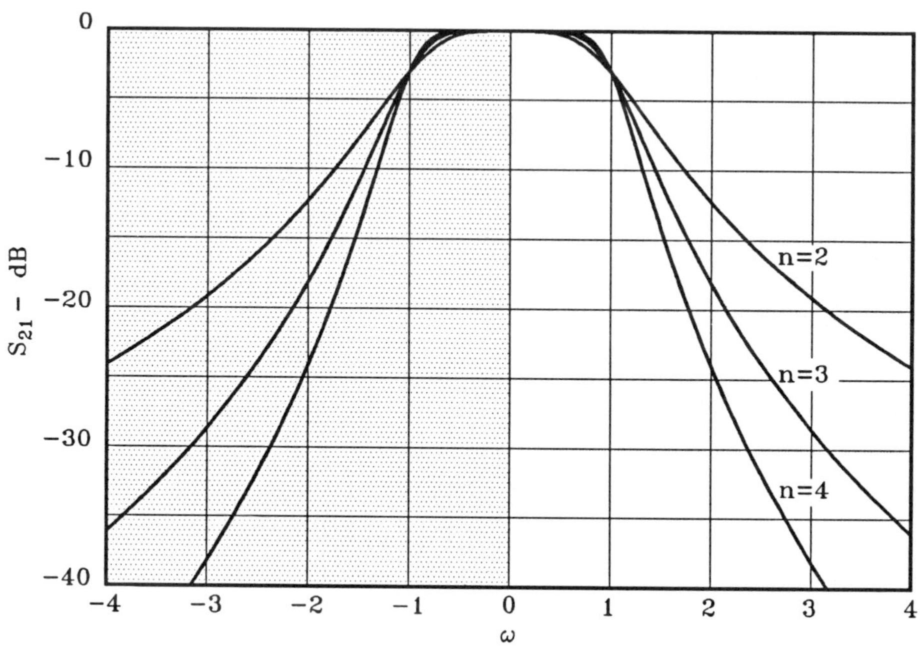

Figure 4.4 Low-pass filter response.

A typical low-pass filter specification includes the cutoff frequency, maximum insertion loss, or VSWR in the passband, along with a stopband frequency or range of frequencies and the rejection or stopband insertion loss. We have three parameters that alter the response and must be determined for each set of specifications: K_m, ϵ^2, and n. Given the following specifications,

Passband frequency:	$\omega = 0$ to $\omega = 1$
Maximum VSWR:	6:1
Stopband frequency:	$\omega = 2$ to $\omega = \infty$
Stopband rejection:	15 dB minimum

we will synthesize a low-pass filter. From the VSWR specification we have

$$|s_{11}| = \frac{VSWR - 1}{VSWR + 1} = \frac{5}{7} = 0.714$$

or

$$G_t = |s_{21}|^2 = 1 - |s_{11}|^2 = 0.490 = -3.10 \text{ dB}$$

With $K_m = 1$ and $\epsilon^2 = 1$, G_t is -3.01 dB at $\omega = 1$ for any value of n which meets the passband requirements. Figure 4.4 shows that at $\omega = 2$, G_t is approximately -12 dB for $n = 2$, -18 dB for $n = 3$, and -24 dB for $n = 4$. We choose $n = 3$, which provides a 3-dB design margin in the stopband, and the gain function is

$$G_t(\omega^2) = \frac{1}{1 + (\omega^2)^3}$$

From (4.8) we have

$$|s_{11}(s)|^2 = 1 - G_t(\omega^2) = \frac{\omega^6}{\omega^6 + 1}$$

Then, substituting $s = j\omega$ to transform into the s-plane,

$$|s_{11}(s)|^2 = \frac{s^6}{s^6 - 1}$$

which must be factored into $s_{11}(s)s_{11}(-s)$. This function is not difficult to factor; however, it will usually be easier to compute the roots of $|s_{11}(s)|^2$ and use these to form $s_{11}(s)$. The function has six zeros, all at 0.0, and the poles are

Poles of $\lvert s_{11}(s)\rvert^2$
1.0
-1.0
0.5 ± j0.866025
-0.5 ± j0.866025

to six decimal places, as shown in Figure 4.5. The figure shows that poles in the left half-plane are mirror images of those in the right half-plane and are all equally spaced about the unit circle. The poles of this function are easy to solve in closed form, which leads to explicit design equations for the Butterworth filter.

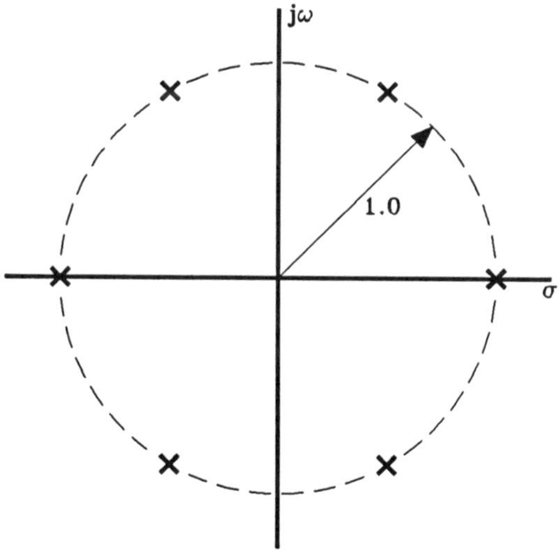

Figure 4.5 Poles of $|s_{11}(s)|^2$.

We select the three poles in the left half-plane, so the denominator of $s_{11}(s)$ will be a strictly Hurwitz polynomial and $s_{11}(s)$ becomes

$$s_{11}(s) = \pm\frac{h(s)}{g(s)} = \pm\frac{s^3}{s^3 + 2s^2 + 2s + 1}$$

The sign of $s_{11}(s)$ must be resolved before we can form $Z(s)$ and synthesize the network. We observe that at $s = 0$, s_{11} is zero or the load and source are perfectly matched, and when $s = \infty$, s_{11} can be either +1 or -1. A reflection coefficient of +1 corresponds to an open circuit, and -1 corresponds to a short circuit; therefore, if we use a plus sign, the input element will be a series inductor, and the negative sign will result in a shunt capacitor at the input. We select the inductor and form $Z(s)$ from (4.11):

$$Z(s) = \frac{2s^3 + 2s^2 + 2s + 1}{2s^2 + 2s + 1}$$

The order of $Z(s)$ is n, which will require n elements in the filter. Then, using the techniques of Sections 4.1 and 4.2, we can synthesize the network shown in Figure 4.6.

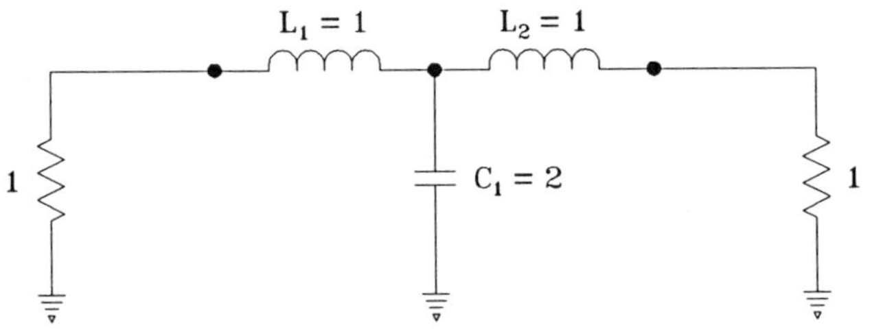

Figure 4.6 Butterworth low-pass filter.

The network in Figure 4.6 has a source and load termination of 1Ω and a cutoff frequency of 1 rad/sec. Most practical RF and microwave filters operate in a 50Ω system and have cutoff frequencies much higher than our circuit. Element values of the synthesized network can easily be scaled to the desired impedance level and frequency using the methods of Section 4.5.

This filter has a relatively high passband VSWR, which, incidentally, resulted in poles on the unit circle and integer polynomial coefficients. We can reduce the VSWR by adjusting the value of ϵ^2 in (4.13), which leads to poles on nonunit circles about the origin of the s-plane and polynomial coefficients that are not integers. ϵ^2 effectively changes the frequency scale of the curves in Figure 4.4, but not their shape, and it is necessary to use another approximation if the design does not meet specifications.

4.4.2 The Chebyshev or Equal-Ripple Low-Pass Filter

The Butterworth filter has relatively poor characteristics near the cutoff frequency. It is flat over a portion of the passband and then slowly makes a transition into the stopband region where it finally approaches an asymptotic value of $6n$ dB per octave. We can improve the characteristics by allowing the response to ripple between a minimum and maximum value in the passband and remain maximally flat in the stopband.

Rewriting (4.13),

$$G_t(\omega^2) = \frac{K_m}{1 + \varepsilon^2 F_n^2(\omega)} \tag{4.14}$$

we see that the function $F_n(\omega)$ must remain within an upper and lower limit from $\omega = 0$ to $\omega = 1$, monotonically increase or decrease to plus or minus infinity for ω greater than 1, and be even or odd so the square will be a function of ω^2. ω^n is one function that has these properties; however, the Chebyshev polynomial $T_n(x)$ also meets the criteria, but has the advantage that there are n zeros for $-1 \le x \le 1$. The Chebyshev polynomial is defined as

$$T_n(x) = \cos(n \cos^{-1} x)$$

and can be computed using the following recurrence formula:

$$T_0(x) = 1 \tag{4.15a}$$

$$T_1(x) = x \tag{4.15b}$$

$$T_n(x) = 2\,x\,T_{n-1}(x) - T_{n-2}(x) \tag{4.15c}$$

The first six values are

$$T_0(x) = 1$$
$$T_1(x) = x$$

$$T_2(x) = 2x^2 - 1$$
$$T_3(x) = 4x^3 - 3x$$
$$T_4(x) = 8x^4 - 8x^2 + 1$$
$$T_5(x) = 16x^5 - 20x^3 + 5x$$
$$T_6(x) = 32x^6 - 48x^4 + 18x^2 - 1$$

Figure 4.7(a) shows a plot of $T_5(x)$, and Figure 4.7(b) shows $T_6(x)$; the upper plot in (a) and (b) is the function, and the lower plot is the function squared. The figure shows that the square of the function ripples between 0 and 1 for $-1 \le x \le 1$, attains zero n times in this range, and monotonically increases for $x < -1$ or $x > 1$.

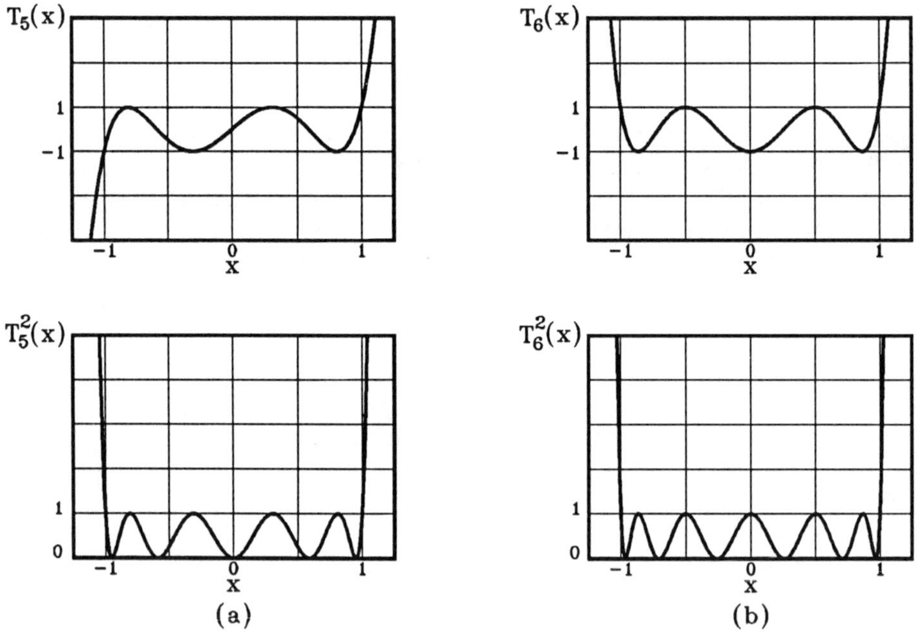

Figure 4.7 (a) Chebyshev polynomial $T_5(x)$, and (b)$T_6(x)$.

We can write the transducer gain for the Chebyshev low-pass filter as

$$G_t(\omega^2) = \frac{K_m}{1 + \varepsilon^2 T_n^2(\omega)} \tag{4.16}$$

The minimum value of $T_n^2(\omega)$ is zero where $G_t(\omega) = K_m$; hence, K_m defines the minimum insertion loss. $T_n^2(\omega)$ ripples between 0 and 1 in the passband from $\omega = 0$ to $\omega = 1$; therefore, $G_t(\omega)$ ripples between K_m and $K_m/(1 + \epsilon^2)$, and we see that ϵ^2 defines the magnitude of the ripple, or

$$Ripple = 10\,\mathrm{Log}\left(\varepsilon^2 + 1\right)\ \mathrm{dB}$$

and solving for ϵ^2,

$$\varepsilon^2 = 10^{(Ripple/10)} - 1 \tag{4.17}$$

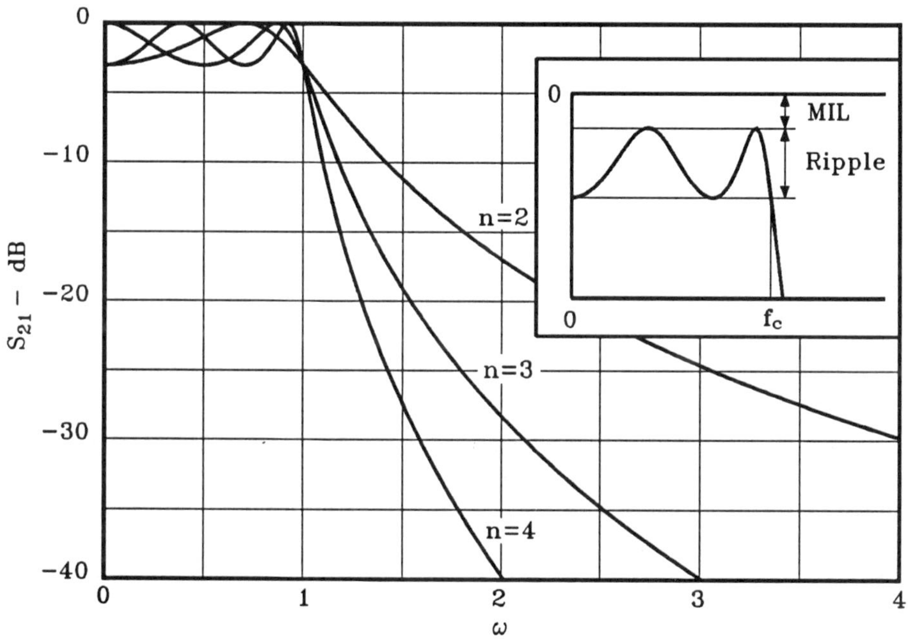

Figure 4.8 Chebyshev low-pass response.

Figure 4.8 shows the response of the Chebyshev low-pass filter of (4.16), with K_m and ϵ^2 set to 1, 0-dB minimum insertion loss, and 3.01-dB ripple. We can compare these plots with those of the Butterworth filter in Figure 4.4 which has similar specifications illustrating the superior stopband rejection of the Chebyshev approximation. The stopband rejection is controlled somewhat by the ripple because of the $\epsilon^2 T_n^2(\omega)$ term in the denominator. Except for small ripple values, the Chebyshev approximation offers improved stopband rejection over the

Butterworth for the same filter order, and can often be used to reduce the number of elements. Figure 4.8 also shows that the total loss at the cutoff frequency is always the sum of minimum insertion loss and ripple since $T_n^2(1) = 1$, or $G_t(1) = K_m/(1 + \epsilon^2)$; however, the loss at zero is either minimum for n odd, $T_n^2(0) = 0$, or maximum for n even, $T_n^2(0) = 1$. The low-pass filter containing series inductors and shunt capacitors is effectively transparent at zero frequency. Thus, the load is connected directly to the source, and any insertion loss must be due to a mismatch, which means that nonzero minimum insertion loss, or even-order Chebyshev filters, *always results in unequal terminations.*

The synthesis procedure for Chebyshev filters is identical to the Butterworth design in the previous section, and as an example we can synthesize a Chebyshev filter using the specifications of Section 4.4.1, except that the maximum passband VSWR is reduced to 2:1.

The VSWR specification requires that

$$|s_{11}| <= 0.333$$

or

$$G_t = |s_{21}|^2 = 1 - |s_{11}|^2 = 0.889 = -0.512 \text{ dB}$$

For 0-dB minimum insertion loss $K_m = 1$, and

$$\varepsilon^2 <= 10^{(0.512/10)} - 1 = 0.125$$

We can set ϵ^2 equal to 0.1, 0.414-dB ripple, to provide some margin for component tolerance, and then evaluate (4.16) at $\omega = 2$ to determine the required order of the filter. $G_t(2) = -7.709$ dB for $n = 2$, which does not meet the rejection specification; however, the loss is 18.363 dB for $n = 3$, and we can write the transducer gain as

$$G_t(\omega^2) = \frac{1}{1 + 0.1\left(4\,\omega^3 - 3\,\omega\right)^2}$$

After expanding and replacing ω^2 with $-s^2$, we can use (4.8):

$$|s_{11}(s)|^2 = \frac{s^6 + 1.50\,s^4 + 0.5625\,s^2}{s^6 + 1.50\,s^4 + 0.5625\,s^2 - 0.625}$$

This function has six poles and six zeros, with numerical values shown in the following table.

Zeros	*Poles*
0.0	0.663910
0.0	-0.663910
0.0 ± j0.866025	0.331955 ± j1.039511
0.0 ± j0.866025	-0.331955 ± j1.039511

The poles and zeros are illustrated in Figure 4.9, which shows the left and right half-plane symmetry about the jω axis. Unlike poles of the Butterworth filter shown in Figure 4.5, the poles of the Chebyshev filter are located on an ellipse in the *s*-plane.

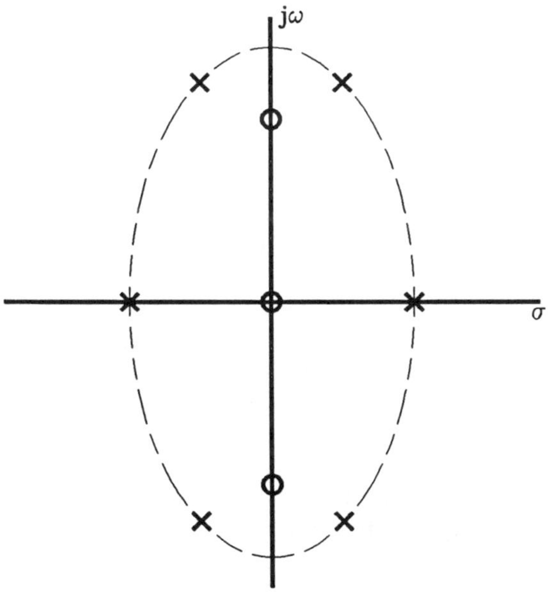

Figure 4.9 Poles and zeros of $|s_{11}(s)|^2$.

We select the three left half-plane poles to ensure that the denominator of $s_{11}(s)$ will be strictly Hurwitz, and then turn our attention to the zeros. The numerator of $s_{11}(s)$ does not have to be strictly Hurwitz, and we are free to select each zero from the left or right half-plane. However, the numerator and denominator must be real polynomials, which means we must choose either real or complex conjugate pairs of roots. After selecting one real and one complex conjugate zero pair, we can factor $|s_{11}(s)|^2$,

$$s_{11}(s) = \pm \frac{s^3 + 0.75s}{s^3 + 1.327821s^2 + 1.631554s + 0.790569}$$

and then use the positive sign to form $Z(s)$:

$$Z(s) = \frac{2s^3 + 1.327821s^2 + 2.381554s + 0.790569}{1.327821s^2 + 0.881554s + 0.790569}$$

which is similar to the expression for the input impedance of the Butterworth filter in Section 4.4.1. Completing the element extraction, we have the filter network shown in Figure 4.10.

The Chebyshev approximation is perhaps the most common type used for filter design. Passband VSWR can easily be controlled, we can achieve good stopband rejection with few elements, and the synthesis procedure is not complicated. However, phase or group delay distortion is greater than that of the Butterworth filter and may prohibit its use in some critical applications, such as communication systems.

4.4.3 The Inverse Chebyshev Low-Pass Filter

Some filter specifications call for high values of rejection at stopband frequencies very close to the passband edge, which requires a large number of elements when realized as a Butterworth or Chebyshev filter. For example, there may be a strong radar or other signal close to the operating band of a receiver which much be suppressed to avoid overloading the preamplifiers. Maximum passband VSWR of 2:1 and 60-dB rejection at $\omega = 2$ requires twelve elements for the Butterworth filter, and seven elements for the Chebyshev design. The inverse Chebyshev only requires four elements to meet the same specifications.

The Chebyshev response is equal ripple in the passband and maximally flat in the stopband. We can invert this function to realize a new approximation which is maximally flat in the passband and equal ripple in the stopband. Subtracting the Chebyshev transducer gain $G_t(\omega)$ in (4.16) from K_m yields

$$G_t(\omega^2) = \frac{K_m \varepsilon^2 T_n^2(\omega)}{1 + \varepsilon^2 T_n^2(\omega)}$$

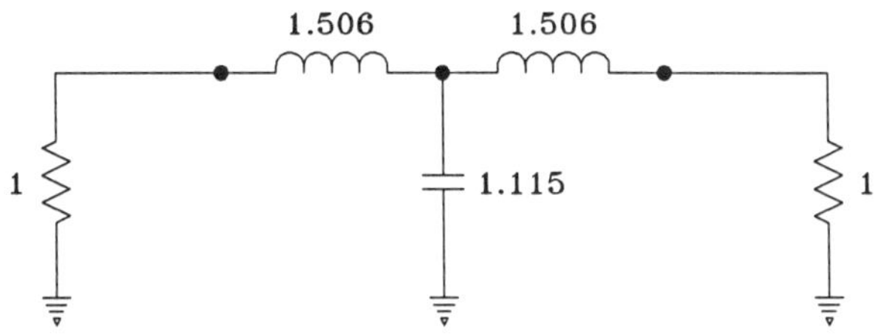

Figure 4.10 Chebyshev low-pass filter.

which is shown in Figure 4.11 for $n = 3$, $K_m = 1$, and $\epsilon^2 = 0.001$. K_m defines the minimum insertion loss. However, the value at the cutoff frequency is $K_m\epsilon^2 / (1 + \epsilon^2)$, or approximately -30 dB, as shown.

The Chebyshev response has been inverted, resulting in an equal-ripple stopband from $\omega = 0$ to $\omega = 1$ and maximally flat passband above $\omega = 1$. We can effectively reverse the frequency scale by substituting $1/\omega$ for ω. $\omega = 0$ is transformed into $\omega = \infty$, $\omega = \infty$ maps into $\omega = 0$, and then the expression for gain becomes

$$G_t(\omega^2) = \frac{K_m\, \varepsilon^2\, T_n^2(1/\omega)}{1 + \varepsilon^2\, T_n^2(1/\omega)}$$

which is also shown in Figure 4.11 for the same value of K_m and ϵ^2. This response is maximally flat in a passband about $\omega = 0$ and equal ripple in a stopband extending from $\omega = 1$ to $\omega = \infty$. In addition, there is a *transmission zero* or attenuation pole between $\omega = 1.0$ and $\omega = 1.5$, and another at $\omega = \infty$. We can write this expression in a form similar to that of the Butterworth and Chebyshev by normalizing such that the cutoff frequency is at $\omega = 1$ where the loss is $K_m/(1 + \epsilon^2)$ and the equal-ripple stopband is between ω_s and infinity. The transducer gain of the inverse Chebyshev low-pass filter is

$$G_t(\omega^2) = \frac{K_m\, T_n^2(\omega_s/\omega)}{\varepsilon^2\, T_n^2(\omega_s) + T_n^2(\omega_s/\omega)} \qquad (4.18)$$

At $\omega = 0$, $G_t(0) = K_m$, which is the minimum insertion loss, and at the cutoff frequency where $\omega = 1$, $G_t(1) = K_m / (1 + \epsilon^2)$, so ϵ^2 defines the passband ripple as given in (4.17). The transducer gain at ω_s is

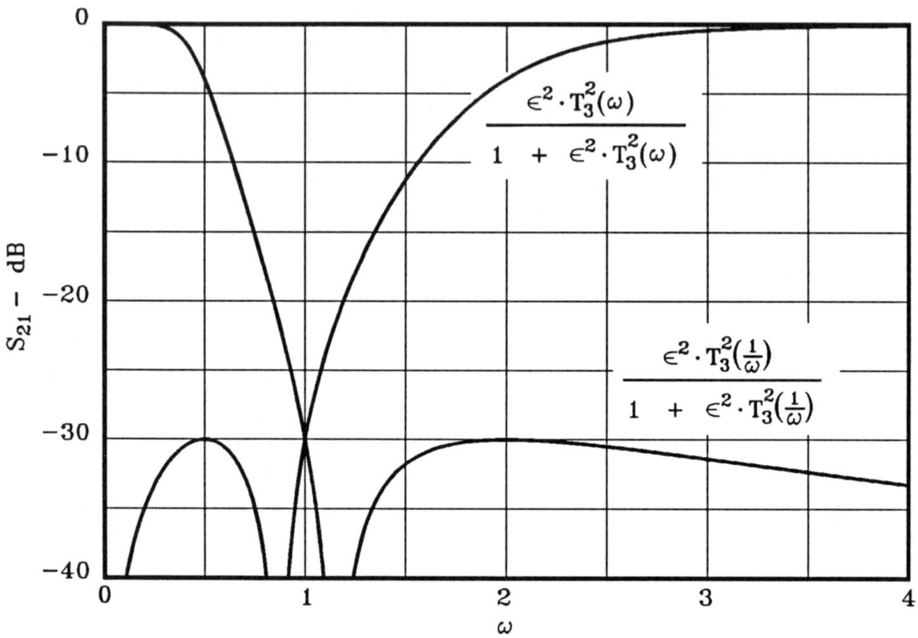

Figure 4.11 Inverse Chebyshev response.

$$G_t\left(\omega_s^2\right) = \frac{K_m\, T_n^2(1)}{\varepsilon^2\, T_n^2\left(\omega_s\right) + T_n^2(1)} = \frac{K_m}{1 + \varepsilon^2\, T_n^2\left(\omega_s\right)}$$

and the minimum stopband loss, L_s, is

$$L_s = -10\,\mathrm{Log}\left\{\frac{K_m}{1 + \varepsilon^2\, T_n^2\left(\omega_s\right)}\right\}\ \mathrm{dB} \tag{4.19}$$

We can compare the inverse Chebyshev filter with the Butterworth and Chebyshev of Sections 4.4.1 and 4.4.2 by designing a low-pass filter using the same specifications. We use the same minimum insertion loss and ripple as the Chebyshev design (K_m = 1 and ϵ^2 = 0.1), which provides a maximum VSWR less than 2:1. At ω_s = 2, L_s is 7.709 dB for n = 2, and 18.363 dB for n = 3, which is identical to that computed for the Chebyshev filter of Section 4.4.2. Figure 4.12 shows the frequency response of this approximation as the lighter of the two plots.

The transmission zero occurs at $\omega = 2.3$ and can be shifted to improve the rejection at $\omega = 2$.

The transmission zeros of $G_t(\omega^2)$ are due to the zeros of the numerator of (4.18), or zeros of $T_3(\omega_s/\omega)$. $T_3(x)$ has zeros at $x = 0$ and $x = \pm\sqrt{3}/2$, or at $\omega = \infty$ and $\omega = \pm 2\omega_s/\sqrt{3}$. If we force the transmission zero to occur at $\omega = 2$, then ω_s becomes

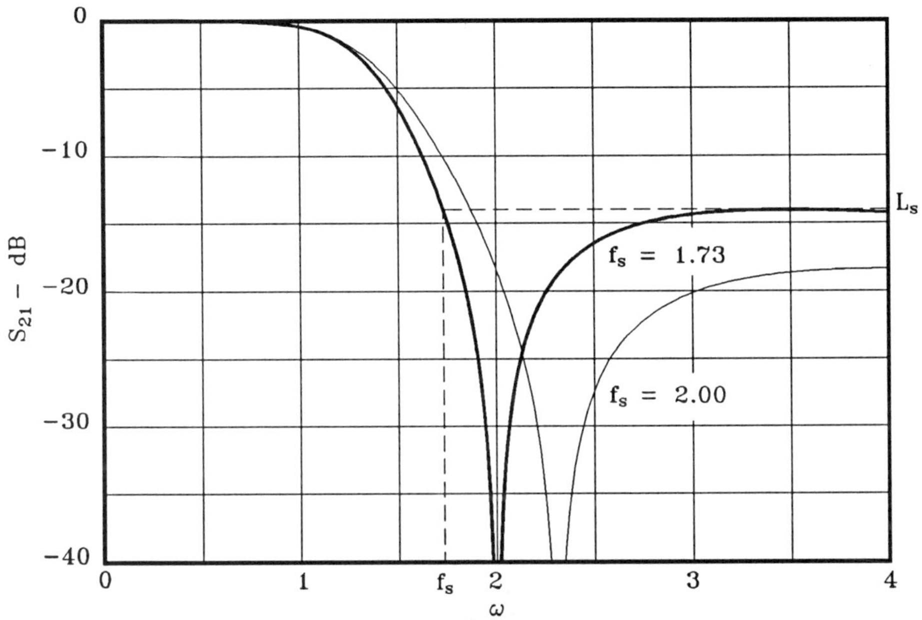

Figure 4.12 Inverse Chebyshev low-pass filter.

$\sqrt{3}$ and we must recompute L_s, which is now 14.031 dB, as shown in the darker plot of Figure 4.12. We can write this transducer gain as

$$G_t(\omega^2) = \frac{T_3^2(\sqrt{3}/\omega)}{0.1\, T_3^2(\sqrt{3}) + T_3^2(\sqrt{3}/\omega)}$$

or

$$G_t(\omega^2) = \frac{\omega^4 - 8\,\omega^2 + 16}{0.9\,\omega^6 + \omega^4 - 8\,\omega^2 + 16}$$

Substituting $s = j\omega$, we can write $|s_{11}(s)|^2$ as

$$|s_{11}(s)|^2 = \frac{0.9s^6}{0.9s^6 - s^4 - 8s^2 - 16}$$

which has six zeros at 0.0 and six poles at

$\pm(2.053741)$
$\pm(0.499824 \pm j1.342831)$

Selecting the left half-plane poles, we can write

$$s_{11}(s) = \pm\frac{s^3}{s^3 + 3.053390s^2 + 4.106039s + 4.216370}$$

and using the positive sign,

$$Z(s) = \frac{2s^3 + 3.053390s^2 + 4.106039s + 4.216370}{3.053390s^2 + 4.106039s + 4.216370}$$

The order of the numerator is greater than the denominator; hence, $Z(s)$ has a pole at $\omega = \infty$ and a series inductor is available for extraction. However, before we begin the extraction, which is destined to fail anyway, we must take a closer look at the elements in this type of filter.

The gain function of the Butterworth and Chebyshev filters exhibits a transmission zero at infinity, and the resulting input impedance can be realized using only series inductors and shunt capacitors which also have poles at infinity, as shown in Figure 4.1. The inverse Chebyshev approximation also has a transmission zero at infinity, but, in addition, there are one or more finite frequency transmission zeros. Finite frequency attenuation poles can be implemented using the two circuits shown in Figure 4.13. The parallel resonant network has infinite impedance at the resonant frequency where the inductive and capacitive reactances cancel, and the series resonant circuit has zero impedance at the resonant frequency. Once we determine that $Z(s)$ or $Y(s)$ has a pole at $s = j\omega$, we can solve for the values of L and C and then subtract both elements simultaneously.

Element	*Z(s) or Y(s)*	*Pole*	*Test*	*Value*
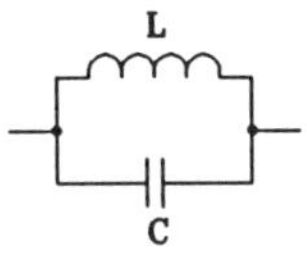	$Z(s) = \frac{\frac{s}{C}}{s^2 + \frac{1}{LC}}$	$Z(s)\Big\|_{s=\pm j\omega}$	$Y(j\omega) = 0$	$\frac{1}{C} = \left\{\frac{(s^2+\omega^2)}{s}\cdot Z(s)\right\}\Big\|_{s=j\omega}$ $L = \frac{1}{\omega^2 \cdot C}$
	$Y(s) = \frac{\frac{s}{L}}{s^2 + \frac{1}{LC}}$	$Y(s)\Big\|_{s=\pm j\omega}$	$Z(j\omega) = 0$	$\frac{1}{L} = \left\{\frac{(s^2+\omega^2)}{s}\cdot Y(s)\right\}\Big\|_{s=j\omega}$ $C = \frac{1}{\omega^2 \cdot L}$

Figure 4.13 Element extraction test and value.

We know that this filter has a transmission zero at $\omega = 2$, and $Z(s)$ or $Y(s)$ can be evaluated at $s = j2$ to determine whether a series or parallel resonant circuit is present. Replacing s with j2, we have

$$Z(j2) = 0.0 + j0.973833$$

which reveals that we do not have an attenuation pole at $\omega = 2$. However, if we subtract an element with a reactance of j0.973833, $Z(j2)$ will be zero and we can extract a series resonant circuit. A series inductor has an impedance of $j\omega L$, and at $\omega = 2$, the required value of L is 0.486916. $Z(s)$ has a pole at $\omega = \infty$, and a series 0.655010H inductor is available, so we can partially extract 0.486916H. The input impedance becomes

$$Z(s) = \frac{0.513255s^3 + 1.054093s^2 + 2.053019s + 4.216370}{3.053390s^2 + 4.106039s + 4.216370}$$

which is zero at $s = j2$, and we can evaluate the expression shown in Figure 4.13 to determine the value of L in the series resonant circuit. The expression is indeterminate at $s = j2$, since both numerator and denominator are zero. However, the denominator of $Y(s)$ contains $(s^2 + \omega^2)$ because of the series resonant circuit, which can be factored out and used to cancel the identical term in the numerator. After removing the common term and multiplying the denominator by s, we have

$$\frac{1}{L} = \left.\frac{3.053390s^2 + 4.106039s + 4.216370}{0.513254s^2 + 1.054093s}\right|_{s=j2}$$

or

$$L = 0.256718$$

and

$$C = 0.973833$$

Removing the L-C network results in an input impedance of

$$Z(s) = \frac{0.513255s + 1.054093}{1.054093}$$

and we can extract a series 0.486916H inductor to reveal the 1Ω termination resistor. The inverse Chebyshev low-pass filter is complete as shown in Figure 4.14.

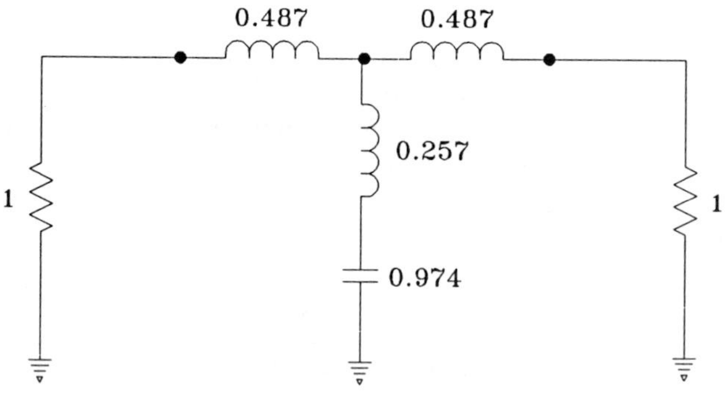

Figure 4.14 Inverse Chebyshev low-pass filter.

The inverse Chebyshev approximation is useful when the stopband is close to the cutoff frequency or discrete frequencies must be suppressed. However, as the stopband edge is moved closer to the cutoff frequency, the equal-ripple stopband loss is decreased, as illustrated in Figure 4.12. It is sometimes possible to combine more than one type of filter to achieve more desirable characteristics. For example, an inverse Chebyshev can be combined with a Butterworth filter to improve the stopband rejection at higher frequencies, although the order of the filter can also be increased with the same effect.

4.4.4 The Elliptic Low-Pass Filter

The inverse Chebyshev filter of Section 4.4.3 turned out to be maximally flat in the passband and equal ripple in the stopband. We can achieve an equal-ripple response in both the passband and stopband by requiring the function $F_n(\omega)$ in (4.14) to have Chebyshev characteristics in the passband and inverse Chebyshev characteristics in the stopband. The solution to this problem is in terms of elliptic integrals and functions; hence, it is known as an *elliptic filter*. Explicit expressions for $F_n(\omega)$ are given by Baher and Rhodes [6,10] and will not be reproduced here; however, we will compare the elliptic filter with the inverse Chebyshev filter.

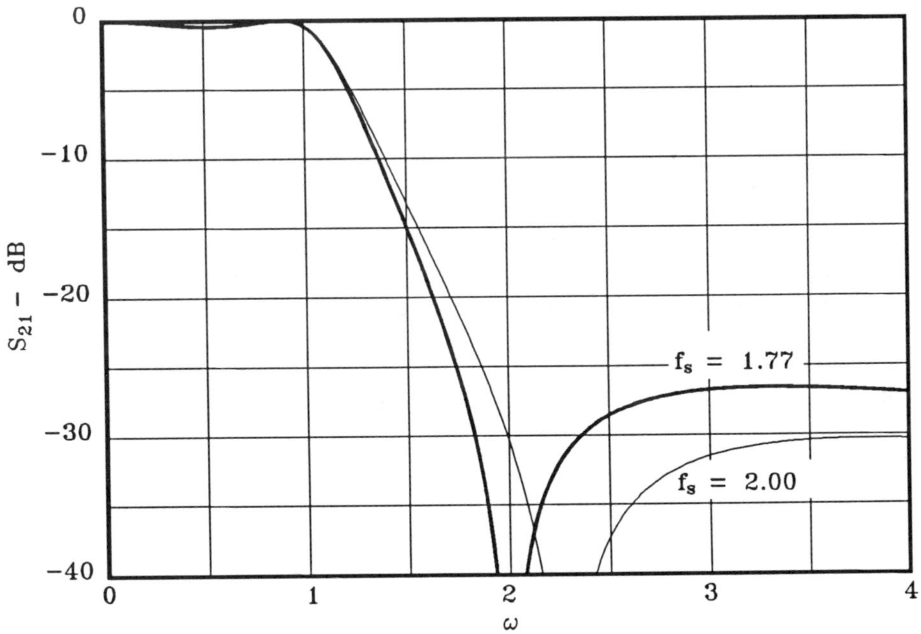

Figure 4.15 Elliptic low-pass filter.

The response of an elliptic filter designed to the same specifications as the inverse Chebyshev filter is shown in Figure 4.15. The minimum equal-ripple stopband loss is 30.325 dB when $\omega_s = 2$, and 26.584 dB when ω_s is moved to 1.772 to provide an attenuation pole at $\omega_s = 2$. When compared with the inverse Chebyshev response shown in Figure 4.12, we see that the elliptic filter provides an improvement of more than 12 dB in the stopband rejection. The topology of the elliptic filter is identical to that of the inverse Chebyshev shown in Figure 4.14; only the element values change. The series inductors are 1.313H, and values of the inductor and capacitor in the series *L-C* network are 0.279H and 1.313F, respectively.

The amplitude response of elliptic filters are optimum when compared with other filter types of the same degree and are often used when the filter requirements are stringent. Elliptic filters are usually classified as one of four types: a, b, c, and s [11]. Unfortunately not all elliptic filters can be realized using only positive value inductors and capacitors; some networks may require negative elements, ideal transformers, or coupled coils. The type s filter is the odd order case which is symmetric, has $(n - 1)/2$ transmission zeros, and equal terminations for 0-dB minimum insertion loss. Type s can usually be realized in a form useful to RF and microwave designers except for certain combinations of passband ripple and stopband frequency. When an undesirable topology is encountered, it is often possible to change the specifications or order to achieve a realizable network.

4.4.5 Phase Approximation

Certain communication systems cannot tolerate phase distortion of the signal and require filters that have linear phase in the passband. Specifications for these systems are often given in terms of *group delay*, which is negative of the derivative of phase with respect to frequency. The group delay of a network with perfectly linear phase is a constant, and any deviation from phase linearity shows up as a deviation from the constant. Thus, group delay is often specified as a maximum value or maximum peak-to-peak ripple about an average value.

We can write $s_{21}(s)$ as

$$s_{21}(s) = \frac{K_m}{Q(s)} \tag{4.20a}$$

Then, separating $Q(s)$ into its even and odd components,

$$s_{21}(s) = \frac{K_m}{Q_e(s) + Q_o(s)} \tag{4.20b}$$

and since the even terms are real and the odd terms imaginary in $j\omega$, we can write the phase of $s_{21}(s)$ as

$$\phi(s) = -\tanh^{-1}\left\{\frac{Q_o(s)}{Q_e(s)}\right\} \tag{4.20c}$$

The group delay is the negative of the derivative of phase with respect to frequency, or

$$t_d(s) = -\frac{d\phi(s)}{ds} \tag{4.21}$$

and if we let $\phi(s) = -s$, the group delay is constant for all frequencies, and the derivatives of $t_d(s)$ are also zero at $\omega = 0$, which is the condition for maximally flat group delay response. Substituting $\phi(s) = -s$ into (4.20c) and then taking the hyperbolic tangent of both sides and using an identity for the hyperbolic tangent, we can write

$$\tanh s = \frac{\sinh s}{\cosh s} = \frac{Q_o(s)}{Q_e(s)}$$

The series expansion for the hyperbolic sine and cosine functions is

$$\sinh s = s + \frac{s^3}{3!} + \frac{s^5}{5!} + \cdots + \frac{s^{2n+1}}{(2n+1)!}$$

$$\cosh s = 1 + \frac{s^2}{2!} + \frac{s^4}{4!} + \cdots + \frac{s^{2n}}{(2n)!}$$

and the continued fraction expansion of ($\sinh s/\cosh s$) is

$$\tanh s = \cfrac{1}{\cfrac{1}{s} + \cfrac{1}{\cfrac{3}{s} + \cfrac{1}{\cfrac{5}{s} + \cdots}}}$$

Truncating the expansion with the nth term, recombining to form a rational approximation to $\tanh s$, and then equating numerator and denominator with $Q_o(s)$ and $Q_e(s)$ yields

$$Q_n(s) = B_n(s)$$

where $B_n(s)$ is a *Bessel polynomial* which can be computed from the recurrence formula:

$$B_0(s) = 1$$

$$B_1(s) = s + 1 \tag{4.22}$$

$$B_n(s) = (2n-1)B_{n-1}(s) + s^2 B_{n-2}(s)$$

Finally, we can write $s_{21}(s)$ as

$$s_{21}(s) = \frac{K_m}{B_n(s)} \tag{4.23}$$

As an example of the linear phase approximation, we can design an $n = 3$ low-pass filter. From (4.22) we have

$$B_3(s) = s^3 + 6s^2 + 15s + 15$$

and using (4.23),

$$s_{21}(s) = \frac{K_m}{s^3 + 6s^2 + 15s + 15}$$

At zero frequency we have $s_{21}(0) = K_m/15$, or $K_m = 15$ for 0-dB minimum insertion loss, and we can express $|s_{21}(s)|^2$ as

$$|s_{21}(s)|^2 = s_{21}(s)\, s_{21}(-s) = \frac{225}{-s^6 + 6s^4 - 45s^2 + 225}$$

and

$$|s_{11}(s)|^2 = \frac{s^6 - 6s^4 + 45s^2}{s^6 - 6s^4 + 45s^2 - 225}$$

Solving for the poles and zeros, multiplying the left half-plane roots to form $s_{11}(s)$, and then using $+s_{11}(s)$, we have

$$Z(s) = \frac{2s^3 + 10.406405s^2 + 21.708204s + 15}{1.593595s^2 + 8.291796s + 15}$$

which is the same topology as the Butterworth filter shown in Figure 4.6. The element values are

$L_1 = 1.255\text{H}$
$C_1 = 0.553\text{F}$
$L_2 = 0.192\text{H}$

The linear phase approximation was first proposed by W.E. Thomson [12,13] and is known as a Thomson, Bessel, or Bessel-Thomson filter. Figure 4.16 shows the amplitude response of the Thomson, Butterworth, and Chebyshev low-pass filters, which have identical topologies. As shown in the figure, insertion loss of the Bessel filter gradually increases outside the passband, and the stopband rejection is not as good as that of the Butterworth or Chebyshev filters, although the slope does eventually reach $6n$ dB per octave.

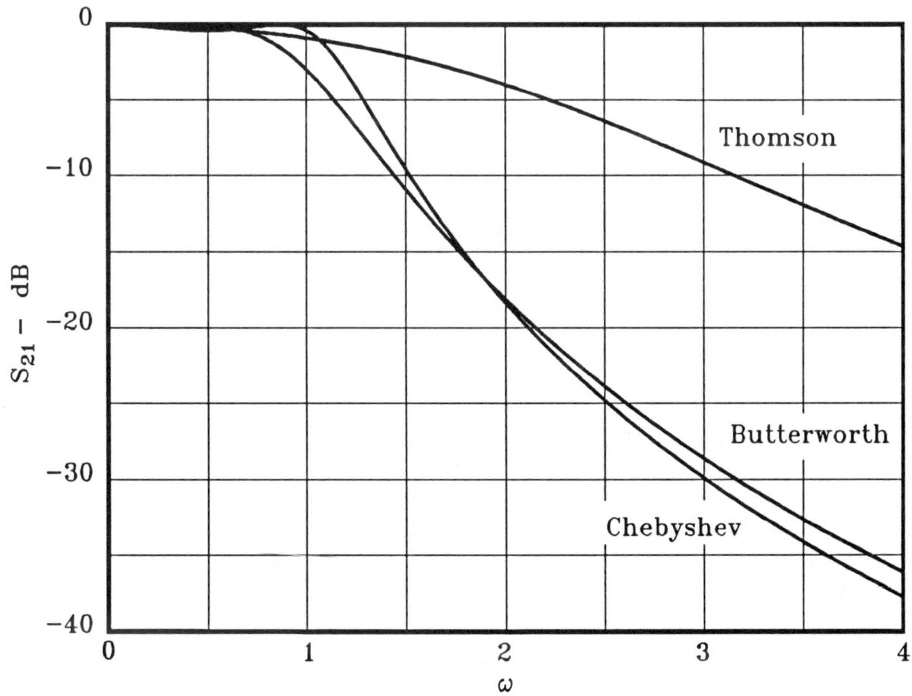

Figure 4.16 $N = 3$ low-pass filter response.

However, the group delay response of the Bessel filter is far superior to any of the filter designs based on amplitude approximation, as shown in Figure 4.17. The group delay is essentially constant at 1.0 sec from $\omega = 0$ to $\omega = 1$, compared with deviations of between 40% and 100% for the other four filters. Figure 4.18 shows the time domain response of the five filters to an ideal 1V step function. As shown, the Thomson filter has the fastest rise time of any other filter and virtually no overshoot or ringing.

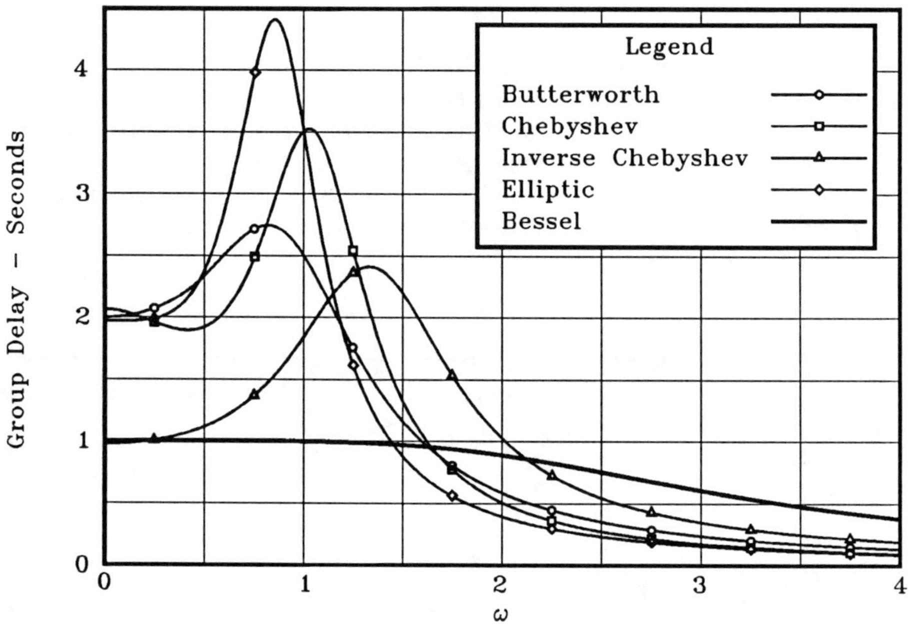

Figure 4.17 Low-pass filter group delay.

The Thomson filter can be characterized as a *Gaussian*-type filter; that is, a filter whose impulse response exhibits no overshoot or ringing and has symmetry about the time its amplitude is maximum. The Gaussian filter [14], which is based on an amplitude approximation to the Gaussian response curve, offers slightly better time domain characteristics, but less stopband rejection than the Bessel filter. A third type of filter that can be classified as Gaussian is based on an equal ripple error to the linear phase response [14]. This filter has better frequency domain characteristics than either the Bessel or Gaussian filters, but less than ideal time domain response.

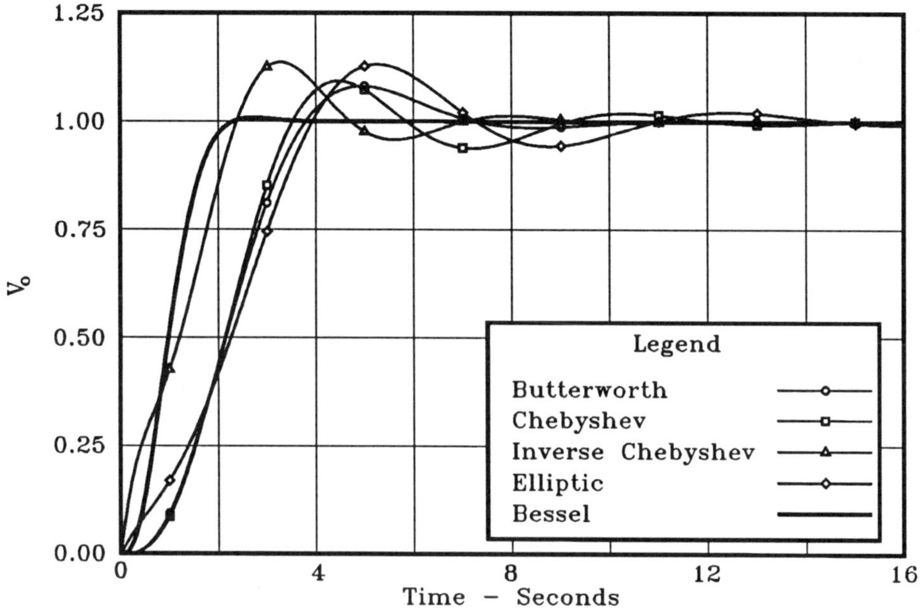

Figure 4.18 Low-pass filter step response.

Occasionally, simultaneous linear phase and high stopband rejection specifications will be encountered that cannot be realized using any of the filter types discussed in this chapter. One possible solution is to design a transitional filter; that is, a filter with Gaussian characteristics in the passband, and Butterworth or Chebyshev characteristics in the stopband. It is also possible to introduce finite frequency transmission zeros into $|s_{21}(s)|^2$ in such a manner that the passband phase and delay is unaffected, but the stopband becomes equal ripple, similar to the inverse Chebyshev and elliptic filters. In some cases, the transitional or Gaussian filter with equal-ripple stopband will also prove inadequate, and it will be necessary to design the filter based on the amplitude requirements and then add a phase equalizer. The phase equalizer is an all-pass network designed to compensate the phase deviation of the filter without changing the amplitude response [10].

4.5 FREQUENCY AND IMPEDANCE SCALING

The low-pass filters of Section 4.4 had cutoff frequencies of 1 rad/sec and 1Ω source and load terminations. These values are not practical for most applications. However, this normalization generally results in numeric values within an order of magnitude of 1.0, which helps improve numerical stability. The final step is to scale the element values, which results in a network with the desired impedance and frequency parameters.

The relationship between input reflection coefficient and input impedance was derived in Section 1.3, and Equation (4.11a) is a special case when we assume $Z_o^* = Z_o$, or the reference impedance is real. We set $Z_o = 1$ to write (4.11b), and, in effect, have formed a normalized input impedance:

$$Z_n(s) = \frac{Z(s)}{Z_0}$$

It is necessary to multiply $Z_n(s)$ by Z_o to scale to the desired value, which is easily accomplished by multiplying the impedance of each element by Z_o. Figure 4.1 shows the impedance of the inductor and capacitor, and when multiplied by Z_o,

$$L = Z_0 L_n$$

$$C = \frac{C_n}{Z_0}$$

For the terminating resistors,

$$R = Z_0 R_n$$

The subscript n represents the normalized element value, and Z_o is a real reference impedance.

We developed the approximations for a cutoff frequency of 1 or a normalized frequency of

$$\omega_n = \frac{\omega}{\omega_r}$$

where ω_r is the reference equal to the cutoff frequency for the low-pass filters. The frequency variable is actually $s_n = j(\omega/\omega_r)$, and the input impedance of (4.11) should properly be written as $Z_n(s_n)$. Since $s = j\omega$, $s_n = (s/\omega_r)$ and we can transform $Z_n(s_n)$ into $Z_n(s)$ by replacing each occurrence of s_n with (s/ω_r). Noting that the reactance of the normalized element at s_n must equal that of the actual element at s, we can solve for the element value. For example, from Figure 4.1, the reactance of the inductor is

$$s_n L_n = sL$$

Substituting (s/ω_r) for s_n,

$$\frac{sL_n}{\omega_r} = sL$$

or

$$L = \frac{L_n}{\omega_r}$$

Similarly, the capacitor value is

$$C = \frac{C_n}{\omega_r}$$

Combining impedance and frequency scaling, we can write the element values as

$$L = \frac{Z_0 L_n}{\omega_r} \tag{4.24a}$$

$$C = \frac{C_n}{\omega_r Z_0} \tag{4.24b}$$

and

$$R = Z_0 R_n \tag{4.24c}$$

As a practical example, we will scale the Butterworth filter shown in Figure 4.6 to operate in a 50Ω system with a cutoff frequency of 1.0 GHz. Thus,

$$Z_0 = 50$$

$$\omega_r = 2\pi f_c = 2\pi 10^9 \simeq 6.283 \cdot 10^9$$

The normalized L_1 and L_2 value is 1.0, and

$$L = \frac{Z_0 L_n}{\omega_r} = \frac{50 \cdot 1.0}{6.283 \cdot 10^9} = 7.958 \text{ nH}$$

The normalized capacitor is 2.0, and the design value is

$$C = \frac{C_n}{\omega_r Z_0} = \frac{2.0}{6.283 \cdot 10^9 \cdot 50} = 6.366 \text{ pF}$$

4.6 BANDPASS FILTER SYNTHESIS

The bandpass filter is perhaps the most common type of filter and is designed to pass a range of frequencies and reject all others. This filter has a passband extending from a lower frequency f_1 to an upper frequency f_2, where, in the ideal case, signals are transmitted from source to load without distortion. The bandpass filter has an upper stopband extending from f_2 to infinity and a lower stopband extending from zero to f_1. Ideally, the bandpass filter provides total isolation between source and load in both stopbands.

The bandpass filter finds numerous applications in systems where extraneous signals cause performance degradation. For example, strong off-frequency signals may overload a receiver preamplifier to the point where the desired weaker signal cannot be received at all. Up and down conversion can also generate many unwanted signals; the local oscillator (LO) may contain many harmonics above and below the desired frequency, and even if the LO is free from harmonics, the mixer generates products equal to the sum and difference of the RF and LO, as well as higher-order products. Extraneous input signals are generally not a problem with satellite communication systems which use highly directional antennas. However, receiver sensitivity is affected to a large extent by the IF bandwidth, and bandpass filters are often used to reduce bandwidth and suppress unwanted noise. It also turns out that the bandpass filter is the most efficient structure to use for matching networks and is frequently found in many types of RF and microwave components.

4.6.1 Low-Pass to Bandpass Transformation

Figure 4.4 shows the response of the Butterworth low-pass approximation, which in fact appears to be a bandpass filter centered about zero frequency. A passband extends from -1 to +1, and insertion loss monotonically increases in the two stopbands below -1 and above +1. All low-pass approximations developed in Section 4.4 have symmetry about zero as a result of the transducer gain, which is a function of ω^2, and the most obvious approach to the bandpass approximation problem is to modify the low-pass approximation so that the response is shifted to positive frequencies.

Consider the transform

$$s \Rightarrow K_f\left\{\frac{s}{\omega_0} + \frac{\omega_0}{s}\right\} \tag{4.25a}$$

or

$$\omega \Rightarrow K_f\left\{\frac{\omega}{\omega_0} - \frac{\omega_0}{\omega}\right\} \tag{4.25b}$$

A low-pass frequency of -1 should transform into a bandpass frequency of ω_1, and +1 into ω_2, providing two equations in two unknowns:

$$-1 = K_f\left\{\frac{\omega_1}{\omega_0} - \frac{\omega_0}{\omega_1}\right\}$$

$$+1 = K_f\left\{\frac{\omega_2}{\omega_0} - \frac{\omega_0}{\omega_2}\right\}$$

which has the solution

$$\omega_0 = \sqrt{\omega_1\omega_2}$$

$$K_f = \frac{\omega_0}{\omega_2 - \omega_1} \tag{4.25c}$$

and (4.25b) becomes

$$\omega \Rightarrow \frac{\omega}{\omega_2 - \omega_1} - \frac{\omega_1\omega_2}{\omega(\omega_2 - \omega_1)} \tag{4.25d}$$

The factor ω_o is known as the *center frequency* and corresponds to $\omega = 0$ from the low-pass response. The inverse of K_f (i.e., upper minus lower passband edge divided by the center frequency) is called the *fractional bandwidth* and usually denoted by w. We can write (4.25d) in terms of the fractional bandwidth as

$$\omega \Rightarrow \frac{1}{w}\left(\frac{\omega}{\omega_0} - \frac{\omega_0}{\omega}\right) \tag{4.25e}$$

The transformation in (4.25) relates the low-pass to bandpass frequencies in the following manner:

Low-Pass	*Bandpass*
$-\infty$	0
-1	ω_1
0	ω_o
+1	ω_2
$+\infty$	$+\infty$

which confirms we have the necessary mapping and must now decide how to implement it. We have two choices, we can either apply (4.25) directly to $G_t(\omega^2)$, then repeat the synthesis, or we can apply (4.25) to the synthesized elements and then compute the modified values. The latter is the simpler approach, and we will investigate the effect of this transformation on the low-pass elements.

It will simplify matters somewhat if we normalize to the upper passband edge ω_2 and introduce two new variables:

$$\omega_a = 1 - \omega_{1n} \tag{4.26a}$$

and

$$\omega_b = \frac{1 - \omega_{1n}}{\omega_{1n}} \tag{4.26b}$$

where ω_{1n} represents the normalized lower passband edge, ω_1/ω_2. Substituting these variables, the low-pass to bandpass transformation becomes

$$s \Rightarrow \frac{s}{\omega_a} + \frac{1}{s\omega_b} \tag{4.26c}$$

As shown in Figure 4.1, the impedance of the series inductor is sL, and when (4.26) is applied, the input impedance becomes

$$Z(s) = \left\{\frac{s}{\omega_a} + \frac{1}{s\omega_b}\right\}L = s\frac{L}{\omega_a} + \frac{1}{s}\frac{L}{\omega_b}$$

which is the impedance of a series inductor equal to (L/ω_a) and a series capacitor equal to (ω_b/L). Similar calculations show that the low-pass capacitor becomes a parallel *L-C* combination when transformed into a bandpass element. Figure 4.19 shows the transformed topology and element values for each of the four low-pass elements. A bandpass filter is easily designed by synthesizing a low-pass filter and then transforming the elements. This process can be most easily illustrated with an example. A bandpass filter is required with the following specifications:

Passband:	1.25 GHz to 1.75 GHz
Maximum VSWR:	2.0:1
Stopband rejection:	15-dB minimum at 1 GHz and 2 GHz

In Section 4.4.2 we analyzed the 2.0:1 VSWR specification and determined the maximum passband ripple to be 0.512 dB. We can use the same values for the bandpass filter, $\epsilon^2 = 0.1$ or approximately 0.414-dB ripple. The next step is to determine the filter order necessary to meet the stopband requirements; however, since we intend to use the low-pass approximation, we must relate the bandpass stopband to the low-pass stopband. Canceling the (2π) terms from the numerator and denominator of (4.25d) and then substituting numerical values for f_1 and f_2, we have

$$\omega \Rightarrow \frac{f}{f_2 - f_1} - \frac{f_1 f_2}{f(f_2 - f_1)} = 2f - \frac{4.375}{f}$$

A bandpass frequency of 1.0 GHz transforms into $\omega = -2.375$, and 2.0 GHz into $\omega = 1.8125$ for the low-pass approximation. The low-pass approximation is symmetrical about zero, and the bandpass rejection specification becomes 15 dB at $\omega = 1.8125$ and $\omega = 2.375$ for the low-pass filter. We can set $K_m = 1$ and $\epsilon^2 = 0.1$ and then evaluate (4.16) at $\omega = 1.8125$ to determine the required order. $n = 2$ only provides 6.131-dB rejection; however, $n = 3$ provides 15.414 dB at $\omega = 1.8125$ and 23.362 dB at $\omega = 2.375$. The $n = 3$ Chebyshev low-pass filter shown in Figure 4.10 has 0.414-dB ripple and 0-dB minimum insertion loss, and will meet the bandpass specifications when transformed.

We normalize to the upper passband edge of 1.75 GHz; then ω_{1n} is 1.25/1.75, or 5/7; and computing ω_a and ω_b from (4.26), we have $\omega_a = 2/7$ and $\omega_b = 2/5$. Figure 4.19 shows that the series 1.506H inductor transforms into a series *L-C* network with

$$L = \frac{1.506}{\omega_a} = \frac{7(1.506)}{2} = 5.271\text{H}$$

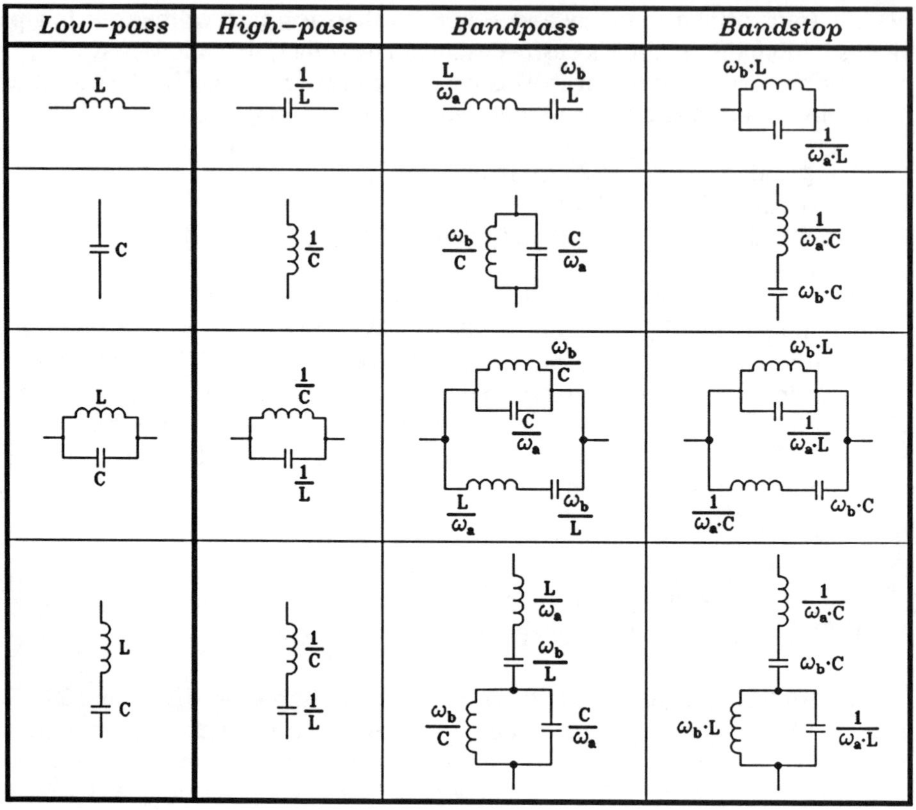

Figure 4.19 Frequency transformations from low-pass.

$$C = \frac{\omega_b}{1.506} = \frac{2}{5(1.506)} = 0.266\text{F}$$

and the shunt 1.115F capacitor transforms into a parallel *L*-*C* as

$$L = \frac{\omega_b}{1.115} = \frac{2}{5(1.115)} = 0.359\text{H and}$$

$$C = \frac{1.115}{\omega_a} = \frac{7(1.115)}{2} = 3.903\text{F}$$

The final steps are to scale the normalized element values to 50Ω and an upper passband frequency of 1.75 GHz:

$$Z_0 = 50$$

$$\omega_r = 2\pi 1.75 \cdot 10^9$$

The complete bandpass filter network is shown in Figure 4.20 along with the response from 1.0 to 2.0 GHz. The filter has an equal-ripple passband extending from 1.25 to 1.75 GHz as required, although the passband is not symmetrical about the center frequency of 1.5 GHz. The low-pass response was symmetrical about zero; however, the mapping of (4.25) transformed zero into ω_o, or approximately 1.48 GHz. The bandpass response does appear symmetrical when plotted using a logarithmic frequency scale.

4.6.2 Bandpass Approximation

Transforming the low-pass network into a bandpass network provides a somewhat limited selection of topologies. The low-pass filter consists of alternating series inductors and shunt capacitors and has only two possible topologies: a series input inductor resulting from the selection of $+s_{11}$ to form $Z(s)$, or its dual, a shunt input capacitor which results when $-s_{11}$ is used. The transformed bandpass network has only two possibilities as well: a series L-C combination, or shunt-connected parallel L-C at the input. However, we can apply the transform of (4.26) directly to $G_t(\omega^2)$ and form an input impedance for the bandpass network, which can be used to synthesize all possible topologies.

The approximation for the Chebyshev low-pass filter was given in Section 4.4.2 as equation (4.16) and is

$$G_t(\omega^2) = \frac{K_m}{1 + \varepsilon^2 T_n^2(\omega)}$$

Using the numerical values from the design of Section 4.6.1 and the $n = 3$ Chebyshev polynomial from (4.15), we have

$$G_t(\omega^2) = \frac{1}{1 + 0.1\left(4\omega^3 - 3\omega\right)^2}$$

$\omega_a = 2/7$, $\omega_b = 2/5$, and the low-pass to bandpass transform (4.26) becomes

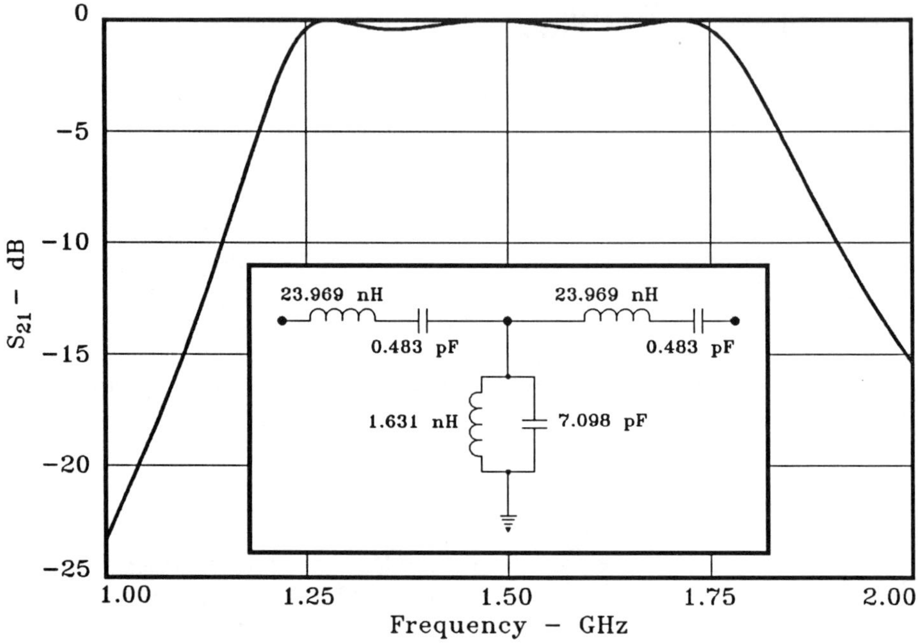

Figure 4.20 $N = 3$ bandpass filter response.

$$\omega \Rightarrow \frac{\omega}{\omega_a} - \frac{1}{\omega\,\omega_b} = \frac{7\omega^2 - 5}{2\omega}$$

When we replace ω with its transformed value and then expand it, the transducer gain becomes

$$\frac{\omega^6}{2941.225\,\omega^{12} - 12965.4\,\omega^{10} + 23549.4\,\omega^8 - 22554.75\,\omega^6 + 12015\,\omega^4 - 3375\,\omega^2 + 390.625}$$

and we can use (4.8) to compute the input reflection coefficient, $|s_{11}(s)|^2$:

$$\frac{2941.225s^{12} + 12965.4s^{10} + 23549.4s^8 + 22555.75s^6 + 12015s^4 + 3375s^2 + 390.625}{2941.225s^{12} + 12965.4s^{10} + 23549.4s^8 + 22554.75s^6 + 12015s^4 + 3375s^2 + 390.625}$$

$|s_{11}(s)|^2$ has twelve poles and zeros which are shown in Table 4.1.

Selecting the three different pairs of complex conjugate zeros and left half-plane poles, we expand to form $s_{11}(s)$ and then use $+s_{11}(s)$ to write the input impedance $Z(s)$ as

$$\frac{s^6 + 0.189689s^5 + 2.240063s^4 + 0.280203s^3 + 1.600045s^2 + 0.096780s + 0.364431}{0.189689s^5 + 0.035982s^4 + 0.280203s^3 + 0.025701s^2 + 0.096780s}$$

Table 4.1 Bandpass Filter Poles and Zeros

Zeros	*Poles*
0.0 ± j0.730443	0.039203 ± j0.708328
0.0 ± j0.730443	-0.039203 ± j0.708328
0.0 ± j0.845154	0.094844 ± j0.839816
0.0 ± j0.845154	-0.094844 ± j0.839816
0.0 ± j0.977879	0.055641 ± j1.005331
0.0 ± j0.977879	-0.055641 ± j1.005331

After extracting a 5.271787H series inductor, 0.265565F series capacitor, and a 0.358751H shunt inductor,

$$Z'(s) = \frac{0.004631s^2 + 0.000878s + 0.003308}{0.018072s^3 + 0.003428s^2 + 0.013787s}$$

and we have a series and shunt capacitor available for extraction. Thus far the synthesized network is identical to the transformed network shown in Figure 4.20; however, we can choose the series capacitor, which will result in a different topology. Extracting a 4.168027F series capacitor, the input impedance becomes

$$Z''(s) = \frac{0.001230s + 0.000233}{0.075325s^2 + 0.014288s + 0.057465}$$

and we see that there is only one element available for extraction. We extract the shunt 61.259894F capacitor, then a 0.021397H series inductor, and find that the termination is 0.004060. Finally, the element values are scaled to 50Ω and 1.75 GHz, which results in the bandpass filter shown in Figure 4.21.

This filter has exactly the same response as the one shown in Figure 4.20; however, the element values and load termination have changed considerably. Although this topology does meet the passband and stopband specifications, it

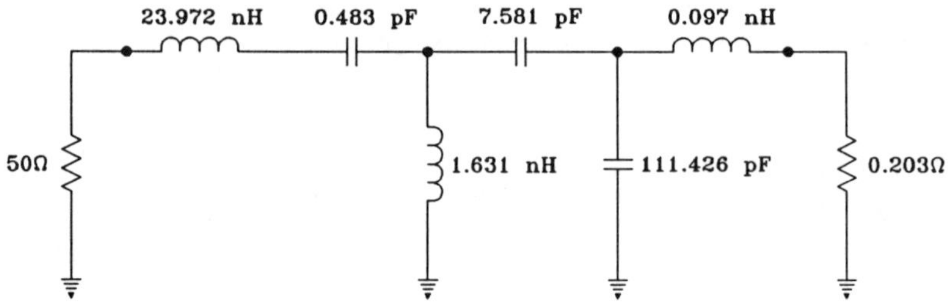

Figure 4.21 Cheybyshev bandpass filter.

would not be acceptable for this design because the output termination is much lower than 50Ω; nevertheless, it could be very useful for some applications. For example, the real and imaginary component of the input impedance of a power transistor is typically only a few ohms. A reactive element can be added to bring the impedance to resonance, and then a bandpass filter structure can be designed to transform 50Ω into the required value.

Transforming $G_t(\omega^2)$ is substantially more complicated than just transforming low-pass elements as illustrated above; however, it does provide considerable flexibility in selecting a topology. We formed $Z(s)$ from $+s_{11}(s)$ and then extracted a series inductor. However, there is also a series capacitor that could be selected. When $-s_{11}(s)$ is used, we have the choice between a shunt inductor and shunt capacitor which forms a dual network with exactly the same electrical characteristics. Considering only the input element, we have a total of four different topologies, although only two are unique because a series or shunt L-C network is the same, whether the inductor or capacitor is extracted first. Continuing this process, we consider all possible choices for the second element of each of the four networks, then all choices for the third element of all networks with different first and second elements, and so on until we have defined all possible topologies. Table 4.2 shows the total number of topologies and the number of unique topologies for the bandpass filter from $n = 1$ to $n = 9$. As shown, we have 102 unique networks for $n = 5$ and over 11,000 networks for $n = 9$!

Many networks will not have the desired load termination as in the above example. However, the techniques of Section 4.9.1 can be applied to transform the load at the expense of an additional element, thus creating another class of topologies that are nonminimum. When faced with an immense number of networks, it becomes difficult to select the one best suited to the application. However, if we can define desirable attributes of the topology, a computer can be used to sort through all possibilities and narrow the choices to only a few networks.

Table 4.2 Bandpass Network Topologies

N	*Total Topologies*	*Unique Topologies*
1	4	2
2	12	6
3	40	14
4	140	38
5	504	102
6	1848	282
7	6864	786
8	25740	3000
9	97240	11492

For example:

- The network should have element values within a specified range.
- If constrained optimization is necessary to limit the range of element values, the optimized response should be as close as possible to the target.
- Element sensitivity should be at a minimum; that is, changes in element values should have minimal effect on the response, resulting in higher production yields.
- The network should have a minimum number of certain types of elements. Inductors are usually labor-intensive, larger, and not as well controlled as capacitors; however, they do sometimes provide a means of tuning a circuit.
- The network should have certain topology constraints:
    - When accurately characterized, the equivalent circuit for most lumped elements is a combination of resistance, inductance, and capacitance. Topologies can be selected that are useful in absorbing these parasitic elements, resulting in structures that are useful at much higher frequencies.
    - Shunt capacitors at each node can be used to absorb the parasitic capacitance associated with bonding pads.

- Series inductance at the input and output port can be used to absorb the series inductance associated with connector pins or transistor leads.

- No DC path from the input port to ground, a DC path from the output to ground which provides a DC block at the input, and a convenient means to ground the gate of an FET when used as an input matching network.

- No DC path from the output port to ground and a DC path from the input to ground which can be used to provide bias when used as an output matching network. The RF ground can be realized using a bypass capacitor and bias injected at this low impedance point, resulting in little effect on the RF performance and only one additional element.

Although transforming $G_t(\omega^2)$ rather than directly transforming the low-pass elements has the advantage of a broader selection of topologies, it has the disadvantage that numerical accuracy becomes much more difficult to maintain. When the numeric values of the $n = 3$ bandpass filter are compared with those of the equivalent $n = 3$ low-pass filter in Section 4.4.2, the problem becomes evident. The low-pass filter has six poles and zeros that are widely separated, as shown in Figure 4.9. However, the $n = 3$ bandpass filter has 12 poles and zeros that are closely grouped, as shown in Table 4.1. Also, the coefficients of $Z(s)$ range from 2.4 to 0.8 for the low-pass filter compared with 2.2 to 0.03 for the bandpass filter. The number of poles and zeros increases rapidly as n is increased (20 each for $n = 5$), and the grouping of roots, as well as the range of coefficients, is affected by the filter bandwidth. Double precision is usually required for all computations. However, even that is not sufficient for high-order or narrow-bandwidth designs, and other methods must be employed.

One approach is to compute the low-pass poles and zeros and then apply the transform of (4.25) directly to these roots, or, alternatively, transform the low-pass $Z(s)$. This approach has the advantage that there are one-half the number of roots and they are well separated, so the accuracy is enhanced. However, the coefficients of $Z(s)$ are unchanged, and small numerical errors at each stage of the extraction process tend to be compounded such that the resulting element values are not valid. Better numerical accuracy can be maintained if the poles and zeros are not multiplied to form polynomials, but instead carried as products throughout the synthesis procedure. Coefficients are computed directly from these products when required, reducing errors due to polynomial multiplication and addition. Perhaps the most accurate method is to transform $j\omega$ into a plane where the roots are well separated and coefficients of the input impedance function are well behaved

[15,16]. Once each element has been extracted, its value is obtained by transforming back into the s-plane.

Depending on the scope of the synthesis program, some of the accuracy enhancement techniques may be implemented, or perhaps none at all. Usually a program intended strictly for filter synthesis will be the most accurate for large-order or small-bandwidth filters. However, some techniques cannot be applied in all situations, and a general-purpose program intended for the design of a wide variety of filter types and matching networks may not be as accurate. Loss of numerical accuracy is easy to ascertain for odd-order filters which are symmetrical. The values of symmetric elements, as well as the input and output terminations, should be equal for 0-dB minimum insertion loss. Generally, the element values in the first half of the network will be the most accurate and can be used to correct errors in the second half of the network. Provided the input and output terminations are equal, (4.12b) can be used to compute $s_{22}(s)$ and the network synthesized from the output side to determine element values in the second half of any network.

The numerator of $s_{21}(s)$ for the bandpass filter of this section is odd, and, from (4.12b), we need only replace s in the numerator of $s_{11}(s)$ with $-s$ to form $s_{22}(s)$. The numerator of $s_{11}(s)$ is even; hence, the input and output reflection coefficients are equal, and $Z_{in} = Z_{out}$ for equal reference impedance at both ports. Generally we only know the exact value of the output termination in the special case where the input and output terminations are equal. However, for the purposes of an example, we assume the value of 0.004060 shown in Figure 4.21 is correct and multiply $Z(s)$ by 0.004060 to change the normalization at the output port from 1 to 0.004060Ω. Examining Z_{out}, we see that a series inductor equal to 0.021403H is available which agrees to five decimal places with the value computed using $s_{11}(s)$. Had we transformed Z_{in} and Z_{out} from the low-pass equivalent and used a known value for the output termination, the last element could be calculated with a high degree of confidence, even for much higher-order filters.

4.7 HIGH-PASS FILTER SYNTHESIS

The high-pass filter is designed to pass all frequencies above the cutoff frequency f_c and reject all frequencies below. Looking at Figure 4.4, we see that the low-pass filter can be transformed into a high-pass filter by reversing the frequency scale. The low-pass to high-pass transform is

$$s \Rightarrow \frac{1}{s} \tag{4.27}$$

which maps ∞ into 0 and 0 into ∞, reversing the low-pass frequencies as desired. We used this transform during the development of the inverse Chebyshev approximation in Section 4.4.3, and the frequency reversing effect is shown in Figure 4.11.

The low-pass elements shown in Figure 4.1 are transformed into high-pass elements by replacing s with $1/s$ and we see that low-pass inductors become high-pass capacitors, and capacitors are transformed into inductors, as shown in Figure 4.19. Thus, design of the high-pass filter is accomplished very simply by replacing s with $1/s$ in $G_t(\omega^2)$ or designing a low-pass structure and then transforming the elements. We illustrate the latter by transforming the $n = 3$ Chebyshev low-pass filter shown in Figure 4.10 into a high-pass filter with a cutoff frequency of 1.5 GHz, stopband extending from 0 to 1.0 GHz, and 50Ω source and load impedance.

The normalized high-pass stopband frequency is (1.0/1.5), which is inverted for use with the low-pass approximation. We evaluate the transducer gain of Section 4.4.2 at $\omega = 1.5$ and determine that the insertion loss of the high-pass filter is 9.590 dB at 1.0 GHz. The low-pass filter in Figure 4.10 is transformed by replacing the 1.506H series inductors with (1/1.506)F series capacitors, and the shunt 1.115F capacitor with a (1/1.115)H inductor, and then normalizing to 1.5 GHz and 50Ω. The high-pass filter is shown in Figure 4.22.

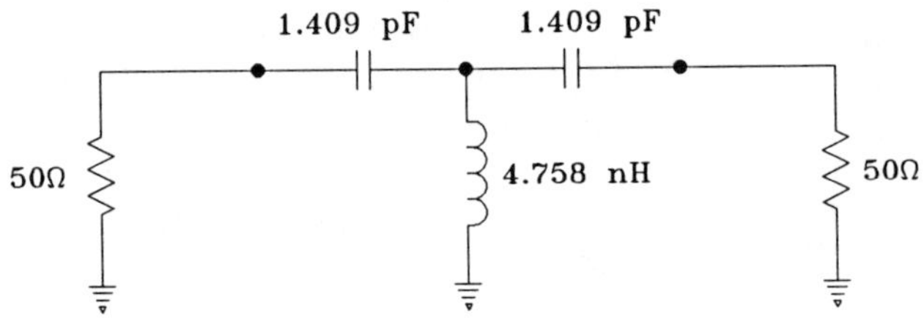

Figure 4.22 Cheybyshev high-pass filter.

Figure 4.23 shows the response of the high-pass filter at positive and negative frequencies from -4.0 to +4.0 GHz. As with all approximations that are functions of ω^2, the response is symmetrical about zero.

4.8 BANDSTOP FILTER SYNTHESIS

The bandstop filter is intended to reject a range of frequencies and pass all others, essentially the complement of the bandpass filter. This is another common filter type often used in communications systems to suppress strong signals causing interference. Bandstop filters can also be purchased at many consumer electronics outlets and used to attenuate strong local television and FM signals causing interference with other stations.

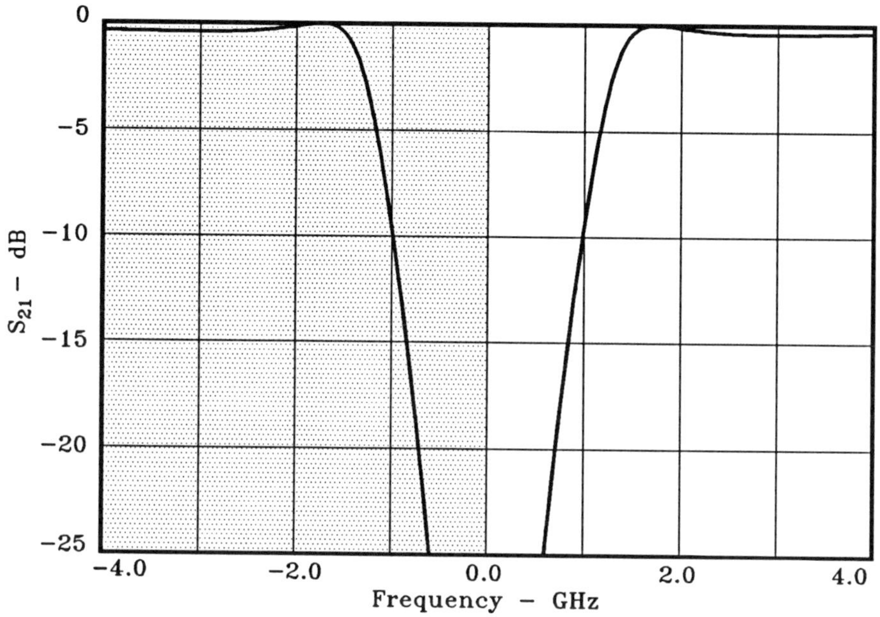

Figure 4.23 $N = 3$ **high-pass filter response.**

The high-pass response shown in Figure 4.23 has the characteristics of a bandstop filter centered about zero. There is an equal-ripple passband extending from negative infinity to -1.5 GHz, another passband from 1.5 GHz to positive infinity, and a stopband extending from -1.5 to +1.5 GHz. Since we used the transform given by (4.26) to shift the low-pass response of Figure 4.4 to positive frequencies resulting in a bandpass approximation, and the transformation of (4.27) to reverse the low-pass response forming a high-pass response, it seems reasonable that we can combine the two for the bandstop approximation, easily accomplished by inverting (4.26); that is,

$$s \Rightarrow \frac{1}{\dfrac{s}{\omega_a} + \dfrac{1}{s\omega_b}} = \frac{s\omega_a}{s^2 + \dfrac{\omega_a}{\omega_b}} \qquad (4.28)$$

Applying (4.28) to the low-pass series inductor, we have

$$sL \Rightarrow \left\{ \frac{s\omega_a}{s^2 + \dfrac{\omega_a}{\omega_b}} \right\} \cdot L = \frac{s(\omega_a L)}{s^2 + (\omega_a / \omega_b)}$$

which has the form of the parallel *L-C* network shown in Figure 4.13. Equating coefficients, we find that

$$C_{bs} = \frac{1}{\omega_a L_l}$$

and

$$L_{bs} = \omega_b L_l$$

where the subscripts *bs* and *l* denote the bandstop and low-pass element values, respectively. Similarly, it can be shown that the low-pass capacitor is transformed into a series *L-C*, as shown in Figure 4.19.

We can design a bandstop filter by simply transforming the elements of an appropriate low-pass network. As an example, we can convert the $n = 3$ Chebyshev low-pass filter of Figure 4.10 into a bandstop filter with a passband extending from 0 to $f_1 = 1.25$ GHz, another from $f_2 = 1.75$ GHz to infinity, and a stopband from $f_s = 1.35$ to 1.65 GHz. The low-pass filter was designed for $K_m = 1$ and $\epsilon^2 = 0.1$, which provides 0-dB MIL and 0.414-dB ripple in the passband of the bandstop filter. In order to predict the stopband characteristics, we must compute the equivalent low-pass frequency for 1.35 and 1.65 GHz and then evaluate the low-pass approximation. Replacing *s* in (4.28) with $j\omega$, we have

$$\omega \Rightarrow \frac{\omega\,\omega_a}{\omega_{1n} - \omega^2} \tag{4.29}$$

where (ω_a/ω_b) has been reduced to ω_{1n}. Normalizing to f_2, $\omega_{1n} = (1.25/1.75)$, $\omega_a = (0.5/1.75)$, and $\omega_b = (1.25/1.75)$. At $f_s = 1.35$ GHz, $\omega = (1.35/1.75)$, which has an equivalent low-pass frequency of 1.849315 when substituted into (4.29); then, evaluating $G_t(\omega^2)$ at 1.849315, we find the stopband rejection to be 16.02 dB. Similarly, $f_s = 1.65$ GHz has an equivalent low-pass frequency of -1.542056 and 10.45-dB stopband rejection.

The series 1.506H inductors in the low-pass filter of Figure 4.10 are transformed into series-connected parallel *L-C* networks. From Figure 4.19,

$$L_{bs} = \omega_b L_l = (0.5/1.25)\cdot(1.506) = 0.602400\text{H}$$

$$C_{bs} = \frac{1}{\omega_a L_l} = \frac{1}{(0.5/1.75)\cdot(1.506)} = 2.324037\text{F}$$

and the shunt series L-C is calculated from the 1.115F shunt capacitor as

$$L_{bs} = \frac{1}{\omega_a C_l} = \frac{1}{(0.5/1.75)\cdot(1.115)} = 3.139013\text{H}$$

$$C_{bs} = \omega_b C_l = (0.5/1.25)\cdot(1.115) = 0.446000\text{F}$$

Figure 4.24 shows the complete bandstop filter normalized to 1.75 GHz and 50Ω.

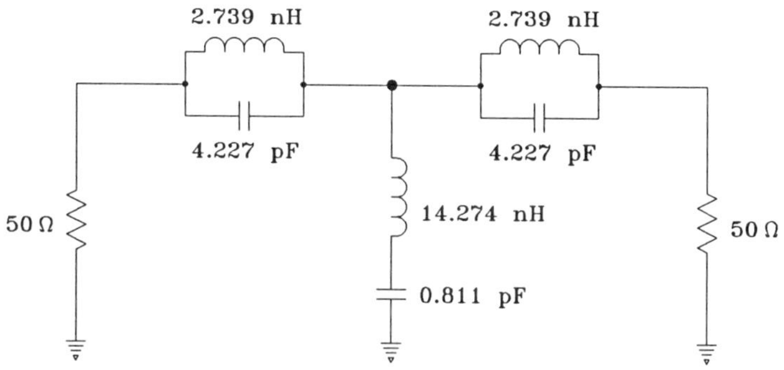

Figure 4.24 Cheybyshev bandstop filter.

The response of the filter from 1.0 to 2.0 GHz is shown in Figure 4.25. Substituting into (4.29), we find that a bandstop frequency of 0 maps into the low-pass approximation as $\omega = 0$, and 1.25 GHz maps into $\omega = +1$, which reveals that the lower passband is a mapping of the positive frequency portion of the low-pass passband. Similarly, 1.75 GHz maps into $\omega = -1$ and infinity maps into $\omega = 0$, which indicates that the upper passband is equivalent to the negative portion of the low-pass passband, but reversed. Thus, the bandstop filter has an insertion loss at zero and infinity equal to the DC value of the equivalent low-pass filter, minimum in the case of odd-order Chebyshev, and maximum for even-order equal-ripple filters. The low-pass to bandstop transformation given by (4.18) does not affect the terminations, and even-order equal ripple or nonzero minimum insertion loss results in structures with unequal terminations.

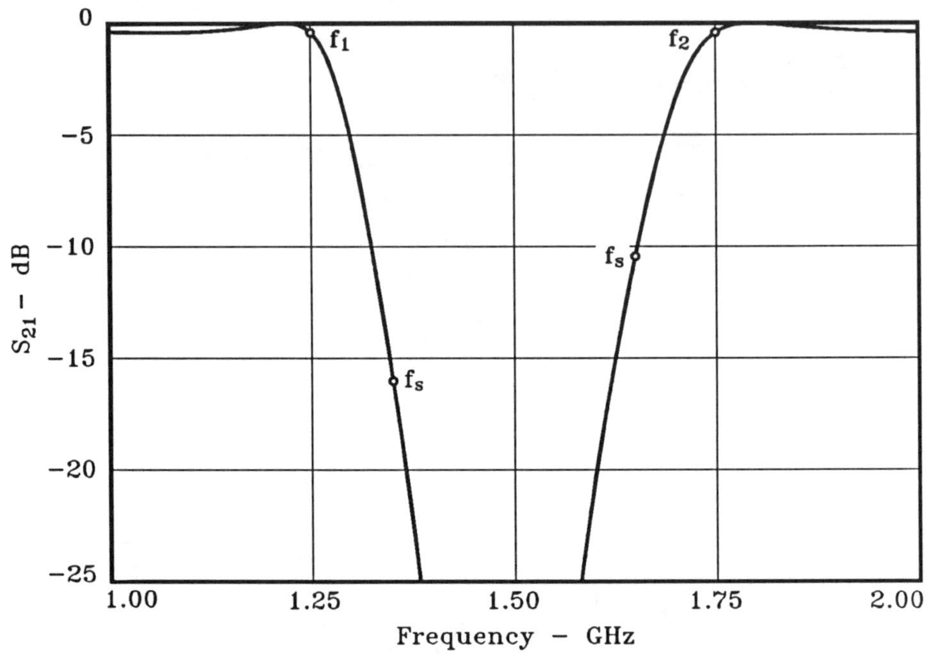

Figure 4.25 N =3 bandstop filter response.

The transformation of (4.28) can be applied directly to $G_t(\omega^2)$ to derive the bandstop approximation function. However, we end up with finite frequency transmission zeros and must use the techniques of Section 4.4.3 to extract these elements. All transmission zeros are located at ω_o, and the expressions given in Figure 4.13 can easily be used to determine element values, although the synthesis becomes more complicated than necessary. The bandstop filter has only two possible topologies, a series-connected parallel L-C as the input element, or its dual, a shunt-connected series L-C input element. Either of these configurations can be realized by simply transforming the appropriate low-pass network.

4.9 TERMINATION ADJUSTMENT

Perhaps the single most difficult task associated with synthesis is designing a network with a prescribed output termination. The termination at one port will always be correct since we can reference $Z(s)$ to the desired value; however, the termination at the other port cannot be specified. We observed during the development of the various low-pass filter approximations in Section 4.4 that the output termination is dependent on the insertion loss at zero frequency since the structure is transparent and any mismatch must be a result of unequal source and load terminations. This observation is valid for the other filter types as well, since they are all transformed from the low-pass approximation and the terminations are invariant under the mappings. However, we have an additional problem with structures containing both low-pass and high-pass elements (poles at $s = \infty$ and $s = 0$, respectively). The bandpass filter shown in Figure 4.20 was designed by transforming the elements of a low-pass filter, while the filter shown in Figure 4.21 was synthesized using the bandpass approximation. Both filters have exactly the same response and number of elements; however, there is almost a 250:1 difference in the output termination! As it turns out, we can use this characteristic to our advantage.

4.9.1 Network Transformations

If we examine the topology of the two bandpass filters in Figures 4.20 and 4.21, we see that the only difference is the orientation of the last two capacitors. The structure in Figure 4.20 has a shunt capacitor followed by a series capacitor, while the filter shown in Figure 4.21 has the opposite arrangement, a series capacitor followed by a shunt capacitor. Close inspection of the inductor and termination resistor following the two capacitors shows that the impedance has been reduced by the same factor. We can conclude that, at least in this case, the shunt-series capacitor topology is equivalent to a series-shunt configuration followed by an ideal transformer, as shown in Figure 4.26.

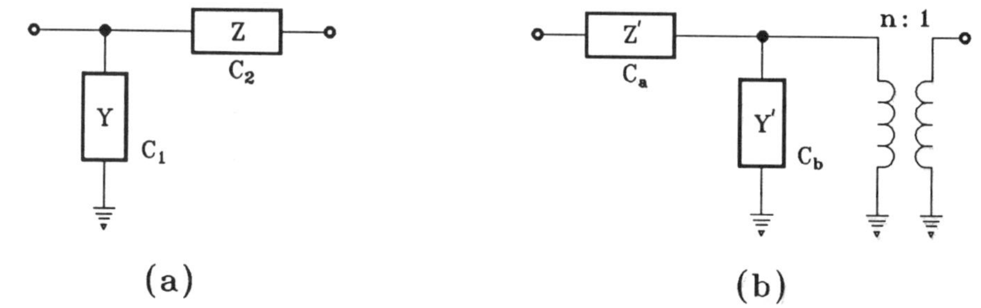

Figure 4.26 (a) Original network; (b) transformed network.

Using *ABCD*-parameters for the series Z, shunt Y, and ideal transformer developed in Section 2.1, the original network is

$$\begin{bmatrix} 1 & 0 \\ Y & 1 \end{bmatrix} \begin{bmatrix} 1 & Z \\ 0 & 1 \end{bmatrix} = \begin{bmatrix} 1 & Z \\ Y & (1 + ZY) \end{bmatrix} \tag{4.30a}$$

and the transformed network is

$$\begin{bmatrix} 1 & Z' \\ 0 & 1 \end{bmatrix}\begin{bmatrix} 1 & 0 \\ Y' & 1 \end{bmatrix}\begin{bmatrix} n & 0 \\ 0 & 1/n \end{bmatrix} = \begin{bmatrix} n(1 + Z'Y') & Z'/n \\ nY' & 1/n \end{bmatrix} \tag{4.30b}$$

Equating the B, C, and D parameters from the original and transformed networks yields

$$\begin{aligned} Z' &= nZ \\ Y' &= (1/n)Y \\ 1/n &= 1 + ZY \end{aligned} \tag{4.30c}$$

Substituting the appropriate capacitor values from the figure, we have

$$\frac{1}{sC_a} = n\frac{1}{sC_2}$$

$$sC_b = \frac{1}{n}sC_1$$

$$\frac{1}{n} = 1 + \frac{sC_1}{sC_2}$$

or

$$\begin{aligned} C_a &= \frac{C_2}{n} \\ C_b &= \frac{C_1}{n} \\ \frac{1}{n} &= 1 + \frac{C_1}{C_2} \end{aligned} \tag{4.31}$$

which can be used to calculate the element values of the transformed network. From the original network in Figure 4.20, $C_1 = 7.098$, $C_2 = 0.483$, and

$$\frac{1}{n} = 1 + \frac{C_1}{C_2} = 1 + \frac{(7.098)}{(0.483)} = 15.604938$$

$$C_a = \frac{C_2}{n} = (0.483)(15.604938) = 7.537185$$

$$C_b = \frac{C_1}{n} = (7.098)(15.604938) = 110.763852$$

We can remove the ideal transformer by multiplying the impedance of all elements connected to the secondary by n^2:

$$L' = n^2 L = (1/15.604938)^2(23.969) = 0.098430$$

$$R_L' = n^2 R_L = (1/15.604938)^2(50.000) = 0.205327$$

which is within about 1% of the values shown in Figure 4.21. Had we used higher precision arithmetic, the transformed element values would exactly equal those in the synthesized network of Figure 4.21.

The general form of (4.30) allows any type of elements, since we have derived the expressions in terms of impedance and admittance. However, closer inspection reveals that all elements must be the same type, either inductors or capacitors. Equation (4.30c) relates the turns ratio of the ideal transformer to $(1 + ZY)$, and since n is frequency-independent, s in Z and Y must cancel in ZY, which is true when both elements are capacitors or inductors. Similarly, the impedance and admittance of the original and transformed networks are related through n, which constrains the elements in both networks to be the same.

When we have two adjacent shunt-series inductors or capacitors, (4.30) provides a simple means to reduce the output termination. However, we are not restricted to one single transformation. The transformer turns ratio is given in (4.30c) as the reciprocal of $(1 + ZY)$; thus, n is bounded by

$$\frac{1}{1 + ZY} <= n <= 1$$

and as we reduce ZY, n becomes larger until it equals one when ZY is zero. We have three alternatives for adjusting ZY: we can reduce Z, reduce Y, or reduce both simultaneously. Figure 4.27(a) shows the case where we reduce Y_1 by splitting it into Y_a and Y and then transforming Y and Z_2 into Z_b, Y_c, and an ideal transformer. The network consisting of Y_a, Z_b, and Y_c has a pi configuration and is known as a *pi* transformation. We can also split Z_2 into Z and Z_c, as shown in Figure 4.27(b), and then transform Y_1 and Z into Z_a, Y_b, and an ideal transformer. The figure shows that Z_c has been multiplied by n^2 and moved to the primary side of the transformer to illustrate the tee configuration.

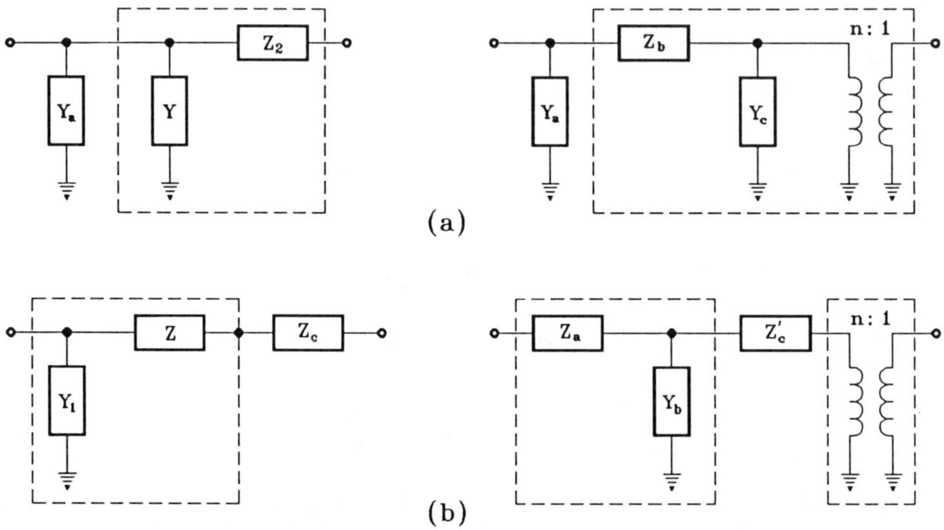

Figure 4.27 Partial network transformation: (a) Pi configuration; (b) Tee configuration.

Given n, C_1, and C_2, we can easily solve (4.31) for the capacitor values in either the tee or pi configuration. From the last expression in (4.31), we can solve for the input capacitor,which, together with C_2, provides the required n,

$$C = \left(\frac{1}{n} - 1\right)C_2$$

Figure 4.27(a) shows that the admittance of C_1 must equal the sum of the admittance of C_a and C, or

$$C_a = C_1 - C$$

and from the first and second expressions in (4.31),

$$C_b = \frac{C_2}{n}$$

$$C_c = \frac{C}{n}$$

Figure 4.28 shows the three capacitive transformations that can be used to reduce the output termination, and Figure 4.29 shows the equivalent inductive transformations. Two additional transformations are possible if we simultaneously reduce Z and Y; however, this results in four elements in a pi-tee configuration useful only in the rare instance where both the tee and pi transforms produce nonrealizable element values.

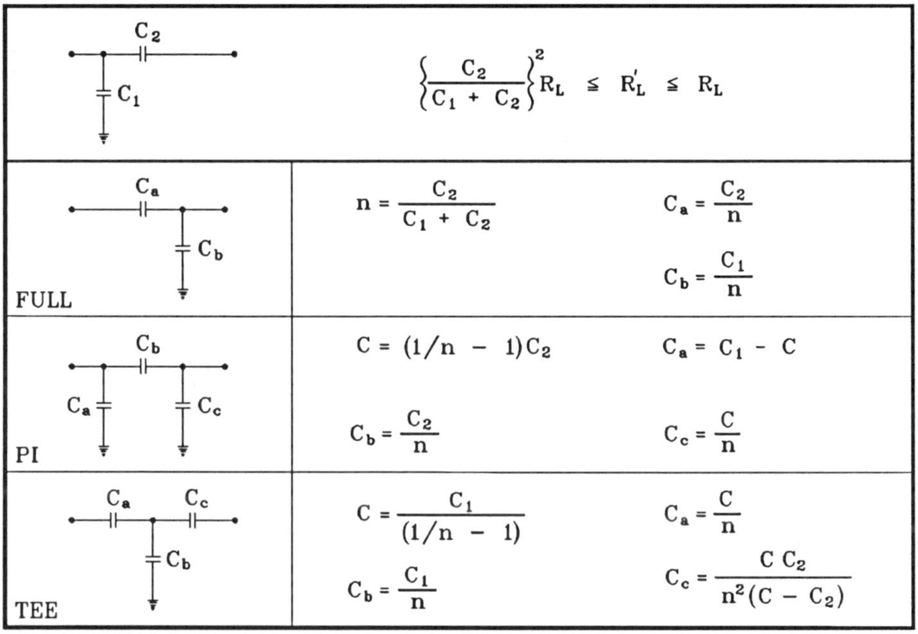

Figure 4.28 Capacitor transformations to reduce R_L.

We can use similar techniques to derive an equivalence for a dual network of the one shown in Figure 4.26(a); that is, a series impedance followed by a shunt admittance. As one might expect, the transformations shown in Figure 4.30 for capacitive elements and Figure 4.31 for inductive elements result in higher output terminations.

The network transformations shown in Figures 4.28 through 4.31 are attributable to E.L. Norton and are called *Norton transformations* [14,17]. The transforms are easily applied to adjust the output termination of a synthesized network by first computing the required impedance ratio as

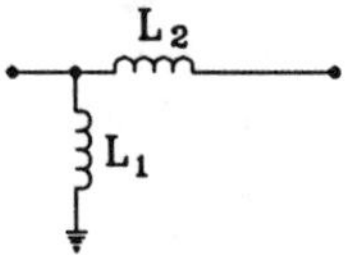	$\left\{\frac{L_1}{L_1 + L_2}\right\}^2 R_L \le R_L' \le R_L$	
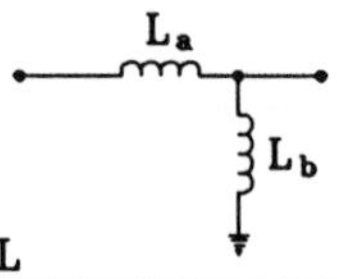 FULL	$n = \frac{L_1}{L_1 + L_2}$	$L_a = n\,L_2$ $L_b = n\,L_1$
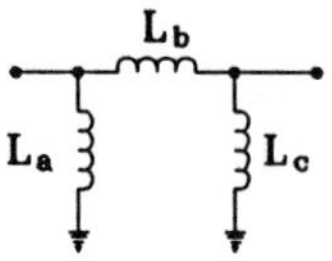 PI	$L = \frac{L_2}{(1/n - 1)}$ $L_b = n\,L_2$	$L_a = \frac{L\,L_1}{L - L_1}$ $L_c = n\,L$
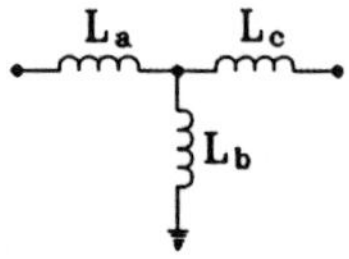 TEE	$L = (1/n - 1)L_1$ $L_b = nL_1$	$L_a = n\,L$ $L_c = n^2(L_2 - L)$

Figure 4.29 Inductor transformations to reduce R_L.

$$n^2 = \frac{R_L'}{R_L}$$

where the primed variable is the desired, or transformed, value. We then use Figures 4.28 through 4.31 to identify two elements that have the appropriate effect and compute the upper and lower bounds from the element values. Assuming the desired output is within range, we select either a pi or tee configuration, compute new element values, and then eliminate the ideal transformer by multiplying the impedance of all remaining elements by n^2:

$$L' = n^2 L$$

$$C' = \frac{C}{n^2} \tag{4.32}$$

$$R' = n^2 R$$

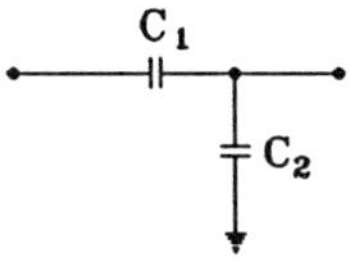	$R_L \leq R_L' \leq \left\{\frac{C_1 + C_2}{C_1}\right\}^2 R_L$	
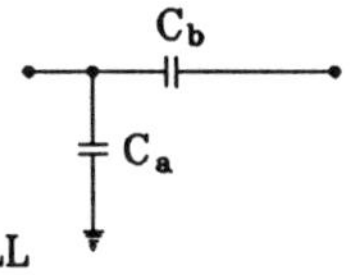 FULL	$n = \frac{C_1 + C_2}{C_2}$	$C_a = \frac{C_2}{n}$ $C_b = \frac{C_1}{n}$
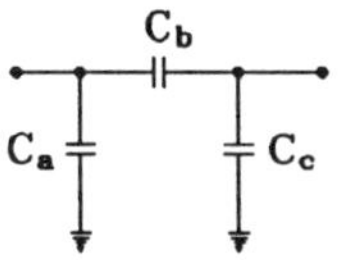 PI	$C = (n - 1)C_1$ $C_b = \frac{C_1}{n}$	$C_a = \frac{C}{n}$ $C_c = \frac{C_2 - C}{n^2}$
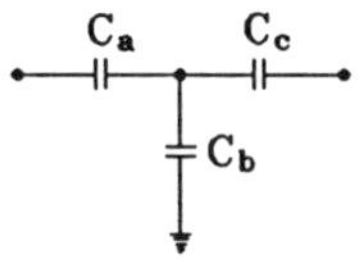 TEE	$C = \frac{C_2}{n - 1}$ $C_b = \frac{C_2}{n}$	$C_a = \frac{C\,C_1}{C - C_1}$ $C_c = \frac{C}{n}$

Figure 4.30 Capacitor transformations to increase R_L.

The transform may not be capable of completely converting the output to the desired value. In that case, we can search the network for another pair of elements, or apply the full transform and then search for a second transform to complete the adjustment. Note that if the full transform is used, n^2 in (4.32) must be computed from the element values rather than the actual and desired terminations.

Typically, we have a number of choices for adjusting the output termination. In most cases there will be several pairs of elements that can be used, and we always have the choice of a pi or tee configuration. Even when one transform is insufficient, there may be several combinations that will work, and in some cases the application of one transform creates a second which did not exist in the original network. The best approach is to calculate all possibilities and then select the network that has the minimum number of elements and practical component values.

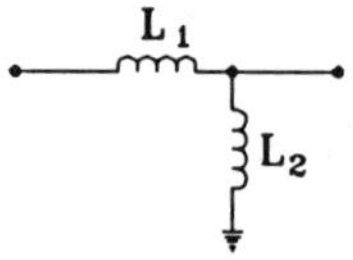	$R_L \le R_L' \le \left\{\frac{L_1 + L_2}{L_2}\right\}^2 R_L$	
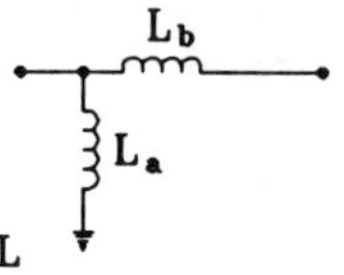 FULL	$n = \frac{L_1 + L_2}{L_2}$	$L_a = n L_2$ $L_b = n L_1$
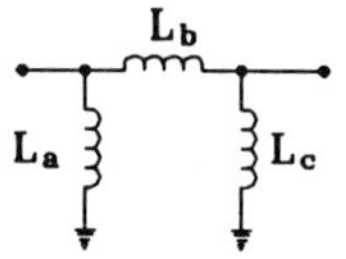 PI	$L = \frac{L_1}{n - 1}$ $L_b = n L_1$	$L_a = n L$ $L_c = \frac{n^2 L L_2}{L - L_2}$
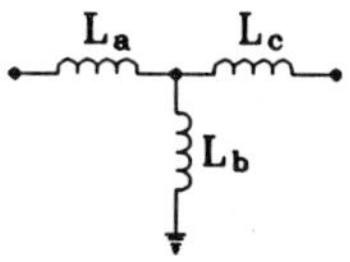 TEE	$L = (n - 1)L_2$ $L_b = n L_2$	$L_a = L_1 - L$ $L_c = n L$

Figure 4.31 Inductor transformations to increase R_L.

4.9.2 Transformer Networks

The Norton transformations cannot always be successfully applied to adjust the output termination. Matching network design requires the network to be referenced to the real component of the load, which can be very high or low relative to the characteristic impedance of the system, usually 50Ω, and the structure may not contain sufficient transformations. We have the option of adding more elements which may provide the desired adjustment. However, when a low-pass or high-pass approximation is used, the network consists of alternating series and shunt elements of different types that do not provide any transformations, and we must resort to other means of solving this problem.

Matthaei [18] references a quasi-low-pass impedance-transforming filter which was later used by Levy [19] to develop a particularly useful distributed equivalent. The quasi-low-pass mapping is

$$\omega \Rightarrow \left(\frac{2\omega^2 - \omega_2^2 - \omega_1^2}{\omega_2^2 - \omega_1^2}\right) \tag{4.33}$$

which increases the degree of the filter from n to $2n$ because of the ω^2 term. The center frequency of this mapping is

$$\omega_o = \frac{\omega_1 + \omega_2}{2}$$

which corresponds to $\omega = 0$ for the original low-pass response, and the fractional bandwidth is the same one given for the bandpass filter:

$$w = \frac{\omega_2 - \omega_1}{\omega_o} = \frac{2(\omega_2 - \omega_1)}{\omega_2 + \omega_1}$$

If we apply this mapping to the Chebyshev low-pass approximation in (4.16) and then plot the response as shown in Figure 4.32, we find some interesting properties. There is an equal-ripple passband extending from ω_1 to ω_2 and a stopband above ω_2. However, the region below ω_1 is not a passband as it would be for the low-pass approximation, nor is it a stopband which would be the case for a bandpass response. Instead, $G_t(\omega^2)$ gradually decreases and attains a finite value at $\omega = 0$, which is well below the passband insertion loss.

Examination of the mapping in (4.33) shows that it has a value of -1 at $\omega = \omega_1$, increases until it reaches +1 at $\omega = \omega_2$, and then continues to increase as ω increases. Below ω_1, (4.33) decreases toward $\omega = 0$, where it has a value of

$$\left(\frac{\omega_1^2 + \omega_2^2}{\omega_1^2 - \omega_2^2}\right)$$

This function has transmission zeros only at infinity. Thus it has the same topology as the low-pass filter (i.e., series inductors and shunt capacitors). The structure is transparent at DC, and the insertion loss at $\omega = 0$ is due entirely to a mismatch between the source and load impedance. At zero, the input reflection coefficient can be written as

$$s_{11} = \frac{Z - Z_o}{Z + Z_o} = \frac{r - 1}{r + 1}$$

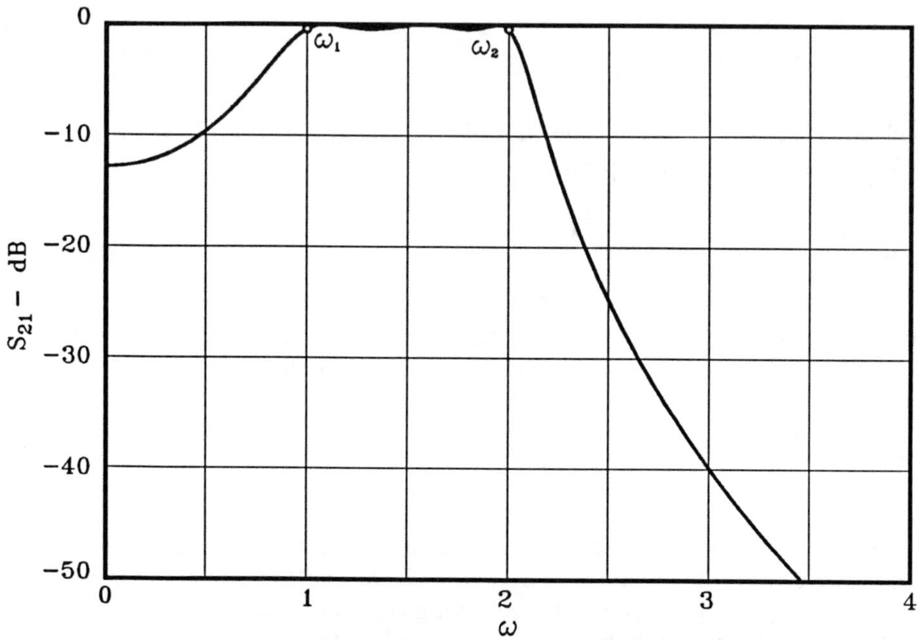

Figure 4.32 Quasi-low-pass response.

where r is Z/Z_o, or the ratio of load to source impedance. We can use the relationship between s_{11} and s_{21} given in (2.38a) for a lossless network to compute $|s_{21}|^2$ due to the mismatch:

$$|s_{21}|^2 = 1 - |s_{11}|^2 = 1 - \left(\frac{r-1}{r+1}\right)^2 = \frac{4r}{(r+1)^2}$$

which must be equal to the quasi-low-pass approximation at $\omega = 0$:

$$\frac{4r}{(r+1)^2} = \frac{K_m}{1 + \varepsilon^2 T_n^2\left\{\dfrac{\omega_1^2 + \omega_2^2}{\omega_1^2 - \omega_2^2}\right\}}$$

Setting $K_m = 1$ and then solving for ϵ^2, we have

$$\varepsilon^2 = \frac{(r-1)^2}{4rT_n^2\left\{\dfrac{\omega_1^2+\omega_2^2}{\omega_1^2-\omega_2^2}\right\}} \tag{4.34}$$

which can be used to compute the passband ripple required to realize the source and load over a specified bandwidth.

As an example, we can design a low-pass impedance-matching filter which transforms 50Ω into 5Ω across a passband extending from 1 to 2 GHz. Normalizing to 2 GHz, $\omega_1 = 0.5$, $\omega_2 = 1$, $r = 5/50$, or 0.1, and (4.34) becomes

$$\varepsilon^2 = \frac{(0.1-1)^2}{(4)(0.1)T_n^2\left\{\dfrac{(0.5)^2+1}{(0.5)^2-1}\right\}} = \frac{2.025}{T_n^2(-5/3)}$$

which we evaluate for $n = 1$, and we find that $\epsilon^2 = 0.729$, or approximately 2.38-dB passband ripple. However, $\epsilon^2 = 0.098$ for $n = 2$, and the required ripple is 0.404 dB, or less than 2.0:1 VSWR across the 1- to 2-GHz passband. Applying (4.33) to the low-pass Chebyshev approximation in (4.16) for $n = 2$, $K_m = 1$, and $\epsilon^2 = 0.097576$; expanding to form $|s_{21}|^2$; and then using (2.38a), we have

$$|s_{11}(s)|^2 = \frac{s^8 + 2.5s^6 + 2.203125s^4 + 0.800781s^3 + 0.102600}{s^8 + 2.5s^6 + 2.203125s^4 + 0.800781s^3 + 0.153267}$$

which has eight poles and zeros. The zeros are all on the $j\omega$ axis, and we select complex conjugate pairs of zeros and left half-plane poles, and then multiply to form $\pm s_{11}(s)$. $Z(s)$ can be computed from either $+s_{11}$ or $-s_{11}$, which has been used in previous examples to determine the input element type. However, in this case the input reflection coefficient at $s = 0$ is due to the mismatch between 50Ω and 5Ω, or approximately -0.8. Thus we use $-s_{11}$ to compute $Z(s)$ as

$$Z(s) = \frac{0.402938s^3 + 0.162360s^2 + 0.328710s + 0.035590}{s^4 + 0.402938s^3 + 1.412358s^2 + 0.328710s + 0.355902}$$

$Z(0)$ is 0.1, which confirms that the ratio of load to source impedance is correct. We can extract the elements as follows:

- Shunt capacitor, $C = 2.481771$F.
- Series inductor, $L = 0.675419$H.
- Shunt capacitor, $C = 6.754166$F.

- Series inductor, L = 0.248178H.
- Output resistor, R = 0.1.

The complete impedance-transforming filter is shown in Figure 4.33(a) where the element values have been scaled to 50Ω and 2.0 GHz.

Depending on the operating frequency and construction techniques, the element values may turn out to be nonrealizable. Chip capacitors often used at these frequencies contain parasitic inductance which causes series and parallel resonance. Although we do have the option of using two shunt capacitors at each node, which may help, we now have six elements instead of four. Another alternative is to use a different type of capacitor that is physically smaller and can be used at higher frequencies. However, these components are usually costly and labor-intensive. Small value inductors are also difficult to realize. The inductor can often be constructed using a very short wire segment. However components will physically be very close to each other, causing undesirable coupling. In addition, the inductance of a connector pin or transistor lead may be much greater than the series inductor, which will degrade the performance.

One of the major advantages of network synthesis over tables of element values or closed-form design equations is flexibility; we can usually design a number

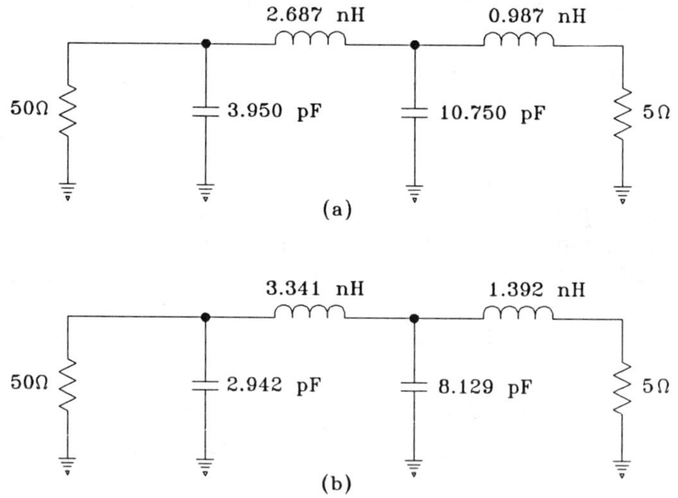

Figure 4.33 1 - 2-GHZ impedance-matching filter: (a) K_m = 1.00; (b) K_m = 0.95.

of networks with similar electrical characteristics. In this case, we do not have a choice of topologies or zero locations. However, we can add some minimum insertion loss, $K_m < 1$, which will move the zeros off the jω axis and provide some adjustment for the element values. For example, we can change K_m from 1.00 to 0.95, approximately 0.22-dB minimum insertion loss, and then select the right half-plane zeros to synthesize the network shown in Figure 4.33(b), which has lower value capacitors and higher value inductors when compared with the original design. Unfortunately the VSWR increased from 1.85:1 to 2.12:1. However, it may turn out that this network is better than the first when built and tested. We also have the option of increasing the order, along with the minimum insertion loss, which will reduce the VSWR at the expense of two additional elements. Alternatively, we could synthesize a transformer based on the Butterworth approximation, which may provide better component values.

4.10 SINGLY TERMINATED NETWORKS

The networks we have dealt with thus far have all been *doubly terminated*; that is, they have real terminations at each port. We now turn our attention to the *singly terminated* network [20], which is terminated at one port by either a zero impedance voltage source or infinite impedance current source, as shown in Figure 4.34. The power absorbed by the voltage-terminated network shown in Figure 4.34(a) is

$$P = |V_s|^2 \operatorname{Re}\{Y_{out}\}$$

However, the network is lossless, and all the power must be absorbed by the input termination, R_1:

$$|V_s|^2 \operatorname{Re}\{Y_{out}\} = |V_1|^2 G_1$$

and we can write the voltage transfer function as

$$\left|\frac{V_s}{V_1}\right|^2 = \frac{G_1}{\operatorname{Re}\{Y_{out}\}}$$

Thus, the power absorbed by R_1 is controlled by $\operatorname{Re}\{Y_{out}\}$ and if the network is designed to have equal ripple or maximally flat transmission characteristics, $\operatorname{Re}\{Y_{out}\}$ must exhibit the same response. Similarly, we can write the current transfer function for the network in Figure 4.34(b) as

$$\left|\frac{I_s}{-I_1}\right|^2 = \frac{R_1}{\operatorname{Re}\{Z_{out}\}}$$

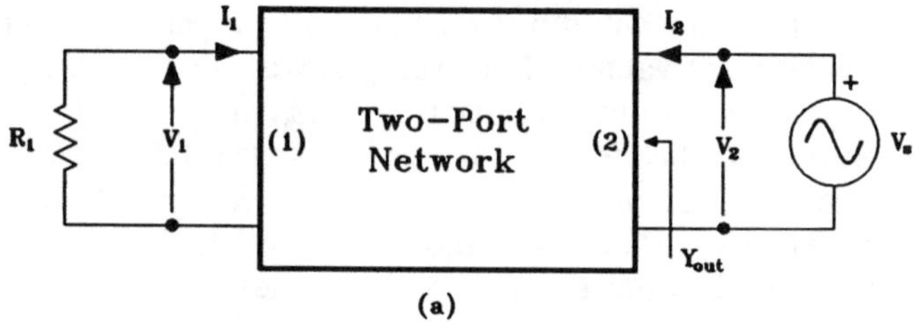

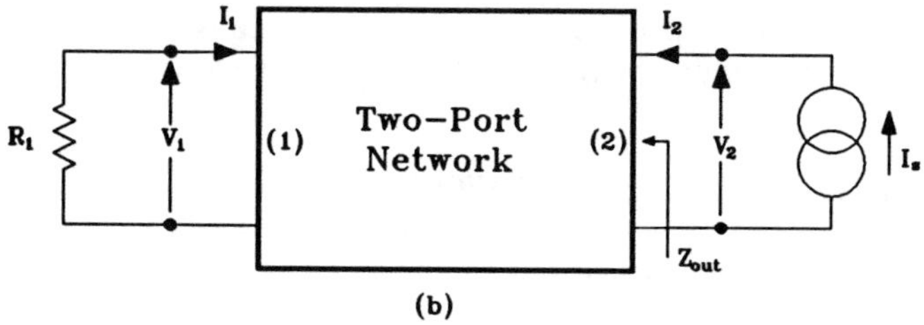

Figure 4.34 Singly terminated networks: (a) voltage; (b) current source.

which shows that $\text{Re}\{Z_{\text{out}}\}$ must possess the same response characteristics as the network when the output is driven by an infinite impedance current source.

We can set the active sources to zero and write the input impedance of the networks in Figure 4.34 as

$$Z(s) = \frac{N(s)}{D(s)} = \frac{N_e(s) + N_o(s)}{D_e(s) + D_o(s)}$$

where the subscripts *e* and *o* refer to the even and odd components, respectively. The average power absorbed by the one-port network is

$$P = \frac{1}{2}|I_1|^2 \text{Re}\{Z(j\omega)\}$$

which must equal zero since all the elements are lossless and the network is terminated in either an open or short. In general, the current is not zero and we can write

$$\text{Re}\{Z(j\omega)\} = \frac{N_e(s)D_e(s) - N_o(s)D_o(s)}{D_e^2(s) + D_o^2(s)} = 0$$

or

$$N_e(s)D_e(s) - N_o(s)D_o(s) = 0$$

Since the even and odd parts of the numerator or denominator cannot be zero simultaneously, $N_e(s)$ and $D_o(s)$ must both equal zero and

$$Z(s) = \frac{N_o(s)}{D_e(s)}$$

or both $N_o(s)$ and $D_e(s)$ are zero, and

$$Z(s) = \frac{N_e(s)}{D_o(s)}$$

which leads to the conclusion that the input impedance of the singly terminated network is either a ratio of even to odd or odd to even polynomials.

Using (4.11a) with $\text{Re}\{Z(j\omega)\} = 0$, it can be shown that $|s_{11}| = 1$, or

$$s_{11}(s)\,s_{11}(-s) = 1$$

for the singly terminated networks, and from (4.12a) we can write

$$\frac{h(s)\,h(-s)}{g(s)\,g(-s)} = 1$$

Assuming $h(s)$ does not equal $g(s)$, which results in the trivial case where $s_{11} = 1$ at all frequencies, we have

$$h(s) = \pm g(-s) \text{ and}$$

$$h(-s) = \pm g(s)$$

or

$$s_{11}(s) = \pm\frac{g(-s)}{g(s)} \tag{4.35}$$

Thus we can form the input reflection coefficient of a singly terminated network directly from the poles of a doubly terminated network.

The $n = 3$ Chebyshev low-pass filter of Section 4.4.2 has

$$g(s) = s^3 + 1.327821s^2 + 1.631554s + 0.790569$$

and we can use (4.35) to write the input reflection coefficient of a singly terminated network with similar characteristics as

$$s_{11}(s) = \pm\frac{-s^3 + 1.327821s^2 - 1.631554s + 0.790569}{s^3 + 1.327821s^2 + 1.631554s + 0.790569}$$

At $s = 0$, s_{11} can be either $+1$, corresponding to an open circuit input impedance, or -1, which corresponds to a short circuit input. We select $-s_{11}$, which is the zero impedance termination shown in Figure 4.34(a), and then use (4.11a) to compute the input impedance:

$$Z(s) = \frac{2s^3 + 3.263108s}{2.655642s^2 + 1.581138}$$

which does in fact equal zero at $s = 0$. $Z(s)$ also meets the positive real requirement that the highest and lowest power of the numerator and denominator can differ at most by one. It also turns out that the poles and zeros of $Z(s)$ and $Y(s)$ must be on the $j\omega$ axis, and each successive term in the numerator and denominator must differ by exactly a power of 2; there can be no missing terms. $Z(s)$ contains both low-pass and high-pass elements. However, the synthesized network must be consistent with the low-pass approximation, and we select only elements with poles at infinity:

- Series inductor, $L = 0.753114$H.
- Shunt capacitor, $C = 1.281475$ F.
- Series inductor, $L = 1.310658$ H.
- Output resistor, $R = 0$.

The output admittance of this network is shown in Figure 4.35(a) and confirms the fact that $\mathrm{Re}\{Y_{out}\}$ is equal ripple and has a low-pass characteristic. Thus power absorbed by R_1 also has a low-pass response when driven from the output port by an ideal voltage source. We can compare the low-pass network with a high-pass equivalent using Figure 4.19 to transform the elements into:

- Series capacitor, $C = 1.327821$ F.

- Shunt inductor, L = 0.780350 H.
- Series capacitor, C = 0.762975 F.
- Output resistor, R = 0.

The output admittance of the high-pass network shown in Figure 4.35(b) also exhibits equal-ripple Re$\{Y_{out}\}$ characteristics, but in a passband extending from ω = 1 to infinity.

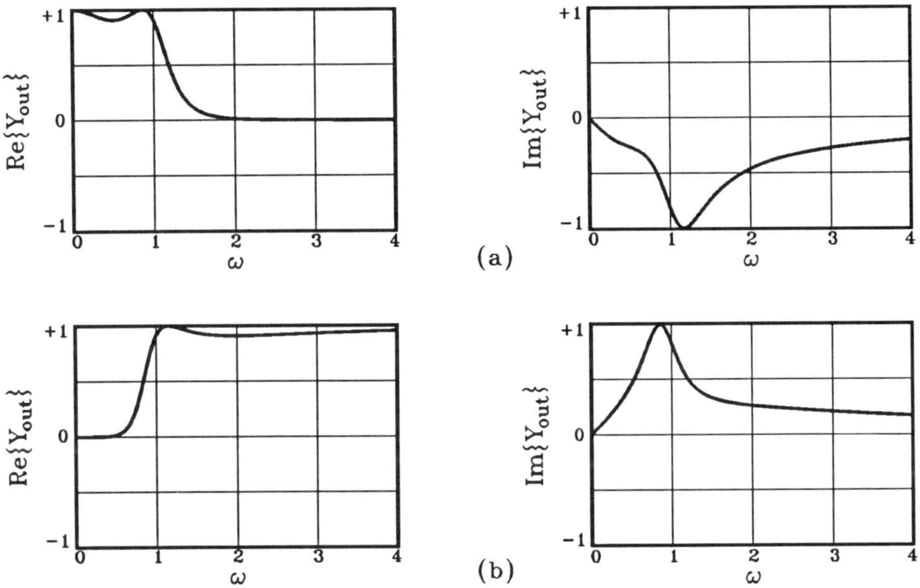

Figure 4.35 Singly terminated Y_{out}: (a) low-pass; (b) high-pass.

Comparing the output admittance characteristics of both networks in Figure 4.35, we see that the low-pass conductance ripples about +1 mho between ω = 0 and ω = 1, and then decreases to 0 as ω increases, while, simultaneously, G_{out} of the high-pass network is 0 at ω = 0, and ripples about +1 between ω = 1 and infinity. The susceptance of the two networks have opposite characteristics: the low-pass susceptance is zero at ω = 0, dips to -1 mho around the cutoff frequency, and then increases to zero again as ω approaches infinity, while the high-pass susceptance also approaches zero at ω = 0 and infinity, but has a +1-mho peak in the vicinity of the cutoff frequency. The output admittance of the two networks can be added together with the result that Re$\{Y_{out}\}$ ripples about +1 mho and Im$\{Y_{out}\}$ cancels almost to zero at all frequencies. We can define a *crossover* frequency where the output conductance of each network is equal. Below this frequency the low-pass conductance dominates, and above it the high-pass conductance dominates. The conductance must add to +1; hence, the conductance of each network should ideally be 0.5 at the crossover frequency.

Referring to Figure 4.35, we see that the cutoff frequency of both networks is 1; however, the conductance of the low-pass network is 0.5 at ω = 1.19 and the high-pass conductance 0.5 at ω = 0.84. We can multiply the low-pass elements by 1.19 and the high-pass elements by 0.84, which shifts the low-pass response down and the high-pass response up such that the conductance is 0.5 at ω = 1 for both networks. The admittance is added by paralleling the output port of the two networks as shown in Figure 4.36, forming a three-port network.

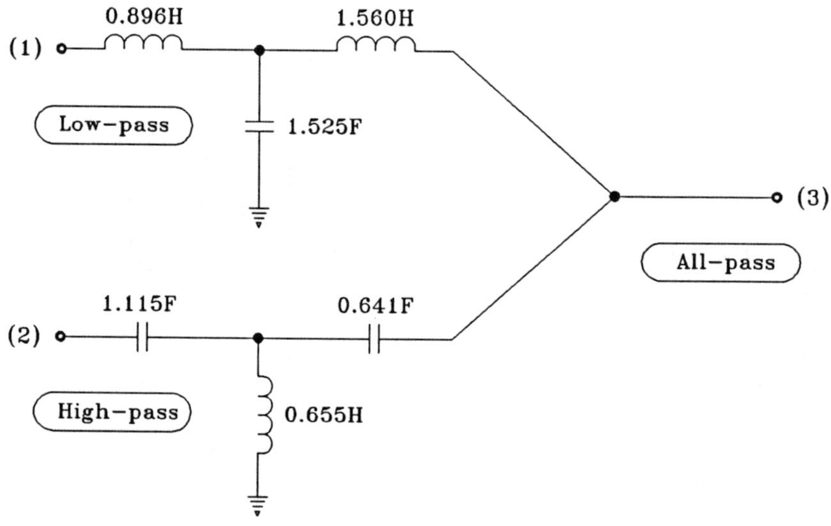

Figure 4.36 Normalized diplexer network.

The three-port network has one port which is the input of the low-pass network, a second port which is the input of the high-pass network, and a third port which is common to the output of both networks. The admittance of the common output port shown in Figure 4.37 is (1 + j0) at ω = 0, 1, and ∞, while the conductance is within ±0.125 of 1.000, and the susceptance ±0.125 across the full frequency range shown. Complete analysis of this three-port reveals that the response between ports 1 and 3 is equal-ripple low-pass with a cutoff frequency of approximately $\omega = 0.84$, and the response between ports 2 and 3 is equal-ripple high-pass with a cutoff frequency of approximately 1.19. At $\omega = 0$, ports 1 and 3 are directly connected while port 2 is isolated; at $\omega = 1$, power applied to any port is equally split between the other two ports; and at $\omega = \infty$, ports 2 and 3 are directly connected while port 1 is isolated. This structure forms a *diplexer*, a network capable of combining two separate frequency bands into a single broadband output while preserving the input impedance at the respective ports. The network is also reciprocal; a broadband input signal may be split into a low-frequency channel and a high-frequency channel. The VSWR at ports 1 and 2 of the diplexer shown in Figure 4.36 is approximately 2:1 in the respective passbands due to the 0.414-dB ripple specified for the low-pass approximation, and the VSWR at the common output port is approximately 1.2:1. Most designs require better performance, and passband ripples are typically specified as 0.1 dB or less, and, in some cases, simple auxiliary networks are used at the output to improve the VSWR.

The dual of the diplexer network can be synthesized using the open circuit output termination shown in Figure 4.34(b). In this case, the networks are the dual of those shown in Figure 4.36 (i.e., shunt-connected input and output elements in each network). $\mathrm{Re}\{Z_{out}\}$ is constant for these networks, and the output ports must

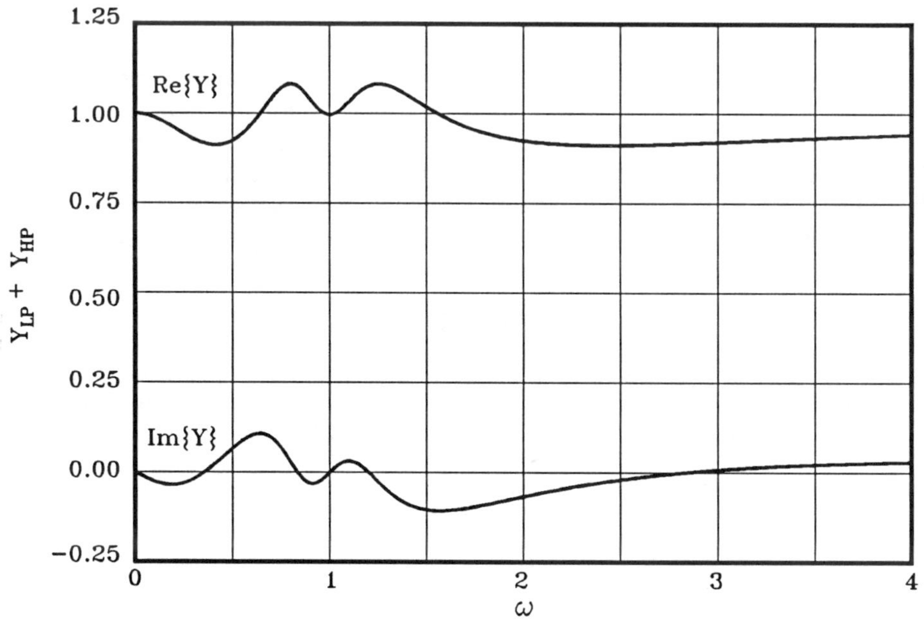

Figure 4.37 Sum of low-pass and high-pass admittance.

be connected in series for constant broadband output impedance. This configuration is inconvenient at RF and microwave frequencies and is seldom used. Regardless of the output termination, both networks should be the same order, which ensures that the slope of the real and imaginary components of Y_{out} or Z_{out} is such that the real parts add to 1, and the imaginary parts cancel at all frequencies. The most convenient approach is to use the low-pass approximation, transform the synthesized network, and then scale element values for a normalized crossover frequency of 1.

4.11 CONCLUSION

The basic fundamentals of modern network synthesis have been presented in this chapter, which when applied to RF and microwave circuit design offer significant advantages over other techniques. A single set of electrical specifications can result in numerous topologies, all of which meet the design requirements. By proper placement of the zeros, we can generate identical sets of topologies with different element values. The most significant difficulty associated with synthesis is the lack of control over the output termination. Several solutions have been presented. We can change the order of the approximation, transform the network, synthesize an impedance-matching network, or use a combination of techniques.

Topologies in this chapter have been limited to simple ladder structures containing inductors and capacitors easily realized at RF and low microwave frequencies; however, similar techniques can be used to synthesize networks consisting of resistors and capacitors, resistors and inductors, coupled coils, and lattice structures. One particularly useful network at lower frequencies uses operational amplifiers, resistors, and capacitors to achieve the desired characteristic [7,21]. The transfer function of certain operational amplifier circuits is related to the approximations developed in this chapter, and component values can be computed directly from the poles of s_{11}.

REFERENCES

[1] W.N. Tuttle, "Simplified Procedures for Approximation and Predistortion in Filter Synthesis by Digital Computer," *IEEE Trans. on Circuit Theory*, Vol. CT-15, December 1968, pp. 488-492.

[2] J.W. Craig, *Design of Lossy Filters*, MIT Report 19, The MIT Press, Cambridge, Massachusetts, 1970.

[3] Otto Brune, "Synthesis of a Finite Two Terminal Network Whose Driving-Point Impedance Is a Prescribed Function of Frequency," *Journal Math and Physics*, Vol. 10, No. 3, 1931, pp. 191-236.

[4] M.E. Van Valkenburg, *Introduction to Modern Network Synthesis*, John Wiley & Sons, Inc., New York, 1960.

[5] F.F. Kuo, *Network Analysis and Synthesis*, John Wiley & Sons, Inc., New York, 1966.

[6] H. Baher, *Synthesis of Electrical Networks*, John Wiley & Sons, Inc., New York, 1984.

[7] G.C. Temes and J.W. LaPatra, *Introduction to Circuit Synthesis and Design*, McGraw-Hill Book Company, New York, 1977.
[8] T.T. Ha, *Solid-State Microwave Amplifier Design*, John Wiley & Sons, Inc., New York, 1981.
[9] V. Belevitch, *Classical Network Theory*, Holden-Day, Inc., San Francisco, 1968.
[10] J.D. Rhodes, *Theory of Electrical Filters*, John Wiley & Sons, Inc., New York, 1976.
[11] Thomas R. Cuthbert, Jr., *Circuit Design Using Personal Computers*, John Wiley and Sons, New York, 1983.
[12] W.E. Thomson, "Delay Networks Having Maximally-Flat Frequency Characteristics," *Proc. IEE*, 96, November 1949, pp. 487-490.
[13] W.E. Thomson, "Network With Maximally Flat Delay," *Wireless Engr.*, 29, October 1952, pp. 256-263.
[14] A.I. Zverev, *Handbook of Filter Synthesis*, John Wiley & Sons, Inc., New York, 1967.
[15] H.J. Orchard and G.C. Temes, "Filter Design Using Transformed Variables," *IEEE Trans. on Circuit Theory*, Vol. CT-15, December 1968, pp. 385-408.
[16] G. Szentirmai, *Computer-Aided Filter Design*, IEEE Press, New York, 1973.
[17] G.A. Vincent, "Impedance Transformation Without a Transformer," *Frequency Technology*, Vol. 7, No. 9, September 1969, pp. 15-21.
[18] G.L. Matthaei, "Short-Step Chebyshev Impedance Transformers," *IEEE Trans. on Microwave Theory and Techniques*, Vol. MTT-14, No. 8, August 1966, pp. 372-383.
[19] R. Levy, "Synthesis of Mixed Lumped and Distributed Impedance-Transforming Filters, *IEEE Trans. on Microwave Theory and Techniques*, Vol. MTT-20, No. 3, March 1972, pp. 223-233.
[20] G.L. Matthaei, L. Young, and E.M.T. Jones, *Microwave Filters, Impedance-Matching Networks, and Coupling Structures*, McGraw-Hill Book Company, New York, 1964.
[21] G.C. Temes and S.K. Mitra, *Modern Filter Theory and Design*, John Wiley & Sons, Inc., New York, 1973.
[22] L. Weinberg, *Network Analysis and Synthesis*, McGraw-Hill Book Company, New York, 1962.
[23] N. Balabanian, *Network Synthesis*, Prentice-Hall, Inc., Englewood Cliffs, New Jersey, 1958.
[24] E.A. Guillemin, *Synthesis of Passive Networks*, John Wiley & Sons, Inc., New York, 1960.
[25] J.E. Storer, *Passive Network Synthesis*, McGraw-Hill Book Company, New York, 1957.
[26] S.K. Mitra, *Analysis and Synthesis of Linear Active Networks*, John Wiley & Sons, Inc., New York, 1969.

Chapter 5

Distributed Network Synthesis

Synthesis techniques for the lumped element networks of Chapter 4 are valid at all frequencies. However, as we increase the design frequency, it becomes more difficult to realize the lumped elements necessary to fabricate the circuit. Internal parasitics, which may be negligible at lower frequencies, cause significant deviations from the anticipated characteristics, resulting in hardware performance that deviates markedly from that expected. In addition, we assume voltage and current is not a function of distance from the terminals of a lumped element. That is, the impedance is independent of physical length. When the size of the element is an appreciable fraction of a wavelength, the voltage and current is a function of distance from the connecting points, and the element can no longer be considered lumped. It is difficult to define a particular size where the lumped assumption begins to deteriorate; most designers have a rule of thumb based on experience with specific elements, types of circuits, and fabrication methods. However, if we assume one one-hundredth of a wavelength, 3.6°, then the maximum size of a lumped element is approximately 59 mils or 1.5 mm in air at 2 GHz. In the case of a capacitor that has a dielectric constant of 40 or higher, the size for 0.01λ is reduced below 10 mils or 0.3 mm. Even if components can safely be considered lumped at the design frequency, the high-frequency response may be degraded, which can result in unacceptable out-of-band characteristics.

We can avoid these problems by designing with distributed elements; that is, cascaded transmission lines and transmission lines terminated with an open or short. The distributed element consists of two conductors separated by a dielectric material [1], and the terminal or port parameters can be described in terms of the voltage and currents as a function of distance along the conductors. The characteristics of the distributed element are dependent on physical properties and are useful for a number of applications from very low to very high frequencies. Transmission lines can be constructed using a variety of different form factors. We can make a transmission line simply by twisting a pair of insulated wires together. An inexpensive 300Ω transmission line used with television antennas is made from parallel wires separated by a dielectric material, and common 50Ω or 75Ω coaxial cable consists of a center conductor surrounded by a dielectric material and an outer conductor. However, the most common implementations of the transmission lines used for RF and microwave circuit design are *stripline* [2,3] and *microstrip* [3-5]. Stripline is formed by printing a conductor pattern on one side of an insulating substrate while the reverse side is fully metalized. A second substrate, which is metalized on only one side, is placed above the first forming a sandwich. This arrangement results in a center conductor which is suspended in a dielectric material between two ground planes. The two substrates can be fused together and connectors attached, forming a compact structure convenient for filters or other planar networks. Microstrip is similar to stripline except that only one substrate is used, as shown in Figure 5.1. In this case, the conductor is suspended above a single ground plane providing easy access for attaching active devices or other components, which accounts for the popularity of this method. Other fabrication techniques such as slot line, coplanar waveguide, coplanar strips, finline, and suspended substrate are sometimes used [3-5]. However, microstrip is by far the most common.

The characteristic impedance and propagation velocity of the transmission lines in both media are determined by the conductor linewidth, substrate thickness, and dielectric constant. Figure 5.1 shows the width-to-height ratio of an isolated line versus impedance for three values of relative dielectric constant, ϵ_r, at zero frequency. The lower value of 2.2 is usually associated with relatively soft substrate material which is easily machined. Substrates with a dielectric constant of 10 can be purchased either as a soft material or a much harder material known as *alumina*, or aluminum oxide. Alumina substrates are difficult to machine. However, line definitions are usually superior when compared with the softer materials. Alumina is the most common substrate material used in the manufacturing of microwave integrated circuits (MIC). Gallium arsenide, $\epsilon_r = 12.9$, is frequently used in the manufacture of monolithic microwave integrated circuits (MMIC) [6].

The mode of propagation is transverse electromagnetic (TEM) for stripline. At low frequencies, the microstrip propagation mode also resembles TEM. However, due to the air-dielectric interface in microstrip, some of the field is in air and the mode is quasi-TEM at higher frequencies. As the conductor width or substrate thickness is increased, we reach a point where other modes are supported, which places an upper limit on these dimensions. On the other end of the scale, there is a minimum linewidth that can reliably be maintained using normal production methods. These constraints limit the range of realizable impedance values, which complicates the design procedure somewhat. The quasi-TEM mode

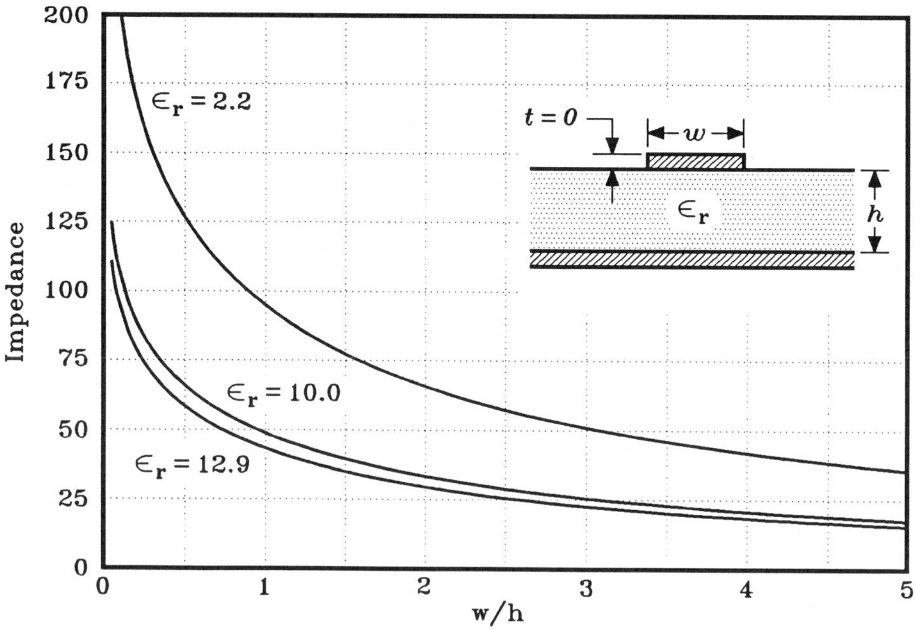

Figure 5.1 Microstrip impedance versus width to height ratio.

of microstrip causes other problems as well. Since a portion of the field is in air, the velocity of propagation is different than it would be if the conductor were completely surrounded by the dielectric material, and we must compute an *effective dielectric constant*. Quasi-TEM characteristics also result in an effective dielectric constant and characteristic impedance which are functions of frequency. This behavior is known as *dispersion* and must be considered in the design of broadband circuits.

Severe constraints are placed on the design of microstrip circuits due to the limited range of linewidths. For the purpose of examples in this chapter, we shall assume a 25-mil (0.635 mm) alumina substrate with $\epsilon_r = 10$. Using simple and relatively inexpensive etching procedures, we can achieve minimum linewidths between 3 and 4 mils, 0.076 to 0.102 mm. Figure 5.1 shows that this corresponds to a maximum impedance of approximately 100Ω. The maximum linewidth should be much less than one-quarter wavelength [6]; in fact, it should be on the order of one-eighth wavelength or less. Assuming an upper operating frequency of 16 GHz and an effective dielectric constant of 7.5, the maximum linewidth is approximately 34 mils (0.855 mm), which imposes a lower limit of 40Ω.

Although distributed elements may deviate from the ideal, most high-frequency circuits use the transmission line, shorted stub, and open stub. These design techniques are also applied to reduce the cost of products, even at frequencies where lumped elements can be used. The distributed element circuit is much less expensive and more reproducible than an equivalent lumped element design. Most of the circuitry is simply etched on a printed circuit board and requires minimal labor and materials. Digital designers are also beginning to use these methods for the design of high-speed circuits where propagation delay and reflected signals can cause significant problems. We can apply the network synthesis methods of Chapter 4 to the design of distributed-element circuits and realize the same advantages. A number of topologies and associated element values can be synthesized that provide equivalent electrical performance.

5.1 DISTRIBUTED EQUIVALENCE

The impedances of the shorted and open transmission line or stub were derived in Section 2.1.2 and were given by equations (2.9d) and (2.9e), respectively:

$$Z_{SC} = jZ_C \mathrm{Tan}\,\theta$$

$$Z_{OC} = \frac{Z_C}{j\mathrm{Tan}\,\theta}$$

If we change variables such that

$$S = j\,Tan\,\theta$$

we can write the impedance as

$$Z_{SC} = SZ_C \tag{5.1a}$$

and

$$Z_{OC} = \frac{Z_C}{S} \tag{5.1b}$$

Comparing (5.1) with the *s*-plane impedance for the inductor and capacitor shown in Figure 4.1,

$$Z = sL$$

$$Z = \frac{1}{sC}$$

we see that the s-plane inductor corresponds to the S-plane shorted stub, $L = Z_C$, and the capacitor corresponds to the open stub with $C = 1/Z_C$. Thus, we can synthesize an equivalent distributed network simply by replacing inductors with shorted stubs and capacitors with open stubs. However, this equivalence is only valid for electrical lengths less than 90° (i.e., Tanθ positive). When the variable $S = j\text{Tan}\theta$ is a common factor, the distributed network is defined as *commensurate*. That is, all transmission lines have the same electrical length.

The complex variable mapping was proposed by Richards [7] and is known as *Richards' variable*. In its complete form, the mapping is

$$S = \text{Tanh}(s\tau) = \Sigma + j\Omega \tag{5.2a}$$

and if we let $s = j\omega$, then

$$S = \text{Tanh}(s\tau) = \text{Tanh}(j\omega\tau) = j\text{Tan}(\omega\tau) = j\text{Tan}\,\theta \tag{5.2b}$$

A number of different notations can be found in the literature for the distributed mapping S; t, p, λ, and Ω are common. However, we shall use S to denote the term $j\Omega$, or $j\text{Tan}\theta$, to emphasize the similarity with the s-plane. The reader should be careful to distinguish between the lower case s used with lumped elements and the upper case S used with distributed elements.

Normally L, C, and s are normalized; hence, the characteristic impedance Z_c and S are both normalized, and we must use the frequency and impedance scaling techniques of Section 4.5 to compute the actual values. We can multiply the impedance of each element by the reference Z_o to scale from 1Ω. Since z_c appears in the numerator of (5.1a) and (5.1b), we can multiply both by Z_o to change the impedance level of the filter. The original filter is designed for a cutoff frequency of $\omega = 1$, or $s = j$, and S must have the same value. Thus, if we divide Ω by $\text{Tan}\theta_c$, where θ_c is the electrical length at the actual cutoff frequency, $(\text{Tan}\theta/\text{Tan}\theta_c)$ will equal 1 and $S = j$ at $f = f_c$. Combining the impedance and frequency scaling, we have

$$Z_{CS} = \frac{z_c Z_o}{\text{Tan}\,\theta_c} = \frac{Z_o L}{\text{Tan}\,\theta_c} \tag{5.3a}$$

for the actual characteristic impedance of the shorted stub, and

$$Z_{CO} = z_c Z_o \text{Tan}\,\theta_c = \frac{Z_o \text{Tan}\,\theta_c}{C} \tag{5.3b}$$

for the open stub. L and C are both normalized values. Z_{CS} denotes the characteristic impedance of the shorted stub and Z_{CO} the characteristic impedance of the open stub. θ_c is the electrical length of the transmission lines at the cutoff frequency and must be the same for all elements in the network; that is, $\text{Tan}\theta_c$ is the common normalization factor for the commensurate elements.

As an example, we can realize the lumped-element $n = 3$ Chebyshev low-pass filter designed in Section 4.4.2 and shown in Figure 4.10 as a distributed network. We assume a reference impedance of 50Ω and a cutoff frequency of 0.5 GHz where the electrical length is 45°. Both series inductors have a value of 1.506 and

$$Z_{CS} = \frac{Z_o L}{\text{Tan}\theta_c} = \frac{(50)(1.506)}{\text{Tan}45°} = 75.3000\Omega$$

for the equivalent series shorted stubs, and

$$Z_{CO} = \frac{Z_o \text{Tan}\,\theta_c}{C} = \frac{(50)(\text{Tan}45°)}{1.115} = 44.8431\,\Omega$$

for the shunt open stub. The complete low-pass filter is shown in Figure 5.2(a) and the response in Figure 5.3.

The response shown in Figure 5.3 as the darker of the two plots confirms that the distributed equivalent low-pass filter has the same characteristics as the original lumped-element network. The passband ripple is 0.414 dB and the cutoff frequency is the design value of 0.5 GHz. However, we also observe from the figure that the response is repeated a number of times. A second passband appears from 1.5 to 2.5 GHz and a third begins at 3.5 GHz. Since ω maps into $\text{Tan}\theta$, the frequency range from $\omega = 0$ to ∞ maps into electrical lengths from $\theta = 0°$ to $90°$. The tangent function is periodic and attains a value of ± 1 for odd multiples of 45° or $\lambda/8$ wavelength, $\pm\infty$ at odd multiples of 90° or $\lambda/4$, and 0 at all multiples of 180° or $\lambda/2$. The transducer gain is a function of ω^2 and the transformed equivalent $\text{Tan}^2\theta$. Therefore, we have an infinite number of passbands centered about frequencies where $\theta = n180°$ and extending from $n180° - 45°$ to $n180° + 45°$, $n = 0,1,2,...,\infty$. In this example, $\theta = 45°$ at 0.5 GHz; hence, we have a second passband centered at (4)•(0.5) or 2.0 GHz, another at (8)•(0.5) or 4.0 GHz, and so forth.

The periodic nature of the distributed response often results in undesired out-of-band characteristics. However, we can decrease the electrical length at the cutoff frequency, which causes the additional passbands to be moved up in frequency. For example, if we let $\theta_c = 22.5°$ at 0.5 GHz in the above example,

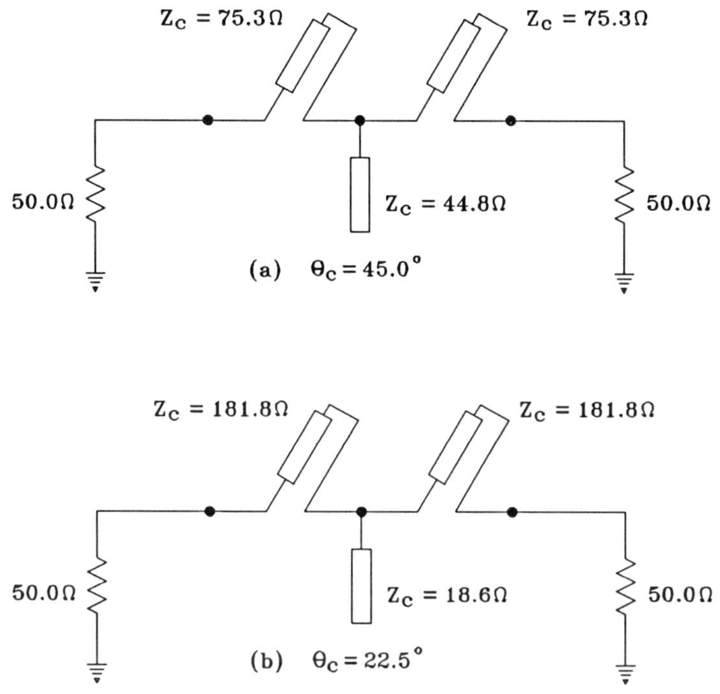

Figure 5.2 Low-pass filter: (a) θ_c = 45°; (b) θ_c = 22.5°.

$$Z_{CS} = \frac{Z_o L}{\text{Tan}\theta_c} = \frac{(50)(1.506)}{\text{Tan}(22.5)^\circ} = 181.7903\Omega$$

and

$$Z_{CO} = \frac{Z_o \text{Tan}\,\theta_c}{C} = \frac{(50)(\text{Tan}(22.5)^\circ)}{1.115} = 18.5746\Omega$$

as shown in Figure 5.2(b). The response is shown in Figure 5.3 as the lighter of the two plots. The cutoff frequency of 0.5 GHz and the 0.414-dB ripple has been preserved; however, the start of the second passband has been moved from 1.5 to 3.5 GHz. Comparing the two circuits shown in Figure 5.2, we see that the range of element values is significantly expanded when we reduce the electrical length. Typical microstrip impedance values range from 40Ω to 100Ω, and we could realize the element values in Figure 5.2(a), but not 5.2(b). We have the option of changing the ripple, minimum insertion loss (MIL), zero locations, number of elements, and the type of approximation in an effort to find realizable element values. However, the series open and shorted stubs cannot easily be realized on planar microstrip, and it is necessary to find another topology before we concern ourselves with element values.

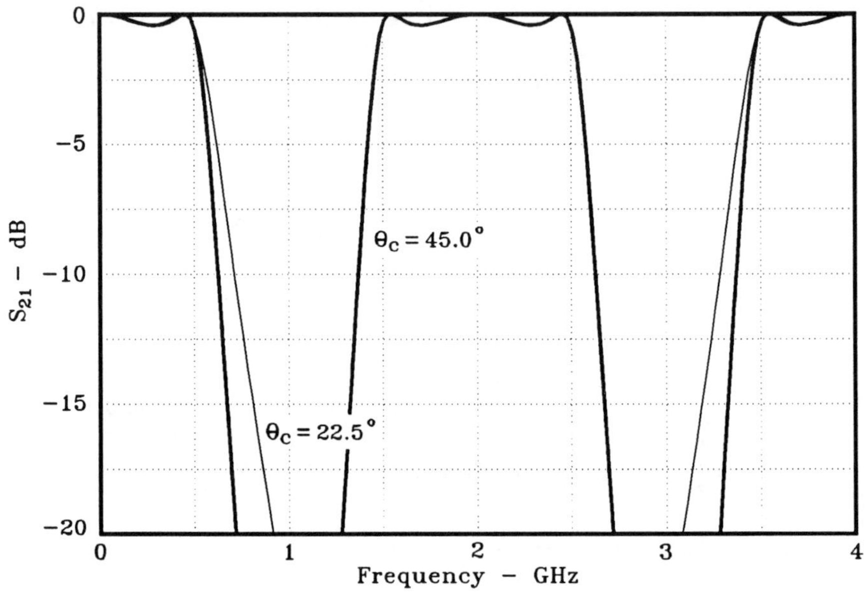

Figure 5.3 Distributed-element low-pass filter.

5.1.1 The Transmission Line Element

The previous section demonstrates the ease with which we can realize a distributed network using only open and shorted stubs. However, we also have the cascaded transmission line, known as a *unit element* (*UE*) in synthesis theory [8], which turns out to be extremely important for physical realizability. Levy [9] has suggested the term *transmission line element*, or *TLE*, as a more direct and descriptive term, and we shall use it here. We derived the expression for the input impedance of a transmission line terminated with an arbitrary load, Z_L, in Section 2.1.2, which is given by (2.9). Assuming the line is lossless, we can substitute (2.9b) into (2.9c), apply Euler's identity, and write the input impedance as

$$Z_{in} = Z_C \cdot \frac{Z_L + jZ_C \mathrm{Tan}\theta}{Z_C + jZ_L \mathrm{Tan}\theta} = Z_C \cdot \frac{Z_L + Z_C S}{Z_C + Z_L S} \tag{5.4a}$$

Let $Z(S)$ be the driving-point impedance of a network consisting of shorted stubs $Z_{si}S$ and open stubs Z_{oi}/S, which is positive real and rational in S. Then, when we cascade a transmission line element, the input impedance becomes

$$Z_{in}(S) = Z_C \cdot \frac{Z(S) + Z_C S}{Z_C + Z(S)S} \tag{5.4b}$$

which is also positive real and a rational function in S. Working backwards, we can evaluate $Z_{in}(S)$ at $S = 1$:

$$Z_{in}(S)\big|_{S=1} = Z_{in}(1) = Z_C \cdot \frac{Z(1) + Z_C \cdot 1}{Z_C + Z(1) \cdot 1} = Z_C \tag{5.5a}$$

which provides the characteristic impedance of the transmission line element directly from the input impedance. How do we test the driving-point impedance for the presence of a TLE? Richards' Theorem [10] states

> *A unit element of characteristic impedance* Z(1) *can always be extracted from a positive real rational impedance* Z(S) *leaving a remainder of*
>
> $$Z'(S) = Z(1) \cdot \frac{Z(S) - SZ(1)}{Z(1) - SZ(S)} \tag{5.5b}$$
>
> *which is also rational and positive real of degree at most equal to that of* Z(S). *Furthermore, if*
>
> $$\mathrm{Ev}\{Z(1)\} = 0$$
>
> *i.e., the even component of* Z(S) *evaluated at* S = *1 equals 0, then*
>
> $$\deg\{Z'(S)\} = \deg\{Z(S)\} - 1$$

In other words, a transmission line element can always be extracted from the positive real impedance function, and if the even component evaluated at $S = 1$ is 0, then the degree is reduced by 1. The remainder in (5.5b) appears to be 1 degree higher than $Z(S)$ due to the additional factor of S in the numerator and denominator. However, since both are rational and 0 for $S = 1$, the term $(S - 1)$ can be factored from the numerator and denominator and then canceled, which lowers the degree back to $Z(S)$. Also, if

$$Z(-1) = -Z(1) \tag{5.5c}$$

the numerator and denominator of (5.5b) are 0 for $S = -1$. Then $(S + 1)$ can be factored and canceled, which lowers the degree of the remainder to 1 less than $Z(S)$. The equality in (5.5c) is equivalent to Richards' requirement that the even component of $Z(S)$ be 0 when evaluated at $S = 1$.

The consequence of Richards' Theorem is that we can add any number of TLEs to our circuit. When (5.5c) does not hold, the degree of the impedance function is not reduced by extracting a TLE, and the element is nonminimum or *redundant*. However, the cascaded transmission line, which has no lumped-element counterpart, is often necessary to provide a distributed circuit topology which is compatible with common fabrication methods.

We usually work with normalized values while synthesizing the network, and it is necessary to scale the transmission line element to the desired impedance level and frequency. When we multiply all impedances in the network by Z_o, the normalized driving-point impedance $z(S)$ becomes $Z_o \bullet z(S)$, and the normalized input impedance of the TLE cascaded with $Z(S)$ given in (5.4b) becomes $Z_o \bullet z_{in}(S)$. Substituting $Z_o \bullet z(S)$ into (5.5b), we find that we must multiply z_c by Z_o to scale $z_{in}(S)$; that is,

$$Z_{in}(S) = \left[Z_o z_c\right] \cdot \frac{\left[Z_o z(S)\right] + \left[Z_o z_c\right] S}{\left[Z_o z_c\right] + \left[Z_o z(S)\right] S} = Z_o z_{in}(S)$$

Thus,

$$Z_C = Z_o z_c \tag{5.5d}$$

and we simply multiply the normalized TLE by Z_o to scale the impedance. At the cutoff frequency, $S = j$, the electrical length of the TLE is 45° (i.e., jTan45° = j), and the input impedance is

$$Z_{in}(j1) = Z_C \cdot \frac{Z(j1) + jZ_C}{Z_C + jZ(j1)}$$

When we assume an arbitrary electrical length and modified Z_c for the TLE, the input impedance can be written as

$$Z'_{in}(j1) = Z'_C \cdot \frac{Z(j1) + jZ'_C \text{Tan}\theta}{Z'_C + jZ(j1)\text{Tan}\theta}$$

Equating $Z_{in}(j1)$ and $Z'_{in}(j1)$, we find that there is no simple real factor that can be applied to the characteristic impedance, making the two equal when θ is not 45°. Therefore, the electrical length of the TLE must be 45° at the cutoff frequency and cannot be scaled.

Returning to the distributed equivalent of the low-pass filter, the input reflection coefficient was computed in Section 4.4.2 as

$$s_{11}(s) = \pm \frac{s^3 + 0.75s}{s^3 + 1.327821s^2 + 1.631554s + 0.790569}$$

and we used the positive sign to form the impedance function, resulting in a network with a series inductor as the first element. We cannot realize a series stub using stripline or microstrip. However, if we select the negative sign to compute the dual impedance function and then convert to the *S*-plane,

$$Z(S) = \frac{1.327821S^2 + 0.881554S + 0.790569}{2S^3 + 1.327821S^2 + 2.381554S + 0.790569} \tag{5.6a}$$

Then, using Figure 4.1, we see that the input element is in the form of a shunt capacitor which has a shunt open stub as a distributed equivalent. The value of the capacitor is (2/1.327821) or 1.506227F and z_c is $1/C$ or 0.663911. Extracting the capacitor, we have a remainder of

$$Z'(S) = \frac{1.327821S^2 + 0.881554S + 0.790569}{1.190778S + 0.790569} \tag{5.6b}$$

Again using Figure 4.1, the input element is a series inductor equal to (1.327821/ 1.190778) or 1.115087H. The distributed equivalent is a series shorted stub. However, we cannot realize this element and elect to remove a transmission line element instead. Evaluating (5.6b) at $S = 1$ and $S = -1$, we find that $Z(1) = z_c =$ 1.514094 and $Z(-1) = -3.090479$. $Z(1)$ does not equal $-Z(-1)$, and the TLE is redundant. Using (5.5b) to compute the remainder, the input impedance becomes

$$Z''(S) = \frac{0.719388S^2 + 0.477608S - 1.196995}{1.327821S^3 + 0.881554S^2 - 1.012380S - 1.196995} \tag{5.6c}$$

which seems to be 1 degree higher than $Z'(S)$. The numerator and denominator are 0 at $S = 1$, and we can factor $(S - 1)$ from both, and then cancel, leaving

$$Z''(S) = \frac{0.719388S + 1.196995}{1.327821S^2 + 2.209375S + 1.196995} \tag{5.6d}$$

which is the same degree as $Z'(S)$. Comparing (5.6d) with (5.6b), we see that the transmission line element essentially inverted $Z'(S)$ and changed the coefficients somewhat. The input element is now in the form of a shunt capacitor or open stub and is acceptable. The capacitor has a value of (1.327821/0.719399) or 1.845766F, and the equivalent z_c is 0.541781. After extracting the capacitor, the input impedance is

$$Z'''(S) = \frac{0.719388S + 1.196995}{1.196995}$$

and our choice of available elements is limited to a 0.600994H series inductor or a redundant transmission line element equal to 1.600994. Removing the TLE and canceling the $(S - 1)$ term in the numerator and denominator,

$$Z''''(S) = \frac{1.916383}{0.719388S + 1.916383}$$

We can remove another shunt capacitor equal to 0.375388F, or an equivalent shunt open stub, $z_c = 2.663907$, which reveals that the output termination is 1.0. The topology and element values are shown in Figure 5.4(a) for the low-pass filter scaled to 50Ω and an electrical length of 45° at 0.5 GHz.

The topology is acceptable for realization using microstrip or stripline and has a frequency response shown by the lighter plot in Figure 5.5. However, the element values of the shunt stubs are not practical; the first two are too low and the last too high. We can realize the first two stubs as a parallel combination with characteristic impedance of each equal twice the original value, as shown in figure 5.4(b). However, the output stub must be handled differently. At the cutoff frequency of 0.5 GHz, the impedance of the stub is $-jZ_c\text{Cot}(45°)$ or -j133.2Ω. If we change the characteristic impedance to 100Ω, we can use (2.9e) to compute the electrical length necessary to provide the same impedance:

$$\theta = \text{Tan}^{-1}\left(\frac{-jZ_c}{Z_{OC}}\right) = \text{Tan}^{-1}\left(\frac{100.0}{133.2}\right) = 36.9°$$

as shown in Figure 5.4(b). The darker plot in Figure 5.5 shows the response of the modified filter. Although the network is no longer commensurate, the response of the two networks are virtually identical up to the cutoff frequency, approximately 0.01 dB difference. However, there are some differences in the second passband, and a significant difference in the third. This structure has transmission zeros at frequencies where the shunt stubs are odd multiples of one-quarter wavelength; that is, the impedance of each stub is zero. For electrical lengths of 45° at 0.5 GHz, the frequencies are 1.0 GHz, 3.0 GHz, and so forth. However, we changed the electrical length of one stub, which adds additional transmission zeros at odd multiples of 1.22 GHz. The second zero is at 3.66 GHz, which places it inside the third passband. We can sometimes use this characteristic to suppress some of the upper passbands by strategically placing the transmission zeros.

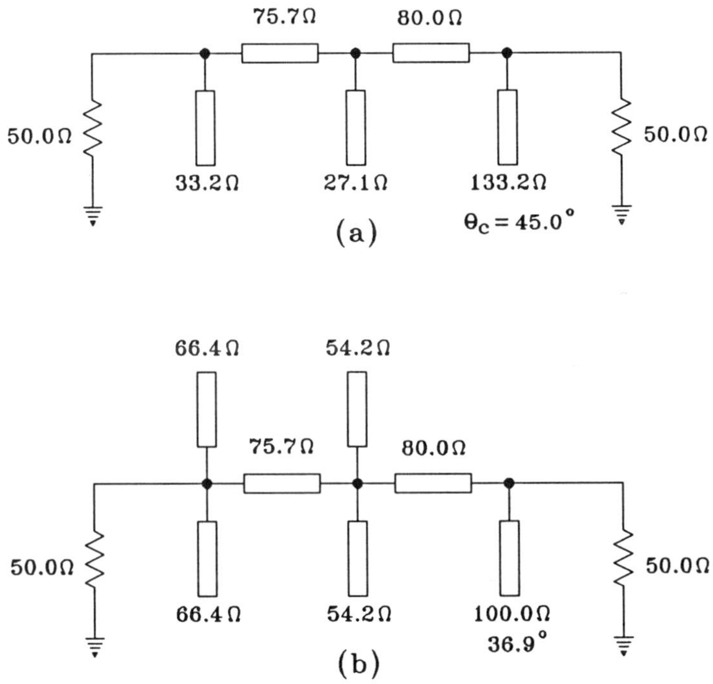

Figure 5.4 Low-pass filter: (a) synthesized; (b) modified.

We cannot scale the TLE to an electrical length different from 45° and maintain an accurate passband response. In some cases, the impedances of the lines will be nonrealizable, and it becomes absolutely necessary to modify the structure if we are to build hardware. In most instances, we can constrain the

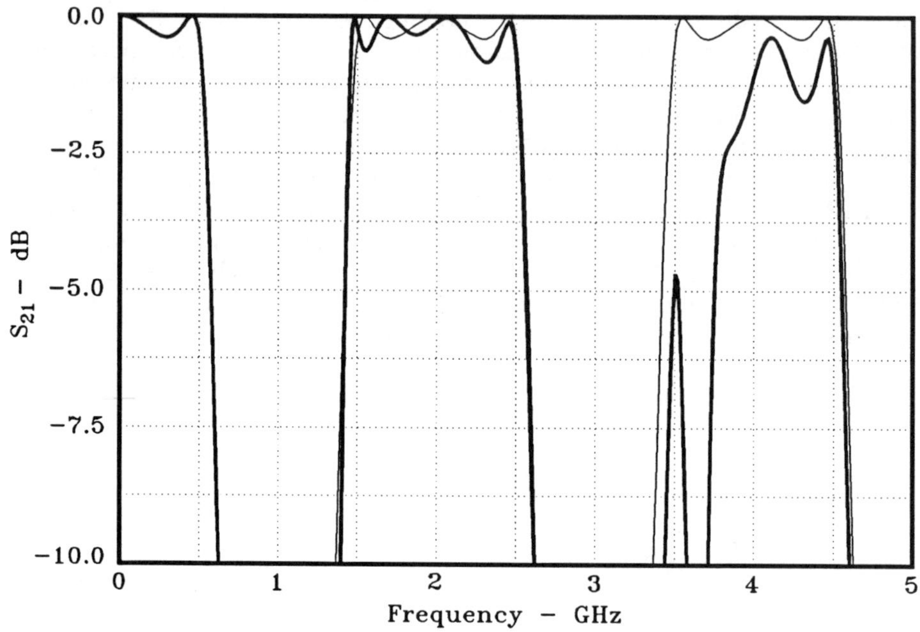

Figure 5.5 Distributed low-pass filter using TLEs.

characteristic impedance and change the electrical lengths of the TLEs such that the cutoff frequency is preserved. Usually, the passband ripple is affected to the extent that another design iteration is required, or constrained minimax optimization becomes necessary. We also have the option of using quarter-wavelength lines, discussed in Section 3.1.3, to change the internal impedance level of the filter. The impedance-transforming properties are only valid at a single frequency. Nevertheless, this method is often useful for adjusting the element values of narrow-band filters. However, before resorting to approximation techniques, we have a number of network transformations and distributed approximations that can be applied in an effort to synthesize a realizable network.

5.1.2 The Double-Length TLE

Levy [9,11] used a double-length TLE (i.e., two identical cascaded TLEs) for the synthesis of an impedance-transforming filter. The use of double-length TLEs resulted in convenient element values, compared with the original network, and it is worthwhile examining this special case of two cascaded TLEs. Figure 5.6 shows the network input impedance, represented as $Z(S)$, and the input impedance to the second TLE, represented as $Z_1(S)$. Equation (5.5b) can be used to extract the first TLE from the input impedance, leaving a remainder of

$$Z_1(S) = Z_a \frac{Z(S) - SZ_a}{Z_a - SZ(S)} \tag{5.7a}$$

We can then evaluate (5.7a) at $S = 1$ to determine the value of Z_b. This expression is indeterminate since the numerator and denominator are both zero because of the common $(S - 1)$ term, which must be factored and canceled. However, we can apply L'Hopital's Rule and write

$$Z_b = Z_1(1) = Z_a \left. \frac{[Z(S) - SZ_a]'}{[Z_a - SZ(S)]'} \right|_{S=1} \tag{5.7b}$$

where the prime denotes a derivative with respect to S. Replacing $Z(S)$ with $N(S)/D(S)$, substituting in (5.7b), and then taking derivatives and expanding, we have

$$Z_b = Z_1(1) = Z_a \frac{Z_a D(1) + Z_a D'(1) - N'(1)}{N(1) - Z_a D'(1) + N'(1)} \tag{5.7c}$$

Thus, we can compute the value of both TLEs directly from the original input impedance polynomials. After extracting the shunt input capacitor in Section 5.1.1, we evaluated (5.6b) at $S = 1$ to determine the value of the TLE. Using (5.6b), $Z_a = 1.514094$, $N(1) = 2.999944$, $D(1) = 1.981347$, $N'(1) = 3.537196$, and $D'(1) = 1.190778$. Substituting into (5.7c),

$$Z_b = (1.514094) \frac{(1.514094)(1.981347) + (1.514094)(1.190778) - 3.537196}{2.999944 - (1.514094)(1.190778) + 3.537196}$$

$$= (1.514094) \frac{1.265699}{4.734190} = 0.404797$$

which agrees with the value of the second TLE computed from (5.6d) after the first TLE was extracted and the $(S - 1)$ terms canceled.

Now, for two identical TLEs we require that Z_a equals Z_b. Applying this constraint to (5.7c) (i.e., $Z_1(1) = Z(1) = Z_a$),

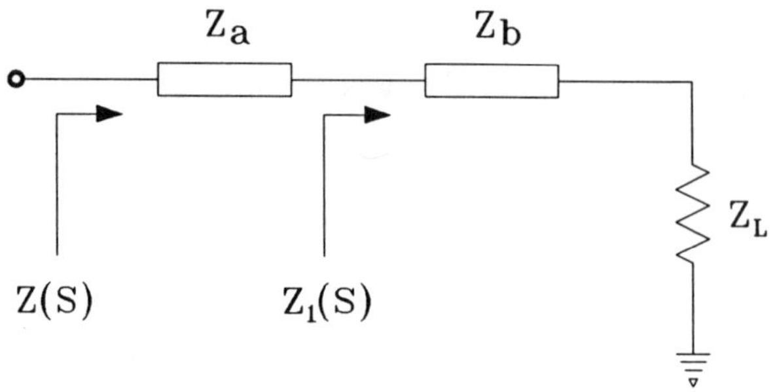

Figure 5.6 Two-cascaded TLEs.

$$Z_a = \frac{2N'(1) + N(1)}{2D'(1) + D(1)} = \frac{N(1)}{D(1)} \tag{5.8a}$$

and we can write the condition for two equal TLEs as

$$Z_a = \frac{N(1)}{D(1)} = \frac{N'(1)}{D'(1)} \tag{5.8b}$$

Thus, when the ratio of the original numerator and denominator polynomials evaluated at $S = 1$ is equal to the ratio of both derivatives, also evaluated at $S = 1$, we can extract two identical TLEs. The condition in (5.8b) is rarely encountered during the synthesis process. However, we can sometimes use partial element extraction to force the double-length TLE. For example, when we extract a shunt capacitor (shunt open stub), the remainder is

$$Z_1(S) = \frac{N(S)}{D(S) - CSN(S)} \tag{5.9a}$$

where $N(S)$ and $D(S)$ are the numerator and denominator polynomials, respectively, of the original driving-point impedance. Imposing (5.8b) on (5.9a), we have

$$\left. \frac{N(S)}{D(S) - CSN(S)} \right|_{S=1} = \left. \frac{N'(S)}{[D(S) - CSN(S)]'} \right|_{S=1} \tag{5.9b}$$

Then expanding and solving for C,

$$C = \frac{N(1)D'(1) - N'(1)D(1)}{N^2(1)} \tag{5.9c}$$

If a shunt capacitor with sufficient value is available, we can extract a partial value determined from (5.9c), two identical TLEs, and then the remainder of the capacitor. Using a similar approach, we can compute the partial values for the other elements, although the shunt open stub is the most useful configuration.

5.2 DISTRIBUTED NETWORK TRANSFORMATIONS

Once we extracted the shunt open stub from the impedance function of the low-pass filter in Section 5.1.1, the remainder given by (5.6b) revealed that a series shorted stub or redundant TLE were the only choices. However, after removing the redundant TLE, we found that, in addition to another TLE, a shunt open stub was present. It appears that somehow the TLE has transformed the series shorted stub into a shunt open stub. In fact, when we compared equations (5.6b) and (5.6d), it was obvious that the effect of the TLE was to invert the impedance function and modify the coefficients. Before pursuing this equivalence, we need the two-port representation of the TLE in terms of Richards' variable.

After factoring Cosθ from the **ABCD** matrix for the lossless transmission line shown in Figure 2.9, we have

$$Ch_{TLE} = \mathrm{Cos}\,\theta \begin{bmatrix} 1 & jZ_C \dfrac{\mathrm{Sin}\,\theta}{\mathrm{Cos}\,\theta} \\ \dfrac{j}{Z_C}\dfrac{\mathrm{Sin}\,\theta}{\mathrm{Cos}\,\theta} & 1 \end{bmatrix} = \mathrm{Cos}\,\theta \begin{bmatrix} 1 & Z_C S \\ \dfrac{S}{Z_C} & 1 \end{bmatrix}$$

Then, using the trigonometric identity

$$\mathrm{Cos}\,\theta = \frac{1}{\pm\sqrt{1 + \mathrm{Tan}^2\theta}} = \frac{1}{\pm\sqrt{1 - (j\mathrm{Tan}\,\theta)^2}}$$

we can write the **ABCD** matrix as

$$Ch_{TLE} = \frac{1}{\pm\sqrt{1 - S^2}} \begin{bmatrix} 1 & Z_C S \\ \dfrac{S}{Z_C} & 1 \end{bmatrix} \tag{5.10}$$

where the sign applied to the radical depends on the quadrant in which θ falls. Normally, (5.10) is written without the sign uncertainty, and we shall do likewise. Most applications of (5.10) result in cancellation of the radical. However, the reader should be aware that it may be necessary to test θ when the radical appears in an expression.

After removing the first open stub in Section 5.1.1, we had a choice of the series shorted stub followed by a TLE, which we know from Richards' Theorem is always available. Instead, we removed a TLE followed by a shunt open stub. The first option is shown in Figure 5.7(a), and we can use (2.1), (2.2), (5.1), and (5.10) to write the combined **ABCD** matrix as

$$Ch_a = \begin{bmatrix} 1 & Z_1 S \\ 0 & 1 \end{bmatrix} \frac{1}{\sqrt{1 - S^2}} \begin{bmatrix} 1 & Z_2 S \\ \dfrac{S}{Z_2} & 1 \end{bmatrix}$$

$$= \frac{1}{\sqrt{1 - S^2}} \begin{bmatrix} \left(1 + \dfrac{Z_1}{Z_2} S^2\right) & (Z_1 + Z_2) S \\ \dfrac{S}{Z_2} & 1 \end{bmatrix} \tag{5.11a}$$

The second option shown in Figure 5.7(b) has a combined **ABCD** matrix given by

$$Ch_b = \frac{1}{\sqrt{1 - S^2}} \begin{bmatrix} 1 & Z_a S \\ \dfrac{S}{Z_a} & 1 \end{bmatrix} \begin{bmatrix} 1 & 0 \\ \dfrac{S}{Z_b} & 1 \end{bmatrix}$$

$$= \frac{1}{\sqrt{1 - S^2}} \begin{bmatrix} \left(1 + \dfrac{Z_a}{Z_b} S^2\right) & Z_a S \\ \left(\dfrac{1}{Z_a} + \dfrac{1}{Z_b}\right) S & 1 \end{bmatrix} \tag{5.11b}$$

We observe that all four elements of (5.11a) and (5.11b) have exactly the same form in S; only the coefficients differ. Comparing B from both **ABCD** matrices, we have

$$Z_a = Z_1 + Z_2 \tag{5.11c}$$

and from A,

$$Z_b = \frac{Z_2}{Z_1} Z_a \tag{5.11d}$$

Thus, the two networks are equivalent at all frequencies, and the element values of one can easily be computed from the element values of the other. Similarly, we can assume the equivalence shown in Figure 5.7(c) and 5.7(d), and write the combined matrix for Figure 5.7(c):

$$\boldsymbol{Ch}_c = \frac{1}{\sqrt{1 - S^2}} \begin{bmatrix} 1 & (Z_1 + Z_2)S \\ \dfrac{S}{Z_1} & \left(1 + \dfrac{Z_2}{Z_1} S^2\right) \end{bmatrix} \tag{5.12a}$$

and for Figure 5.7(d):

$$\boldsymbol{Ch}_d = \frac{1}{\sqrt{1 - S^2}} \begin{bmatrix} 1 & Z_b S \\ \left(\dfrac{1}{Z_a} + \dfrac{1}{Z_b}\right) S & \left(1 + \dfrac{Z_b}{Z_a} S^2\right) \end{bmatrix} \tag{5.12b}$$

Comparing B from (5.12a) and (5.12b),

$$Z_b = Z_1 + Z_2 \tag{5.12c}$$

Then, by examining D from both matrices,

$$Z_a = \frac{Z_1}{Z_2} Z_b \tag{5.12d}$$

and we see that these networks are also identical at all frequencies.

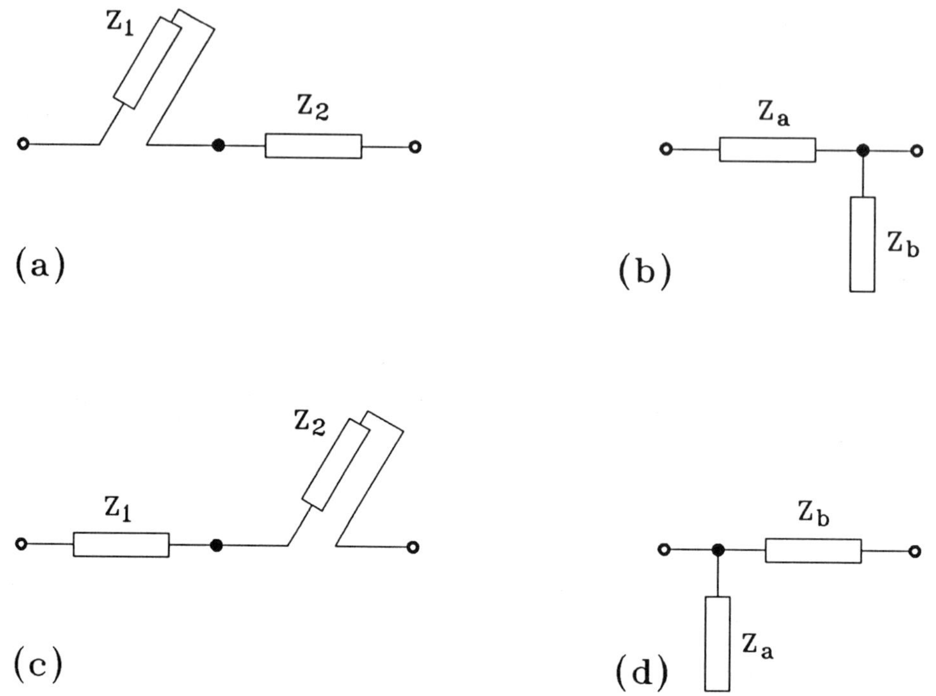

Figure 5.7 Shorted and open stub equivalent network.

Equivalences in the form of (5.11) and (5.12) were first noticed by K. Kuroda [12] and are known as *Kuroda identities* or *Kuroda transforms*. These identities play an important role in transforming a nonrealizable topology into one that can be realized, as well as altering element values to more desirable levels. For example, the original distributed equivalent filter designed in Section 5.1 and shown in Figure 5.2(a) is not realizable in microstrip or stripline because of the series stubs. However, we can add 50Ω transmission lines at the input and output ports without affecting the amplitude response. In fact, we often measure the amplitude response of a filter using 50Ω cables connected to the signal source and power meter or network analyzer. Figure 5.8(a) shows the filter augmented with 50Ω transmission line elements at the input and output. The first TLE-stub combination is in the form of Figure 5.7(c), and we can use (5.12) to transform into the equivalent network shown in Figure 5.7(d). Using (5.12c),

$$Z_b = Z_1 + Z_2 = 50.0 + 75.3 = 125.3\Omega$$

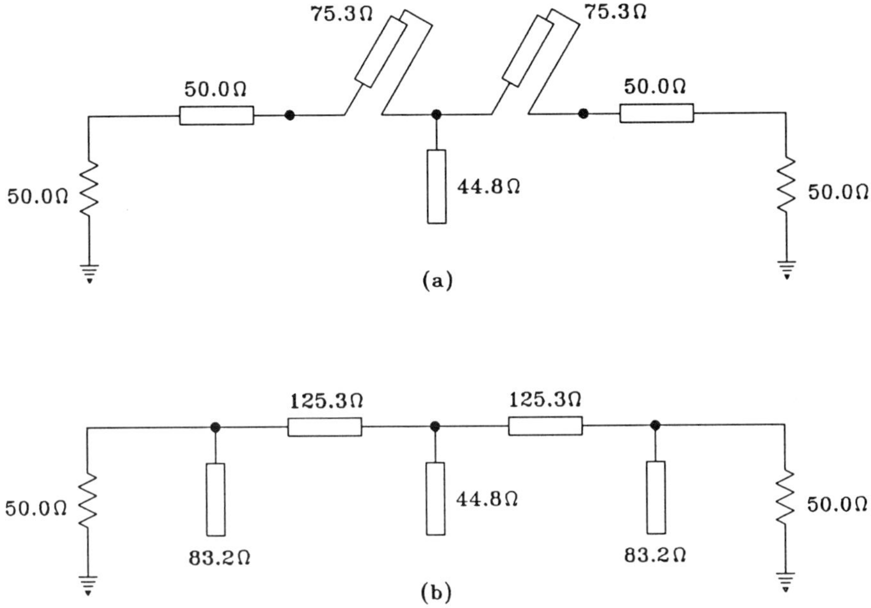

Figure 5.8 Low-pass filter: (a) augmented; (b) transformed.

and from (5.12d),

$$Z_a = \frac{Z_1}{Z_2} Z_b = \frac{50.0}{75.3} 125.3 = 83.2\Omega$$

The last stub-TLE combination is in the form of Figure 5.7(a), and we can compute element values for the equivalent shown in Figure 5.7(b). Using (5.11c),

$$Z_a = Z_1 + Z_2 = 75.3 + 50.0 = 125.3\Omega$$

then (5.11d),

$$Z_b = \frac{Z_2}{Z_1} Z_a = \frac{50.0}{75.3} 125.3 = 83.2\Omega$$

and we have the equivalent low-pass filter shown in Figure 5.8(b). Comparing the transformed filter in Figure 5.8(b) with the synthesized filter in Figure 5.4(a), we see that the topology is identical, but the element values are different. How can the two filters be equivalent when the element values are so different? An analysis of the transformed filter reveals that the amplitude response is identical to that of the original filter shown in Figure 5.3. However, the phase response is different. Inserting transmission line elements having the same characteristic impedance as the source and load has the effect of adding phase shift but preserving the amplitude response. In this case, the low-pass filter was based on an amplitude approximation, and additional phase shift is not detrimental to the performance.

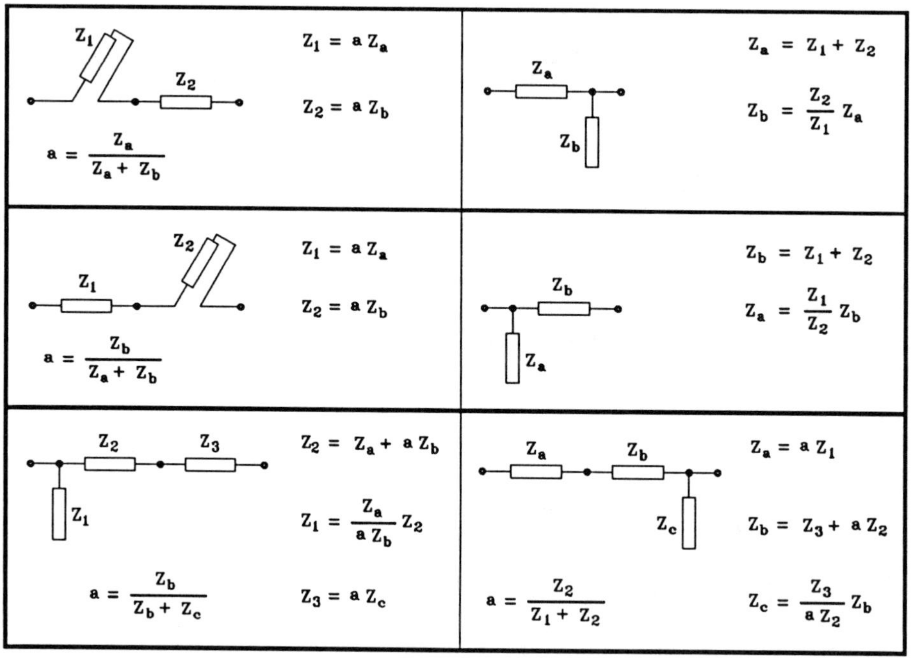

Figure 5.9 Low-pass Kuroda identities.

We can easily derive a large number of Kuroda identities by computing the **ABCD** matrices for combinations of distributed elements, searching for those that have matching forms in all four elements, and then solving for the element values of one network in terms of the other. References 13 through 19 provide tables of a few Kuroda identities, including some coupled transmission line [2-5] equivalents which are convenient for physical realization of certain types of filters. Figure 5.9

shows the Kuroda identities using a low-pass element in combination with one and two transmission line elements. Equivalent circuits are shown in each row, along with the appropriate equations used to compute element values. For example, the first row corresponds to Figures 5.7(a) and (b), and the second row to Figures 5.7(c) and (d). The third row is derived by applying a transform from the first or second row twice. The shunt open stub is moved between the two transmission line elements where it becomes a series shorted stub, and then moved again to restore the open stub at the opposite end of the network.

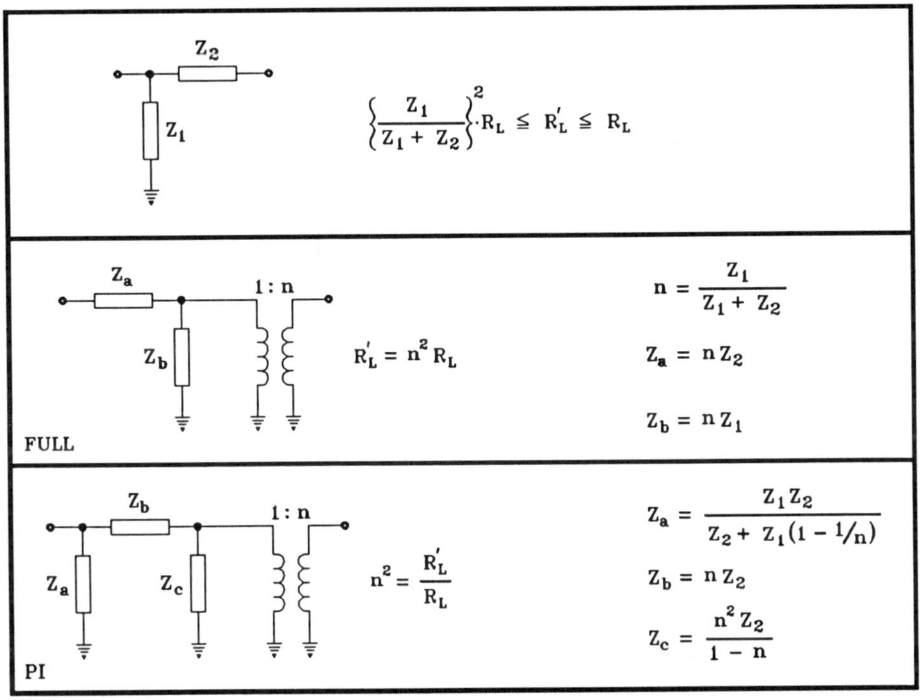

Figure 5.10 Kuroda identities using shunt shorted stub and TLE to reduce R_L.

Figure 5.10 shows the Kuroda identities for a shunt shorted stub and TLE combination. Comparing the equivalent networks in the second and third row with the original network in the top row, we see that the shorted stub is not converted into a different element type by the transformation. In fact, only the low-pass elements shown in Figure 5.9 are changed. The figure also shows that it is necessary to add an ideal transformer to the equivalent circuit. In this case, the turns ratio is less than 1 and we can transform the termination R_L into any value R'_L within the limits given by the expression associated with the original network in the top row. The second row shows the full transform where $R'_L = n^2 \bullet R_L$, and the third row shows the pi equivalent, which results from a partial transform.

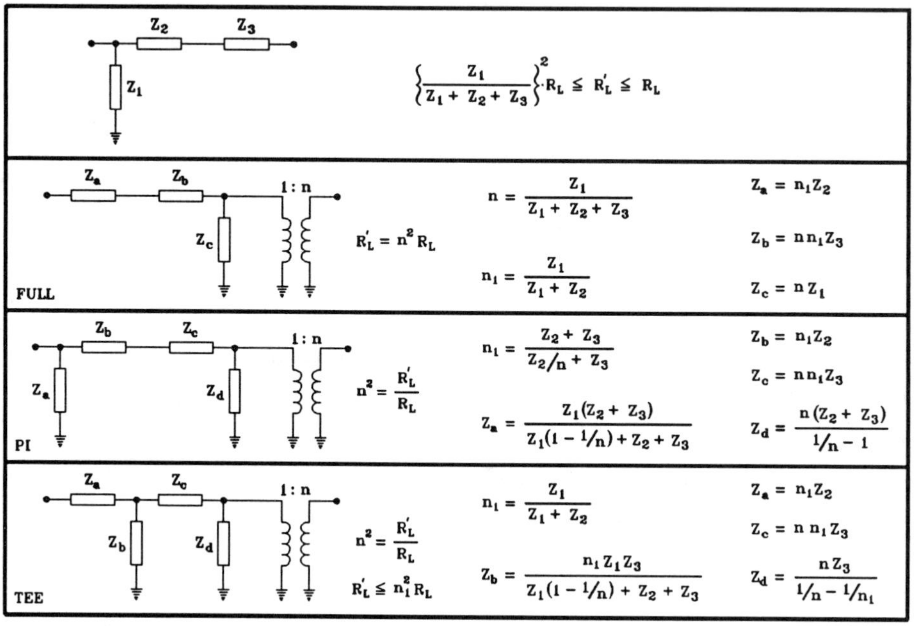

Figure 5.11 Kuroda identities using shunt shorted stub and two TLEs to reduce R_L.

Figure 5.11 shows the identities for a network consisting of one shunt shorted stub and two transmission line elements. The equivalent networks are similar to the ones in Figure 5.10, except that we also have a tee configuration shown in the bottom row. However, the termination for the tee has an additional constraint; that is, it must be less than $n_1^2 \bullet R_L$. When the termination exactly equals $n_1^2 \bullet R_L$, Z_d is infinite, and the portion of the network formed by Z_a and Z_b corresponds to the full transform in Figure 5.10. Terminations greater than $n_1^2 \bullet R_L$ can be realized using the pi configuration of Figure 5.10, which results in a shorted stub at either end of Z_a rather than Z_c, as shown.

Figures 5.12 and 5.13 are similar to the identities shown in Figures 5.10 and 5.11 except that the shunt shorted stub is located at the output port of the original network, which results in turns ratios greater than 1. The Kuroda identities shown in Figures 5.14 through 5.17 are derived in the same manner as those in Figures

5.10 through 5.13, except that the shunt shorted stub is replaced with a series open stub. The effect of the open series stub on R_L is opposite that of the shunt shorted stub; that is, as we move the element from the output port toward the input, the termination is reduced.

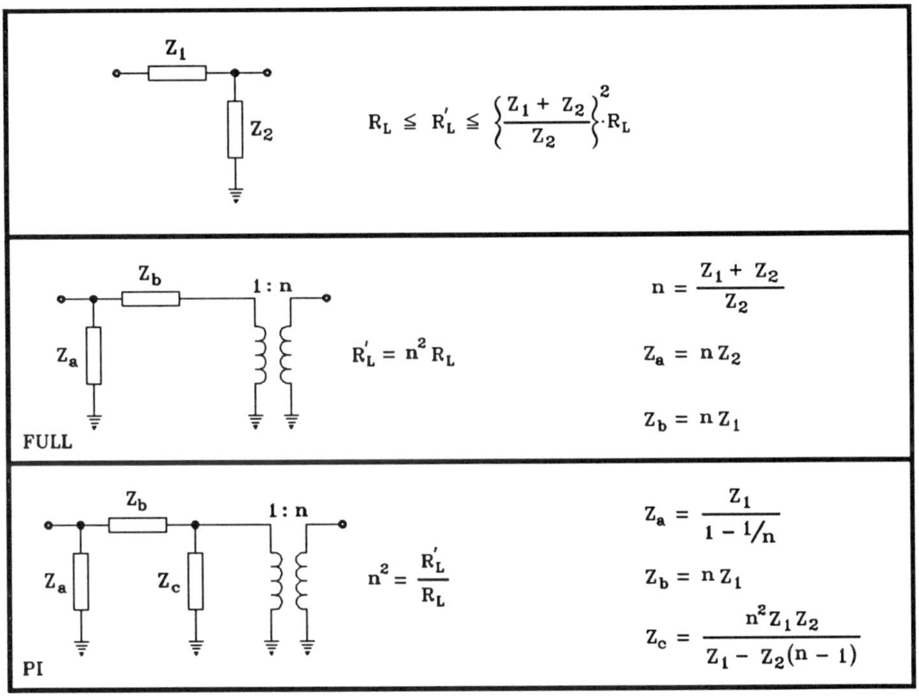

Figure 5.12 Kuroda identities using shunt shorted stub and TLE to increase R_L.

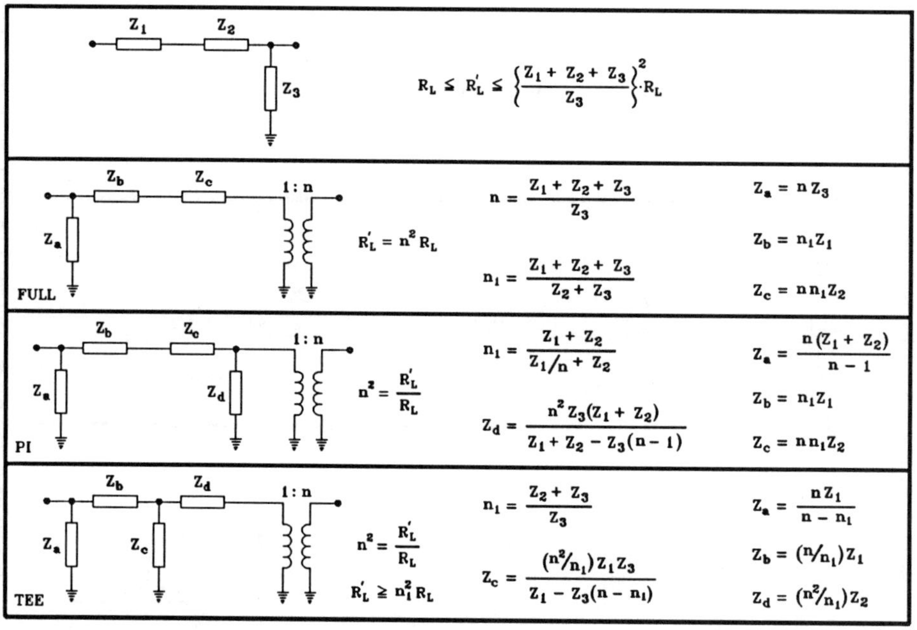

Figure 5.13 Kuroda identities using shunt shorted stub and two TLEs to increase R_L.

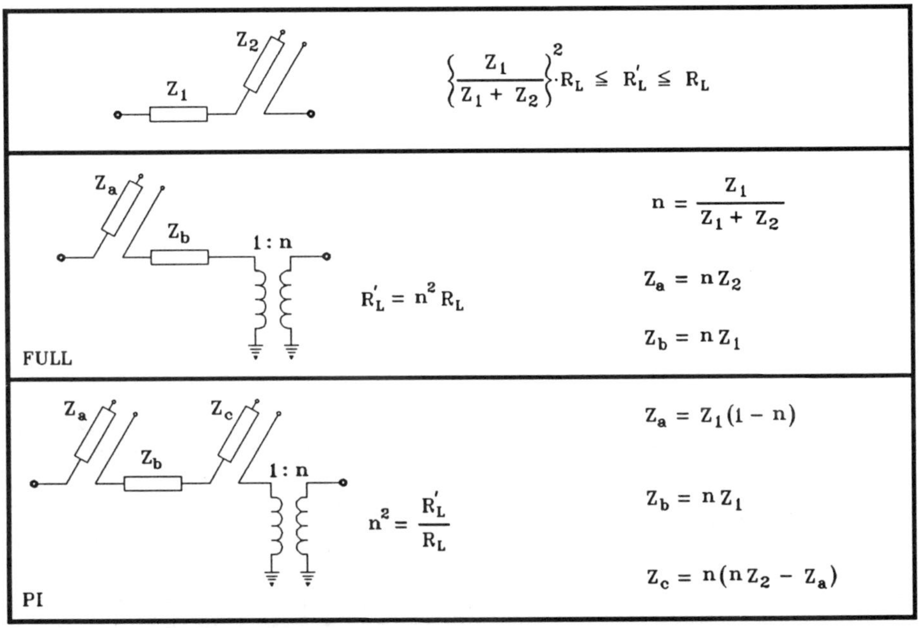

Figure 5.14 Kuroda identities using series open stub and TLE to reduce R_L.

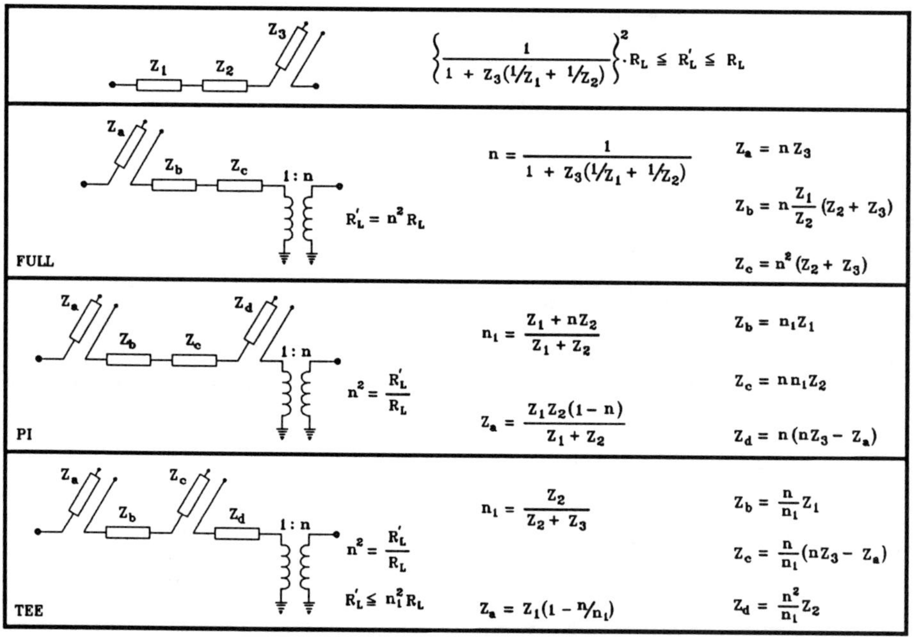

Figure 5.15 Kuroda identities using series open stub and two TLEs to reduce R_L.

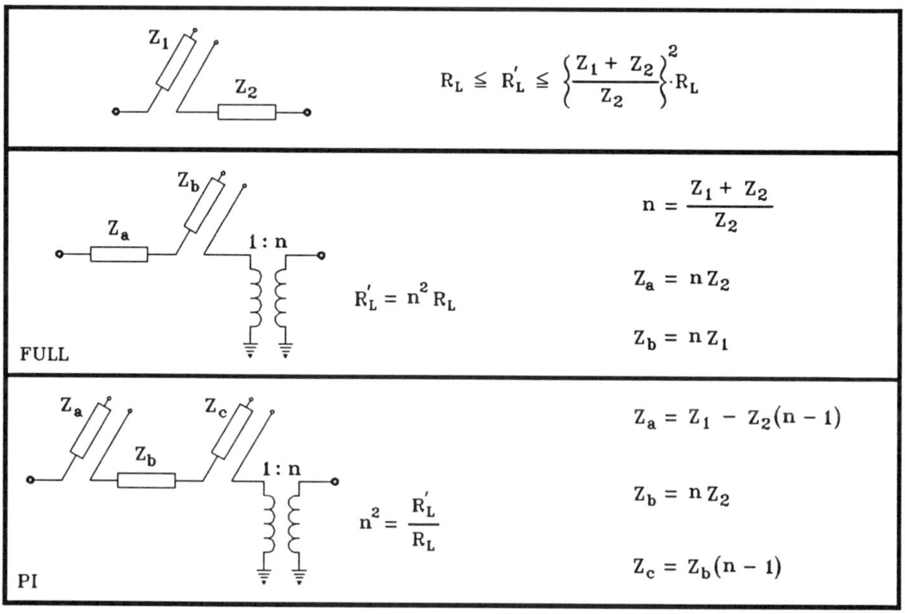

Figure 5.16 Kuroda identities using series open stub and TLE to increase R_L.

Figures 5.18 through 5.21 show the Kuroda identities for networks consisting of series and shorted stubs. These identities are the distributed equivalent of the Norton transformations shown in Figures 4.28 through 4.31 and can easily be derived directly from the lumped-element expressions. We simply replace capacitors with open stubs, and inductors with shorted stubs. The normalized capacitance becomes the reciprocal of the characteristic impedance of the open stubs, and the normalized inductance is replaced by the impedance of shorted stubs.

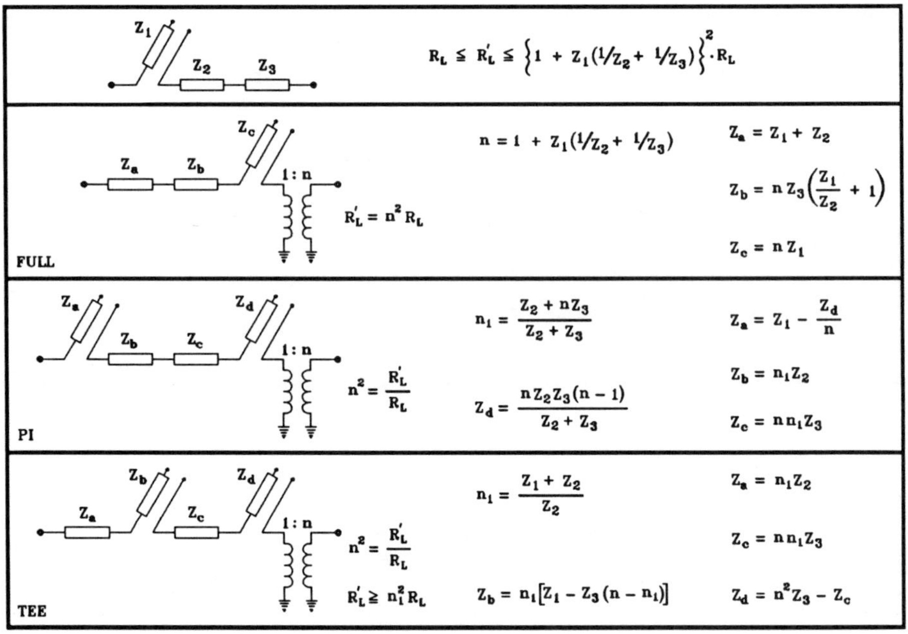

Figure 5.17 Kuroda identities using series open stub and two TLEs to increase R_L.

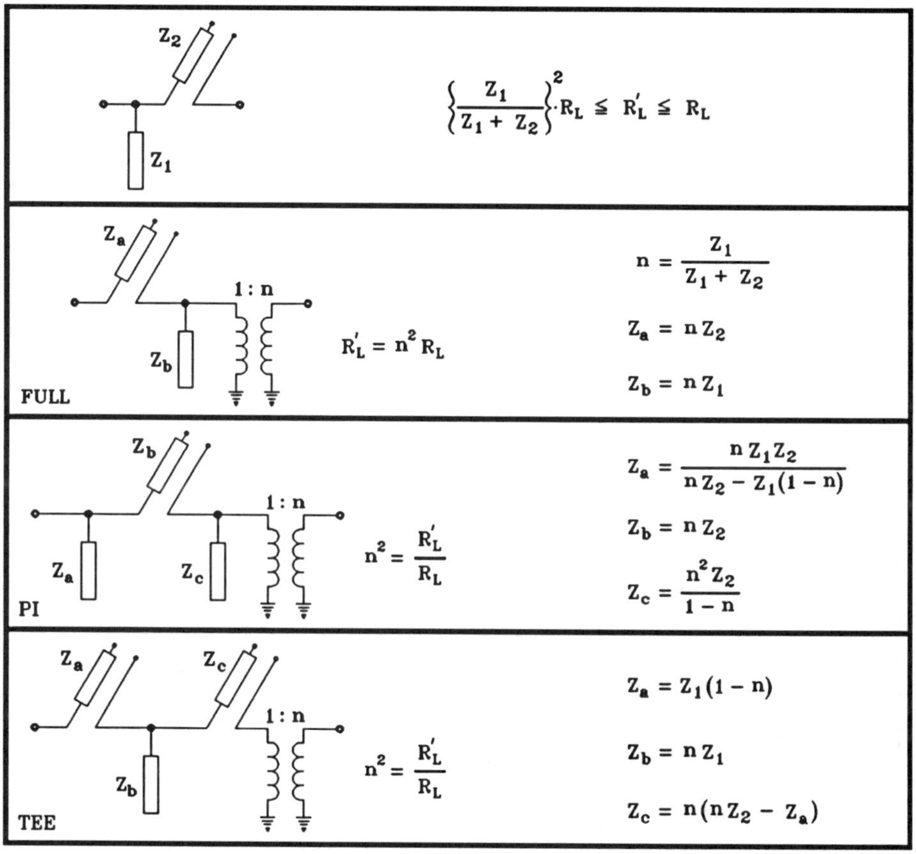

Figure 5.18 Kuroda identities using open stubs to reduce R_L.

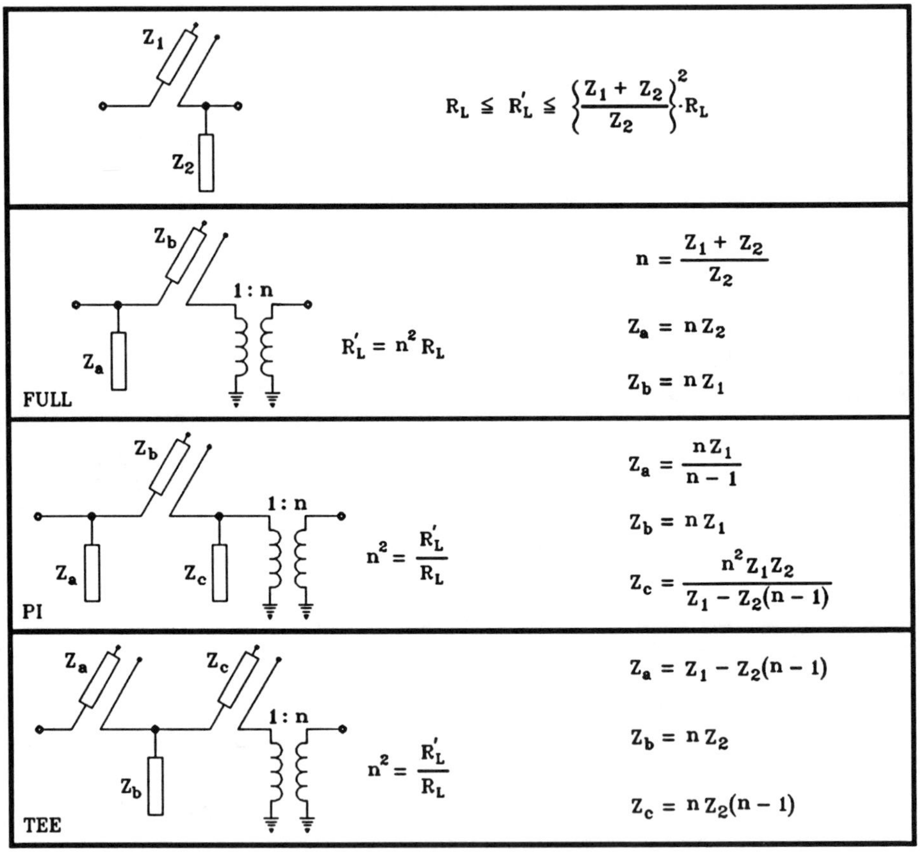

Figure 5.19 Kuroda identities using open stubs to increase R_L.

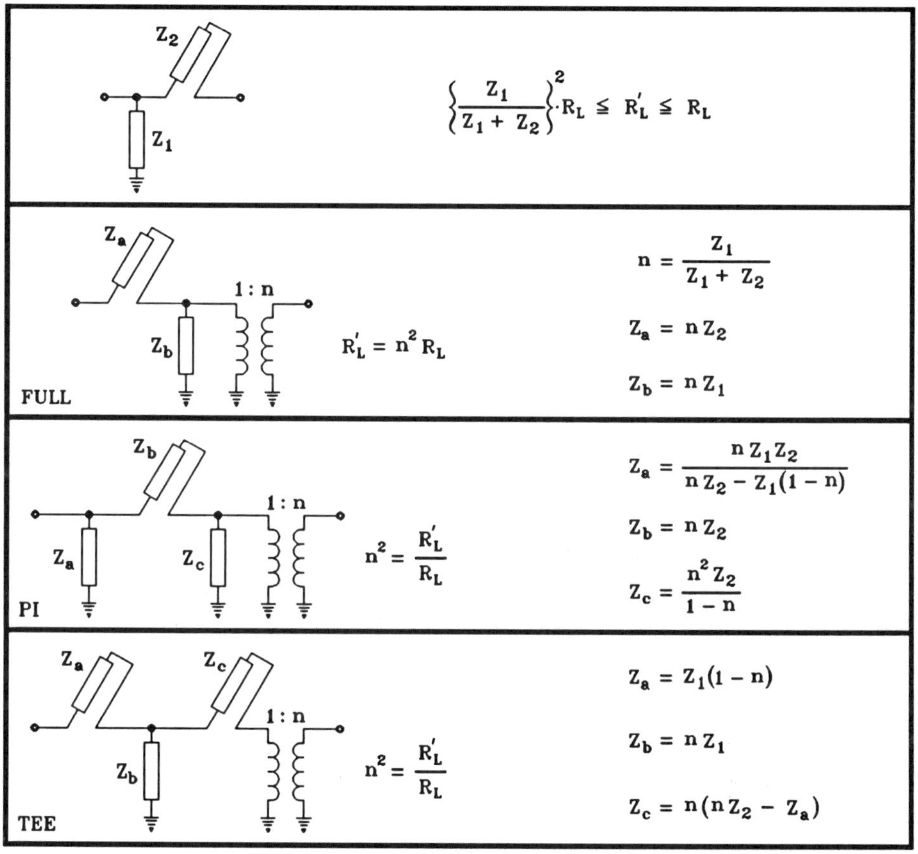

Figure 5.20 Kuroda identities using shorted stubs to reduce R_L.

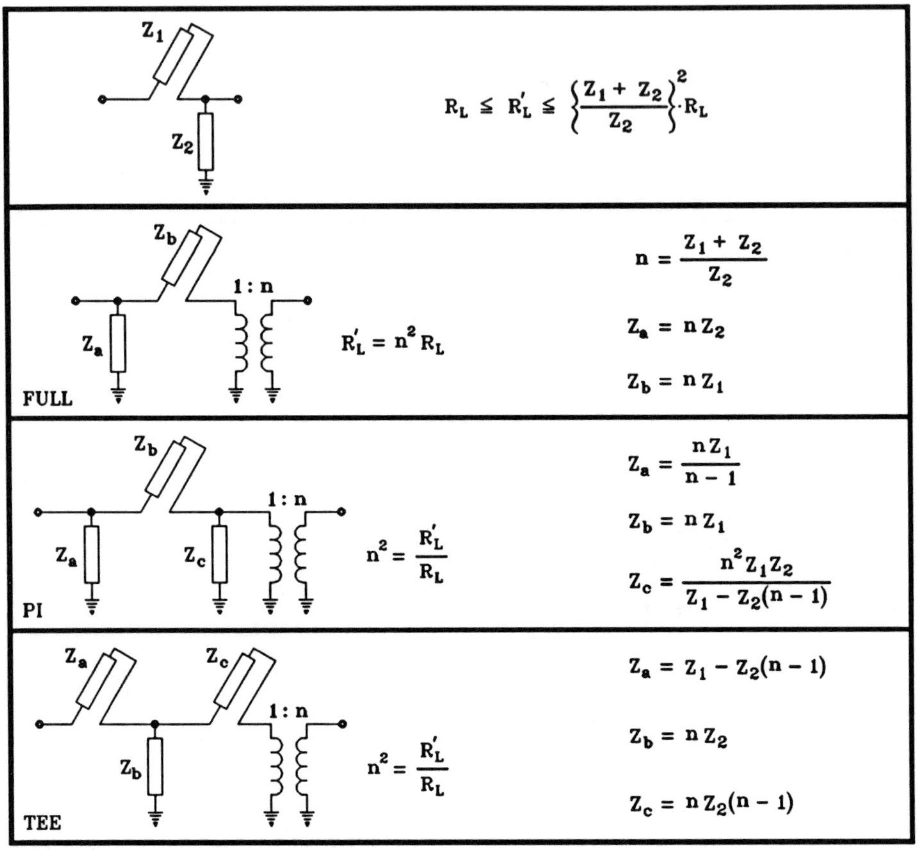

Figure 5.21 Kuroda identities using shorted stubs to increase R_L.

We can synthesize a number of useful distributed-element networks using the techniques discussed thus far. Starting with a lumped-element approximation, we can transform into the S-plane and then synthesize a distributed equivalent including as many redundant transmission line elements as necessary to provide a realizable topology. We also have the option of applying Kuroda identities to the distributed equivalent network, synthesized network, augmented network, or some combination. However, synthesis based on a distributed-element approximation offers several advantages over these methods.

5.3 DISTRIBUTED APPROXIMATIONS

Redundant transmission line elements are useful for transforming a nonrealizable network into a realizable one by altering the topology and element values. However, since the redundant TLE does not reduce the order of the impedance function, it does not contribute to the response (except to add phase shift) and is essentially wasted. Examining the **ABCD** matrix of the transmission line element given by (5.10), we see that the parameters are functions of S and should be useful as filter elements. Horton and Wenzel [20] show that the transmission line element provides almost the same attenuation characteristics as an open or shorted stub in the vicinity of the cutoff frequency for narrow to moderate bandwidth filters. As the bandwidth is increased, the TLE is less effective and the stub-type element provides superior attenuation. In either case, the nonredundant transmission line element does contribute to the response and can be used to reduce the number of elements as well as provide realizable topologies.

Series open circuit and shunt short circuit stubs are high-pass elements that produce transmission zeros at $S = 0$, while series short circuit and shunt open circuit stubs are low-pass elements that produce transmission zeros at $S = \infty$. The transmission line element is different in that it does not have a transmission zero; that is, some power is transmitted from input to output at all frequencies. Using Figure 1.11, we can compute $s_{21}(S)$ from the **ABCD** matrix given in (5.10):

$$s_{21}(S) = \frac{\sqrt{1 - S^2}}{1 + \frac{1}{2}(z_c + 1/z_c)S} \tag{5.13}$$

Then the transducer gain function for the distributed network must have the form

$$|s_{21}(j\Omega)|^2 = \frac{K_m S^{2m}(1 - S^2)^n}{P_{n+m+q}(S^2)} \tag{5.14}$$

where

$P(S)$ is a strictly Hurwitz polynomial.
n is the number of transmission line elements.
m is the number of high-pass elements.
q is the number of low-pass elements.
K_m is a constant selected such that $G_t \leq 1$.

The numerator of (5.14) contains the required number of transmission zeros due to the m high-pass elements, and terms of the form $(1 - S^2)$ due to the n transmission line elements. The denominator is a function of all elements and must be derived to provide the desired type of approximation.

5.3.1 Cascaded Transmission Line Networks

When we restrict the network to a cascade of TLEs, the number of high-pass and low-pass elements is zero, and (5.14) reduces to

$$|s_{21}(j\Omega)|^2 = \frac{K_m(1 - S^2)^n}{P_n(S^2)} = \frac{K_m(1 + \Omega^2)^n}{p_n\Omega^{2n} + p_{n-1}\Omega^{2(n-1)} + \cdots + p_1\Omega^2 + p_0} \tag{5.15a}$$

Substituting the trigonometric identities

$$\mathrm{Cos}^2\theta = \frac{1}{1 + \Omega^2} \tag{5.15b}$$

and

$$\mathrm{Sin}^2\theta = \frac{\Omega^2}{1 + \Omega^2} \tag{5.15c}$$

into (5.15a) and expanding and collecting terms, we can write

$$|s_{21}(j\Omega)|^2 = \frac{K_m}{q_n \mathrm{Sin}^{2n}\theta + q_{n-1}\mathrm{Sin}^{2(n-1)}\theta + \cdots + q_1\mathrm{Sin}^2\theta + q_0} \tag{5.15d}$$

where the coefficients q_i are combinations of the original p_i coefficients. Thus, we need only determine the coefficients of the polynomial as a function of $\mathrm{Sin}^2\theta$ to realize the desired approximation.

One such approximation is the distributed equivalent of the maximally flat or Butterworth response given by (4.13):

$$G_t(\Omega^2) = \frac{K_m}{1 + \varepsilon^2\left(\dfrac{\mathrm{Sin}^2\theta}{\mathrm{Sin}^2\theta_c}\right)^n} \tag{5.16a}$$

where θ_c is the electrical length of the transmission line elements at the cutoff frequency. The term $\mathrm{Sin}^2\theta_c$ is the frequency normalization discussed in Section 5.1; that is, $(\mathrm{Sin}^2\theta/\mathrm{Sin}^2\theta_c)$ is 1 at the cutoff frequency. In contrast with the lumped to distributed transformation, the distributed approximation is valid for any electrical length less than 90°, although the network must also be commensurate (i.e., the same electrical length for all elements). At $\Omega = 0$ or $\theta = 0°$, all derivatives are

0; at $\Omega = \infty$ where $\theta = \pi/2$, one derivative is 0. $\mathrm{Sin}^2\theta$ monotonically increases from 0 at $\theta = 0°$ to 1 at $\theta = 90°$, decreases to 0 again at $\theta = 180°$; then repeats with a period of 180°. At $\Omega = 0$, $G_t(\Omega)$ is K_m, $K_m/(1 + \epsilon^2)$ at the cutoff frequency where $(\mathrm{Sin}^2\theta/\mathrm{Sin}^2\theta_c)$ equals 1, and

$$G_t(\infty) = \frac{K_m}{1 + \dfrac{\varepsilon^2}{\mathrm{Sin}^{2n}\theta_c}} \tag{5.16b}$$

at $\Omega = \infty$ or frequencies where the electrical length is an odd multiple of 90°. The response of the distributed, low-pass, maximally flat filter consisting of cascaded transmission line elements is similar to the lumped equivalent shown in Figure 4.4, except that the maximum attenuation is not infinite and the response repeats. We can adjust the frequency and maximum attenuation by selecting the appropriate value for the electrical length at the cutoff frequency. For example, assuming $n = 3$, $K_m = 1$, $\epsilon^2 = 1$, and an electrical length of 60° at a cutoff frequency of 2 GHz, the maximum attenuation frequency is 2.0(90°/60°), or 3.0 GHz. Substituting into (5.16b), the attenuation is 5.28 dB. If we shorten the electrical length to 30° at 2.0 GHz, the maximum attenuation frequency becomes 2.0(90°/30°), or 6.0 GHz, and the loss is equal to 18.13 dB. The loss at the original frequency of 3.0 GHz has increased as well and is 9.54 dB. Unfortunately, the range of TLE impedance values is expanded as we reduce the electrical length, and it is usually necessary to make a compromise between attenuation and realizability.

The term $(\mathrm{Sin}\theta/\mathrm{Sin}\theta_c)$ is 0 for $\Omega = 0$, monotonically increases to 1 at the cutoff frequency Ω_c, and then continues to increase to its maximum value of $(1/\mathrm{Sin}\theta_c)$ at $\Omega = \infty$. The discussion of the Chebyshev polynomial in Section 4.4.2 revealed that $T_n^2(x)$ ripples between 0 and 1 in the range $-1 \le x \le 1$ and monotonically increases outside this range. Therefore, we can write the distributed equal ripple or Chebyshev approximation as

$$G_t(\Omega^2) = \frac{K_m}{1 + \varepsilon^2 T_n^2\left(\dfrac{\mathrm{Sin}\,\theta}{\mathrm{Sin}\,\theta_c}\right)} \tag{5.17a}$$

which is identical to (4.16), with ω replaced by $(\mathrm{Sin}\theta/\mathrm{Sin}\theta_c)$. Maximum attenuation occurs at $\theta = 90°$, $\Omega = \infty$, where

$$G_t(\infty) = \frac{K_m}{1 + \varepsilon^2 T_n^2\left(\dfrac{1}{\mathrm{Sin}\,\theta_c}\right)} \tag{5.17b}$$

As an example of a network consisting of cascaded transmission line elements, we can design a low-pass filter with a passband extending to 1.0 GHz, voltage standing wave ratio (VSWR) less than 2.0:1, and 15-dB minimum attenuation at 3.0 GHz. Following the example in Section 4.4.2, which has similar specifications, we set $K_m = 1$ and $\epsilon^2 = 0.1$ or 0 dB MIL and 0.414 dB ripple. For maximum attenuation at 3.0 GHz, the electrical length at the cutoff frequency is 90°(1.0/3.0), or 30°, and we can apply (5.17b) to compute the required number of TLEs. The attenuation is 1.46 dB for $n = 1$, 7.71 dB for $n = 2$, and 18.36 dB for $n = 3$. Thus, the design requires three transmission line elements. We can substitute $T_3(x)$ from Section 4.4.2 and write (5.17a) as

$$G_t(\Omega^2) = \frac{1.0}{1 + (0.1)(4x^3 - 3x)^2} = \frac{1}{1.6x^6 - 2.4x^4 + 0.9x^2 + 1}$$

where x represents $(\mathrm{Sin}\theta/\mathrm{Sin}\theta_c)$. Using (5.15c), x^2 can be written as

$$x^2 = \frac{\mathrm{Sin}^2\theta}{\mathrm{Sin}^2\theta_c} = \left(\frac{\Omega^2}{1 + \Omega^2}\right) \Big/ \mathrm{Sin}^2(30°) = \frac{4\Omega^2}{1 + \Omega^2}$$

and the transducer gain as a function of Ω becomes

$$G_t(\Omega^2) = \frac{1}{(1.6)\left(\frac{4\Omega^2}{1 + \Omega^2}\right)^3 - (2.4)\left(\frac{4\Omega^2}{1 + \Omega^2}\right)^2 + (0.9)\left(\frac{\Omega^2}{1 + \Omega^2}\right) + 1}$$

or

$$G_t(\Omega^2) = \frac{\Omega^6 + 3\Omega^4 + 3\Omega^2 + 1}{68.6\Omega^6 - 28.2\Omega^4 + 6.6\Omega^2 + 1}$$

Transforming into the S-plane, $S = j\Omega$; then subtracting the transducer gain from 1, we have

$$|S_{11}|^2 = \frac{67.6S^6 + 31.2S^4 + 3.6S^2}{68.6S^6 + 28.2S^4 + 6.6S^2 - 1}$$

and the corresponding poles and zeros are

Zeros	*Poles*
$\pm(0.000000)$	$\pm(0.315050)$
$\pm(0.000000 \pm j0.480384)$	$\pm(0.253042 \pm j0.564976)$

All zeros are on the $j\Omega$ axis. We select one real and one complex conjugate pair to form the numerator and the left half-plane poles to form the denominator.

$$s_{11}(S) = \pm\frac{8.221922S^3 + 1.897367S}{8.282512S^3 + 6.801061S^2 + 4.494677S + 1}$$

Before continuing with the synthesis, we can confirm that this expression is correct. At $\Omega = 0$, $s_{11} = 0$, and at $\Omega = \infty$, $s_{11} = (8.221922/8.282512)$, or 0.992685. Subtracting s^2_{11} from 1, we have $s^2_{21} = 0.014577 = -18.363241$ dB, which is the same value as that computed from (5.17b). Using the positive sign, we compute the input impedance as

$$Z(S) = \frac{S^3 + 0.412075S^2 + 0.387293S + 0.060590}{0.003671S^3 + 0.412075S^2 + 0.157370S + 0.060590}$$

then test for available elements. Since the highest and lowest powers of the numerator and denominator are the same, there are no stubs. However, we can always use Richards' Theorem to extract a TLE. Evaluating $Z(S)$ at $S = 1$, we have $Z(1) = 2.935049$. At $S = -1$, $Z(-1) = -2.935037$, and the transmission line element is nonredundant. Extracting the 2.935049 TLE, then canceling the common $(S^2 - 1)$ factor from the numerator and denominator,

$$Z'(S) = \frac{0.031624S^2 + 0.614776S + 0.177835}{S^2 + 0.401300S + 0.177835}$$

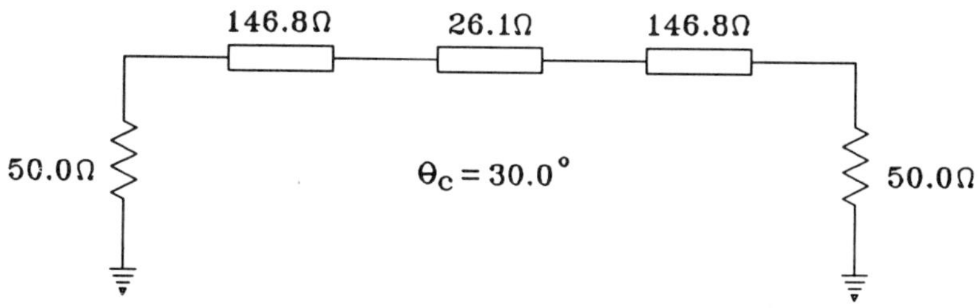

Figure 5.22 Chebyshev TLE low-pass filter.

Applying Richards' Theorem again, $Z(1) = 0.521953$, $Z(-1) = -0.521957$, and we find a second nonredundant transmission line element. Extracting this element, we find a third TLE with a value of 2.935049 and a termination of 1.0. Figure 5.22 shows the low-pass filter scaled to 50Ω, and Figure 5.23 shows the corresponding response.

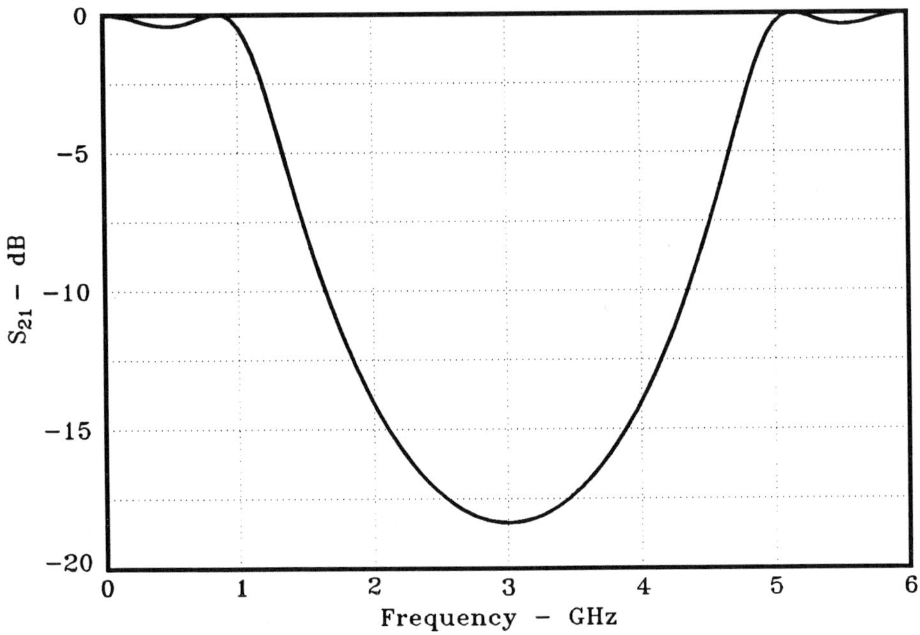

Figure 5.23 TLE low-pass filter response.

The response is that predicted: 0.414-dB ripple up to a cutoff frequency of 1.0 GHz and 18.36-dB maximum attenuation at 3.0 GHz. However, the element values are somewhat extreme for microstrip or stripline. Had we selected $-s_{11}$ to form $Z(S)$, the input impedance would be inverted and the normalized impedances of the transmission line elements would be inverted as well; that is, 17.0Ω, 95.8Ω, and 17.0Ω scaled to 50Ω. In addition to extreme element values, we also have discontinuities caused by the abrupt change in linewidth between the alternating high and low impedance transmission lines [2-5]. Thus, it will be necessary to change the number of elements, ripple, and electrical length in an effort to find a combination that is realizable and meets specifications, and then model the structure to correct for the effects of the discontinuities.

As this example illustrates, the transmission line element does exhibit filter characteristics and provides a very simple topology. No opens or shorts are required. Since the maximum attenuation is finite and dependent on K_m, ϵ, and n, it is often not possible to simultaneously achieve the desired response and realizable elements. However, if we add a low-pass stub which causes a transmission zero at the quarter-wave frequency, the maximum attenuation is infinite and independent of the other parameters.

5.3.2 Networks With TLEs and Low-Pass Stubs

When we allow both transmission line elements and low-pass stubs, (5.14) can be written as

$$|s_{21}(j\Omega)|^2 = \frac{K_m(1 - S^2)^n}{P_{n+q}(S^2)} \tag{5.18a}$$

Then, as demonstrated in Section 5.1.1, we can write

$$|s_{21}(j\Omega)|^2 = \frac{K_m}{1 + \varepsilon^2 F_k^2\left(\dfrac{\operatorname{Sin}\theta}{\operatorname{Sin}\theta_c}\right)} \tag{5.18b}$$

where $k = n + q$ (i.e., the total number of transmission line elements and low-pass stubs). Horton and Wenzel [20] show that when

$$F_k^2 = \left(\frac{\operatorname{Sin}\theta}{\operatorname{Sin}\theta_c}\right)^{2n}\left(\frac{\operatorname{Tan}\theta}{\operatorname{Tan}\theta_c}\right)^{2q} \tag{5.19}$$

the response exhibits maximally flat response about $\theta = 0°$, has q transmission zeros, and q zero-derivatives at $\theta = 90°$ or $\Omega = \infty$. Therefore, we can write the Butterworth or maximally flat approximation as

$$|s_{21}(j\Omega)|^2 = \frac{K_m}{1 + \varepsilon^2\left(\dfrac{\operatorname{Sin}\theta}{\operatorname{Sin}\theta_c}\right)^{2n}\left(\dfrac{\operatorname{Tan}\theta}{\operatorname{Tan}\theta_c}\right)^{2q}} \tag{5.20a}$$

Finally, substituting (5.15c), $S_c = j\Omega_c$ and $S = j\Omega$ into (5.20a), we have

$$|s_{21}(j\Omega)|^2 = \frac{K_m}{1 + \varepsilon^2\left(\frac{S^2}{S_c^2}\right)^{n+q}\left(\frac{1 - S_c^2}{1 - S^2}\right)^n} \tag{5.20b}$$

Horton and Wenzel also provide an appropriate function for the equal ripple approximation:

$$F_k = \mathrm{Cos}\left(n\phi_n + q\phi_q\right) \tag{5.21a}$$

where

$$\mathrm{Cos}\,\phi_n = \frac{\mathrm{Sin}\,\theta}{\mathrm{Sin}\,\theta_c} = \frac{S\sqrt{1 - S_c^2}}{S_c\sqrt{1 - S^2}} \tag{5.21b}$$

and

$$\mathrm{Cos}\,\phi_q = \frac{\mathrm{Tan}\,\theta}{\mathrm{Tan}\,\theta_c} = \frac{S}{S_c} \tag{5.21c}$$

Expanding (5.21a), we have

$$F_k = \mathrm{Cos}(n\phi_n)\mathrm{Cos}(q\phi_q) - \mathrm{Sin}(n\phi_n)\mathrm{Sin}(q\phi_q) \tag{5.21d}$$

Then, replacing ϕ with $\mathrm{Cos}^{-1}(x)$ and substituting (5.21b) and (5.21c), we can write (5.21d) as

$$F_k = T_n\left(\frac{S\sqrt{1 - S_c^2}}{S_c\sqrt{1 - S^2}}\right)T_q\left(\frac{S}{S_c}\right) - U_n\left(\frac{S\sqrt{1 - S_c^2}}{S_c\sqrt{1 - S^2}}\right)U_q\left(\frac{S}{S_c}\right) \tag{5.21e}$$

where $T_k(x)$ is the Chebyshev polynomial of the first kind (i.e., $\mathrm{Cos}(k \bullet \mathrm{Cos}^{-1}x)$) and $U_k(x)$ is the kth degree unnormalized Chebyshev polynomial of the second kind [21]. The unnormalized Chebyshev polynomial of the second kind is defined as

$$U_k(x) = \mathrm{Sin}\left(k\,\mathrm{Cos}^{-1}x\right) \tag{5.22a}$$

and the normalized Chebyshev polynomial of the second kind is related to the Chebyshev polynomial of the first kind by

$$U_{n-1}'(x) = \frac{1}{1 - x^2}\left[x T_n(x) - T_{n+1}(x)\right] \tag{5.22b}$$

The first six values are

$$U_0'(x) = 1$$

$$U_1'(x) = 2x$$

$$U_2'(x) = 4x^2 - 1$$

$$U_3'(x) = 8x^3 - 4x \tag{5.22c}$$

$$U_4'(x) = 16x^4 - 12x^2 + 1$$

$$U_5'(x) = 32x^5 - 32x^3 + 6x$$

$$U_6'(x) = 64x^6 - 80x^4 + 24x^2 - 1$$

and we can compute the unnormalized polynomials using

$$U_n(x) = \sqrt{1 - x^2} \cdot U_{n-1}'(x) \tag{5.22d}$$

Thus, the final form of the distributed equal ripple approximation for structures consisting of transmission line elements and low-pass stubs is

$$|s_{21}(S)|^2 = \frac{K_m}{1 + \varepsilon^2\left[T_n\left(\frac{S\sqrt{1 - S_c^2}}{S_c\sqrt{1 - S^2}}\right)T_q\left(\frac{S}{S_c}\right) - U_n\left(\frac{S\sqrt{1 - S_c^2}}{S_c\sqrt{1 - S^2}}\right)U_q\left(\frac{S}{S_c}\right)\right]^2} \tag{5.23}$$

We can modify the Chebyshev three-element low-pass filter of Section 5.1.1 by replacing one of the TLEs with a shunt open stub. The number of TLEs is two and the number of low-pass elements one. Therefore, $n = 2$, $q = 1$, $K_m = 1$, and $\epsilon^2 = 0.1$ for this design. The commensurate electrical length is 30° at 1.0 GHz for maximum attenuation at 3.0 GHz. Computing

$$S_c = j\mathrm{Tan}\,30° = j\frac{1}{\sqrt{3}}$$

we have

$$\frac{S}{S_c} = -j\sqrt{3}S \qquad (5.24a)$$

and

$$\frac{S\sqrt{1 - S_c^2}}{S_c\sqrt{1 - S^2}} = -j\frac{2S}{\sqrt{1 - S^2}} \qquad (5.24b)$$

In this case, we wish to compare the response characteristics of the two topologies. Usually the design specifications call for specified attenuation at certain frequencies. We can easily determine the response at any frequency by evaluating (5.23). For example, at 2.0 GHz, $\theta = 30°(2.0/1.0)$, or 60°, and

$$S = j\,\mathrm{Tan}\,60° = j\sqrt{3}$$

Substituting S into (5.24) and then into (5.23),

$$|s_{21}(S)|^2 = \frac{1}{1 + (0.1)\left[T_2(\sqrt{3})\,T_1(3) - U_2(\sqrt{3})\,U_1(3)\right]^2}$$

We can use (4.15) to evaluate $T_n(x)$ and (5.22) for $U_n(x)$. The transducer gain at 2.0 GHz is

$$|s_{21}(S)|^2 = \frac{1}{1 + (0.1)\left[(5)(3) - (j2\sqrt{6})(j2\sqrt{2})\right]^2} = 0.011867 = -19.256690\,dB$$

which is approximately 5 dB better than the previous design using three TLEs.

Using (4.15) and (5.24) to compute the Chebyshev polynomials of the first kind results in

$$T_2\left(\frac{\mathrm{Sin}\,\theta}{\mathrm{Sin}\,\theta_c}\right) = \frac{-7S^2 - 1}{1 - S^2}$$

and

$$T_1\left(\frac{\text{Tan}\,\theta}{\text{Tan}\,\theta_c}\right) = -j\sqrt{3}S$$

Then, from (5.22) and (5.24) the Chebyshev polynomials of the second kind are

$$U_2\left(\frac{\text{Sin}\,\theta}{\text{Sin}\,\theta_c}\right) = -j\frac{4S\sqrt{1 + 3S^2}}{1 - S^2}$$

and

$$U_1\left(\frac{\text{Tan}\,\theta}{\text{Tan}\,\theta_c}\right) = \sqrt{1 + 3S^2}$$

Substituting these expressions into (5.23), then expanding, we have

$$|s_{21}(S)|^2 = \frac{(1 - S^2)^2}{-58.198454S^6 - 26.656406S^4 - 5.285641S^2 + 1} \tag{5.25a}$$

Subtracting $|s_{21}(S)|^2$ from 1 provides the input reflection coefficient squared:

$$|s_{11}(S)|^2 = \frac{S^6 + 0.475209S^4 + 0.056456S^2}{S^6 + 0.458026S^4 + 0.090821S^2 - 0.017183} \tag{5.25b}$$

Next, we compute the poles and zeros, select poles from the left half-plane and one real and one complex conjugate pair of zeros to form $\pm s_{11}(S)$. The input reflection coefficient is 0 at $S = 0$ and ± 1 at $S = \infty$. The input element is either a series shorted stub for $+s_{11}$ or a shunt open stub for $-s_{11}$. Selecting $-s_{11}$, the input impedance is

$$Z(S) = \frac{0.399298S^2 + 0.155144S + 0.065541}{S^3 + 0.399298S^2 + 0.392747S + 0.065541} \tag{5.25c}$$

Referring to Figure 4.1, the input element is a shunt capacitor (shunt open stub) with a value of (1/0.399298), or 2.504393. After extracting the capacitor,

$$Z'(S) = \frac{0.399298S^2 + 0.155144S + 0.065541}{0.010756S^2 + 0.228606S + 0.065541} \tag{5.25d}$$

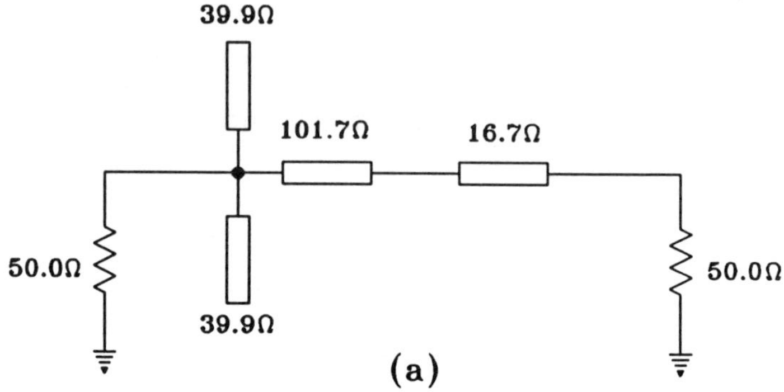

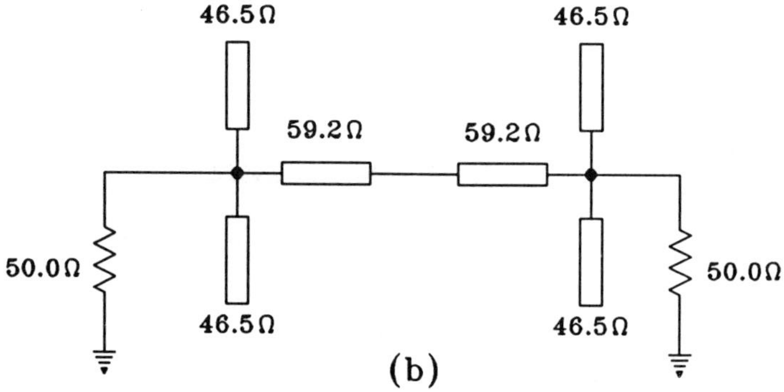

Figure 5.24 Low-pass filter: (a) original; (b) double TLE.

we see that there are no available elements with poles at zero or infinity. Evaluating $Z'(S)$ at $S = 1$, we have $Z(1) = 2.033374$, and at $S = -1$, $Z(-1) = -2.033341$, which confirms that a nonredundant TLE is available. Using (5.5b), factoring $(S^2 - 1)$ from numerator and denominator, and then canceling, the remainder becomes

$$Z''(S) = \frac{0.044474S + 0.133270}{0.399298S + 0.133270} \tag{5.25e}$$

Testing this expression, we find another nonredundant TLE with a value of 0.333748. Removing the second TLE, the remainder is 1 and the load termination is $R = 1$. Figure 5.24(a) shows the topology and element values scaled to 50Ω, and Figure 5.25 shows the response. Comparing this response with that of the TLE filter shown in Figure 5.23 demonstrates the superior stopband characteristics.

Both filters have the same number of nonredundant elements, but the addition of one transmission zero has improved the response considerably. The single 19.95Ω open shunt stub can be realized as two parallel 39.9Ω stubs. However, the last TLE is 16.7Ω which is too low to be constructed on standard microstrip or stripline material. Even if we extract the elements in a different order, the values fall outside the target range of 40Ω to 100Ω. In fact, neither circuit can be built and we must redesign the filter and use different materials or construction techniques.

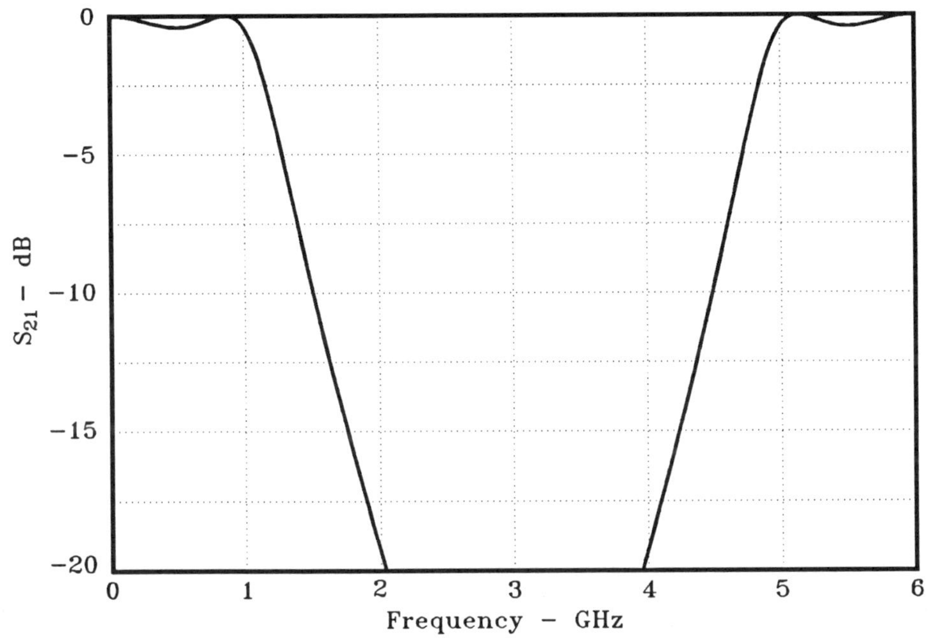

Figure 5.25 Response using two TLEs and one open stub.

However, before giving up on this design, we can try the double-length TLE discussed in Section 5.1.2. The input element is a shunt capacitor, and using the driving-point impedance given in (5.25c) we can compute $N(1) = 0.619984$, $D(1) = 1.857587$, $N'(1) = 0.953741$, and $D'(1) = 4.191344$, and then substitute into (5.9c) yielding

$$C = \frac{(0.619984)(4.191344) - (0.953741)(1.857587)}{(0.619984)^2} = 2.151280$$

The full value of the shunt capacitor is 2.504393F as previously calculated, and since the value required for a double-length TLE is less, we can use partial element extraction. Removing the 2.151280F shunt capacitor from (5.25c) leaves a remainder of

$$Z'(S) = \frac{0.399298S^2 + 0.155144S + 0.065541}{0.140997S^3 + 0.065540S^2 + 0.251750S + 0.065541} \tag{5.26a}$$

Evaluating $Z'(S)$ at $S = 1$ and $S = -1$, we determine that, in addition to the remainder of the capacitor, a 1.183563Ω nonredundant TLE is available. The derivative of the numerator divided by the derivative of the denominator evaluated at $S = 1$ is also 1.183563, which verifies that the TLE is double-length. After removing the TLE,

$$Z''(S) = \frac{0.197512S^2 + 0.091811S + 0.077572}{0.232419S + 0.077572} \tag{5.26b}$$

Computing the value at $S = 1$ and $S = -1$ confirms that another 1.183563 nonredundant TLE is available in addition to a series shorted stub. Extracting the second TLE and a 2.151280F shunt capacitor reduces the impedance to 1.0, and we have the topology shown in Figure 5.24(b).

The networks shown in Figure 5.25 have identical electrical characteristics at all frequencies; that is, they were synthesized from the same driving-point impedance. However, the circuit using the double-length TLE has a narrow range of element values around 50Ω and is easily realized on microstrip or stripline. The discontinuity due to the change in linewidth between the two TLEs has been eliminated. Loss and discontinuities at the junction of the stubs and the TLEs [2-5] cause some differences between the actual and ideal response. However, these effects can be modeled and corrected using optimization. We also have fringing off of the ends of the open stubs [2-5], which can be modeled as a small capacitor and a resistor to account for radiated energy. The physical length of the stubs can be reduced to correct for the excess end capacitance and does not usually cause a significant deviation from the ideal response.

We can use the partial element extraction technique to adjust the element values of most distributed-element networks. In fact, this approach is often necessary to achieve realizable element values or specified load terminations. We can achieve the same results by applying one or more Kuroda transforms to the original network. The element values of Figure 5.24(b) can be computed from the network in Figure 5.24(a) by splitting the shunt open stub into two elements, and then transforming one across both TLEs as shown in the bottom row of Figure 5.9. In this case, the computed value for a double-length TLE resulted in equal stubs at each end of the network. If we extract more or less of the input stub, the impedance of one stub will increase while the other will decrease. Similarly, the impedance of the two TLEs is high-low for an input stub, equal when the stubs are equal, and low-high when the stub is extracted at the output.

5.3.3 Networks With TLEs and High-Pass Stubs

The approximations considered thus far have resulted in a low-pass response. We can transform the low-pass approximation into a high-pass one simply by inverting the frequency variable as was done in Section 4.7 for lumped element networks. Writing the general form of the low-pass approximation given in (5.18a) in terms of the frequency variable Ω,

$$|s_{21}(\Omega)|^2 = \frac{K_m(1 + \Omega^2)^n}{P_{n+q}(\Omega^2)} \tag{5.27a}$$

where n is the number of TLEs and q is the number of low-pass elements. To obtain the high-pass approximation, we invert Ω, resulting in

$$|s_{21}(1/\Omega)|^2 = \frac{K_m(1 + \Omega^2)^n}{\Omega^{2n} P_{n+q}(1/\Omega^2)} \tag{5.27b}$$

Then, expanding the denominator,

$$|s_{21}(1/\Omega)|^2 = \frac{K_m(1 + \Omega^2)^n}{\Omega^{2n}\left[p_{n+q}(1/\Omega^2)^{n+q} + p_{n+q-1}(1/\Omega^2)^{n+q-1} + \cdots + p_1(1/\Omega^2) + p_0\right]} \tag{5.27c}$$

and multiplying the denominator by the Ω^{2n} factor,

$$|s_{21}(1/\Omega)|^2 = \frac{K_m(1 + \Omega^2)^n}{p_{n+q}(1/\Omega^2)^q + p_{n+q-1}(1/\Omega^2)^{q-1} + \cdots + p_1(\Omega^2)^{n-1} + p_0(\Omega^2)^n} \tag{5.27d}$$

Finally, multiplying the numerator and denominator by Ω^{2q}, we can write

$$|s_{21}(1/\Omega)|^2 = \frac{K_m \Omega^{2q}(1 + \Omega^2)^n}{p_{n+q} + p_{n+q-1}\Omega^2 + \cdots + p_1 \Omega^{2(n+q-1)} + p_0 \Omega^{2(n+q)}} \tag{5.27e}$$

Comparing (5.27e) with (5.27a), we see that the transformation results in an additional term of Ω^{2q} in the numerator. This factor provides q transmission zeros at $S = 0$. We conclude that the low-pass elements have been transformed into high-pass elements (i.e., the network consists of high-pass elements and transmission line elements). It is also apparent from (5.27e) that the denominator polynomial is the same as the original low-pass approximation, except that the coefficients have been reversed. However, the transformation from low-pass to high-pass inverts the tangent function, and the high-pass electrical length is transformed as well. For example, at the cutoff frequency, the low-pass filter of Section 5.3.2 had an electrical length of 30° where Tan30° $= 1/\sqrt{3}$. If we create a high-pass approximation by reversing the denominator coefficients, then multiply the numerator by Ω^{2q}, $\text{Tan}\theta_c = \sqrt{3}$, or $\theta_c = 60°$. Using the notation of (5.14), the high-pass approximation becomes

$$|s_{21}(j\Omega)|^2 = \frac{K_m S^{2m}(1 - S^2)^n}{P_{n+m}(S^2)} \tag{5.28}$$

where n is the number of TLEs and m is the number of high-pass elements.

The low-pass approximations contain two frequency-dependent terms, $(\text{Sin}\theta/\text{Sin}\theta_c)$ and $(\text{Tan}\theta/\text{Tan}\theta_c)$. We can write these expressions as a function of Richards' variable S and then invert S to obtain the equivalent high-pass terms. From (5.21b),

$$\frac{\text{Sin}\,\theta}{\text{Sin}\,\theta_c} = \frac{S\sqrt{1 - S_c^2}}{S_c\sqrt{1 - S^2}} \Rightarrow \frac{(1/S)\sqrt{1 - (1/S_c)^2}}{(1/S_c)\sqrt{1 - (1/S)^2}} = \frac{\sqrt{1 - S_c^2}}{\sqrt{1 - S^2}} = \frac{\text{Cos}\,\theta}{\text{Cos}\,\theta_c} \tag{5.29a}$$

and from (5.21c),

$$\frac{\text{Tan}\,\theta}{\text{Tan}\,\theta_c} = \frac{S}{S_c} \Rightarrow \frac{(1/S)}{(1/S_c)} = \frac{S_c}{S} = \frac{\text{Tan}\,\theta_c}{\text{Tan}\,\theta} \tag{5.29b}$$

Substituting (5.29) into (5.20), we can write the maximally flat high-pass approximation as

$$|s_{21}(j\Omega)|^2 = \frac{K_m}{1 + \varepsilon^2\left(\dfrac{S_c^2}{S^2}\right)^m\left(\dfrac{1 - S_c^2}{1 - S^2}\right)^n} \tag{5.30}$$

Then, substituting (5.29) into (5.23), the equal ripple high-pass approximation becomes

$$|s_{21}(S)|^2 = \frac{K_m}{1 + \varepsilon^2\left[T_n\left(\frac{\sqrt{1 - S_c^2}}{\sqrt{1 - S^2}}\right)T_m\left(\frac{S_c}{S}\right) - U_n\left(\frac{\sqrt{1 - S_c^2}}{\sqrt{1 - S^2}}\right)U_m\left(\frac{S_c}{S}\right)\right]^2} \tag{5.31}$$

We can design a three-element high-pass filter using the same specifications as those of the equal ripple low-pass filter in Section 5.3.2; that is, $K_m = 1.0$, $\epsilon^2 = 0.1$, $n = 2$, $m = 1$, $\theta_c = 30°$, and 50Ω terminations. $S_c = j\mathrm{Tan}30°$, or $j/\sqrt{3}$, and we can apply (5.29) and (4.15) to compute the Chebyshev polynomials of the first kind:

$$T_n\left(\frac{\mathrm{Cos}\,\theta}{\mathrm{Cos}\,\theta_c}\right) = T_2\left(\frac{\sqrt{1 - S_C^2}}{\sqrt{1 - S^2}}\right) = \frac{5 + 3S^2}{3 - 3S^2}$$

and

$$T_m\left(\frac{\mathrm{Tan}\,\theta_c}{\mathrm{Tan}\,\theta}\right) = T_1\left(\frac{S_c}{S}\right) = \frac{j}{\sqrt{3}S}$$

Then, using (5.22) and (5.29), we compute the Chebyshev polynomials of the second kind, which are

$$U_n\left(\frac{\mathrm{Cos}\,\theta}{\mathrm{Cos}\,\theta_c}\right) = U_2\left(\frac{\sqrt{1 - S_c^2}}{\sqrt{1 - S^2}}\right) = \frac{-j4\sqrt{1 + 3S^2}}{3 - 3S^2}$$

and

$$U_m\left(\frac{\mathrm{Tan}\,\theta_c}{\mathrm{Tan}\,\theta}\right) = U_1\left(\frac{S_c}{S}\right) = \frac{\sqrt{1 + 3S^2}}{\sqrt{3}S}$$

Substituting the Chebyshev polynomials into (5.31), we have

$$|s_{21}(S)|^2 = \frac{S^2(1 - S^2)^2}{S^6 - 2.833333S^4 - 0.300000} \tag{5.32a}$$

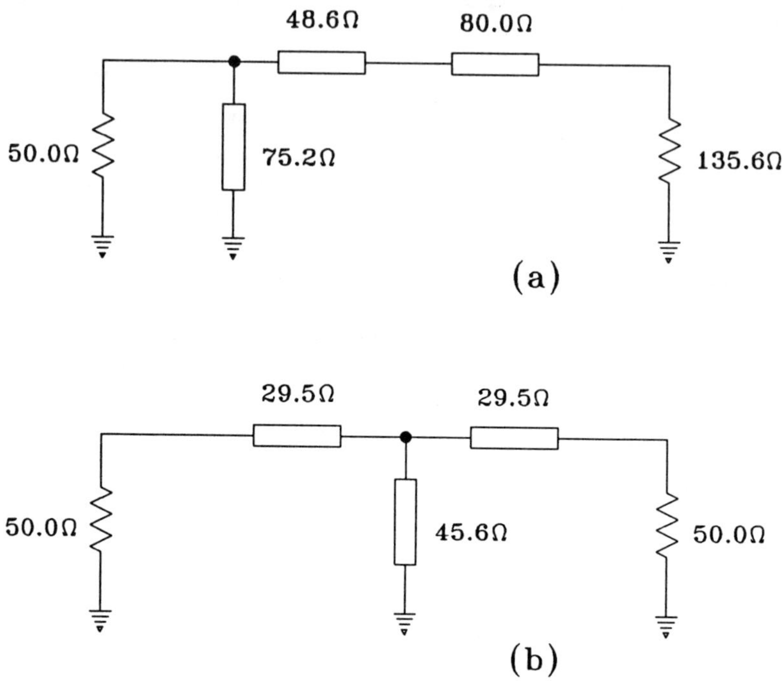

Figure 5.26 High-pass filter: (a) original; (b) transformed.

and after subtracting $|s_{21}(S)|^2$ from one,

$$|s_{11}(S)|^2 = \frac{0.833333\,S^4 + S^2 + 0.300000}{-S^6 + 2.833333\,S^4 + 0.300000} \tag{5.32b}$$

The four zeros fall on the jΩ axis, and we select one complex conjugate pair and the left half-plane poles to form $\pm s_{11}(S)$. At $S = 0$, s_{11} is $+1$, corresponding to an open circuit (series open stub), or -1, which corresponds to a short circuit (shunt shorted stub). We select the shunt shorted stub and compute the input impedance as

$$Z(S) = \frac{S^3 + 1.562332\,S^2 + 1.646648\,S}{S^3 + 3.388074\,S^2 + 1.646648\,S + 1.095445} \tag{5.32c}$$

Evaluating (5.32c) at $S = 1$ and $S = -1$, we find that there is a nonredundant TLE available with a value of 0.590306, or approximately 29Ω. The driving-point

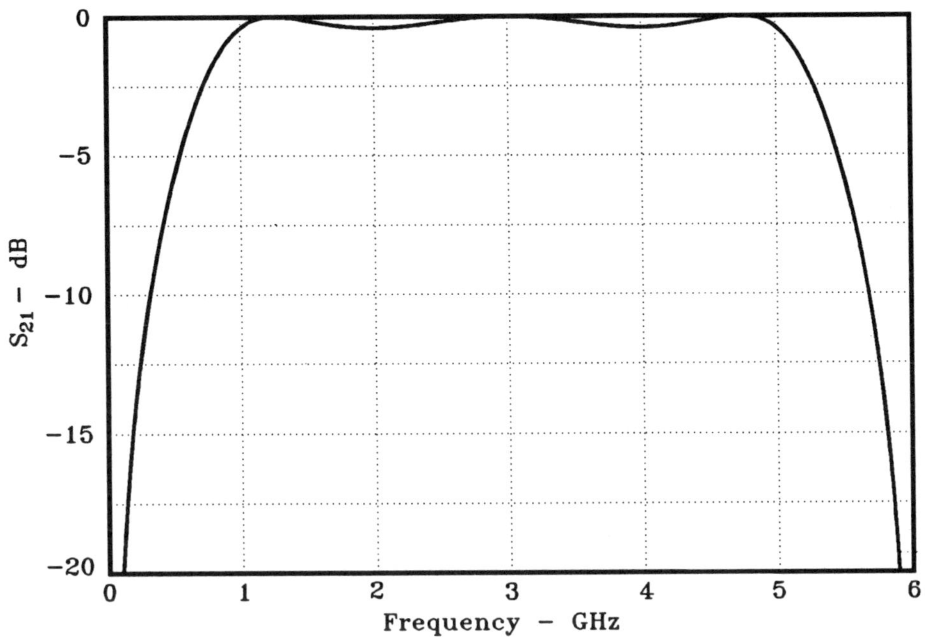

Figure 5.27 High-pass filter response.

admittance has a pole at $S = 0$, which indicates that a shunt shorted stub is also available, and the value is (1.646648/1.095445), approximately 75Ω. The characteristic impedance of the stub is within the range used for other examples, and we extract this element, leaving a remainder of

$$Z'(S) = \frac{1.503177S^2 + 2.348461S + 2.475203}{1.503177S^2 + 4.092874S + 0.912871} \tag{5.32d}$$

Testing this function, we find that the only choice is a nonredundant TLE equal to 0.972026. After extracting the TLE,

$$Z''(S) = \frac{1.420253S + 2.405961}{1.503177S + 0.887334} \tag{5.32e}$$

and once again, the only choice is a nonredundant TLE, which has a value of 1.600584. Removing the second TLE results in a remainder of 2.711449, or an output termination of 135.572440Ω. The complete high-pass filter topology is shown in Figure 5.26(a) and the response in Figure 5.27.

Although the high-pass filter shown in Figure 5.26(a) contains realizable element values, the output termination is not 50Ω. Applying the Kuroda transform shown in Figure 5.10 results in an output termination of 50Ω, but the two TLEs are nonrealizable, as illustrated in Figure 5.26(b). However, we have a relatively narrow range of impedance values, 50Ω to 29.5Ω, and could easily build this filter using a thinner substrate, perhaps 15 mils (0.381 mm). We have several other options as well. We can use the Kuroda transform shown in Figure 5.11 to transform a portion of the input stub to the output, as shown in Figure 5.28. This configuration corresponds to the double-length TLE discussed in Section 5.1.2 and does not improve the realizability, although it might be entirely acceptable using a lower dielectric constant material. We can also change the specifications, the number or elements, ripple, or electrical length. However, when we specify more than one high-pass element, the network must consist of alternating shunt shorted and series open stubs either adjacent to one another or separated by one or more TLEs. This fact is easily demonstrated using the Kuroda identities. If the network contains two shunt shorted stubs separated by any number of TLEs, we can transform the circuit such that both stubs are adjacent and then combine them into a single element.

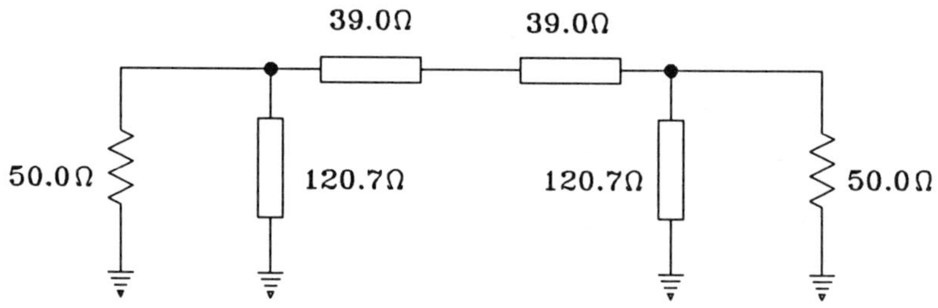

Figure 5.28 Transformed high-pass filter.

We have other alternatives which can be considered when this, or any other, distributed-element network has element values outside the acceptable range. There are coupled-line equivalences for combinations of TLEs and open or shorted stubs [13-16,22,23] which often provide acceptable dimensions as well as a solution to the problem of realizing series stubs in microstrip or stripline. We can also change the substrate thickness and/or dielectric constant when the element values fall within a narrow range, as in Figure 5.26(b). Although the network may be nonrealizable using one fabrication technique, another might produce a simple and cost-effective product. For example, Wenzel [13,14,23] and Horton and Wenzel

[20] show several mechanical configurations that provide compact filter structures using series stubs. Another solution, although only approximate, is to use constrained optimization. The synthesized network is commensurate; that is, all elements have the same electrical length. When we constrain the impedance to realizable values and then allow the electrical length of each element to vary independently, the number of variables essentially doubles and optimization may converge to an acceptable solution.

5.3.4 Bandpass and Bandstop Approximations

We could derive the bandpass and bandstop approximations by defining a low-pass transformation as was done in Section 4.6.1 for the lumped-element bandpass, and in Section 4.8 for the lumped-element bandstop approximation. However, the characteristics of Richards' variable make this step unnecessary. Examining Figure 5.25, which shows the distributed low-pass filter response, we see that the plot also has the appearance of a bandstop filter centered about 3.0 GHz. As θ increases from 0, $|s_{21}(\text{Tan}\theta)|^2$ ripples between K_m and $K_m/(1 + \epsilon^2)$ until it reaches the cutoff frequency where $\theta = \theta_c$. Above cutoff, the response monotonically increases to infinity at $\theta = 90°$, where the impedance of the low-pass series shorted stubs becomes infinite and the impedance of the shunt open stubs becomes 0. Between $\theta = 90°$ and 180°, the tangent function changes quadrants and we can use the trigonometric identity $\text{Tan}(90 - \phi) = -\text{Tan}(90 + \phi)$ to show that the response, which is a function of Ω^2, is symmetrical about 90°. Thus, the center frequency of the distributed bandstop filter is

$$f_0 = \frac{f_1 + f_2}{2} \tag{5.33a}$$

and the fractional bandwidth

$$w = \frac{f_2 - f_1}{f_0} \tag{5.33b}$$

where f_1 is the lower cutoff frequency and f_2 the upper. Comparing the center frequency in (5.33a) with that of the lumped-element networks from Chapter 4, we see that the center frequency of a distributed network is the arithmetic mean of the two cutoff frequencies, whereas f_o is the geometric mean for lumped-element networks. We can write the electrical length θ_c in terms of f_1 and f_2 as

$$\theta_c = \frac{90 f_1}{f_0} = 180 - \frac{90 f_2}{f_0} = 90\left\{1 - \frac{w}{2}\right\} \tag{5.33c}$$

Then we use the maximally flat low-pass approximation in (5.20) or the equal ripple approximation given by (5.23) to design a bandstop filter that has commensurate one-quarter wavelength elements at the center frequency. For example, Figure 5.25 shows a bandstop filter that has a stopband extending from 1.0 to 5.0 GHz. Substituting into (5.33a),

$$f_0 = \frac{1.0 + 5.0}{2} = 3.0$$

then into (5.33c),

$$\theta_c = \frac{(90)(1.0)}{3.0} = 30°$$

which is the same value used to design the original low-pass filter in Section 5.1.2.

Similarly, when we examine the high-pass filter response shown in Figure 5.27, we see that the high-pass approximations can be used to design one-quarter wavelength bandpass filters. Given f_1 and f_2, we can compute the center frequency from (5.33a) and θ_c from (5.33c) and then use either (5.30) or (5.31) to form the appropriate gain function. Although this method does provide approximations for the bandpass and bandstop functions, we are restricted to one particular electrical length and two types of elements. In the case of a bandpass network, only TLEs and high-pass elements are possible, while the bandstop network can contain only TLEs and low-pass elements. At lower frequencies, elements that are one-quarter wavelength at the center frequency can become very large. Also, it turns out that the design of certain types of impedance-matching networks discussed in Chapter 6 requires a bandpass structure with a specified electrical length as well as combinations of TLEs, low-pass, and high-pass elements.

We can derive the generalized form of the distributed bandpass approximation by combining several approximations into one. Mokari-Bolhassan and Ku [24] developed synthesis procedures for bandpass structures containing TLEs, low-pass, and high-pass elements. This approximation takes the form

$$|s_{21}(\Omega)|^2 = \frac{K_m}{1 + \varepsilon^2 \mathrm{Cos}\left(n\phi_n + q\phi_q + m\phi_m\right)} \tag{5.34}$$

where n is the number of TLEs, q is the number of low-pass elements, and m is the number of high-pass elements. The expressions provided for ϕ_n, ϕ_q, and ϕ_m represent the special case of an electrical length equal to one-eighth wavelength at the center frequency and must be modified to provide arbitrary electrical lengths. Levy [9] used Matthaei's [25] lumped-element quasi-low-pass mapping, discussed in Section 4.9.2, to derive a distributed equivalent with n TLEs, q low-pass elements, and arbitrary electrical length. Levy's expressions for ϕ_n is

$$\alpha_n = \mathrm{Cos}\,\phi_n = \frac{\Omega^2\left(\Omega_2^2 + \Omega_1^2 + 2\right) - \left(\Omega_2^2 + \Omega_1^2 + 2\Omega_1^2\Omega_2^2\right)}{\left(\Omega_2^2 - \Omega_1^2\right)\left(1 + \Omega^2\right)} \tag{5.35a}$$

and

$$\alpha_q = \mathrm{Cos}\,\phi_q = \frac{2\Omega^2 - \left(\Omega_2^2 + \Omega_1^2\right)}{\Omega_2^2 - \Omega_1^2} \tag{5.35b}$$

for ϕ_q. Ω_1 is Tanθ_1 and Ω_2 is Tanθ_2, the tangent of the electrical length at the lower and upper bandedges, respectively. Examining (5.35), we find that the value is -1 at $\Omega = \Omega_1$ and monotonically increases to +1 at $\Omega = \Omega_2$. Below Ω_1, the value is less than -1 and greater than +1 for Ω above Ω_2. If we assume a form for the high-pass elements that is the reciprocal of the low-pass elements; that is,

$$\mathrm{Cos}\,\phi_m = \frac{A}{\Omega^2} + B$$

then solve for the constants A and B such that the value is -1 at Ω_1 and +1 at Ω_2, ϕ_m can be written as

$$\alpha_m = \mathrm{Cos}\,\phi_m = \frac{\Omega^2\left(\Omega_2^2 + \Omega_1^2\right) - 2\Omega_1^2\Omega_2^2}{\Omega^2\left(\Omega_2^2 - \Omega_1^2\right)} \tag{5.35c}$$

Substituting (5.35) into the cosine function of (5.34), we have

$$F_k(\Omega^2) = \mathrm{Cos}\left(n\,\mathrm{Cos}^{-1}\alpha_n + q\,\mathrm{Cos}^{-1}\alpha_q + m\,\mathrm{Cos}^{-1}\alpha_m\right) \tag{5.36a}$$

Then, using the trigonometric identities for Cos(A + B) and Sin(A + B),

$$\begin{aligned} F_k(\Omega^2) = \;& \cos\left(n\cos^{-1}\alpha_n\right)\cdot\cos\left(q\cos^{-1}\alpha_q\right)\cdot\cos\left(m\cos^{-1}\alpha_m\right) \\ & -\cos\left(n\cos^{-1}\alpha_n\right)\cdot\sin\left(q\cos^{-1}\alpha_q\right)\cdot\sin\left(m\cos^{-1}\alpha_m\right) \\ & -\sin\left(n\cos^{-1}\alpha_n\right)\cdot\sin\left(q\cos^{-1}\alpha_q\right)\cdot\cos\left(m\cos^{-1}\alpha_m\right) \\ & -\sin\left(n\cos^{-1}\alpha_n\right)\cdot\cos\left(q\cos^{-1}\alpha_q\right)\cdot\sin\left(m\cos^{-1}\alpha_m\right) \end{aligned} \tag{5.36b}$$

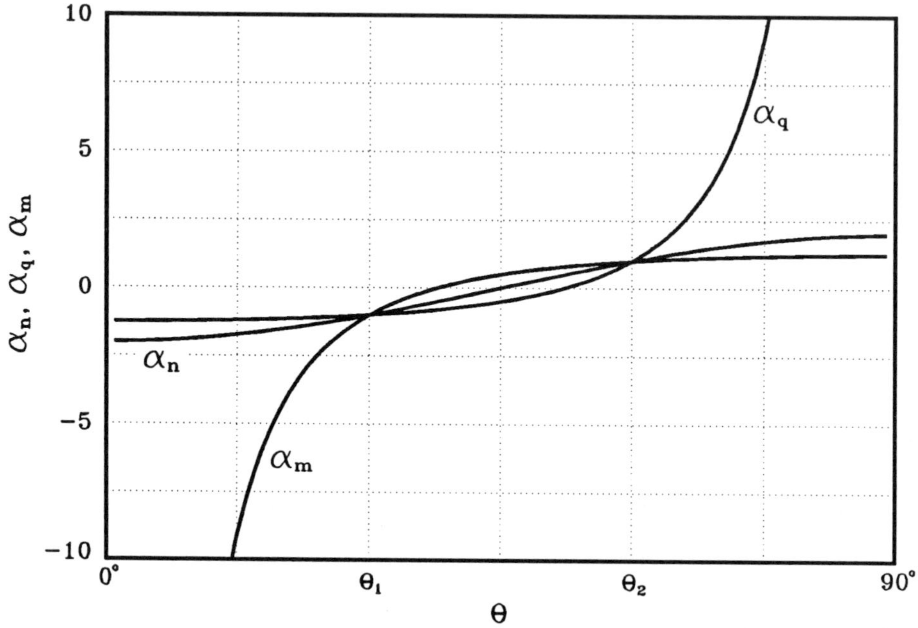

Figure 5.29 α_n, α_q, and α_m versus angle.

We recognize Cos(n • Cos^{-1}x) as the unnormalized Chebyshev polynomial of the first kind, and Sin(n • Cos^{-1}x) as the unnormalized Chebyshev polynomial of the second kind. Writing (5.36b) in terms of the Chebyshev polynomials yields

$$F_k(\Omega^2) = T_n(\alpha_n)\cdot T_q(\alpha_q)\cdot T_m(\alpha_m) - T_n(\alpha_n)\cdot U_q(\alpha_q)\cdot U_m(\alpha_m) \\ -U_n(\alpha_n)\cdot U_q(\alpha_q)\cdot T_m(\alpha_m) - U_n(\alpha_n)\cdot T_q(\alpha_q)\cdot U_m(\alpha_m) \tag{5.36c}$$

Mokari-Bolhassan and Ku [24] show that an approximation in the form of (5.34) requires $n + q + m$ to be even. Examination of (5.36c) provides some insight into the reason for this restriction. Figure 5.29 shows a plot of α_n, α_q, and α_m as a function of angles between 0° and 90°; that is, Ω between 0 and infinity. As shown, all three functions are less than -1 below θ_1, equal to -1 at θ_1, +1 at θ_2, and greater than +1 above θ_2. Figure 4.7 shows the unnormalized Chebyshev polynomials of the first kind, which ripple between ± 1 for $-1 \leq x \leq +1$, are equal to +1 at $x = 1$ and increase for $x > 1$. At $x = -1$, even polynomials attain a value of +1 and increase in value for $x < -1$. However, odd polynomials equal -1 at $x = -1$ and decrease in value for $x < -1$. The unnormalized Chebyshev polynomials

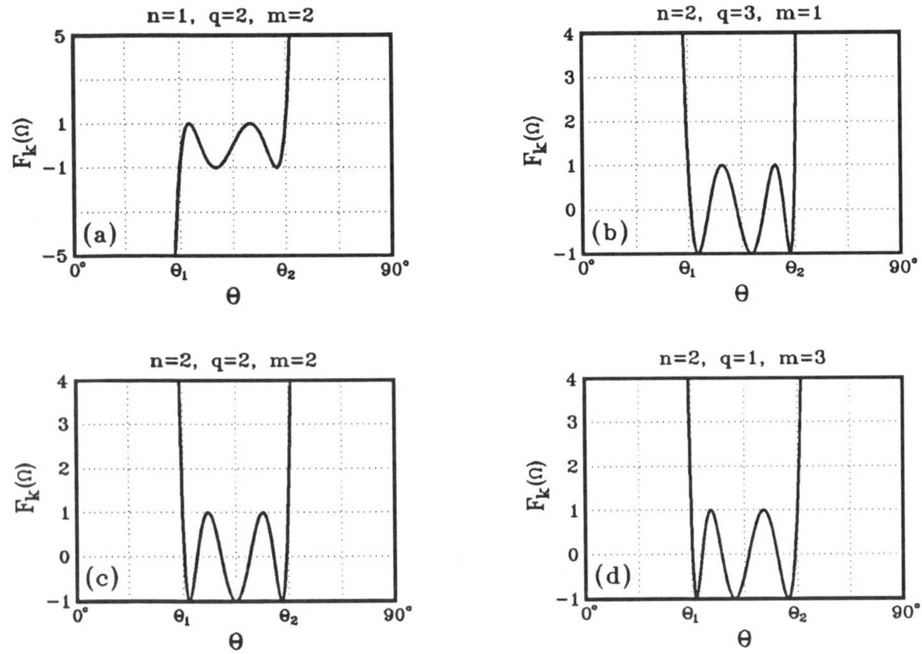

Figure 5.30 $F_k(\Omega)$ versus phase angle.

of the second kind defined in (5.22) ripple between ± 1 for $-1 \leq x \leq 1$, and equal 0 at $x = \pm 1$. Below $x = -1$, $U_n(x)$ is imaginary and decreases for even polynomials or increases for odd polynomials. Above $x = +1$, $U_n(x)$ is also imaginary, but increases in value for all orders. We can apply these characteristics to (5.36c) and deduce that for $n + q + m$ odd, $F_k(\Omega)$ is negative for θ less than θ_1. Hence, $|s_{21}(\Omega)|^2$ is also negative over a portion of the frequency range and cannot be synthesized using positive element values.

Figure 5.30 shows the response of the $F_k(\Omega)$ function for $\theta_1 = 30°$, $\theta_2 = 60°$, and four combinations of n, q, and m. Figure 5.30(a) illustrates the odd-order case where $F_k(\Omega)$ is negative for θ less than θ_1. Figure 5.30(c) reveals that the ripple is symmetric about the center frequency when the number of each type of element is equal, and Figures 5.30(b) and 5.30(d) show that the ripples tend to shift toward one end of the passband when the number of high-pass and low-pass elements are unequal. Figure 5.31 illustrates the effects of changing bandwidth and electrical length. The response centered about 15° and 45° has the same bandwidth (i.e., 2:1), and we see that increasing the electrical length has the effect of expanding the response about the center frequency. Similarly, when we compare the 1.14:1 bandwidth centered at 75° with either of the 2:1 bandwidth curves, we find that

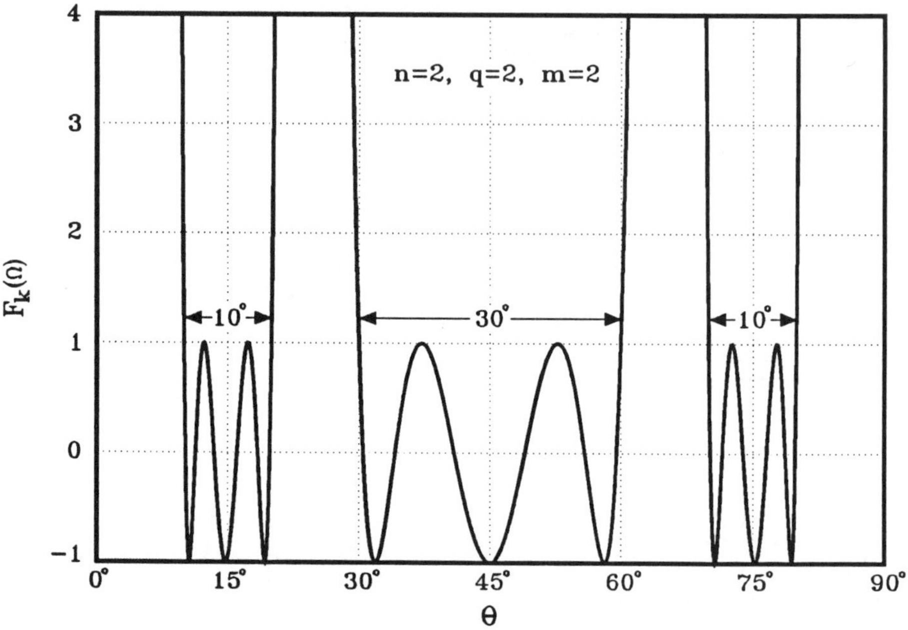

Figure 5.31 $F_k(\Omega)$ for various bandwidth and electrical length.

reducing the bandwidth compresses the passband response. In both cases, the response is skewed slightly to one side of the center frequency. However, the basic shape is determined primarily by the combination of elements rather than bandwidth or electrical length.

Since $F_k(\Omega)$ ripples between +1 and -1 in the passband, we can write the generalized bandpass approximation as

$$|s_{21}(\Omega)|^2 = \frac{K_m}{1 + \dfrac{\varepsilon^2}{2}\cdot\left[1 + F_k(\Omega^2)\right]} \tag{5.37}$$

which is valid for $n + q + m$ even, and $\theta_2 < 90°$. As an example, we will design a stripline bandpass filter using a dielectric material with $\epsilon_r = 10$. The passband should extend from 500 to 750 MHz, exhibit 0.05-dB ripple, and have a response provided by two TLEs, one high-pass element, and one low-pass element. Using

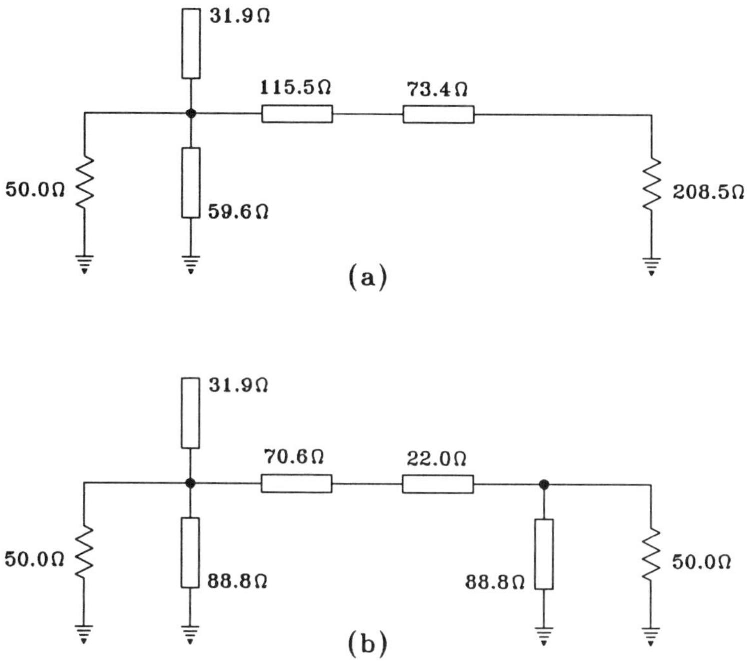

Figure 5.32 Bandpass filter: (a) original; (b) transformed.

$$Z'(S) = \frac{1.131953S^3 + 0.651802S^2 + 0.719348S}{1.775464S^4 + 1.131953S^3 + 2.241650S^2 + 0.352231S + 0.198388} \tag{5.39d}$$

We now have the choice of a shunt shorted stub equal to 3.625956, or 181.297810Ω, an open shunt stub with a value of (1/1.568497), or 31.877659Ω, and a TLE equal to 0.439165, or 21.958242Ω. We can extract any value for the shunt open stub greater than or equal to the full value. One convenient value is that discussed in Section 5.1.2 which results in a double-length TLE. Using the numerator and denominator of (5.39d) in (5.9c), we compute the required value as (1/1.196256), 41.797077Ω. After removing a (1/1.196256) shunt open stub, we find a double-length TLE equal to 0.925246, 46.26231Ω, and then shunt shorted and open stubs with the same values as the input elements. Figure 5.33 shows the advantages of dividing the open and shorted shunt stubs between input and output. The terminations are equal, all elements are realizable, and the stubs are separated by the maximum distance, which becomes an important factor as we reduce the electrical length. When the stubs are in close proximity, energy can be coupled

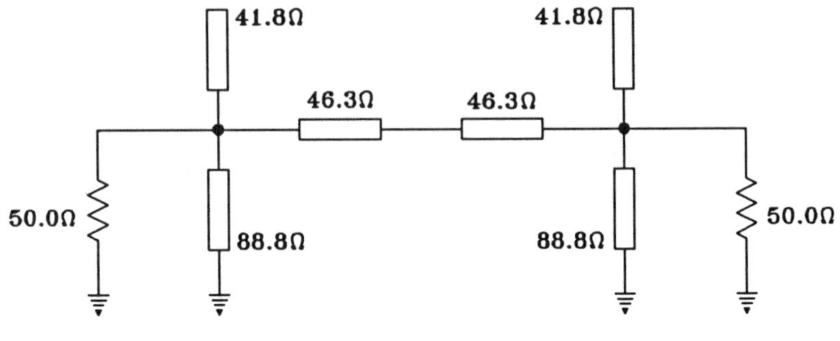

Figure 5.33 Bandpass filter using double-length TLE.

from one element to another, causing significant deviations from the ideal response. Figure 5.34 shows the bandpass filter response.

5.4 IMPEDANCE-TRANSFORMING NETWORKS

Compared with the synthesis of lumped-element networks, distributed-element synthesis is far more difficult. In addition to the common problem of an output termination that is often something other than desired, we must be concerned with element values restricted to a relatively narrow range. Unfortunately, these requirements can be mutually exclusive. For example, the high-pass filter designed in Section 5.3.3 has convenient element values for the original topology, shown in Figure 5.26(a). However, the output termination is 135.6Ω. When we applied Kuroda identities, which reduced the output termination to 50Ω, the element values were no longer realizable in either transformed network, as shown in Figure 5.26(b) or Figure 5.28. In most cases we can achieve acceptable results by using several Kuroda identities, as illustrated in Section 5.3.4, changing one or more synthesis parameters, or both. Occasionally these efforts fail and we must consider a different fabrication method to realize the required elements or break the synthesis problem into two, more manageable, parts: realizability and termination adjustment.

The latter approach provides flexibility by allowing us to concentrate on element values rather than the output termination while synthesizing the network. Once the element values are acceptable, one or more Kuroda identities can be used to adjust the output as much as possible while maintaining realizable elements. We can then synthesize a second network, which is cascaded with the first to transform the termination to the required value. Using the high-pass network in Figure 5.26(a) as an example, we need a network to transform from 135.6Ω into 50Ω. One possibility is to cascade the mirror image of the original

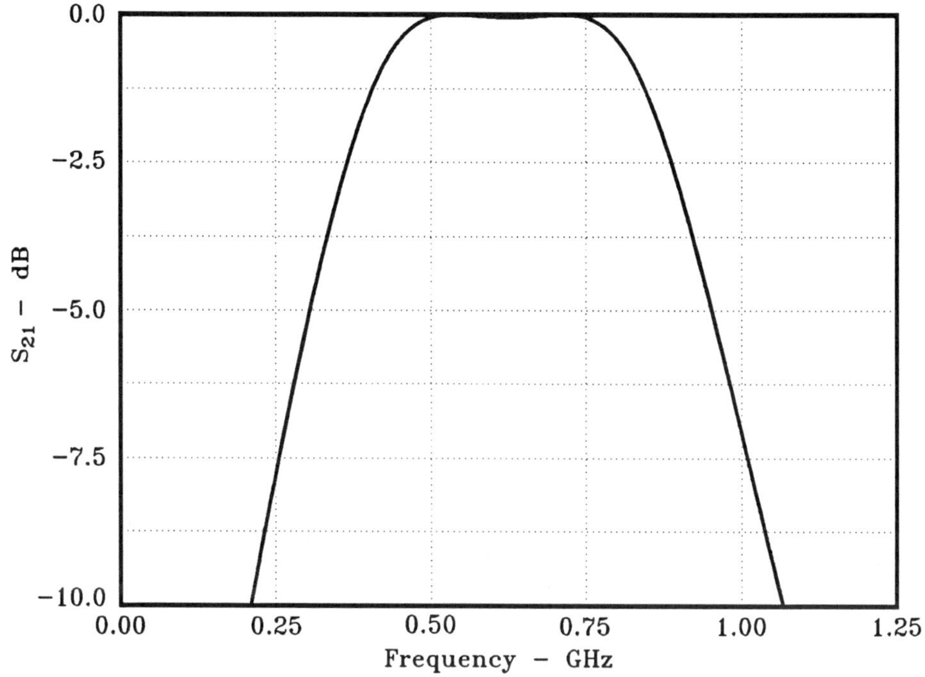

Figure 5.34 Bandpass filter response.

network as shown in Figure 5.35. The original filter shown in Figure 5.26(a) has a 50Ω input and 135.6Ω output. When we cascade the mirror image network, the 135.6Ω output of the first is transformed back to 50Ω by the second network. The response of this combination is shown by the darker trace in Figure 5.36. The total ripple is approximately 0.6 dB, not twice the ripple of each network as might be expected. The design value of 0.414-dB ripple is valid only for the specified terminations, 50Ω and 135.6Ω in this case. Since the filter is lossless, the ripple is caused by a mismatch between the terminations and input or output impedance. Thus, each network is terminated by an impedance that is a function of frequency and is the correct value only at frequencies where the response attains 0 dB. The response of the high-pass filter shown in Figure 5.27 is 0 at approximately 1.1, 3.0, and 4.9 GHz. Examining Figure 5.36, we see that the combined network has similar characteristics.

The response shown by the thinner trace in Figure 5.36 results when we add a 135.6Ω transmission line between the two networks. If the impedance levels at the interface were exactly 135.6Ω, the transmission line would have no effect other

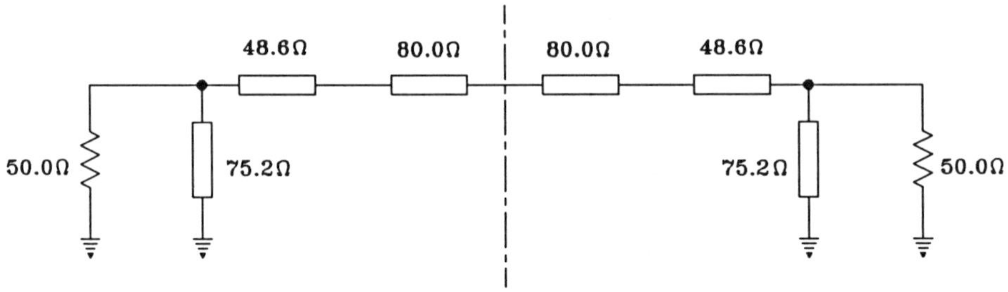

Figure 5.35 High-pass filter cascaded with mirror image.

than to add phase shift. The ripple has increased to approximately 1.6 dB; however, the three 0-dB points from the original response have been preserved. We cannot easily predict the combined response, and the best approach is to design the network using low ripple values so that the impedance levels remain close to a constant value across the passband. Although this method doubles the number of elements and is not an exact solution, it is often useful for realizing networks which would otherwise not be possible.

5.4.1 High-Pass Elements

The technique of cascading a mirror image network to adjust the output termination may require optimization or, at the very least, another design iteration to adjust the ripple. However, examining the combined network in Figure 5.35 provides some insight into the possibility of an exact solution. The network consists of four TLEs and one high-pass element; that is, we can transform either stub to the opposite end and then combine it with the other stub. Therefore, we might be able to synthesize a network having this topology, which provides convenient element values and the required 0.414-dB ripple. Figure 5.37(a) shows the synthesized network which, except for the two center TLEs, is not too different from the mirror image network in Figure 5.37. All elements are realizable and the network has exactly 0.414 dB ripple in a 50Ω system. The original network with $n = 2$ and $m = 1$ is not acceptable, but the $n = 4$, $m = 1$ network is, and the $n = 3$, $m = 1$ case should be investigated as well. The synthesized network shown in Figure 5.37(b) is the best solution to this particular design problem. The circuit has the fewest number of elements, and they are all within bounds. This example illustrates the approach of changing synthesis specifications. Simply adding one additional element to the original design results in a network we can easily build.

When designing an impedance-transforming network, the optimum solution results in the minimum number of elements and ripple. The Kuroda identities

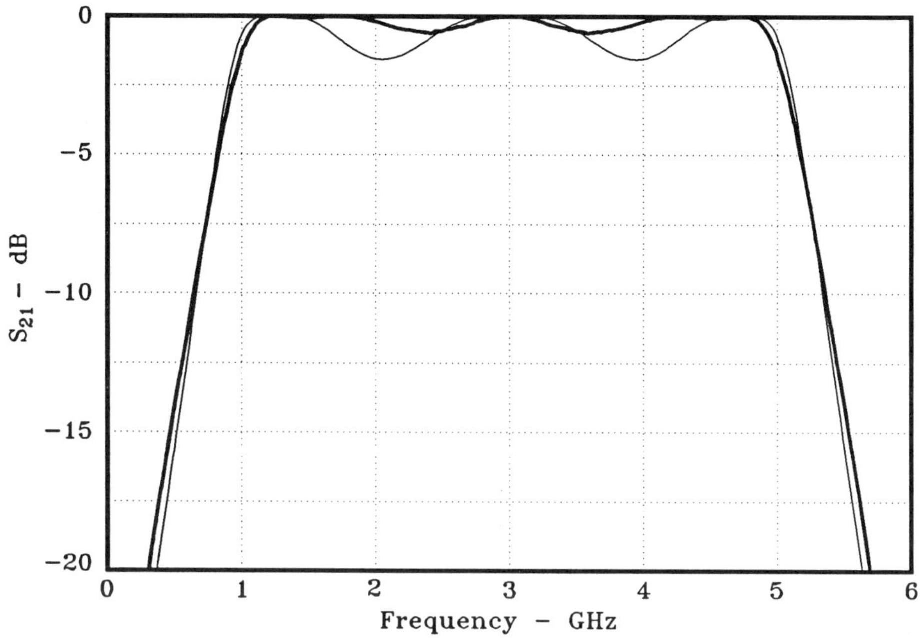

Figure 5.36 Cascaded high-pass network response.

discussed in Section 5.2 demonstrate that there is one particular topology that produces the maximum impedance ratio for a specific combination of elements. For example, Figure 5.12 shows that when the full value of a shorted shunt stub is extracted before a TLE, we have the highest possible output termination. We can examine the impedance-transforming capability of this type of network by plotting impedance versus ripple, as shown in Figure 5.38. The figure shows impedance as a function of ripple for an octave bandwidth 50Ω quarter-wave structure. *SP* denotes the shorted parallel or shunt stub, R_L the output termination, and *TLE* the characteristic impedance of the transmission line. As shown, we can realize the element values in a 50Ω system from about 0.01 to 0.5 dB ripple, although above 0.1 dB the shunt stub will have to be split into two equal parallel stubs. Thus, we can design a network capable of transforming 50Ω into any value between 80Ω and 200Ω across an octave bandwidth simply by adjusting the ripple. Between 50Ω and 80Ω, a single TLE can be used. The ripple required for 80Ω is 0.06 dB and decreases to 0 dB for no transformation. This topology, in fact, corresponds to the quarter-wave transmission line discussed in Section 3.1.3, where the characteristic impedance is the square root of the product of the source and load impedances; 63.2Ω for a 50Ω to 80Ω transformation. When the ripple is too high, we can add

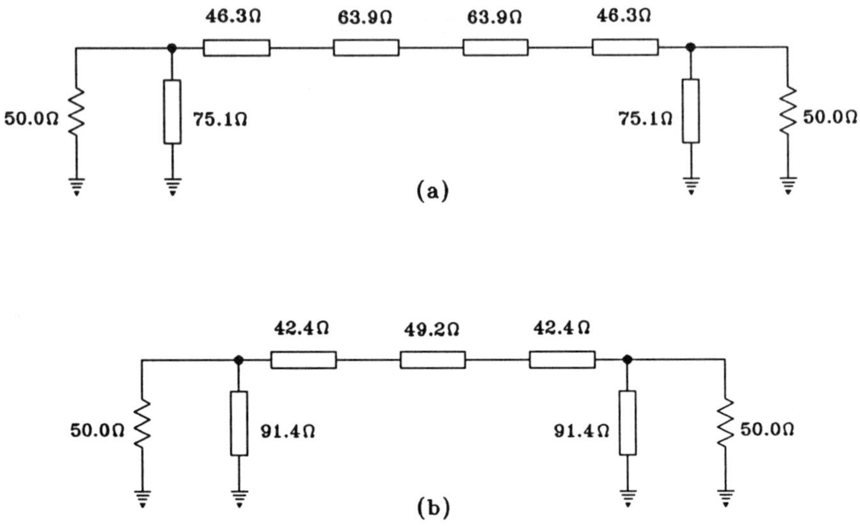

Figure 5.37 Synthesized high-pass filter networks.

another TLE to the network. For example, a network with one shunt shorted stub and two TLEs requires 0.003 dB ripple for a 50Ω to 200Ω octave bandwidth transformation. However, in practice it is sometimes necessary to increase the ripple and split the shunt stub between the input and the output to provide realizable element values.

Carlin and Kohler [26] provide techniques for computing the required ripple for prescribed impedance transformations using topologies consisting of a shorted shunt stub and TLEs. However, the technique is rather complex and uses approximations. Computer-aided synthesis is extremely fast, and it is usually easier to simply synthesize the network for various ripple values. Alternatively, one of the optimization techniques from Chapter 3 can be used. Direct synthesis also has the advantage that we can add one or more low-pass elements to the network, which is sometimes useful for adjusting the element values.

5.4.2 Low-Pass Elements

The lumped-element quasi-low-pass mapping discussed in Section 4.9.2 was used by Levy [9] to develop distributed approximations for a particularly useful impedance-transforming structure consisting of low-pass elements and TLEs. In fact, we also used Levy's derivation for the generalized bandpass network in

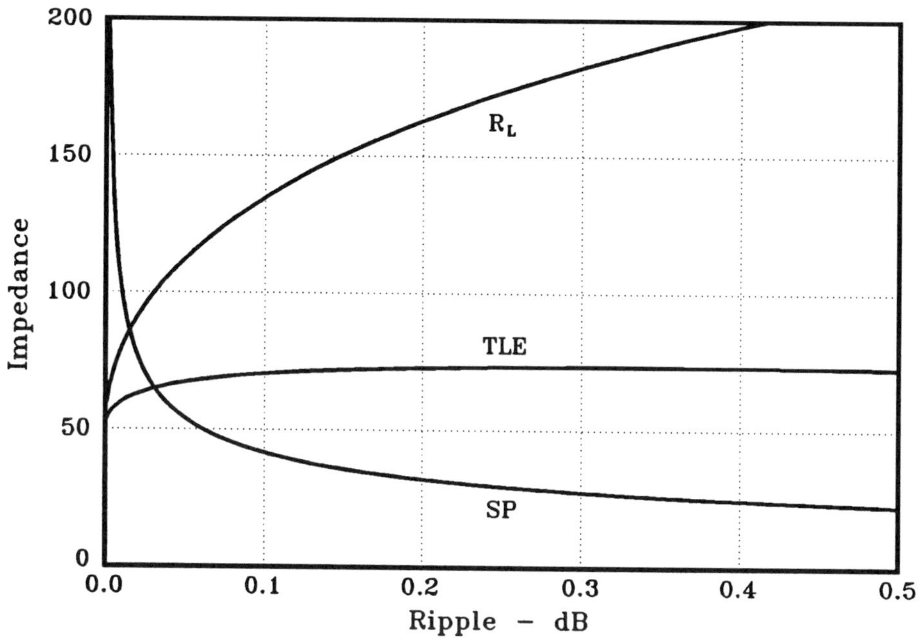

Figure 5.38 Impedance versus ripple for octave bandwidth.

Section 5.3.4, and the quasi-low-pass approximation is a special case. The number of high-pass elements is 0; that is, $m = 0$, and (5.36a) reduces to

$$F_k(\Omega^2) = T_n(\alpha_n)\cdot T_q(\alpha_q) - U_n(\alpha_n)\cdot U_q(\alpha_q) \tag{5.40}$$

where α_n is defined by (5.35a) and α_q by (5.35b). The same restrictions apply to (5.40) as to the original function in (5.36). That is, the total number of elements must be even and θ_2 must be less than 90°. Following the same procedure as in Section 4.9.2, we note that the low-pass network is transparent at zero frequency, and any insertion loss must be due to a mismatch between the source and load terminations. Therefore, we can write

$$\frac{4r}{(r+1)^2} = \frac{K_m}{1+\dfrac{\varepsilon^2}{2}\left[1+F_k(0)\right]} \tag{5.41a}$$

where r is Z_L/Z_S, the ratio of load to source impedances. Setting $K_m = 1$, then solving for ϵ^2, we have

$$\varepsilon^2 = \frac{(r-1)^2}{2r\left[1 + F_k(0)\right]} \tag{5.41b}$$

and

$$Ripple = 10\text{Log}(1 + \varepsilon^2)\ \text{dB} \tag{5.41c}$$

As an example, we will design a 50Ω to 100Ω impedance-transforming network for an octave bandwidth and an electrical length of 45° at the upper band edge. Computing $\theta_1 = \theta_2/2$, we have $\Omega_1 = \text{Tan}\ 22.5° = 0.414214$ and $\Omega_2 = \text{Tan}\ 45° = 1$. Next, we evaluate (5.35a) and (5.35b) to determine α_n and α_q. Letting $n = 1$ and $q = 1$, we compute the Chebyshev polynomials of the first and second kinds. Then we use (5.40) to compute

$$F_k(\Omega^2) = \frac{18.164411\,\Omega^4 - 20.280931\,\Omega^2 + 4.116520}{\Omega^2 + 1}$$

At $\Omega = 0$, $F_k(0) = 4.116520$. Substituting into equation (5.41b),

$$\varepsilon^2 = \frac{(2-1)^2}{(2)(2)(1 + 4.116520)} = 0.048861$$

which is approximately 0.207 dB of ripple. The VSWR is less than 1.6:1, which is acceptable. We could increase n and q if necessary to reduce the ripple. However, $n + q$ must be even. Following the same procedure outlined in Section 5.3.4, we compute

$$|s_{11}(S)|^2 = \frac{S^4 + 1.061467S^2 + 0.281678}{S^4 - 1.191959S^2 + 2.535105}$$

and then form $\pm s_{11}(S)$ from the poles and zeros. Since the output termination is 100Ω, we select $+s_{11}(S)$ and compute

$$Z(S) = \frac{S^2 + 1.045989S + 1.061467}{1.045988S + 0.530734}$$

Evaluating $Z(S)$ at $S = 0$, we find $Z(0) = 2$, which confirms that the expression is correct. The network synthesized from this driving-point impedance is shown in

Figure 5.39(a). The impedance of the open stub is too high by about 5Ω. However, we can change the electrical length, which has an effect on the element values. Reducing θ_2 to 44°, θ_1 is 22°, and the ripple 0.206 dB, slightly lower than the original value. Figure 5.39(b) shows the resulting network, which does have all element values within realizable limits. The response of the quasi-low-pass impedance-transforming network is similar to that shown in Figure 4.32 for the lumped-element equivalent.

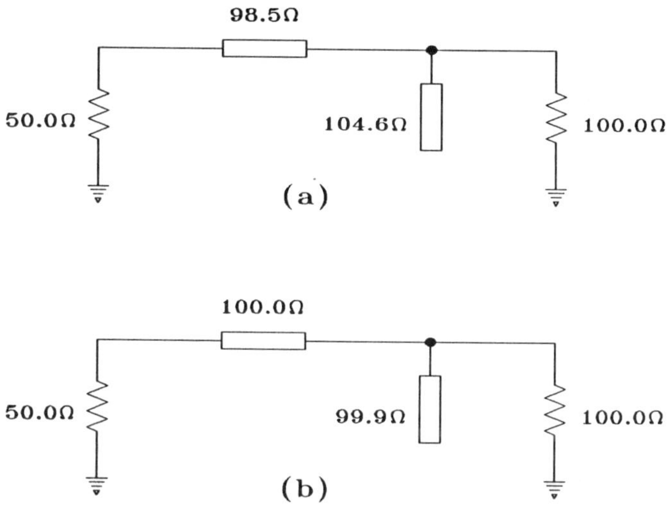

Figure 5.39 Low-pass transformer: (a) θ_2 = 45°; (b) θ_2 = 44°.

Comparing the parameters of the octave bandwidth 50Ω to 100Ω low-pass transformer with those of a similar high-pass network shown in Figure 5.38, we see that the high-pass design is superior. The ripple is only 0.03 dB, and the impedances of both the shorted stub and TLE are approximately 65Ω. However, there are instances where a low-pass structure has an advantage. For example, when the transformer is part of the input matching network of an field-effect transistor (FET), the high-pass circuit approaches a short circuit at lower frequencies, which can cause stability problems. The low-pass network is transparent at low frequencies, and the device is terminated with R_S, usually 50Ω. The advantages of one structure over the other is also a function of the impedance ratio and bandwidth. As we reduce the bandwidth, the impedance level of all elements in the low-pass structure approaches that of R_L for one particular value of θ_2 which is less than 45°. Element values in the high-pass network approach the square root of $R_S \bullet R_L$.

The quasi-low-pass impedance-transforming network is superior for broadband applications. The required ripple for an equivalent high-pass circuit is much higher, as are the element values. For example, Figure 5.40 shows a 50Ω to 100Ω transformer designed for $\theta_2 = 45°$ and a passband extending from 1 to 20 GHz. The ripple is 0.59 dB, which provides less than 2:1 VSWR across a bandwidth of more than four octaves. The element values are higher than desired. However, since the range is from 62.7Ω to 113.7Ω, we could use a lower permittivity substrate material. Reduced ϵ_r will also increase the separation between stubs. The electrical length of 45° at 20 GHz corresponds to approximately 73.8 mils, or 1.9 mm in air. An equivalent high-pass transformer requires more than 6 dB of ripple.

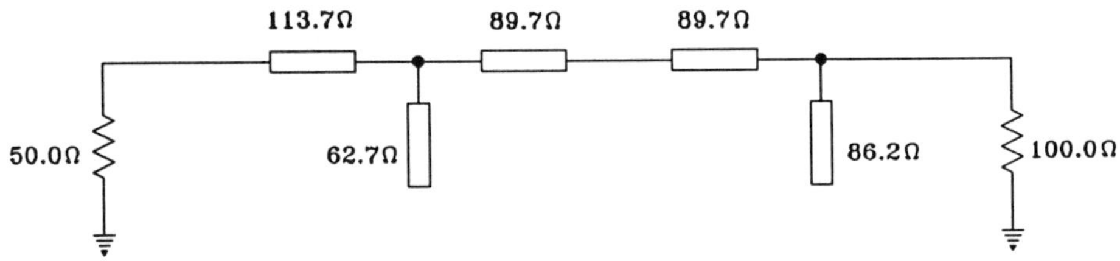

Figure 5.40 1- to 20-GHz low-pass transformer.

5.4.3 Transmission Line Elements

Matthaei [25] originally used the quasi-low-pass mapping to synthesize a structure which he called a *short-step Chebyshev impedance transformer*. The network consists of a cascade of electrically short transmission line elements (i.e., less than or equal to one-eighth wavelength at the center frequency). Matthaei's procedure is also a special case of the generalized bandpass approximation from Section 5.3.4, with both q and m equal to zero. When we restrict the network to TLEs, (5.36a) becomes

$$F_k(\Omega^2) = T_n(\alpha_n) \tag{5.42}$$

where α_n is defined in (5.35a) and $|s_{21}(\Omega)|^2$ in (5.37). Since the transducer gain function has the same form as that in Section 5.4.2, we can use (5.41) to compute the necessary ripple for a given impedance transformer. However, the restriction of an even number of elements and θ_2 less than 90° also applies to this approximation.

The 1- to 2-GHz, 50Ω to 100Ω transformer example from Section 5.4.2 can be designed as a short-step structure which will provide a comparison between the two methods. Using the same electrical length and number of elements, that is, $\theta_1=22.5°$, $\theta_2=45°$, and $n = 2$, we use (5.35a) to compute

$$\alpha_n = \frac{3.828427\,\Omega^2 - 1.828427}{\Omega^2 + 1}$$

and from (5.42),

$$F_k(\Omega^2) = T_2(\alpha_n) = 2\alpha_n^2 - 1$$

At $\Omega = 0$,

$$F_k(0) = (2)(-1.828427)^2 - 1 = 5.686291$$

and we can evaluate (5.41b),

$$\varepsilon^2 = \frac{(2-1)^2}{(2)(2)(1 + 5.686291)} = 0.037390$$

to determine that the required ripple is 0.159 dB, almost 0.05 dB less than the first design. Figure 5.41 shows the topology and element values for the short-step impedance transformer. It appears that this network is a better choice for the design. There are no stubs and the element values are well within realizable limits.

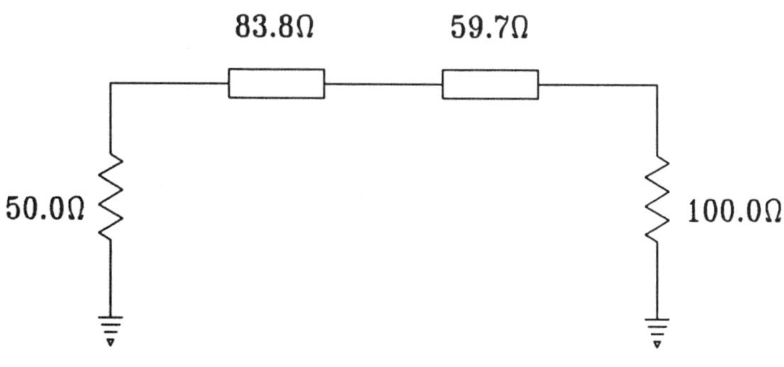

Figure 5.41 Short-step impedance transformer.

However, when we compare the response of the short-step structure, shown as the lighter trace in Figure 5.42, with that of the low-pass network, depicted by the darker trace, we see that there is considerable difference in the frequency characteristics. Both networks are virtually identical up to 2 GHz, that is, within approximately 0.05 dB. However, the low-pass network has a shunt open stub which provides a transmission zero at 4 GHz.

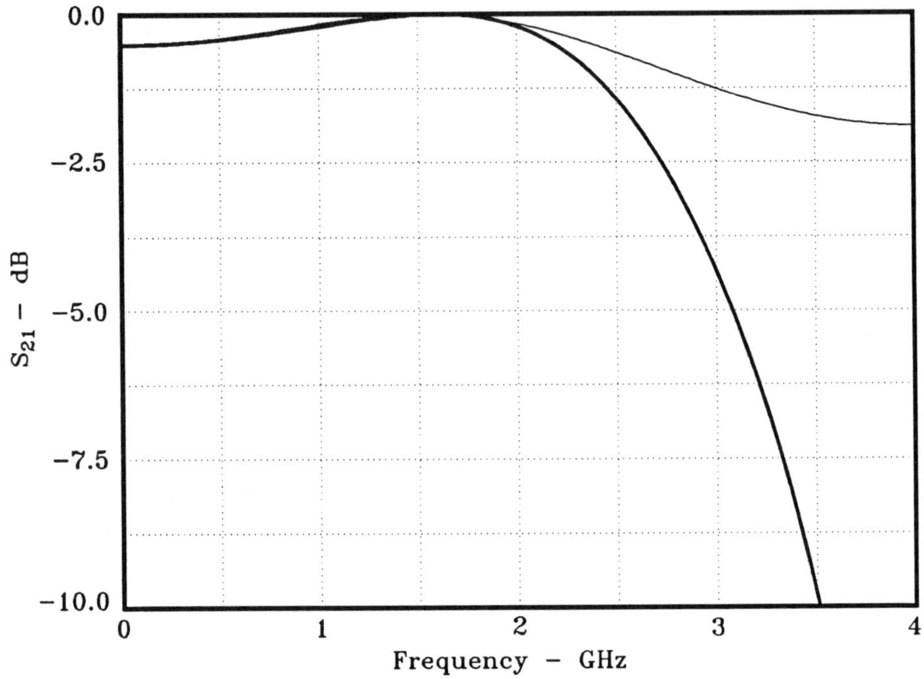

Figure 5.42 Short-step and low-pass transformer response.

Matthaei [25] points out that when the electrical length at the center frequency is $\lambda/8$, the network consists of a cascade of double-length TLEs. Since n is even, we have $n/2$ quarter-wavelength TLEs, which corresponds to the high-pass approximation of Section 5.3.3 with $m = 0$. When m is 0, (5.31) reduces to

$$|s_{21}(S)|^2 = \frac{K_m}{1 + \varepsilon^2 T_n\left(\dfrac{\sqrt{1 - S_c^2}}{\sqrt{1 - S^2}}\right)} \tag{5.43}$$

Then, equating the transducer gain at zero frequency with the mismatch between source and load,

$$\frac{4r}{(r+1)^2} = \frac{1}{1 + \varepsilon^2 T_n^2\left(\sqrt{1+\Omega_1^2}\right)} \tag{5.44a}$$

and

$$\varepsilon^2 = \frac{(r-1)^2}{4rT_n^2\left(\sqrt{1+\Omega_1^2}\right)} \tag{5.44b}$$

where S_c = jTanθ_c from the high-pass approximation becomes jTanθ_1 for the bandpass approximation.

The special case where $n = 1$ is of interest because it corresponds to the quarter-wavelength transmission line transformer discussed in Section 3.1.3. When $n = 1$, $T_1(x) = x$ and (5.44b) reduces to

$$\varepsilon^2 = \frac{(r-1)^2}{4r\left(1 + \mathrm{Tan}^2\theta_1\right)} = \frac{(r-1)^2\mathrm{Cos}^2\theta_1}{4r} \tag{5.45}$$

We can easily calculate the characteristics of a single quarter-wavelength transmission line used to match two real impedances. For example, to match 50Ω to 100Ω from 1 to 2 GHz, the characteristic impedance is the square root of (50) • (100), or 70.7Ω. θ_1 is 90° • (f_1/f_0), or 60°, and

$$\varepsilon^2 = \frac{(2-1)^2\mathrm{Cos}^2 60°}{(4)(2)} = 0.031250$$

corresponding to 0.134 dB ripple. Thus, since the transformation is exact at the center frequency, the insertion loss at 1.5 GHz is 0 and monotonically increases to 0.134 dB at 1.0 and 2.0 GHz.

5.5 CONCLUSION

Mathematically, distributed-element synthesis is not much different from the lumped-element synthesis of Chapter 4. In fact, using Richards' variable we can synthesize a distributed-element network directly from a lumped approximation, although the transmission line elements are redundant. When we base the synthesis on distributed approximations, the network is optimum in the sense that all elements contribute to the response. However, the difficulty arises from the fact that distributed elements depend on physical dimensions and properties of materials. In contrast with a lumped-element network, where we have a range of realizable element values spanning decades, the range of acceptable distributed element impedances may be 3:1 or less.

Fortunately, we have a large number of Kuroda identities which can be applied to change the topology and element values. In addition, redundant TLEs can be added at the input and output and then used to adjust element values, as well as the topology. The distributed synthesis techniques of this chapter are based on a commensurate network; that is, all elements have the same electrical length. When we allow the electrical length of each element to vary independently, optimization can often be used to find an acceptable set of element values.

Network synthesis is an ideal design tool. We simply specify the electrical specifications and follow a relatively uncomplicated procedure, and we are presented with numerous networks, all with identical electrical performance. Perhaps the most difficult task is selecting the particular circuit best suited for the application. However, computer-aided synthesis programs are extremely fast and can incorporate artificial intelligence to provide the optimum solution.

REFERENCES

[1] S. Ramo, J.R. Whinnery, and T. Van Duzer, *Fields and Waves in Communication Electronics*, John Wiley & Sons, Inc., New York, 1965.

[2] H. Howe, Jr., *Stripline Circuit Design*, Artech House, Inc., Dedham, Massachusetts, 1974.

[3] K.C. Gupta, R. Garg, and R. Chadha, *Computer-Aided Design of Microwave Circuits*, Artech House, Inc., Dedham, Massachusetts, 1981.

[4] T.C. Edwards, *Foundations for Microstrip Circuit Design*, John Wiley & Sons, Inc., New York, 1981.

[5] K.C. Gupta, R. Garg, and I.J. Bahl, *Microstrip Lines and Slotlines*, Artech House, Inc., Dedham, Massachusetts, 1979.

[6] R.A. Pucel, *Monolithic Microwave Integrated Circuits*, IEEE Press, New York, 1985.

[7] P.I. Richards, "Resistor-Transmission Line Circuits," *Proc. IRE*, Vol. 30, No. 2, February 1948, pp. 217-220.

[8] H.J. Carlin, "Distributed Circuit Design With Transmission Line Elements," *Proc. IEEE*, Vol. 59, No. 7, July 1971, pp. 1059-1081.

[9] R. Levy, "Synthesis of Mixed Lumped and Distributed Impedance-Transforming Filters," *IEEE Trans. on Microwave Theory and Techniques*, Vol. MTT-20, No. 3, March 1972, pp. 223-233.

[10] P.I. Richards, "General Impedance-Function Theory," *Quart. Appl. Math*, Vol. 6, 1948, pp. 21-29.

[11] R. Levy, "A New Class of Distributed Prototype Filters with Applications to Mixed Lumped/Distributed Component Design," *IEEE Trans. on Microwave Theory and Techniques*, Vol. MTT-18, No. 12, December 1970, pp. 1064-1071.

[12] K. Kuroda, "Derivation Methods of Distributed Constant Filters from Lumped Constant Filters," lectures at Joint Meeting of Konsoi Branch of Institute of Elec. Commun. of Elec., and of Illumin. Engrs. of Japan, October 1952, p. 32 (in Japanese).

[13] R.J. Wenzel, "Exact Design of TEM Microwave Networks Using Quarter-Wave Lines," *IEEE Trans. on Microwave Theory and Techniques*, Vol. MTT-12, No. 1, January 1964, pp. 94-111.

[14] R.J. Wenzel, "Exact Theory of Interdigital Band-Pass Filters and Related Coupled Structures," *IEEE Trans. on Microwave Theory and Techniques*, Vol. MTT-13, No. 5, September 1965, pp. 559-575.

[15] H. Ozaki and J. Ishii, "Synthesis of a Class of Strip-Line Filters," *IRE Trans. on Circuit Theory*, Vol. CT-5, No. 2, June 1958, pp. 104-109.

[16] G.L. Matthaei, L. Young, and E.M.T. Jones, *Microwave Filters, Impedance-Matching Networks, and Coupling Structures*, McGraw-Hill Book Company, New York, 1964.

[17] E.G. Cristal, "Microwave Filters," in *Modern Filter Theory*, G.C. Temes and S.K. Mitra, eds., John Wiley & Sons, Inc., New York, 1973.

[18] H. Baher, *Synthesis of Electrical Networks*, John Wiley & Sons, Inc., New York, 1984.

[19] G.C. Temes and J.W. LaPatra, *Circuit Synthesis and Design*, McGraw-Hill Book Company, New York, 1977.

[20] M.C. Horton and R.J. Wenzel, "General Theory and Design of Optimum Quarter-Wave TEM Filters," *IEEE Trans. on Microwave Theory and Techniques*, Vol. MTT-13, No. 3, May 1965, pp. 316-327.

[21] M. Abramowitz and A. Stegun, eds., *Handbook of Mathematical Functions With Formulas, Graphs, and Mathematical Tables*, National Bureau of Standards, Washington, D.C., 1972.

[22] J.A.G. Malherbe, *Microwave Transmission Line Filters*, Artech House, Inc., Norwood, Massachusetts, 1979.

[23] R.J. Wenzel, "Theoretical and Practical Applications of Capacitance Matrix Transformations to TEM Network Design," *IEEE Trans. on Microwave Theory and Techniques*, Vol. MTT-14, No. 12, December 1966, pp. 635-647.

[24] M.E. Mokari-Bolhassan and W.H. Ku, "Transfer Function Approximations for a New Class of Bandpass Distributed Network Structures," *IEEE Trans. on Microwave Theory and Techniques*, Vol. MTT-25, No. 10, October 1977, pp. 837-847.

[25] G.L. Matthaei, "Short-Step Chebyshev Impedance Transformers," *IEEE Trans. on Microwave Theory and Techniques*, Vol. MTT-14, No. 8, August 1966, pp. 372-383.

[26] H.J. Carlin and W.E. Kohler, "Direct Synthesis of Band-Pass Transmission Line Structures," *IEEE Trans. on Microwave Theory and Techniques*, Vol. MTT-13, No. 3, May 1965, pp. 283-297.

Chapter 6

Impedance Matching and Modeling

The design examples considered in earlier chapters were narrowband, and, in fact, designed for a single frequency. However, most components must operate over bandwidths ranging from a few percent to more than a decade. Even when the specifications call for operation at a single frequency, good design practice dictates that some bandwidth be provided to compensate for component and fabrication tolerance, as well as changes due to temperature and aging. A design that is overly sensitive to component variation usually requires precision components and alignment of each production circuit, which translates into increased cost. Relatively few circuits operate in a benign environment. At the very least, we must contend with temperature rise inside an enclosure due to heat buildup, even when the ambient temperature is constant. Military, airborne, and space applications present the greatest challenges because components are expected to operate over a wide range of environmental conditions for long periods of time.

The network characterization examples of Chapter 2 and the mapping examples of Chapter 3 clearly illustrate the relationship between circuit performance and the matching network. We often find that there is almost an infinite number of choices for an input or output impedance; for example, anywhere on the source or load plane constant gain circle. However, we also observe that certain regions of matching impedance offer advantages over other regions: relative insensitivity to component variations, stability, noise figure, or output power, for example. The single-frequency techniques of previous chapters can be extended to the design of broadband circuits by plotting the appropriate parameters at several frequencies. We then use the individual mappings to define an impedance at each frequency, which provides a locus of impedance points or impedance as a function of frequency.

The choice of the matching impedance has a significant effect on the matching network. At a single frequency, a matching network can sometimes be designed using one element, although a specific topology is usually preferred that can be used to accommodate bias injection or parasitic elements. Figure 3.6 shows the dependence between the two-element single-frequency matching networks and the impedance. The topologies that can be used to match certain impedances are limited, as illustrated in the figure. Broadband matching structures are necessarily more complex than those for single-frequency applications. However, similar topology restrictions apply. For example, if the impedance to be matched is real and greater than Z_O, the Norton transformations of Section 4.9.1 or the Kuroda transforms of Section 5.2 demonstrate that certain topologies are necessary to provide an output impedance higher than the input impedance.

The impedance as a function of frequency not only affects the matching network topology, but also the number of elements. In 1945, Bode [1] analyzed the limitations of matching to an impedance consisting of a parallel resistor and capacitor. Fano [2] published a classical paper in 1950 which provides limitations for matching to other series and parallel R-L and R-C impedance combinations. Youla [3] published another classical paper in 1964 which considers the problem of matching to an arbitrary lumped passive load. Bode and Fano show that the matching limitation takes the form of a gain-bandwidth (GBW) product defined by the topology and element values of the impedance. Thus, for a given impedance, there is a theoretical efficiency for the matching network. At a single frequency where the bandwidth is zero, we can match with maximum efficiency using at most two elements. However, as the bandwidth increases, the efficiency decreases and we can no longer perfectly match the impedance, *even using a matching network with an infinite number of elements!* As a consequence, the definition of an impedance function must be based on the theoretical matching efficiency as well as the performance of the network. Broadband circuit design usually necessitates a compromise between the complexity of the matching networks and the electrical parameters. We may be able to define a frequency-dependent impedance that provides the desired characteristics, but we must also be able to design efficient matching networks.

6.1 SYNTHESIS OF IMPEDANCE-MATCHING NETWORKS

Designers approach the problem of designing impedance-matching networks in various ways. One technique is the *brute force* method, where we assume a topology and initial element values, model the circuit, and then optimize. Given enough time and computer resources, this method can be used successfully although the results are seldom optimum. However, when the specifications are similar to a previous design, brute force can be an effective design procedure. The state of the art for active devices, components, and fabrication techniques is continually advancing. Design requirements change in a corresponding manner: increased operating frequency, broader bandwidth, lower noise figure, and higher output power. Thus, we cannot rely on previous circuits to solve current design problems. Instead, we must develop a unified design approach that allows us to extract the maximum possible performance from a given circuit in an efficient manner.

One approach is an extension of the single-frequency graphical methods of Chapter 3. We plot the impedance as a function of frequency, construct various contours, and then attempt to find a topology that intersects the contours at specific locations [4]. However, this technique has only limited usefulness. Once the number of elements exceeds two or three, the drawing becomes very complex and difficult to interpret. A technique that is practical for the design of complex matching networks is presented by Matthaei, Young, and Jones [5]. Using this method, we first determine whether the load[1] is a parallel or series equivalent circuit by examining the real component of the impedance and admittance. A reactive element is added in parallel or series to resonate the load at the center frequency. Then we compute the susceptance or reactance at the upper and lower band edges and compute a *load decrement* [5] from the real and imaginary components of Z or Y. The load decrement is used to design a filter structure matching the resonated load across the specified frequency range. The resonant load actually forms the first resonator of the filter. This approach can be effectively used for the design of a number of very useful impedance-matching networks; however, the selection of available topologies is somewhat limited. The load decrement method is, in fact, based on closed-form solutions to specific synthesis problems. We can apply similar techniques directly to the general form of network synthesis and realize the advantages illustrated in Chapters 4 and 5.

Figure 6.1(a) shows a typical load consisting of a 50Ω resistor in series with a 20-nH inductor. We could easily match this impedance at a single frequency using the Smith chart. However, we desire a relatively constant low mismatch loss across the entire operating bandwidth, and this requires a different design approach. We can use lumped or distributed synthesis to design a filter with a maximally flat or equal-ripple response across a given passband and, under certain conditions, the structure can be used for impedance matching as well. Figure 6.1(b) shows the $n = 3$ bandpass filter designed in Section 4.6.2 and illustrated in Figure 4.20. Comparing the load with the first two elements of the filter, we see that the resistive components are equal, both reactive elements are the same type, and the value of the series inductor in the filter is 4 nH higher than that of the

[1] In the framework of impedance-matching theory, the term *load* is often used to denote the impedance that will be matched. Although we also use this terminology to designate an output termination (i.e., source and load impedance), the context of its use will generally define the specific meaning.

load. After splitting the 24-nH inductor into a 20-nH inductor in series with a 4-nH inductor, we can replace the first two elements of the filter with the load, as shown in Figure 6.1(c). The response of the load and matching network in Figure 6.1(c) has exactly the same characteristics as the original filter in Figure 6.1(b). The only difference is that the input termination and a portion of the series inductor is provided by the load. The reactive element associated with the load is known as a *parasitic element*, and the process of replacing a portion of the filter with the load is called *parasitic absorption.*

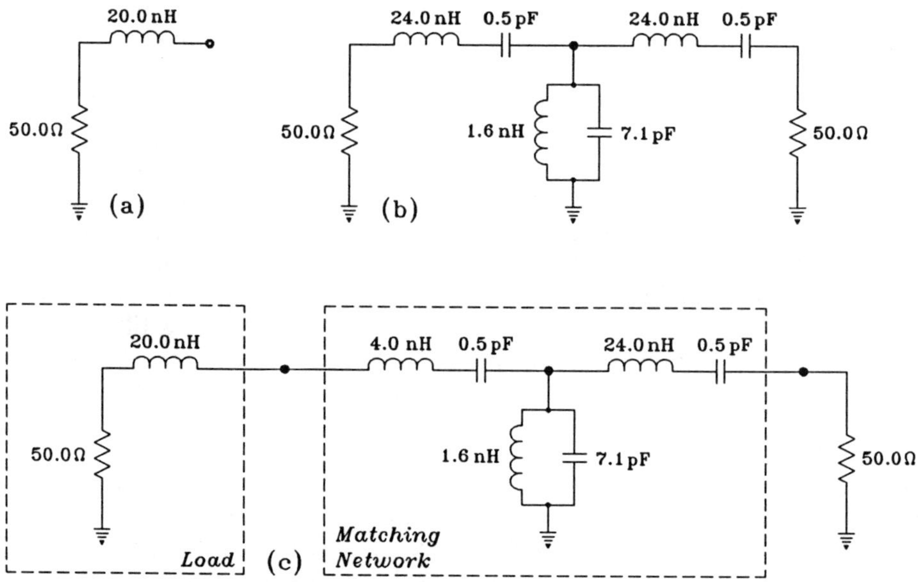

Figure 6.1 Load and impedance-matching network.

Thus, given a load to be matched across a specified frequency range, we need only synthesize a filter for the required bandwidth and center frequency having an input termination equal to that of the load, and an input element of the same type as the load which has a value such that a portion can be replaced by the parasitic element. We can easily meet the first two requirements. The frequency range, along with the number and type of elements, must be specified before we can compute an approximation, and either plus or minus s_{11} can be selected to form an impedance function with the required input element type. The input termination is normalized and we can scale to any value. However, the element values are determined by the coefficients of the s_{11} polynomials and requires further investigation to determine a method for providing the necessary values. We have some flexibility in specifying the number of elements, or order, ripple, minimum insertion loss (MIL), and zero locations, which turn out to be critical parameters in the design of an efficient matching network. Figure 6.2 shows a plot of the value of the series input inductor versus ripple and MIL for a lumped-element Chebyshev $n = 2$, 1.25- to 1.75-GHz bandpass filter. Examining this figure, we find that the inductance increases for larger values of ripple and/or MIL, and below certain limits it is not possible to absorb the 20-nH load parasitic into the filter. This condition is known as ***insufficient parasitic absorption***. The situation where the parasitic equals the input element is called *exact absorption*, and we say that the network has *excess absorption* capability when a portion of the synthesized element is absorbed by the load, leaving the remainder in the matching network, as in the previous example.

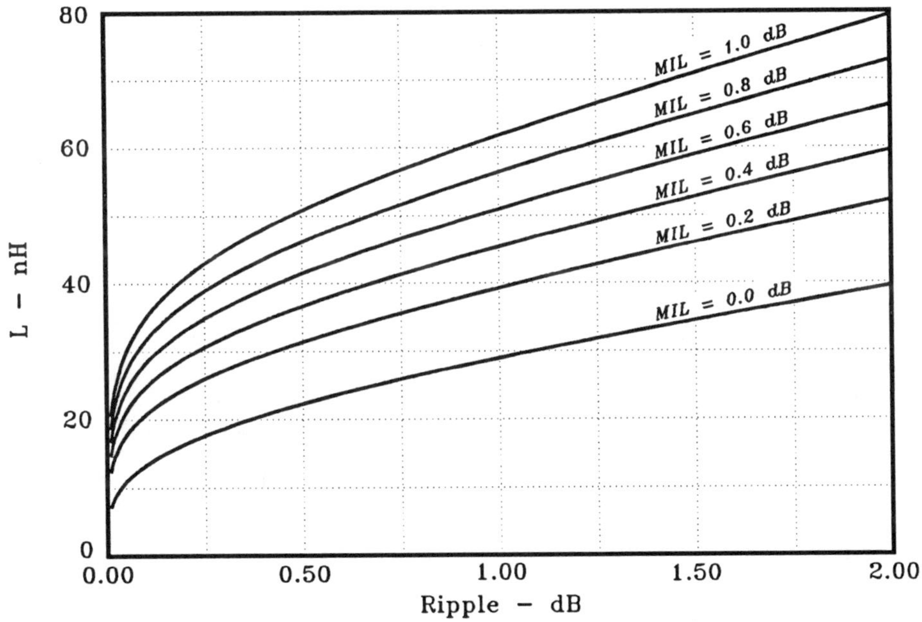

Figure 6.2 Inductance versus ripple and ML.

The most efficient matching network is one that provides maximum power transfer into the load using the minimum number of elements. The network is assumed lossless and the insertion loss is entirely due to a mismatch between the input and output termination. Thus, if we minimize the insertion loss, mismatch is also minimized. Figure 6.2 indicates that the most advantageous choice for the ripple or MIL is exact absorption. When the MIL is held constant, the ripple is minimized, and, similarly, when the ripple is held constant, the MIL is minimized. In addition, the number of elements in the synthesized matching network is reduced by one, since the entire input element is absorbed into the load. Using the load of Figure 6.1(a) as an example, Figure 6.2 shows that when the MIL is 0.0 dB, the ripple is approximately 0.36 dB for exact absorption, 0.08 dB for 0.2-dB MIL, and 0.01 dB for 1.0-dB MIL. The maximum insertion loss is the combined total of the ripple plus MIL and, since the network is lossless, we can use Equation (2.38a) to compute the maximum $|s_{11}|$, input return loss, and voltage standing wave ratio (VSWR). Table 6.1 lists the results of these calculations for each of the constant MIL curves in Figure 6.2, clearly illustrating that one particular combination provides the lowest total loss: 0.284 dB in this case.

Table 6.1 Exact Absorption Values for 20-nH Series Inductor

MIL	*Ripple*	*Total Loss*	*Return Loss*	*VSWR*
0.0 dB	0.363 dB	0.363 dB	10.959 dB	1.790:1
0.2 dB	0.084 dB	0.284 dB	11.986 dB	1.672:1
0.4 dB	0.038 dB	0.438 dB	10.180 dB	1.897:1
0.6 dB	0.021 dB	0.621 dB	8.754 dB	2.150:1
0.8 dB	0.013 dB	0.813 dB	7.677 dB	2.408:1
1.0 dB	0.009 dB	1.009 db	6.833 dB	2.672:1

Figure 6.3 shows the total loss versus ripple for exact absorption of a 20-nH series inductor using the Chebyshev bandpass approximation, with $n = 2, 3, 4$, and 6. The maximum ripple value for the $n = 4$ and $n = 6$ curves requires 0-dB MIL for exact absorption; above these points the network has excess absorption capability. As shown, there is a definite minimum for each curve, which is relatively broad and decreases as the order is increased. Using an $n = 2$ approximation, the synthesized filter has four elements, and the corresponding matching network has three elements. The maximum insertion loss is 0.241 dB, which corresponds to a maximum VSWR of 1.606:1. Table 6.2 lists tabulated values for orders between one and ten. Examining this data, we see that there is a significant reduction in loss between $n = 1$ and $n = 2$, less improvement between $n = 2$ and $n = 3$, and only minimal reduction above $n = 3$. Thus, we can improve the match to a specified load by increasing the order of the approximation used to synthesize the network. However, there is a corresponding increase in the number

of elements and a point of diminishing returns where the improvement does not justify increased circuit complexity. In fact, considering component loss, we may actually *reduce* the power transfer by increasing the number of elements.

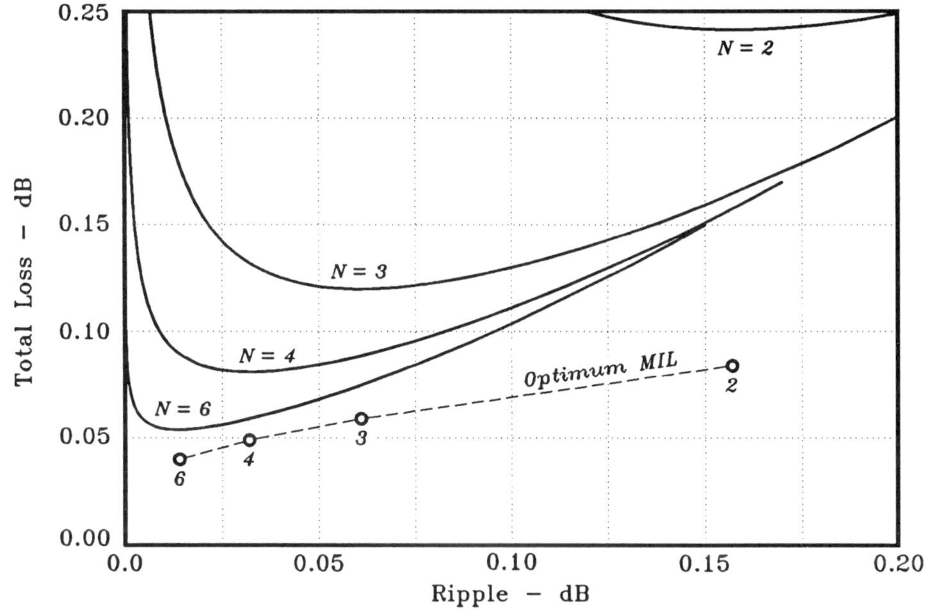

Figure 6.3 Exact absorption and total loss versus ripple.

Table 6.2 Minimum Total Loss as a Function of Approximation Order

N	*Elements*	*Total Loss*	*VSWR*
1	1	1.149 dB	2.863:1
2	3	0.241 dB	1.606:1
3	5	0.120 dB	1.395:1
4	7	0.081 dB	1.315:1
6	11	0.054 dB	1.250:1
10	19	0.040 dB	1.211:1

The only other synthesis parameter we can control is the placement of the zeros in either the left or right half-plane. Figure 6.4 shows series inductance values versus ripple as a function of zero locations for an $n = 3$ Chebyshev bandpass approximation at a constant MIL value of 0.5 dB. The three pairs of complex conjugate zeros are sorted by increasing real values; L denotes a left half-plane location and R a right half-plane location. For example, the curve labeled *L-R-R* represents the case where the zero pair with the smallest real component, or closest to the real axis, is placed in the left half-plane, and the other two zero pairs are located in the right half-plane. As shown, maximum absorption capability occurs when all three pairs are located in the left half-plane. Minimum absorption results when all three are placed in the right half-plane and other combinations fall between the *L-L-L* and *R-R-R* curves.

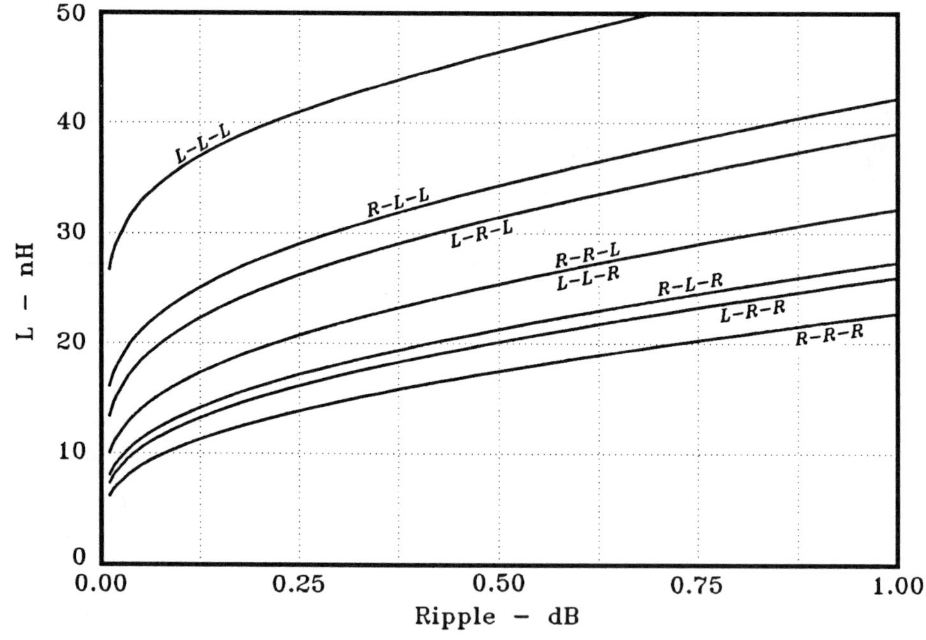

Figure 6.4 Inductance versus ripple and zero location for $n = 3$ and $0.5 =$ db MIL.

The application of network synthesis to the design of impedance-matching networks is not at all complicated and, by carefully selecting the parameters, can provide the optimum performance achievable with a given number of elements. We need only select an approximation with the appropriate passband and the same element type as the load (i.e., low-pass or high-pass). Next, we specify an order, assume a value for the ripple and MIL, place all zeros in the left half-plane, choose the appropriate sign for s_{11}, form the input impedance, and then evaluate the element value. Based on the element value, we can change the ripple or MIL, and then iterate until we find the optimum combination. Fano [2] shows that the minimum loss for low-pass Chebyshev matching networks can be expressed in the form

$$\frac{\text{Tanh}\, na}{\text{Cosh}\, a} = \frac{\text{Tanh}\, nb}{\text{Cosh}\, b}$$

where n is the order of the approximation, and a and b are functions of the ripple and MIL. The expression must be solved by iterative means, although Matthaei, Young, and Jones [5] provide charts which can be used to easily determine the optimum values. We can use the low-pass to high-pass or bandpass transformations in conjunction with the charts to design matching networks for other frequency response characteristics. However, the minimums illustrated in Figure 6.3 are broad, and parasitic absorption monotonically increases for larger values of ripple or MIL. Thus, any of the optimization strategies discussed in Chapter 3 will rapidly converge. The advantage of using optimization is that we can design optimum matching networks using any approximation, and, as it turns out, we can compute the absorption capability directly from the approximation, which simplifies the procedure.

6.1.1 Lumped-Element GBW Limitations

Although computer-aided synthesis is very fast and easy to use, the iterative procedure described in Section 6.1 for finding the exact absorption values of ripple and MIL can be time consuming. Especially since it must be repeated for each order. That is, we assume $n = 1$ and compute the minimum loss; then if it is too high, we try $n = 2$, and so forth. As it turns out, *we do not have to actually synthesize the network to determine the optimum parameters.* Consequently, we can very rapidly determine the insertion loss and number of elements required to match a given load.

Fano [2] derived the limitations for matching to four different load topologies which take the form of equations (6.1a) through (6.1d):

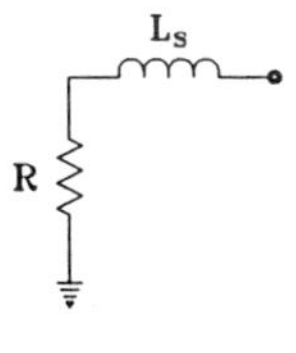

$$\int_0^\infty \ln\left|\frac{1}{s_{11}(\omega)}\right| d\omega <= \pi \cdot \left\{\frac{R}{L_S} - \sum Z_{rhp}\right\} \qquad (6.1a)$$

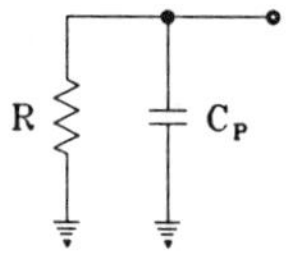

$$\int_0^\infty \ln\left|\frac{1}{s_{11}(\omega)}\right| d\omega <= \pi \cdot \left\{\frac{1}{RC_P} - \sum Z_{rhp}\right\} \tag{6.1b}$$

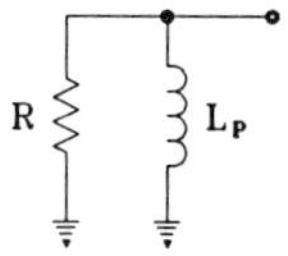

$$\int_0^\infty \frac{1}{\omega^2} \cdot \ln\left|\frac{1}{s_{11}(\omega)}\right| d\omega <= \pi \cdot \left\{\frac{L_P}{R} - \sum \frac{1}{Z_{rhp}}\right\} \tag{6.1c}$$

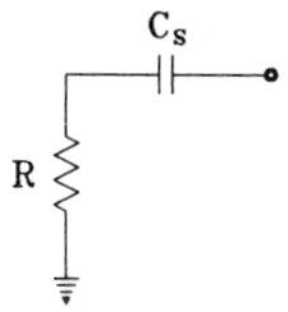

$$\int_0^\infty \frac{1}{\omega^2} \cdot \ln\left|\frac{1}{s_{11}(\omega)}\right| d\omega <= \pi \cdot \left\{RC_S - \sum \frac{1}{Z_{rhp}}\right\} \tag{6.1d}$$

where Z_{rhp} denotes zeros located in the right half-plane, the subscript S denotes series connected elements, and P signifies parallel elements. Equations (6.1a) and (6.1b) represent the low-pass configuration for an inductor and capacitor, respectively. Equations (6.1c) and (6.1d) provide similar expressions for the high-pass load. Examining these equations reveals that the integral of a function must be less than or equal to a constant minus the sum of right half-plane zeros for low-pass loads, or inverted zeros for high-pass loads. Figure 6.5 shows a plot of the low-pass and high-pass integrands for an $n = 3$ Chebyshev, 1.25- to 1.75-GHz bandpass approximation with 0.5-dB MIL. The darker curves represent 0.414-dB ripple and the lighter curves 0.792 dB, ϵ^2 equal to 0.1 and 0.2, respectively. The low-pass function resembles the insertion loss response, which is in fact related since

$$\ln\left|\frac{1}{s_{11}}\right| = -\frac{1}{2}\ln\left|1 - |s_{21}|^2\right|$$

for a lossless network. The maximum function value, or gain, corresponds to the MIL, and ripple reduces the passband gain below the MIL as shown in the figure. The high-pass function is similar, but has an additional slope caused by the $1/\omega^2$ factor.

We can determine the element value limitations by rearranging the low-pass equations, (6.1a) and (6.1b):

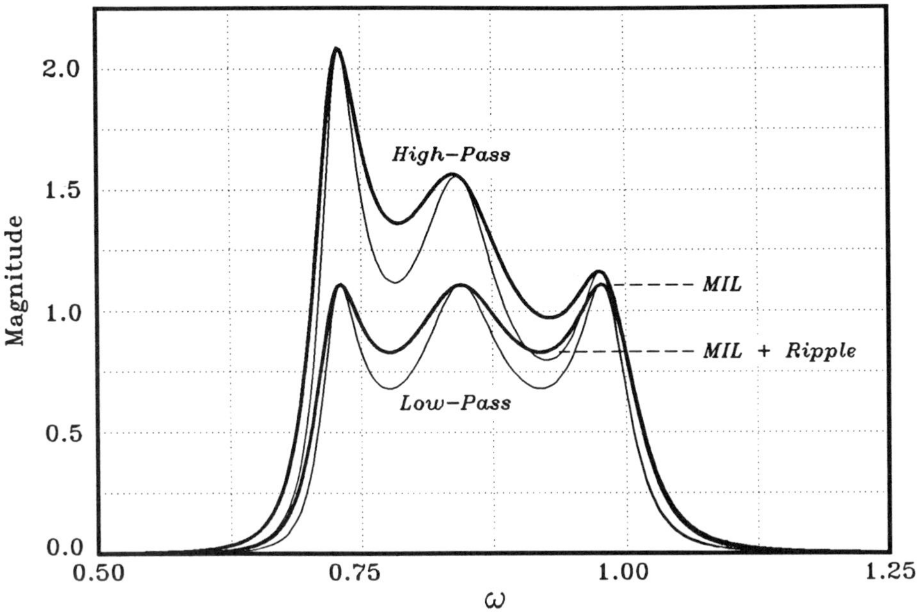

Figure 6.5 Function value versus radian frequency.

$$L_S,\ C_P <= \frac{1}{\frac{I_{LP}}{\pi} + \sum Z_{rhp}} \tag{6.2a}$$

and the high-pass expressions, (6.1c) and (6.1d):

$$L_P,\ C_S >= \frac{I_{HP}}{\pi} + \sum \frac{1}{Z_{rhp}} \tag{6.2b}$$

where I_{LP} is the low-pass integral and I_{HP} the high-pass integral, and the element values are normalized. Maximum parasitic absorption is accomplished by minimizing the denominator of (6.2a) and the right side of (6.2b). When all zeros are placed in the left half-plane, the summation of right half-plane zeros in (6.2a) and (6.2b) is zero and can be eliminated leaving only the low-pass or high-pass integral to be minimized. We can reduce the integral value, or area under the curves in Figure 6.5, by increasing the MIL or ripple. Also, as we increase n, the response above and below the passsband becomes steeper and contributes less to the total area. The number of passband ripples increases as well, which reduces the integral value even further. These results are the same as those obtained by empirical means in Section 6.1. However, it is also readily apparent from Figure 6.5 that as we increase bandwidth, the magnitude must be decreased to maintain a constant integral value. This observation leads to the term *gain-bandwidth product*, which is applied to impedance-matching networks.

As an example, we can compute the absorption capability of the $n = 3$ Chebyshev bandpass filter with 0.414-dB ripple, 0.5-dB MIL, and a passband extending from 1.25 to 1.75 GHz. Numerically integrating the functions shown by the darker curves in Figure 6.5, we have

$$I_{LP} = \int_0^\infty \ln\left|\frac{1}{s_{11}}\right| d\omega = 0.319$$

for the low-pass, and

$$I_{HP} = \int_0^\infty \frac{1}{\omega^2}\cdot\ln\left|\frac{1}{s_{11}}\right| d\omega = 0.447$$

for the high-pass load. Assuming left half-plane zeros, we can substitute into (6.2a)

$$L_S,\ C_P <= \frac{1}{\dfrac{I_{LP}}{\pi} + \sum Z_{rhp}} = \frac{1}{\dfrac{0.319}{\pi} + 0.000} = 9.848$$

and into (6.2b)

$$L_P,\ C_S >= \frac{I_{HP}}{\pi} + \sum\frac{1}{Z_{rhp}} = \frac{0.447}{\pi} + 0.000 = 0.142$$

Scaling to 50Ω and 1.75 GHz, we have $L_S \le 44.782$ nH, $C_P \le 17.913$ pF, $L_P \ge 0.646$ nH, and $C_S \ge 0.258$ pF. The input reflection coefficient has six zeros, which are

Zeros
0.018696 ± j0.725266
0.025371 ± j0.984207
0.044067 ± j0.844005

and if we place the last four in the right half-plane,

$$\sum Z_{rhp} = (0.025371 + j0.984207) + (0.025371 - j0.984207)$$
$$+ (0.044067 + j0.844005) + (0.044067 - j0.844005)$$
$$= 0.138876$$

The low-pass limitation becomes

$$L_S,\ C_P <= \frac{1}{\dfrac{I_{LP}}{\pi} + \sum Z_{rhp}} = \frac{1}{\dfrac{0.319}{\pi} + 0.138876} = 4.159$$

and after scaling the series inductor, we have $L_S \leq 18.912$ nH. Comparing the two series inductance values with the *L-L-L* and *L-R-R* curves in Figure 6.4, we find that the results are the same.

6.1.2 Ideal GBW Limitations

Maximum parasitic absorption requires all zeros to be located in the left half-plane, and the GBW calculations reduce to π/I_{LP} for a low-pass load or I_{HP}/π for a high-pass load. The integrands are well behaved, and although the integration limits are from zero to infinity, the major contribution is within a few octaves above and below the passband. Thus, we can rapidly evaluate the integral using relatively simple numerical integration techniques [6]. However, we are often presented with a number of choices for the load impedance and seek a simple method to evaluate the relative GBW limitations of each.

The optimum value of the total loss shown in Figure 6.3 decreases as we increase the order or number of elements in the matching network. The MIL associated with the optimum is shown as a broken line in the figure and decreases at a much slower rate than the ripple. We can postulate that the MIL is much larger than the ripple as the order approaches infinity. Also, the insertion loss increases by $6n$ dB per octave above and below the passband, which leads to the conclusion that $|s_{11}|$ is constant in the passband and 1 elsewhere for an ideal matching network with an infinite number of elements. Figure 6.6 shows a plot of the ideal low-pass and high-pass integrands. K represents the constant value of the low-pass integrand in the passband; that is,

$$K = \ln\left|\frac{1}{s_{11}}\right| = -\frac{1}{2}\cdot\ln|s_{11}|^2$$

Thus,

$$|s_{21}|^2 = 1 - e^{-2K}$$

and

$$MIL = 10 \cdot \log(1 - e^{-2K}) \text{ dB} \tag{6.3}$$

Integrating, we have

$$\int_{\omega_1}^{\omega_2} \ln\left|\frac{1}{s_{11}}\right| d\omega = K \cdot (\omega_2 - \omega_1) <= \frac{\pi R}{L_S}, \frac{\pi}{RC_P} \tag{6.4a}$$

for low-pass loads, and

$$\int_{\omega_1}^{\omega_2} \frac{1}{\omega^2} \cdot \ln\left|\frac{1}{s_{11}}\right| d\omega = K \cdot \left(\frac{1}{\omega_1} - \frac{1}{\omega_2}\right) <= \frac{\pi L_P}{R}, \pi RC_S \tag{6.4b}$$

for high-pass loads. Then, solving for K in terms of the load elements yields

$$K = \frac{R}{2L_S(f_2 - f_1)} \tag{6.5a}$$

$$K = \frac{1}{2RC_P(f_2 - f_1)} \tag{6.5b}$$

$$K = \frac{2\pi^2 L_P}{R(1/f_1 - 1/f_2)} \tag{6.5c}$$

and

$$K = \frac{2\pi^2 RC_S}{(1/f_1 - 1/f_2)} \tag{6.5d}$$

We can use (6.5) to rapidly compute the loss required to match a load across a given bandwidth using an ideal matching network. For example, the load shown in Figure 6.1(a) is to be matched from 1.25 to 1.75 GHz. We first substitute into (6.5a)

$$K = \frac{50}{(2)(20 \cdot 10^{-9})(1.75 \cdot 10^{9} - 1.25 \cdot 10^{9})} = 2.50$$

and then into (6.3)

$$MIL = 10 \cdot \log(1 - e^{-5.00}) = 0.029 \text{ dB}$$

Comparing this result with the optimum loss shown in Figure 6.3, we see that the improvement is only 0.02 dB as we increase the order from $n = 6$ to $n = \infty$. We can also relate (6.5) to the expressions for the load Q computed from the real and imaginary impedance or admittance components to show that the optimum loss is reduced as we reduce the Q-factor.

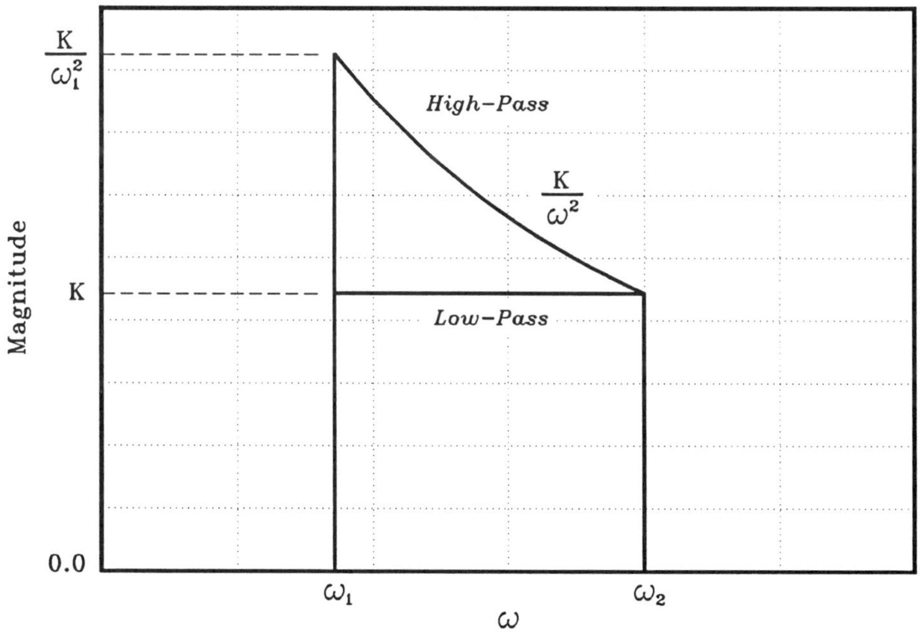

Figure 6.6 Ideal low-pass and high-pass integrands.

The ideal GBW limitations provide valuable design information. Given a choice of load impedances, we can easily compute the loss required to match each and then make a decision based on circuit performance as well as the matching network design. Also, the ideal loss indicates how well we will be able to match to a particular load. The finite-element matching network will exhibit higher loss, and we need not pursue the design if the load requires significant mismatch using an ideal matching network.

6.1.3 Distributed-Element GBW Limitations

We can use Richards' variable, discussed in Section 5.1, to derive the distributed-element GBW limitations. The transformation $\omega \Rightarrow \Omega$ results in ω-plane inductors replaced by Ω-plane shorted stubs, and ω-plane capacitors replaced by Ω-plane open stubs. Transforming (6.1) into the Ω-plane, we have

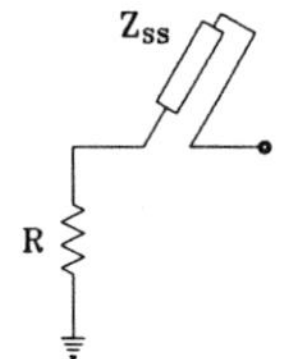

$$\int_0^\infty \ln\left|\frac{1}{s_{11}(\Omega)}\right| d\Omega <= \pi \cdot \left\{\frac{R}{Z_{SS}} - \sum Z_{rhp}\right\} \tag{6.6a}$$

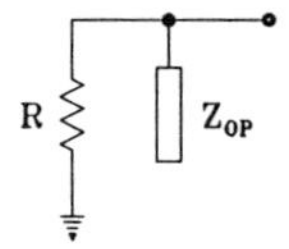

$$\int_0^\infty \ln\left|\frac{1}{s_{11}(\Omega)}\right| d\Omega <= \pi \cdot \left\{\frac{Z_{OP}}{R} - \sum Z_{rhp}\right\} \tag{6.6b}$$

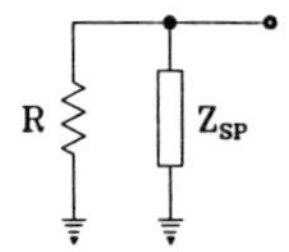

$$\int_0^\infty \frac{1}{\Omega^2} \cdot \ln\left|\frac{1}{s_{11}(\Omega)}\right| d\Omega <= \pi \cdot \left\{\frac{Z_{SP}}{R} - \sum \frac{1}{Z_{rhp}}\right\} \tag{6.6c}$$

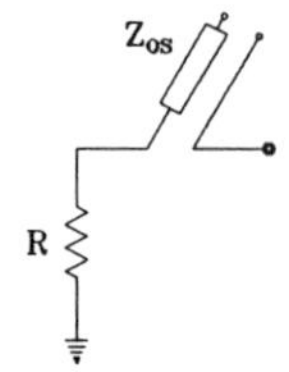

$$\int_0^\infty \frac{1}{\Omega^2} \cdot \ln\left|\frac{1}{s_{11}(\Omega)}\right| d\Omega <= \pi \cdot \left\{\frac{R}{Z_{OS}} - \sum \frac{1}{Z_{rhp}}\right\} \tag{6.6d}$$

where Z represents the characteristic impedance of the distributed element, the first subscript shorted or open, and the second subscript series or parallel connected. The variable Ω is Tanθ, and the integration limits become $\theta = 0$ to $\theta = \pi/2$ if we integrate with respect to θ; that is, $d(\text{Tan}\theta) = \text{Sec}^2\theta \bullet d\theta$.

We designed a 500- to 750-MHz distributed bandpass filter in Section 5.3.4, which can be used to illustrate the distributed GBW calculations. The ripple is 0.05 dB, MIL 0.0 dB, $\theta_1 = 30°$, and $\theta_2 = 45°$, and the $|s_{11}|^2$ polynomials are as shown. Figure 6.7 illustrates the distributed low-pass and high-pass integrands for this approximation plotted versus angle in degrees. The integrands are somewhat different from the corresponding lumped-element functions shown in Figure 6.5. We have two infinitesimally narrow peaks which have infinite magnitude, and the value of the high-pass integrand does not approach zero for small angles. The peaks coincide with the 0-dB insertion loss points; that is, $|s_{11}| = 0$, also shown in the figure. Nonzero minimum insertion loss would result in a passband response similar to that shown in Figure 6.5 for the lumped-element network.

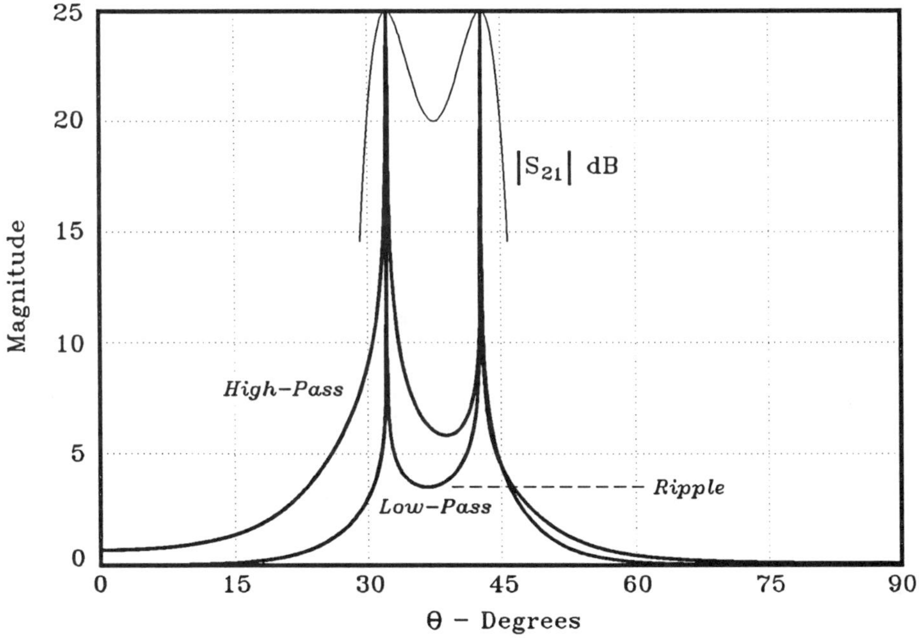

Figure 6.7 Distributed integrands versus angle.

Integrating, we have $I_{LP} = 2.001$ and $I_{HP} = 3.740$. When we rearrange (6.6b) and then substitute, the open shunt stub is

$$Z_{OP} = \frac{R \cdot I_{LP}}{\pi} = \frac{(50)(2.001)}{\pi} = 31.847\Omega$$

Using (6.6c) to compute the shorted shunt stub,

$$Z_{SP} = \frac{R \cdot I_{HP}}{\pi} = \frac{(50)(3.740)}{\pi} = 59.524\Omega$$

and we find the impedance of the shunt open and shorted stubs to be within 0.1% of the values computed in Section 5.3.4. However, these computations are considerably easier since the values are determined directly from the $|s_{21}|^2$ approximation.

We can also use Richards' variable to derive the ideal GBW constraints for distributed loads. Replacing ω with Tanθ, L_S with Z_{SS}, L_P with Z_{SP}, C_S with $1/Z_{OS}$, and C_P with $1/Z_{OP}$ in (6.4), we have

$$\int_{\Omega_1}^{\Omega_2} \ln\left|\frac{1}{s_{11}}\right| d\Omega = K \cdot \left(\Omega_2 - \Omega_1\right) <= \frac{\pi R}{Z_{SS}}, \frac{\pi Z_{OP}}{R} \tag{6.7a}$$

for low-pass loads, and

$$\int_{\Omega_1}^{\Omega_2} \frac{1}{\Omega^2} \cdot \ln\left|\frac{1}{s_{11}}\right| d\Omega = K \cdot \left(\frac{1}{\Omega_1} - \frac{1}{\Omega_2}\right) <= \frac{\pi Z_{SP}}{R}, \frac{\pi R}{Z_{OS}} \tag{6.7b}$$

for high-pass loads. Then solving for K in terms of the load,

$$K = \frac{\pi R}{Z_{SS}\left(\text{Tan}\,\theta_2 - \text{Tan}\,\theta_1\right)} \tag{6.8a}$$

$$K = \frac{\pi Z_{OP}}{R\left(\text{Tan}\,\theta_2 - \text{Tan}\,\theta_1\right)} \tag{6.8b}$$

$$K = \frac{\pi Z_{SP}}{R\left(1/\text{Tan}\,\theta_1 - 1/\text{Tan}\,\theta_2\right)} \tag{6.8c}$$

and

$$K = \frac{\pi R}{Z_{OS}(1/\mathrm{Tan}\,\theta_1 - 1/\mathrm{Tan}\,\theta_2)} \tag{6.8d}$$

6.2 LOAD MODELING

An impedance-matching network is simply a filter with the input termination and at least a portion of the input element provided by the load. We can achieve an optimum design by adjusting the number of elements, ripple, and minimum insertion loss. However, this approach requires that the load be represented as a combination of a resistor and reactive element. Most practical loads cannot accurately be represented as simple two-element models, and the restriction would seem to severely limit the success of this approach. Nevertheless, by carefully modeling the load impedance, we can design matching networks that are nearly optimum, even when the actual load deviates from the ideal model.

Consider *Load A*, which is defined by the tabulated data in Table 6.3. Figure 6.8 shows the data plotted on a Smith chart, revealing that the impedance follows the 50Ω constant resistance contour and has positive reactance which increases with increasing frequency. Therefore, the equivalent load model is a series resistor-inductor combination. We can use the reactance measured at the center frequency to compute the inductance; that is,

$$L = \frac{X_L}{\omega} = \frac{94.248}{(2\pi)(1.50 \cdot 10^9)} = 10.000 \text{ nH}$$

Thus, the load consists of a 50Ω resistor in series with a 10-nH inductor, and

Table 6.3 Load A

Frequency	Γ_L	$Z_L\Omega$
1.00 GHz	$0.532\angle 57.858°$	50.000 + j62.832
1.25 GHz	$0.618\angle 51.854°$	50.000 + j78.540
1.50 GHz	$0.686\angle 46.696°$	50.000 + j94.248
1.75 GHz	$0.740\angle 42.285°$	50.000 + j109.956
2.00 GHz	$0.782\angle 38.512°$	50.000 + j125.664

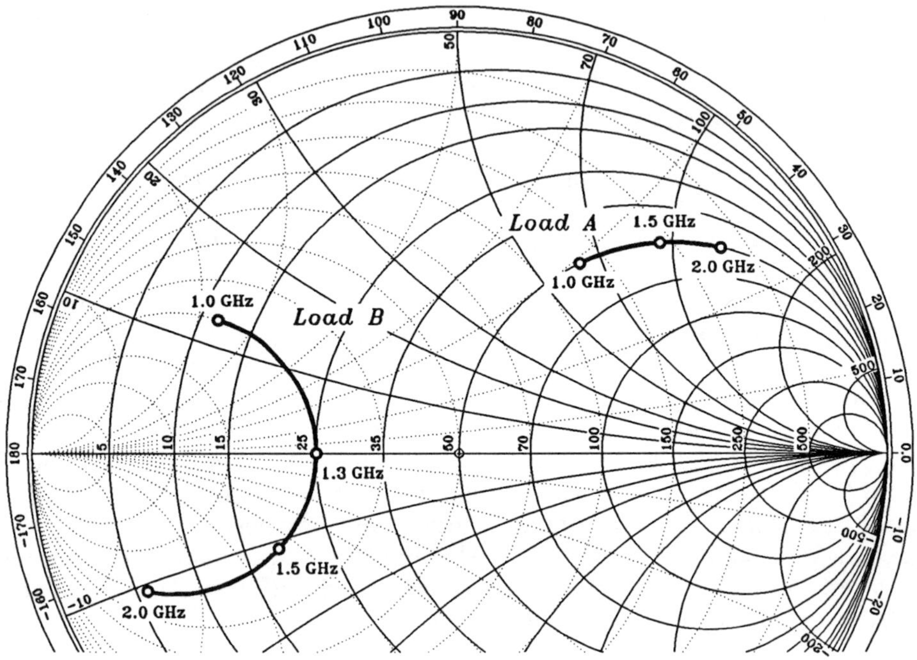

Figure 6.8 Load impedance versus frequency.

we could design the matching network following the procedure used for the load in Figure 6.1(a). However, we must compute the reactance at other frequencies to determine if the element values are optimum. In this case the load model is exact, but when the computed impedance deviates from the desired impedance, we can adjust the load values to minimize the maximum error.

Table 6.4 lists reflection coefficient versus frequency for *Load B*. The plot shown in Figure 6.8 indicates that the load is a parallel equivalent; that is, the reflection coefficient follows the 40-mmho constant conductance contour. However, the susceptance changes signs in the passband and cannot be modeled as a single reactive element without significant error at one end of the band. The susceptance is negative at 1.0 GHz, zero at 1.3 GHz, and positive at 2.0 GHz. Thus, we can model this data as a parallel R-L-C circuit which is resonant at 1.3 GHz. At resonance, the inductive and capacitive susceptances are equal:

$$\frac{1}{\omega_o L} = \omega_o C$$

or

$$L = \frac{1}{\omega_o^2 C} \tag{6.9a}$$

At any frequency, the total susceptance is the sum of the individual components:

$$B_T = B_C + B_L = \omega C - \frac{1}{\omega L}$$

Substituting (6.9a), we have

$$C = \frac{B_T}{2\pi\left(f - f_o^2/f\right)} \tag{6.9b}$$

Using the 2.0 GHz data from Table 6.4 in Equation (6.9b),

$$C = \frac{72.612 \cdot 10^{-3}}{2\pi \cdot 10^9\left[2.00 - (1.30)^2/2.00\right]} = 10.000 \text{ pF}$$

then, from (6.9a),

$$L = \frac{1}{(2\pi \cdot 1.30 \cdot 10^9)^2(10.00 \cdot 10^{-12})} = 1.500 \text{ nH}$$

and the load model is a parallel combination of a 25Ω resistor, 10-pF capacitor, and 1.5-nH inductor.

Checking the admittance of the load model against the values shown in Table 6.4 reveals that this model is also exact. However, we have two reactive elements that must be absorbed by the load. The parallel L-C configuration is easily realized from the lumped-element bandpass approximation. In fact, the low-pass to bandpass transformation shown in Figure 4.19 maps a shunt capacitor into a parallel L-C. As a rule, we cannot achieve exact absorption for both elements, but the ideal GBW constraints can be used to quickly compare the two and determine which is the limiting factor. Using (6.5b) with the resistor-capacitor combination yields

$$K = \frac{1}{(2)(25)(10.000 \cdot 10^{-12})(2.00 \cdot 10^9 - 1.00 \cdot 10^9)} = 2.000$$

Table 6.4 Load B

Frequency	Γ_L	Y_L *mmho*
1.00 GHz	$0.644\angle 150.605°$	40.000 - j43.271
1.25 GHz	$0.348\angle 168.439°$	40.000 - j6.343
1.50 GHz	$0.479\angle -151.784°$	40.000 + j23.512
1.75 GHz	$0.685\angle -151.494°$	40.000 + j49.325
2.00 GHz	$0.800\angle -155.832°$	40.000 + j72.612

and (6.5c) with the R-L elements yields

$$K = \frac{(2\pi^2)(1.500 \cdot 10^{-9})(10^9)}{(25)(1/1.000\ -\ 1/2.000)} = 2.369$$

Substituting these values into (6.3), we find that the inductor requires 0.038-dB MIL and the capacitor 0.080 dB when matched with an ideal network. Thus, the capacitor requires higher loss and determines the optimum parameters for the matching network. That is, the ripple and MIL combination for exact absorption of the capacitor results in excess absorption for the inductor, and the full value cannot be removed from the network. Using the methods of Sections 6.1 and 6.1.1, we compute the optimum loss for the capacitor as 1.56 dB for $n = 1$, 0.41 dB with $n = 2$, and 0.23 dB when $n = 3$. Synthesizing an $n = 3$ lumped-element bandpass filter for a passband extending from 1.0 to 2.0 GHz, with 0.102-dB ripple and 0.133-dB MIL, provides the filter structure shown in Figure 6.9(a). Removing 1.5 nH of the shunt 1.3-nH input inductor and all of the shunt 10.0-pF capacitor, and then applying the Norton transformations to the remaining network results in the matching network shown in Figure 6.9(b). Thus, we can match this load across an octave bandwidth with a maximum VSWR of 1.6:1 using a relatively simple network.

6.2.1 Distributed Models

We can design either lumped or distributed matching networks by using the appropriate type of elements to model the load. The loads shown in Figure 6.8 are high enough in frequency that the dimensions of a distributed-element matching network are practical. In fact, distributed elements are preferable for this

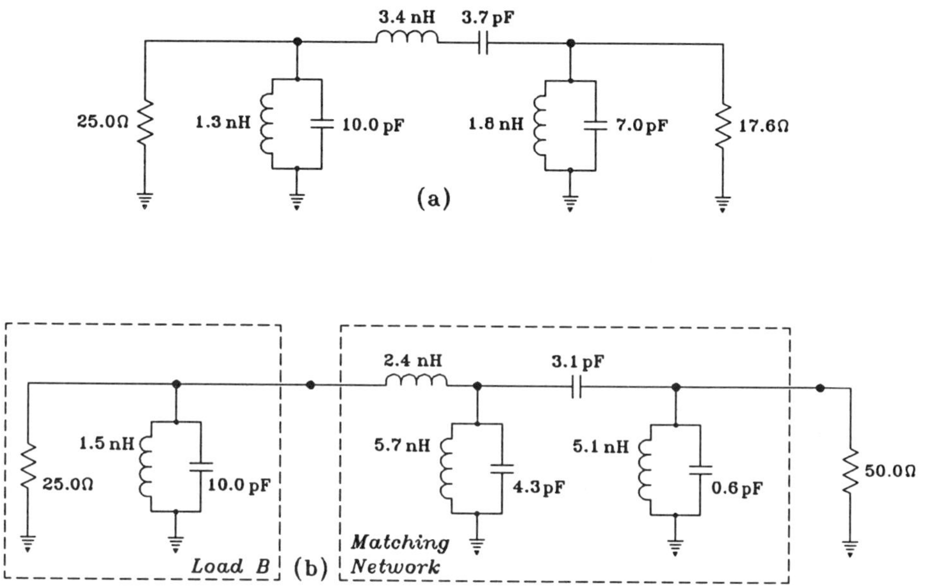

Figure 6.9 Load B: (a) filter; (b) matching network.

application. The circuit is less expensive and the impedance better controlled since the elements are etched directly on a printed circuit board or substrate. Load B has a positive reactance at 1.0 GHz, zero at 1.3 GHz, and negative reactance at 2.0 GHz. A shorted stub exhibits similar characteristics and the load can be modeled as a resistor and shunt shorted stub. The electrical length is 90° at 1.3 GHz, and we can use the susceptance at any frequency to solve for the characteristic impedance; that is,

$$B = \frac{1}{jZ_o \mathrm{Tan}\,\theta}$$

or

$$Z_o = \frac{1}{jB\,\mathrm{Tan}\,\theta} \tag{6.10}$$

Substituting 1.0 GHz data from Table 6.4,

$$Z_O = \frac{1}{j(-j43.271 \cdot 10^{-3})\left[\mathrm{Tan}\left(\frac{1.0}{1.3} \cdot 90°\right)\right]} = 8.765\Omega$$

and at 2.0 GHz,

$$Z_O = \frac{1}{j(j72.612 \cdot 10^{-3})\left[\mathrm{Tan}\left(\frac{2.0}{1.3} \cdot 90°\right)\right]} = 15.545\Omega$$

The wide disparity in Z_O between 1.0 GHz and 2.0 GHz indicates that the load model is not accurate. We must evaluate the significance of model errors and, if possible, minimize the impact on the matching network. The obvious measure of the model error is the differences between the impedance or admittance. However, a more meaningful parameter is the mismatch loss (MML) between the model and a *conjugate* load. Assuming we can design a perfect matching network for the model (i.e., exact absorption using 0-dB ripple and MIL), the input impedance of the network must be the complex conjugate of the model impedance. When the model is replaced with the actual load, any insertion loss is caused by a mismatch between the impedance of the lossless matching network and load, which is the mismatch loss described in Section 2.3.9. Figure 6.10 shows the mismatch loss between a model consisting of a 25Ω resistor in parallel with a shorted quarter-wavelength stub, and the actual load shown in Figure 6.8 as a function of stub impedance Z_O. The mismatch loss is significant at higher frequencies for low Z_O values, and increases at the low end when Z_O is increased. The best choice is Z_O = 13Ω, shown as the darker curve, which provides a maximum loss of 0.12 dB across the entire bandwidth. We can also adjust the quarter-wavelength frequency, which will equalize the ripple across the passband for other values of Z_O. However, the insertion loss of the matching network will add to the mismatch loss caused by load model errors and must be considered as well.

The matching network must be synthesized using the quarter-wavelength bandpass approximation discussed in Section 5.3.4. The electrical length at the center frequency given by (5.33a) is 90°, 90° - θ at f_1, and 90° + θ at f_2. Thus, the response is symmetrical about f_0 and, since f_0 is fixed, we must adjust either f_1 or f_2 to include the full design bandwidth. Rearranging (5.33a), we have

$$f_1 = 2f_0 - f_2 \tag{6.11}$$

The limits for the GBW integrals are from $\Omega = 0$ to $\Omega = \infty$ and we need only integrate the quarter-wavelength approximation from $\theta = 0°$ to $\theta = 90°$; that is, $f = 0$ to $f = f_0$. The ideal GBW constraint given by (6.8c) for the shunt shorted stub reduces to

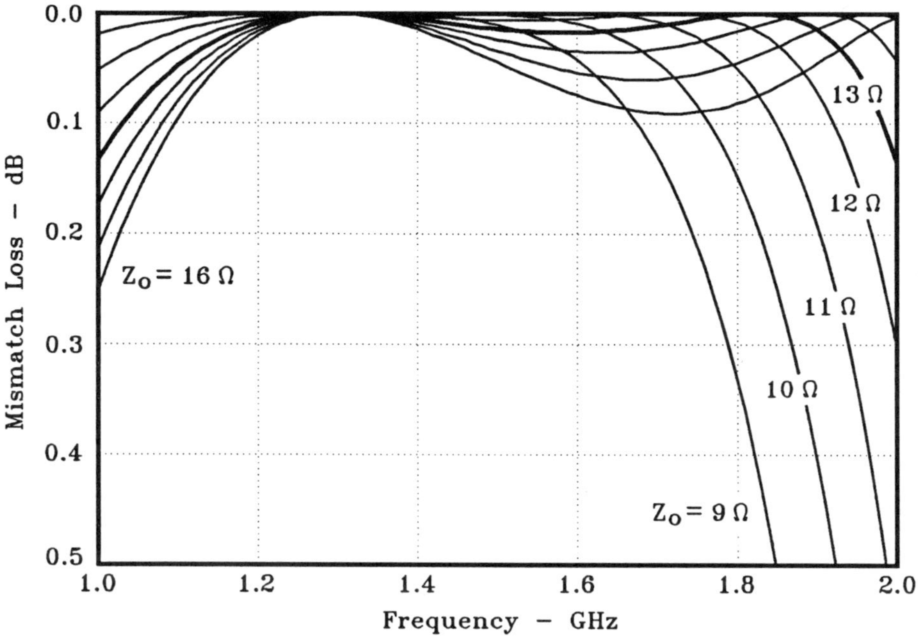

Figure 6.10 Distributed model mismatch loss.

$$K = \frac{\pi Z_{SP}}{R} \cdot \text{Tan}\,\theta_1 \tag{6.12}$$

since $\theta_2 = 90°$ and $1/\text{Tan}\theta_2$ is zero. Using 1.3 GHz as the center frequency,

$$f_1 = (2)(1.3) - 2 = 0.6\,\text{GHz}$$

Then for $Z_O = 9\Omega$,

$$K = \frac{\pi \cdot 9}{25} \cdot \text{Tan}\left(\frac{0.6}{1.3} \cdot 90°\right) = 1.002$$

and for $Z_O = 16\Omega$,

$$K = \frac{\pi \cdot 16}{25} \cdot \text{Tan}\left(\frac{0.6}{1.3} \cdot 90°\right) = 1.781$$

Using (6.3) to compute the minimum insertion loss reveals that the ideal matching network requires 0.629 dB for a Z_O = 9Ω model and 0.125 dB for the Z_O = 16Ω model. Thus, we have a tradeoff between how well the model fits the actual load data and the efficiency of the associated matching network. The optimum is easily determined by adjusting f_0 for minimum mismatch loss at discrete Z_O values, adding the MIL from the ideal matching network, then selecting the combination providing the lowest total loss. Table 6.5 shows calculations for Z_O ranging from 9Ω to 16Ω. Z_O = 13Ω results in the minimum loss, and we can synthesize the distributed--element matching network.

Table 6.5 Matching Efficiency Versus Shunt Stub Z_O

Z_OΩ	*f_0 GHz*	*Load Mismatch (dB)*	*Ideal MIL (dB)*	*Total Loss (dB)*
9.0	1.408	0.123	0.242	0.365
10.0	1.387	0.078	0.218	0.296
11.0	1.364	0.056	0.206	0.262
12.0	1.342	0.046	0.199	0.245
13.0	1.344	0.075	0.149	0.224
14.0	1.344	0.111	0.114	0.225
15.0	1.343	0.150	0.089	0.239
16.0	1.341	0.189	0.070	0.259

The center frequency is 1.34370 GHz, and, using (6.11), the design bandwidth is from f_1 = 0.687400 to f_2 = 2.0 GHz. We let m = 1 (i.e., one high-pass element) and then evaluate the GBW integrals as a function of n, the number of transmission line elements (TLE). The loss is 0.95 dB for n = 1, 0.39 dB for n = 3, and 0.27 dB at n = 5. Using m = 1 and n = 5, the optimum ripple is 0.07506 dB and the MIL 0.19131 dB. The synthesized matching network has a shorted input shunt stub equal to 13Ω, although the element values and output termination are much too high, as shown in Figure 6.11(a). We cannot use low-pass elements with this approximation, and if m is increased beyond 1, the network must contain series open stubs, as discussed in Chapter 5. However, we can increase the loss resulting in excess absorption capability and a portion of the shorted input stub which will be available to adjust the output termination. With the ripple increased from 0.075 to 0.120 dB, the impedance of the shorted shunt stub becomes 12.35526Ω. The remainder is 249.121165Ω after removing the 13Ω portion assigned to the load model. We can use the stub and the last two TLEs to provide a 50Ω output impedance, as shown in Figure 6.11(b). With the exception of the 29Ω input TLE, all elements are realizable and the total loss will be 0.311 dB from the matching network plus 0.149 dB from the load model, or a total loss of approximately 0.46 dB. Figure 6.12 shows the insertion loss versus frequency for the matching network of Figure 6.11(b) terminated with the actual load of Figure 6.8. The maximum loss is almost exactly 0.46 dB in the 1.0- to 2.0-GHz passband, which illustrates one of the advantages of this method: we can predict the efficiency of the matching network without actually designing a circuit.

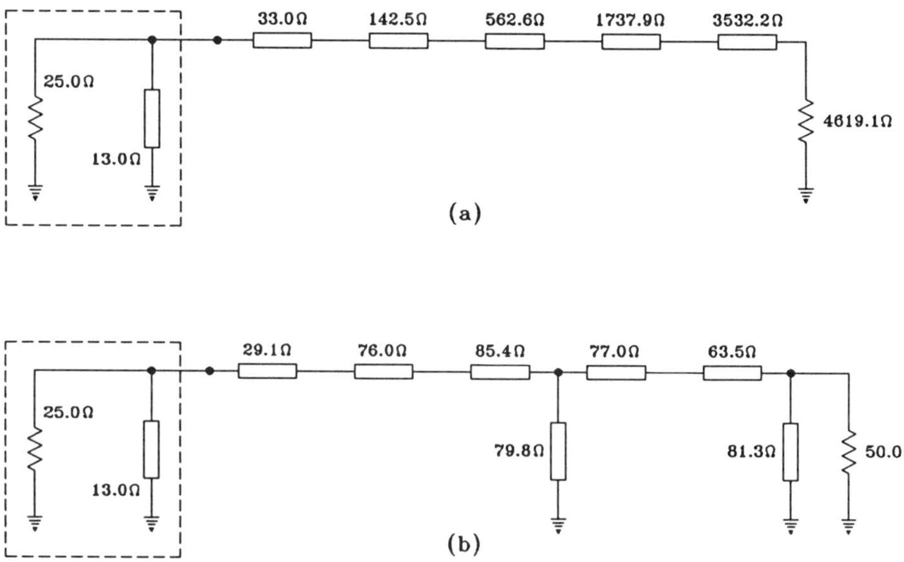

Figure 6.11 Load B matching network; (a) exact loss; (b) excess loss.

We have another option for the design of a distributed-element matching network to match the impedance of Load B in Figure 6.8. The generalized approximation given by (5.34) provides a network with TLEs, low-pass, and high-pass elements with electrical lengths shorter than 90° at f_2. In this case, short electrical lengths are an advantage because of the relatively low frequencies. The load is resonant at 1.3 GHz, which requires two reactive elements in the model (i.e., a shunt shorted and open stub). All elements in the distributed network are commensurate; hence, the electrical length of both stubs must be the same. When the procedure used with the lumped L-C model is followed, the reactances are equal at resonance:

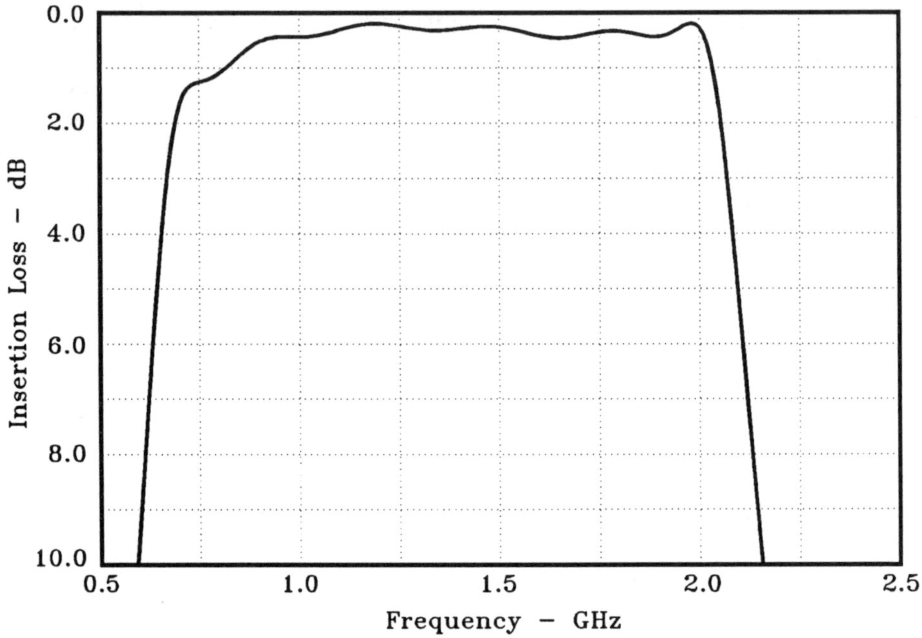

Figure 6.12 Quarter-wavelength matching network.

$$jZ_{SP} \cdot \mathrm{Tan}\,\theta_0 = \frac{jZ_{OP}}{\mathrm{Tan}\,\theta_0}$$

or

$$Z_{SP} = \frac{Z_{OP}}{\mathrm{Tan}^2\theta_0} \tag{6.13}$$

At any frequency, the total susceptance equals the sum of the individual components:

$$B_T = B_{OP} + B_{SP} = j\frac{\mathrm{Tan}\,\theta}{Z_{OP}} - j\frac{1}{Z_{SP}\,\mathrm{Tan}\,\theta}$$

Then, substituting (6.13) and solving for Z_{OP}, we have

$$Z_{OP} = \frac{j}{B_T} \cdot \left(\mathrm{Tan}\,\theta - \frac{\mathrm{Tan}^2\theta_0}{\mathrm{Tan}\,\theta} \right) \tag{6.14}$$

We can assume an electrical length of 45° at 2.0 GHz (i.e., $\theta_2 = 45°$) and then compute the electrical length at resonance as $(f_0/f_2) \bullet \theta_2$ or $\theta_0 = 29.25°$. Using the 2.0 GHz data from Table 6.4 and equation (6.14), we compute Z_{OP} as

$$Z_{OP} = \frac{j}{j\,72.612 \cdot 10^{-3}} \cdot \left(\mathrm{Tan}\,45° - \frac{\mathrm{Tan}^2\,29.25°}{\mathrm{Tan}\,45°} \right) = 9.453\Omega$$

Then, from (6.13),

$$Z_{SP} = \frac{9.453}{\mathrm{Tan}^2\,29.25°} = 30.141\Omega$$

Analyzing the mismatch between this model and data from Table 6.4 shows a maximum insertion loss of 0.033 dB. The model fits the data very well; however, we assumed an angle at f_2 and can minimize the loss by optimizing the electrical length. The optimum value turns out to be $\theta_2 = 37.15361°$, where $Z_{SP} = 32.08093\Omega$, $Z_{OP} = 6.53793\Omega$, and the maximum loss is 0.003 dB. Evaluating equations (6.8b), (6.8c), and (6.3), we find that 0.033-dB MIL is required to match the shorted stub using an ideal matching network and 0.089 dB to match the open stub. The open stub is the dominant parasitic, and we can use the generalized approximation with the GBW integral given by (6.6b) to optimize the required loss. For $n = 3$, $m = 1$, and $q = 2$ (i.e., three TLEs, one high-pass element, and two low-pass elements), the optimum is 0.105-dB ripple and 0.147-dB MIL, or a total loss of 0.252 dB. Figure 6.13(a) shows the matching network synthesized from the optimum approximation. The load absorbed the full value of the open shunt stub, and only a portion of the shorted shunt stub. The remainder is used to adjust the output termination.

The synthesized network has three elements that are not realizable: the 26.6Ω TLE, 28.3Ω shunt open stub, and the 291.1Ω shunt shorted stub. The mismatch loss of the load model is almost zero, and we could change the electrical length of the model, which would have minimal effect on the mismatch loss but alter the element values of the synthesized network. We also have the option of other element combinations or using excess loss as illustrated in the previous example. However, optimization is a powerful design tool and can often be used to successfully adjust the element values. The original matching network in Figure 6.13(a) is optimum as synthesized, and we proceed in steps to avoid deviating too far. The most significant problem appears to be the 291Ω shunt shorted stub which

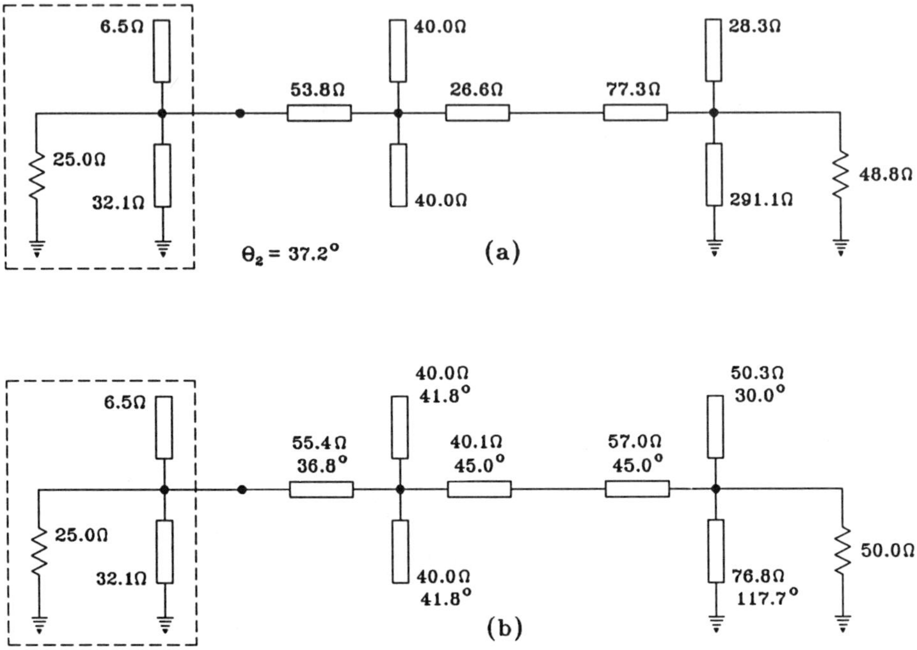

Figure 6.13 Load B matching network: (a) synthesized; (b) optimized.

is almost three times the upper impedance limit. We can replace this element with a 100Ω stub which is 90° at the center frequency or 120° at 2.0 GHz. The reactance is infinite at 1.5 GHz and ±j173 at the band edges. We allow the optimizer to adjust the characteristic impedance and electrical length, change the open stub to 40Ω, optimize the electrical length, then do the same with the 26.6Ω TLE. Next, we optimize only the electrical lengths, then all impedances, and. finally, both impedance and electrical length. Figure 6.13(b) shows the optimum values for characteristic impedance and electrical lengths measured at 2.0 GHz. All elements are realizable and, with the exception of the shorted stub, the electrical lengths are between 30° and 45° at 2.0 GHz. The 77Ω characteristic impedance of the shorted stub requires relatively narrow linewidths when fabricated on microstrip or stripline and is easily folded to conserve space. Thus, the 1.0- to 2.0-GHz matching network shown in Figure 6.13(b) provides a very compact physical structure. Both matching networks are terminated with the actual load in Figure 6.8 and the resulting insertion loss plotted in Figure 6.14. A comparison between the response of the original network, shown as the lighter curve, and the optimized circuit, shown as the darker curve, reveals almost identical performance.

Both networks have approximately 0.25-dB maximum insertion loss in the passband, and the optimized circuit shows slightly better rejection above and below the passband.

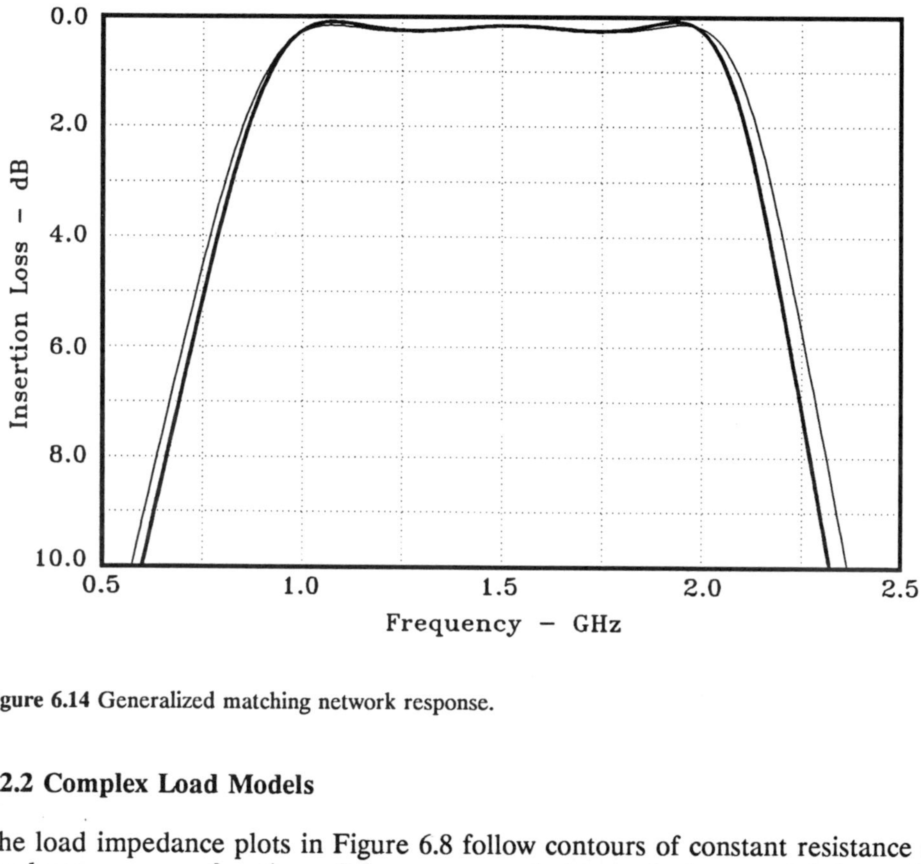

Figure 6.14 Generalized matching network response.

6.2.2 Complex Load Models

The load impedance plots in Figure 6.8 follow contours of constant resistance or conductance as a function of frequency, and the equivalent model is readily determined by inspection. However, most practical loads are combinations of many reactive and resistive elements, and the equivalent model may not be obvious from the impedance plot. For example, Figure 6.15 shows a plot of the load data listed in Table 6.6, which does not follow either a constant resistance or conductance contour. The table lists reflection coefficient, equivalent series impedance, and equivalent parallel impedance as a function of frequency. The series resistance changes from 20.4Ω to 36.2Ω and the parallel resistance from 88.8Ω to 148.9Ω across the 1.0- to 2.0-GHz frequency range, and the form of the load model is not apparent from the figure. We could increase the complexity of the load model.

However, the synthesis of matching networks becomes considerably more difficult. The simple two-element load model is adequate for most purposes if we carefully determine the optimum parameters.

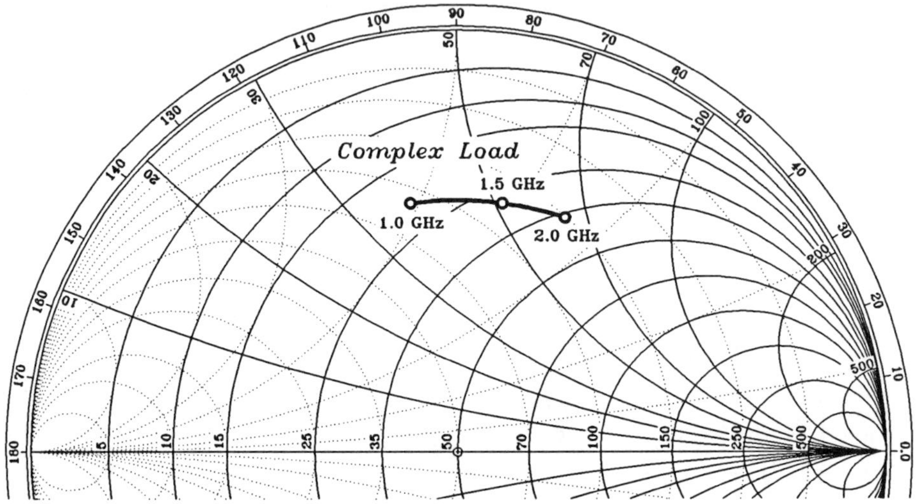

Figure 6.15 Complex load impedance versus frequency.

Table 6.6 Load Data for Figure 6.15

Frequency (GHz)	Γ_L	$Z_S \Omega$	$Z_P \Omega$
1.00	$0.598\angle 100.446°$	20.4 + j37.4	88.8 + j48.5
1.25	$0.596\angle 89.087°$	24.1 + j44.6	106.6 + j57.7
1.50	$0.598\angle 79.798°$	28.0 + j51.4	122.0 + j66.7
1.75	$0.603\angle 72.076°$	32.1 + j57.8	136.3 + j75.6
2.00	$0.609\angle 65.603°$	36.2 + j63.9	148.9 + j84.5

Most load models are simple combinations of two or three elements, and the goal is to minimize the total loss; that is, mismatch loss between the model and actual impedance plus the insertion loss of an ideal matching network. Hence, we can rapidly evaluate all possible models and select the best. The reactive components for the load listed in Table 6.6 are positive over the entire frequency range, and the lumped-element model is either a series or parallel R-L. We could also create R-L-C models by selecting the appropriate resonant frequency. However, using the resistance and reactance at 1.5 GHz, we compute element values for the series R-L model as R_S = 28.0Ω and L_S = 5.454 nH, and then the parallel model as R_P = 122.0Ω and L_P = 7.077 nH. Computing the maximum mismatch loss between the models and load data, we have 0.127 dB for the series model and 0.148 dB for the parallel model. The ideal MIL for the models are calculated from (6.5a) and (6.5c) as 0.026 dB for the series equivalent, and 0.148 dB for the parallel equivalent. Thus, the series model requires a total loss of 0.153 dB and the parallel model 0.193 dB. However, these values are not optimum. Starting with the initial element values calculated at 1.5 GHz, we can analyze the mismatch loss versus frequency to determine the maximum, calculate and add the ideal MIL, which provides the total, and then increment the element values until the optimum combination is found. The best series equivalent is R_S = 27.334Ω and L_S = 5.464 nH, resulting in 0.111 dB of mismatch with the actual load impedance, 0.029 dB MIL from an ideal matching network, and a total loss of 0.140 dB. The optimum parallel model is R_P = 116.089 and L_P = 7.376 nH, which has 0.087 dB MML, 0.029 dB MIL, and 0.116 dB total. Thus, the best choice is the parallel model, and we can synthesize the corresponding matching network. When the lumped-element bandpass approximation is used, the optimum ripple for n = 3 is 0.060 dB and the optimum MIL is 0.058 dB for a total loss of 0.118 dB. The synthesized matching network is shown in Figure 6.16.

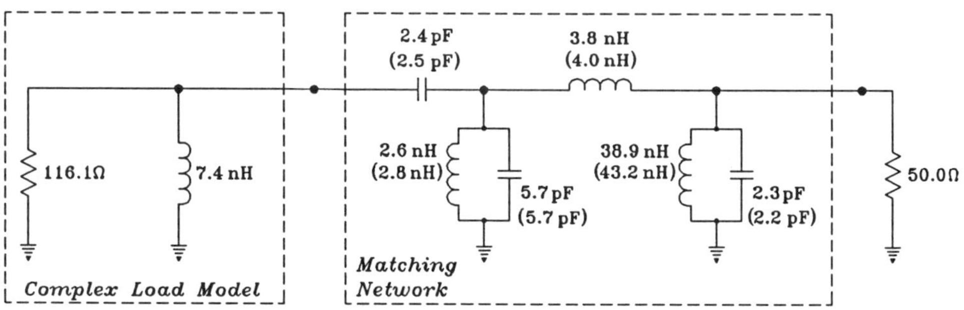

Figure 6.16 Complex load matching network.

We can replace the load model shown in Figure 6.16 with the actual load data from Table 6.6 and then analyze to determine the performance of this matching network. The response is shown as the lighter curve in Figure 6.17, and we see that the maximum loss is approximately 0.2 dB at 1.00 GHz and 1.75 GHz. The loss is greater than the 0.116 dB predicted for an ideal matching network, but almost exactly the 0.205 dB computed by adding the load MML with the ripple and MIL used for the synthesis. The load model mismatch loss exhibits passband ripples, as does the matching network response. We can improve the performance somewhat by optimizing the matching network such that the ripples cancel or at least shift to improve the maximum loss. Optimized element values are shown enclosed by parentheses in Figure 6.16 and the associated response is the darker plot in Figure 6.17. The optimized element values are changed only slightly from the original, but the maximum insertion loss is reduced to 0.1 dB.

It is instructive to examine the input impedance of the synthesized matching network. Figure 6.18 shows s_{11} for the original and optimized network plotted on a Smith chart, with the conjugate load data. The optimized input

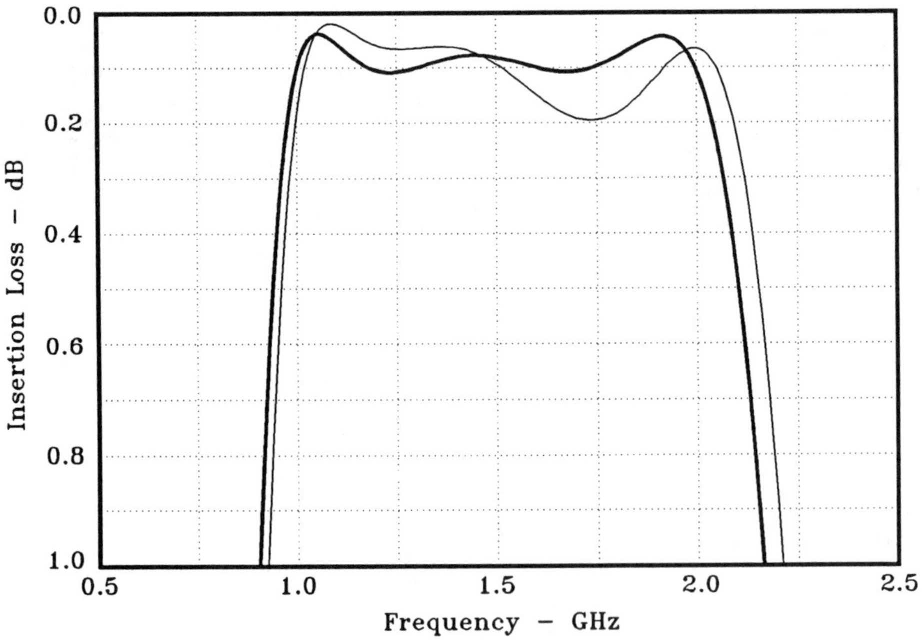

Figure 6.17 Complex load matching network response.

impedance shown in Figure 6.18(b) appears to be better centered around the conjugate load data. However, it is not obvious from either the impedance or reflection coefficient that the insertion loss is equal ripple. When we use the load impedance data as a reference for computing s_{11} of the matching network and then plot the results as shown in Figure 6.19, the behavior is apparent. The synthesized response appears to be shifted slightly high in frequency, but the optimized response is centered about the origin and has two minimum and four maximum points corresponding to the insertion loss ripples shown in Figure 6.17.

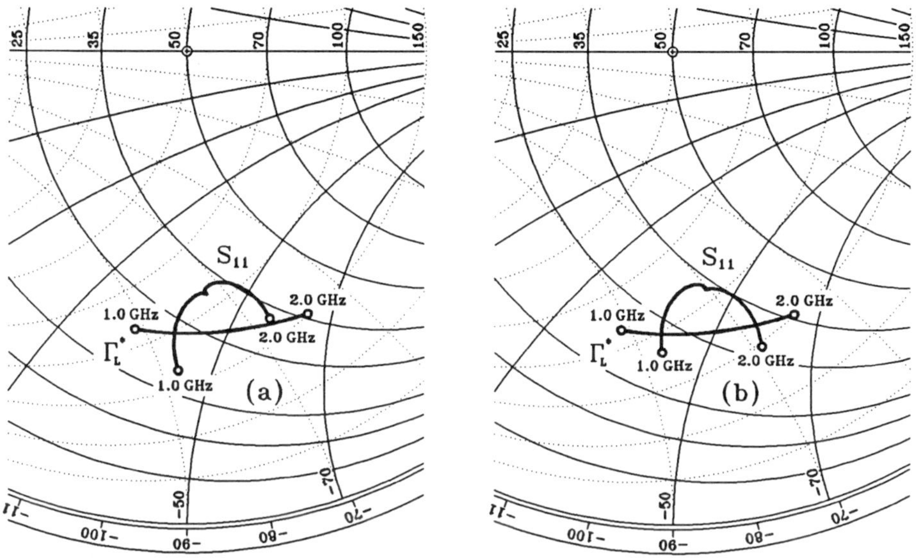

Figure 6.18 Conjugate load and matching network input impedance: (a) synthesized; (b) optimized.

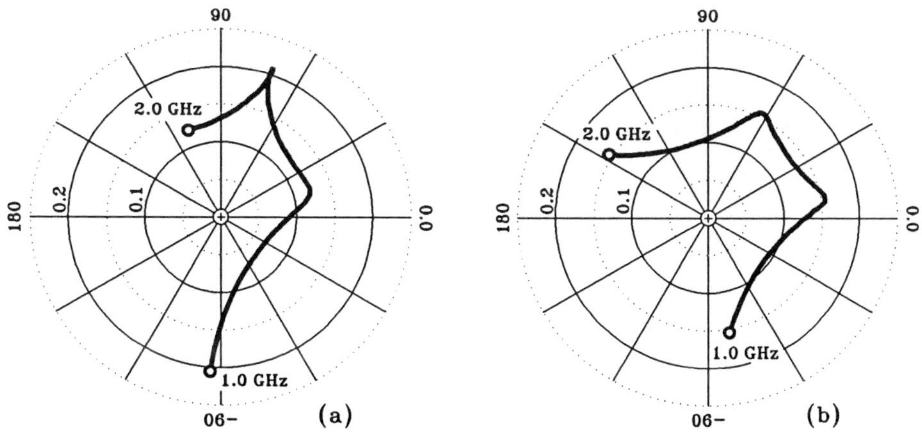

Figure 6.19 Matching network s_{11} referenced to load impedance (a) synthesized; (b) optimized.

6.2.3 Negative-Image Models

The preceding examples clearly demonstrate the advantages of load modeling and matching network synthesis. Given impedance data, it is a relatively simple matter to determine the optimum load model that results in the most efficient matching network. We can easily compute the MML between the load model and impedance data and then add the MIL from an ideal matching network to predict the efficiency of the circuit before it is designed. Once the network is synthesized, we can apply optimization to improve the response although, the circuit is usually optimum as synthesized when the model accurately fits the data. However, in many instances the load impedance is not fixed and we are faced with the task of defining an impedance that provides the specified circuit performance, as well as accurately modeling the load and synthesizing a low-loss matching network.

Figure 6.20 shows the common configuration for a two-port amplifier; that is, input matching network, active device, and output matching network. We can draw similar diagrams for a number of components, including mixers, multipliers, detectors, and antenna interface networks. In each case, the problem is essentially the same: design a matching network that provides a specified impedance at one or more ports. For example, when the active device in Figure 6.20 is unconditionally stable, we can use (2.52) to compute Γ_{MS} and Γ_{ML} for simultaneous input and output match, model the associated load impedances, and then synthesize matching networks. However, G_{MAX} decreases with increasing frequency, and an amplifier designed in this manner usually exhibits a gain slope. When the device is potentially unstable or the design specifications include noise or power, we must construct the appropriate contours and then define the reflection coefficients based on the behavior of several parameters.

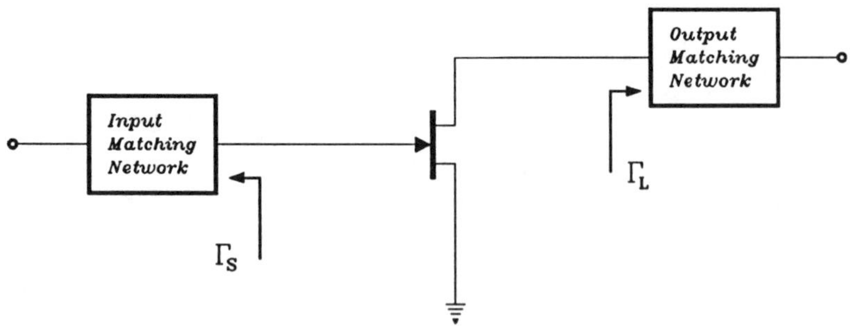

Figure 6.20 Common amplifier configuration.

One approach to the problem of determining load models for matching network design is to assume the device is unilateral (i.e., $s_{12} = 0$) and then model

s_{11} and s_{22} as a resistor and reactive element [7-11]. When designing for noise, we use Γ_0, the reflection coefficient for optimum noise, as the input model. Similarly, the output load can be modeled using the impedance that provides maximum output power. However, this approach does not account for finite s_{12}, which can result in large errors, especially at high frequencies where the magnitude is significant. The gain slope versus frequency is not considered, and, in addition, matching into a potentially unstable device can cause serious stability problems. In fact, it is not necessary, or desirable, to model the device. Instead, the task is to find the impedances that provide specified circuit performance and then design appropriate matching networks [12], as illustrated in Figure 6.21.

Figure 6.21(a) shows the ideal situation at one port of an n-port network. The reflection coefficient Γ is selected to provide the termination necessary for the specified circuit performance. We can postulate an ideal matching network capable of transforming Z_0 into any Γ at all frequencies. The design task is one of replacing the ideal matching network with a circuit that has the same port characteristics as shown in Figure 6.21(b). That is, the actual matching network also transforms Z_0 into Γ, at least across the desired frequency range. Since the ideal matching network is perfect and has an output reflection coefficient of Γ, it exactly matches

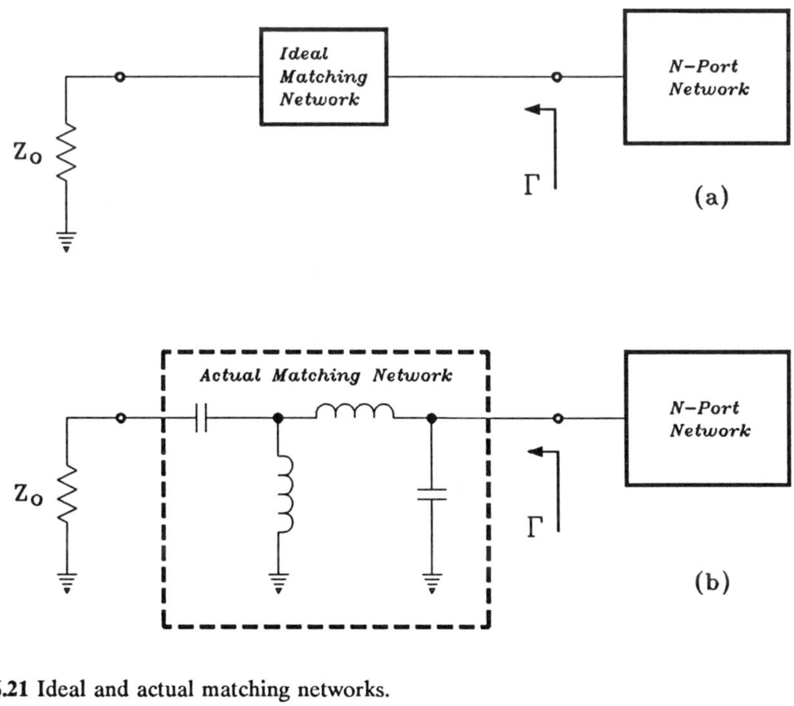

Figure 6.21 Ideal and actual matching networks.

an ideal load with a reflection coefficient of Γ^*, as shown in Figure 6.22(a). The ideal load is *not* the port impedance of the *n*-port network. It is strictly the complex conjugate output impedance of the ideal matching network when the input is terminated with Z_0. The actual matching network has the same output impedance as the ideal network and must also match the ideal load over a limited frequency range, as shown in Figure 6.22(b). Matching network synthesis requires that the load be represented as a resistive termination and one or more reactive elements that can be absorbed from a synthesized network. Thus, the ideal load associated with the actual matching network in Figure 6.22(b) must be constrained to a particular topology with a finite number of elements.

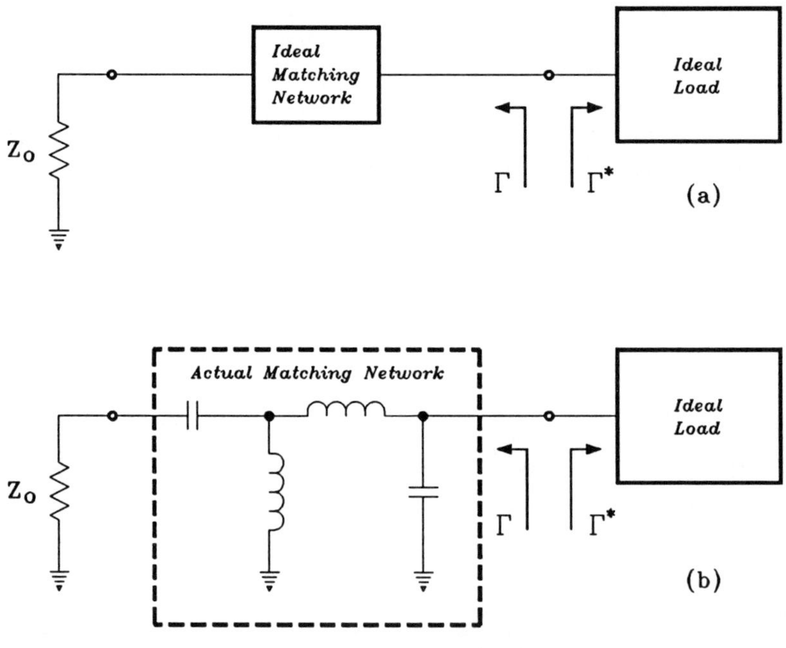

Figure 6.22 Matching networks terminated with ideal loads.

Figure 6.23(a) shows the actual matching network terminated with a load model consisting of elements absorbed from the original synthesized network. Assuming the ripple and MIL are almost zero, the output impedance of the matching network equals the complex conjugate of the load impedance. We can replace the actual matching network with an ideal network that must exactly match the same load, but at all frequencies. One topology that meets these requirements is shown in Figure 6.23(b). The network consists of an ideal transformer and two reactive elements with negative values. Examining this network, we see that positive series inductive reactance exactly cancels the negative series inductive reactance at all frequencies. Similarly, the positive and negative shunt capacitive susceptances cancel. Thus, we can eliminate the reactive elements, and the equivalent matching network consists of an ideal transformer, terminated at the input by Z_0 and the output by R, which provides a perfect match at all frequencies when the appropriate turns ratio is selected. The ideal matching network is the mirror image of the load topology, with positive reactive element values replaced by corresponding negative values and an ideal transformer used to transform Z_0 to R. This circuit is known as a *negative-image matching network* or *model* [13,14] and has some useful properties when applied to matching network synthesis.

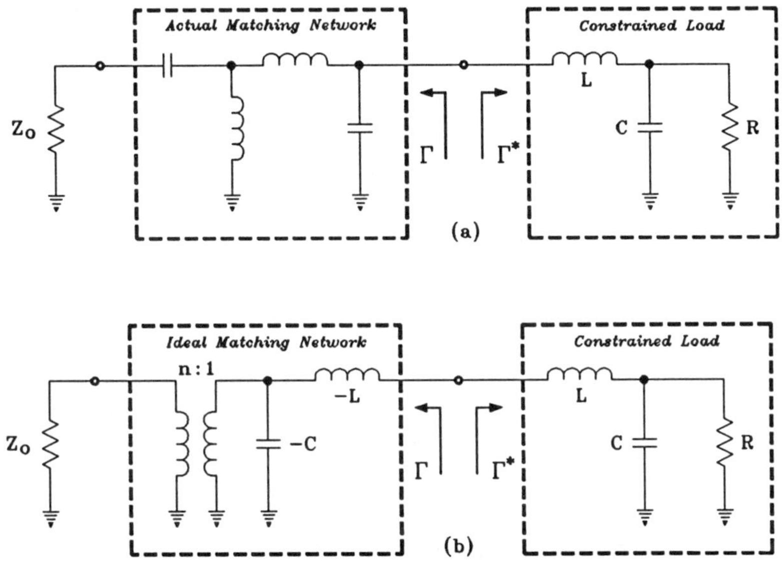

Figure 6.23 Matching networks terminated with constrained loads.

The negative-image matching network has the same characteristics as an actual matching network. That is, they both transform Z_0 into Γ across a specified frequency range. However, the negative-image network is based on the topology and element values of an actual load, and Γ is constrained accordingly. For example, Figure 6.24(a) shows the negative-image model for Load A, and 6.24(b) shows the model for Load B from Section 6.2. Transformer primary and secondary impedance values are shown in the figures instead of the usual turns ratio to emphasize the function of the ideal transformers. When necessary, we can compute the turns ratio from the impedances as described in Section 2.1.3. The output reflection coefficients of both networks are the exact conjugate of the load data plotted in Figure 6.8, and we have 0-dB mismatch loss. When the load

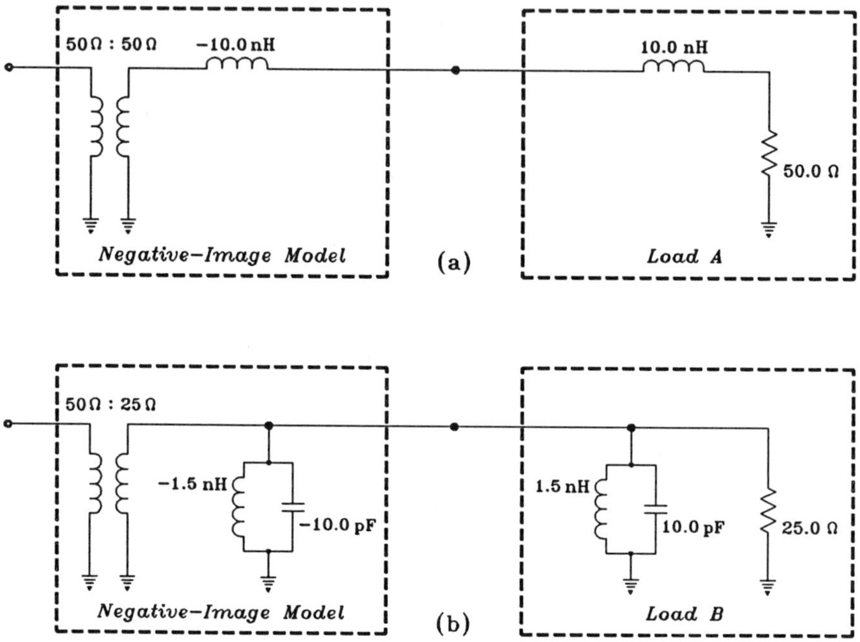

Figure 6.24 Negative-image models: (a) Load A; (b) Load B.

impedance is not well behaved as illustrated by the complex load plotted in Figure 6.15, we can try several ideal matching networks and compare the relative performance. Figure 6.25(a) shows the series equivalent negative-image model using values derived in Section 6.2.2, and Figure 6.25(b) shows the parallel equivalent. The computer model for these networks is extremely simple: an ideal transformer cascaded with a negative element and then terminated with measured load data. We have two optimization variables (i.e., the transformer turns ratio and the inductor value) and can easily optimize the one-port networks to minimize the input reflection coefficient. However, a better approach is to model the negative-image network as a two-port and reference the output port to the measured load data. Using this technique, we can directly calculate the insertion loss, which, in fact, is the mismatch loss between the ideal matching network and actual load. Consequently, minimizing the insertion loss provides a best fit to the load data. The ideal matching network represents the negative-image of the load model that will be used to synthesize a matching network. Therefore, we can use (6.5) or (6.8) to compute the MIL associated with the ideal matching network from the turns ratio and positive element values. Hence, the negative image technique provides a complete description of the required matching network: accurate load model, mismatch loss between the model and actual load data, and MIL required to exactly absorb the load using an ideal matching network.

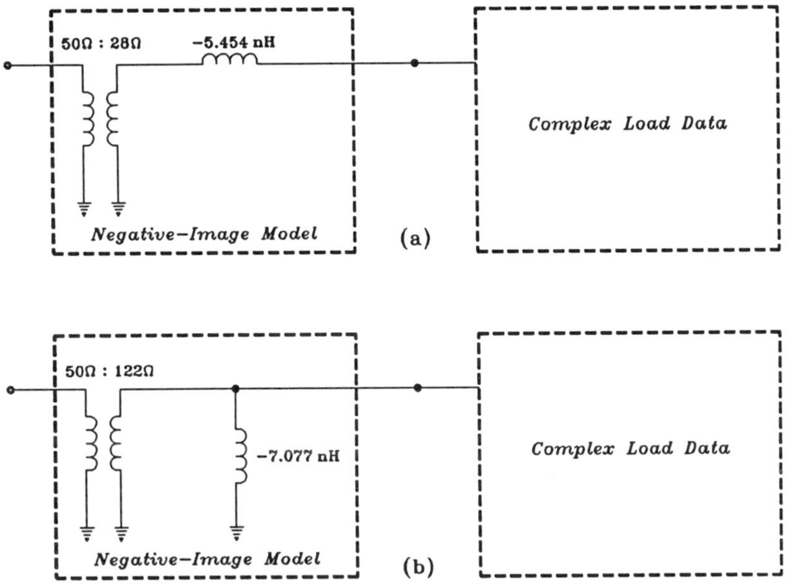

Figure 6.25 Complex load models: (a) series; (b) parallel.

The negative-image matching networks can be used as ordinary two-port elements. That is, we create a computer model of the circuit and place ideal matching networks as needed to provide the required circuit performance. For example, the input and output matching networks shown in Figure 6.20 can initially be modeled using negative-image models. We define optimization goals to conform with the design specifications (i.e., input and output return loss, gain, noise figure, etc.) and then optimize the negative-image parameters. Circuit analysis provides the complete response of the network and we need not continue if the specifications are not met. However, assuming the optimization is successful, we can examine the ideal MIL from each negative-image model to ensure the matching networks can be synthesized, and then design actual matching networks to replace the ideal networks. When the ripple and MIL required to synthesize each matching network are low, the response of the complete circuit will very closely match that provided by the negative-image models.

Amplifier Design Using Negative-Image Matching Networks

The advantages of using negative-images models can best be illustrated with a practical design example. We will use the low-cost NEC NE70000 MESFET chip biased at 3V drain-to-source voltage and 10 mA drain current to design a 4.0- to 8.0-GHz amplifier with 10.0 ± 0.5 dB gain. The *s*-parameters are readily available from the manufacturer's data sheets or the data library provided with most RF and microwave computer-aided design (CAD) software. Table 6.7 lists the *s*-parameters from 2.0 to 10 GHz, and we can apply the methods of Section 2.3 to characterize the device.

Table 6.7 NEC NE70000 MESFET Chip Biased at V_{DS} = 3V and I_{DS} = 10 mA

Frequency (GHz)	S_{11}	S_{21}	S_{12}	S_{22}
2.00	$0.97\angle{-24°}$	$3.02\angle{159°}$	$0.04\angle{77°}$	$0.71\angle{-14°}$
3.00	$0.94\angle{-36°}$	$2.92\angle{149°}$	$0.06\angle{69°}$	$0.70\angle{-20°}$
4.00	$0.91\angle{-48°}$	$2.75\angle{138°}$	$0.07\angle{63°}$	$0.69\angle{-26°}$
5.00	$0.87\angle{-57°}$	$2.64\angle{130°}$	$0.08\angle{57°}$	$0.66\angle{-31°}$
6.00	$0.85\angle{-66°}$	$2.48\angle{123°}$	$0.09\angle{55°}$	$0.66\angle{-36°}$
7.00	$0.81\angle{-74°}$	$2.37\angle{115°}$	$0.10\angle{50°}$	$0.62\angle{-40°}$
8.00	$0.78\angle{-81°}$	$2.23\angle{109°}$	$0.11\angle{45°}$	$0.60\angle{-43°}$
9.00	$0.76\angle{-89°}$	$2.07\angle{103°}$	$0.11\angle{43°}$	$0.59\angle{-47°}$
10.00	$0.73\angle{-96°}$	$1.97\angle{97°}$	$0.12\angle{40°}$	$0.61\angle{-51°}$

Table 6.8 lists the results of the analysis, which indicates that the device is potentially unstable and the maximum stable gain ranges from 16 dB at the low end of the band to 13 dB at the upper end. Since gain is the only specification, we could stabilize the device and then design matching networks without regard for the source and load impedance. However, when noise and output power are design considerations, the use of resistive elements degrades the performance and is usually avoided. Therefore, to fully demonstrate the advantages of negative-image models, the device will not be stabilized, but we will require positive real input and output impedances in a 50Ω system. The maximum stable

gain is 13 dB at 8 GHz, and we can use the excess gain to provide some design flexibility. The design approach will be to select an input impedance from the 11.5-dB constant gain circles, compute s'_{22}, and then select an output impedance from the 1.5-dB mismatch circles and synthesize appropriate matching networks.

Table 6.8 NEC NE70000 Chip, G_{MSG} and K

Frequency (GHz)	G_{MSG} *(dB)*	K
4.00	15.942	0.275
5.00	15.185	0.380
6.00	14.402	0.365
7.00	13.747	0.487
8.00	13.069	0.568

Figure 6.26 shows the input plane stability circles at 4.0, 6.0, and 8.0 GHz, along with the 11.5- and 12.0-dB constant gain circles. The 11.5-dB gain circles are darker than the 12.0-dB, circles and separation between the two indicates the relative sensitivity of gain to changes in source impedance. Ideally, we would select Z_S from the less sensitive gain regions and outside the stability circles. However, we must also consider the behavior as a function of frequency. The input plane impedance must rotate *counterclockwise* with increasing frequency. Z_S or Γ_S represents the termination presented to the input port and is the complex conjugate of an actual impedance. The impedance plot of positive valued reactive elements rotates clockwise on a Smith chart, and the complex conjugate must rotate in the opposite direction. The selection of Z_S has a significant impact on the circuit performance, as well as the synthesized matching networks. For example, if the impedance falls inside the unstable region, the output impedance will be negative real. The impedance must also intersect the desired constant gain circle at each frequency. However, the intersection must be in a similar location at each frequency. That is, if we select a point from one side of the constant gain circle at any frequency, other frequencies must also intersect at the same side. The Z_S plot in Figure 6.26 shows the impedance selected from the left side of the 4.0-, 6.0-, and 8.0-GHz, 11.5-dB constant gain circle. Had we used another point on the 4.0-GHz gain circle, 100Ω - j500Ω for example, the amplifier would have 11.5 dB gain for matched output at the three frequencies, but exhibit a gain peak between 4.0 and 6.0 GHz. This fact is easily demonstrated by constructing numerous constant gain circles between 4.0 and 6.0 GHz. Since the 6.0-GHz impedance is on the left side of the gain circle and the 4.0-GHz point on the right, somewhere in between the impedance must be inside the circles, resulting in much higher gains. However,

the location of the Z_S plot is consistent and the gain at each frequency between 4.0 and 8.0 GHz is approximately the same.

Examination of the constant gain and stability circles in Figure 6.26 reveals that we have an almost infinite number of choices for Z_S. However, it turns out that the matching networks are easier to realize when we apply certain guidelines. The impedance plot should follow either contours of constant resistance or conductance, which simplifies the negative-image model and matching network

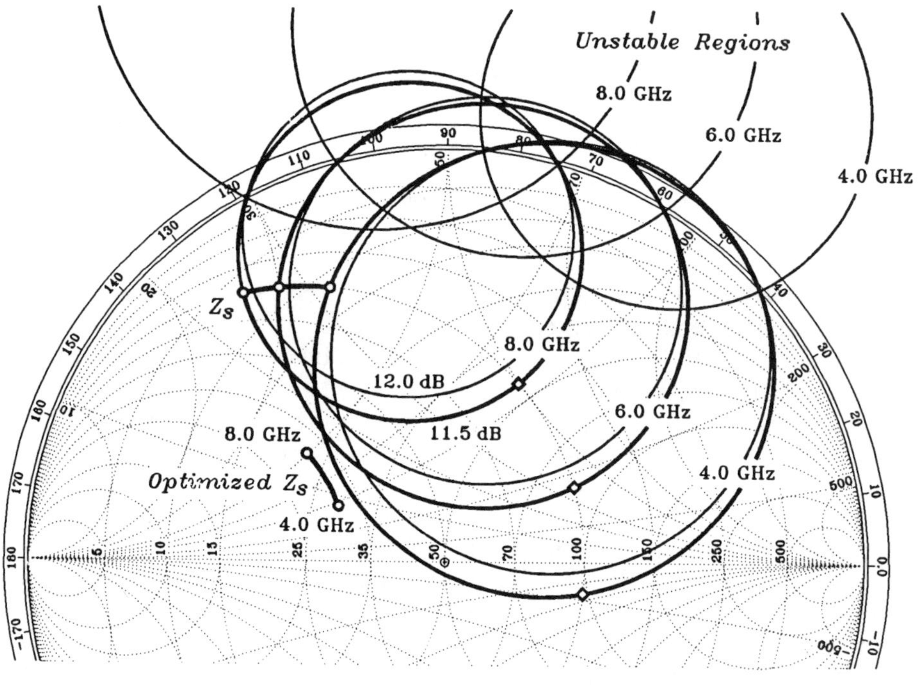

Figure 6.26 Input plane constant gain, stability and source impedance.

synthesis; that is, we have a simple series or parallel equivalent load. In addition, the total rotation of the impedance plot indicates sensitivity of the reactive component to frequency, which can be expressed as a Q-factor. Higher Q loads require increased parasitic absorption capability. Thus, we should minimize the change in reactance, which lowers the Q and improves the efficiency of the matching networks. The synthesized network is referenced to the real part of the load, and we should choose a series or parallel equivalent close to Z_O, which will minimize necessary Norton or Kuroda transformations, as well as provide more convenient element values for distributed networks. An impedance plot that crosses the real axis as illustrated by Load B in Figure 6.8 requires a resonant model (i.e., L-C or a quarter-wavelength distributed element). In the case of a distributed network, we may be required to change the bandwidth because of symmetry of the response, which usually results in increased matching network loss. Also, element values of a distributed network are affected by the electrical length. Angles between 30° and 60° usually provide the most realizable element values. Thus, we can summarize the criteria for selecting a broadband impedance from contours of constant gain or mismatch as follows:

- Z must rotate counterclockwise with increasing frequency.
- Z must intersect the constant gain or mismatch circle at the same relative location at each frequency.
- Z must follow contours of constant resistance for series equivalent models or constant conductance for parallel equivalent models.
- The impedance must be restricted to the stable region.
- The total change in reactance or susceptance should be minimized to maximize the efficiency of the matching network. That is, the spread of Z or Γ over frequency should be minimized.
- The real part of the series or parallel equivalent should be as close to Z_O as practical. Extreme high or low values should be avoided.
- The electrical lengths of distributed models should be between 30° and 60°, if possible. The center frequency of resonant models should be as close as possible to the design center frequency.

We have the choice of either a series or parallel equivalent input model, as shown in Figure 6.26. The series model requires an impedance less than 40Ω, otherwise we cannot intersect the 8.0-GHz, 11.5-dB gain circle. The 35Ω constant resistance circle intersects each gain circle twice, once close to the real axis, and again close to the stability circles. However, if we select the three points close to the real axis, Z_S rotates clockwise with increasing frequency. The other choice rotates in the desired direction, although the impedance is close to the unstable region and rotates through about 35° across frequency. Therefore, we elect to use a parallel equivalent input model. The 10-mmho or 100Ω constant conductance contour also intersects each circle twice. The plot labeled Z_S is selected as the input impedance model. Z_S rotates through about 15° and is in the stable region. The second choice for the 100Ω parallel model, illustrated by the three diamonds in Figure 6.26, is resonant at approximately 4.5 GHz and spread over 70°. At 6.0

GHz, the parallel equivalent is 100.00Ω + j30.08Ω, and the load is the complex conjugate, or a capacitive reactance of 30.08Ω. The high operating frequency dictates that we use distributed elements. Thus, the input model is a 100Ω resistor in parallel with an open stub. We will use optimization to adjust the final values of the negative-image models and can compute initial element values based on an electrical length in the middle of the desirable range (i.e., $\theta_2 = 45°$). At 6.0 GHz, $\theta = 33.75°$ and $Z_{OP} = 20.1\Omega$ for a reactance of -j30.08Ω.

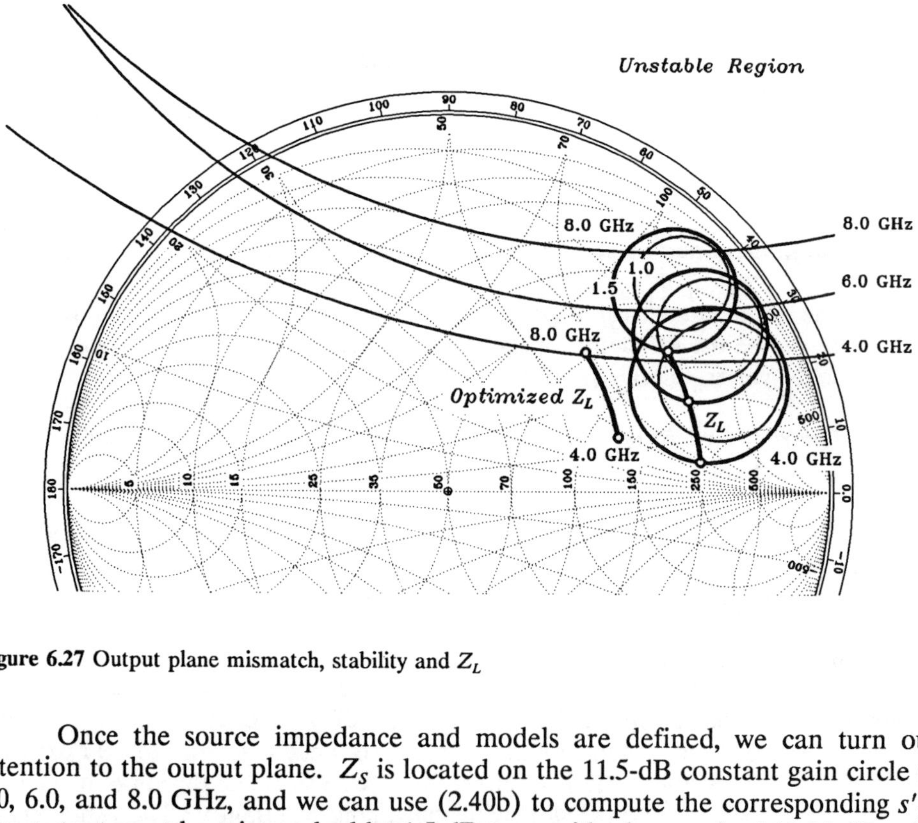

Figure 6.27 Output plane mismatch, stability and Z_L

Once the source impedance and models are defined, we can turn our attention to the output plane. Z_S is located on the 11.5-dB constant gain circle at 4.0, 6.0, and 8.0 GHz, and we can use (2.40b) to compute the corresponding s'_{22}. The output must be mismatched by 1.5 dB to provide the required 10.00-dB gain, and Figure 6.27 shows the 4.0-, 6.0-, and 8.0-GHz, 1.5- and 1.0-dB mismatch circles, along with the corresponding stability circles. We can select the output impedance in exactly the same manner as the input and have a choice of either a series or parallel model. The series resistance must be less than approximately 85Ω to intersect the 8.0-GHz circle. However, this places the 4.0-GHz point in a highly sensitive area where the 1.0- and 1.5-dB circles are in close proximity and also near the unstable region. A better choice is the parallel equivalent shown as Z_L in the figure. The 250Ω resistance is higher than desired, but the impedance is in the less sensitive region and as far as possible from the stability circles. The 6.0-GHz parallel equivalent is 250Ω + j280.50Ω, and we follow the same procedure as that of the input model to compute Z_{OP} = 187.4Ω, with θ_2 = 45°.

Figure 6.28 shows the complete negative-image amplifier model, consisting of the ideal input matching network, FET, and ideal output matching network. We have six optimization variables: the transformer turns ratio, negative Z_{OP}, and θ_2 for each matching network. We could simply set the goal at 10.00 dB gain and then optimize. However, by adding additional constraints and goals, we can ensure that all specifications are met and the matching networks will be realizable. We require that the input and output impedance be positive real and can specify that condition as $|s_{11}| < 1$ and $|s_{22}| < 1$. In addition, the ideal MIL computed from (6.8b) and (6.3) for each model should be as low as possible so that we can synthesize accurate matching networks. Thus, we add an optimization goal for each model requiring the ideal MIL to be less than 0.05 dB. We selected 100Ω for the input model and observe from Figure 6.26 that larger values cause Z_S to move in the direction of the unstable region. Therefore, the input model resistance is constrained between 25Ω and 100Ω, or one-half and twice Z_O. The output resistance is 250Ω, which is higher than desired, but necessary, and we constrain this value between 50Ω and 250Ω. We also limit θ_2 between 30° and 60° for both networks and then optimize.

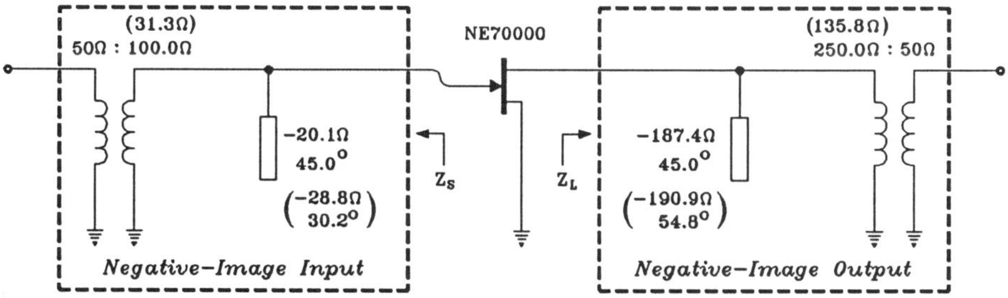

Figure 6.28 Negative-image amplifier model. Optimized values shown in parentheses.

Figure 6.29 shows the gain response of the initial matching networks as the lighter trace. The gain exhibits a slope from 13 dB at the low end of the band to 7 dB at the high end and is 10 dB at 6 GHz, where we computed the element values. The darker trace is the response after optimization, and Figure 6.28 shows the final values in parentheses. The optimized gain is 10.0 ± 0.2 dB, well within specifications, and the ideal MIL for both networks is less than 0.01 dB, which indicates that we should be able to design accurate matching networks. It is interesting to compare the optimized input and output impedances with our original values. Figure 6.26 shows that the optimizer has reduced the input plane gain to approximately 10.5 dB, and that Z_S has moved away from the unstable region. We cannot use the output plane mismatch circles shown in Figure 6.27 because they were computed from the original s'_{22}. However, Z_L must represent a mismatch of about 0.5 dB to compensate for the 10.5-dB input plane gain. We also observe that the resistance and spread over frequency has been reduced.

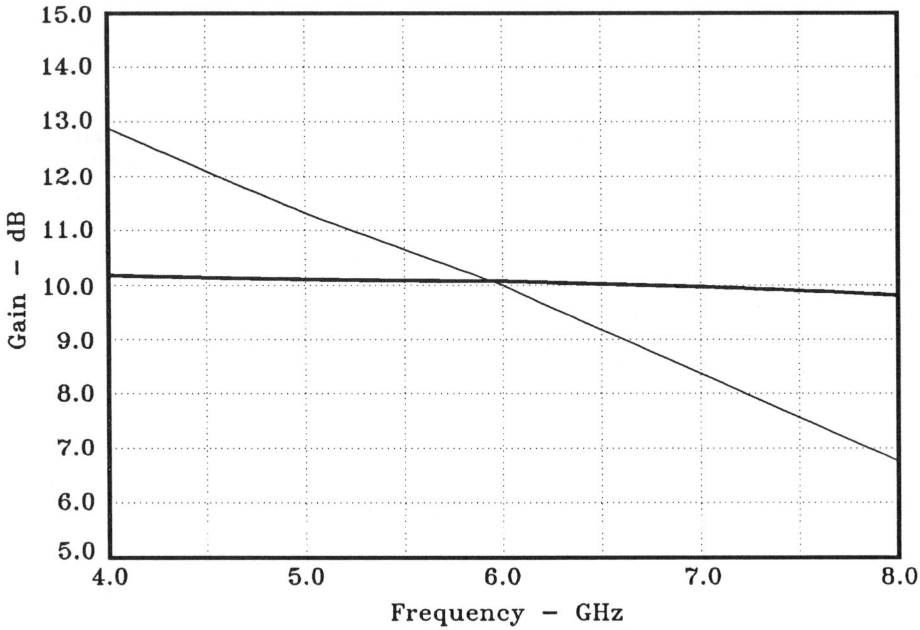

Figure 6.29 Amplifier response with ideal matching networks.

We must now synthesize practical matching networks to replace the ideal ones shown in Figure 6.28. The ideal matching networks exactly match loads that are positive images, and if we can match to the same loads with minimum loss, the synthesized circuits will provide the same impedance characteristics. Figure 6.30 shows the input and output loads formed from the optimized negative-image models of Figure 6.28. The electrical length of both models is less than 90°, and we can use the generalized distributed bandpass approximation from Section 5.3.4. We let $n = 2$, $q = 1$, and $m = 1$ and then compute the optimum loss as 0.003 dB for the input model shown in Figure 6.30(a). The loss provides exact absorption for the 28.8Ω open stub, although the termination cannot be transformed to 50Ω.

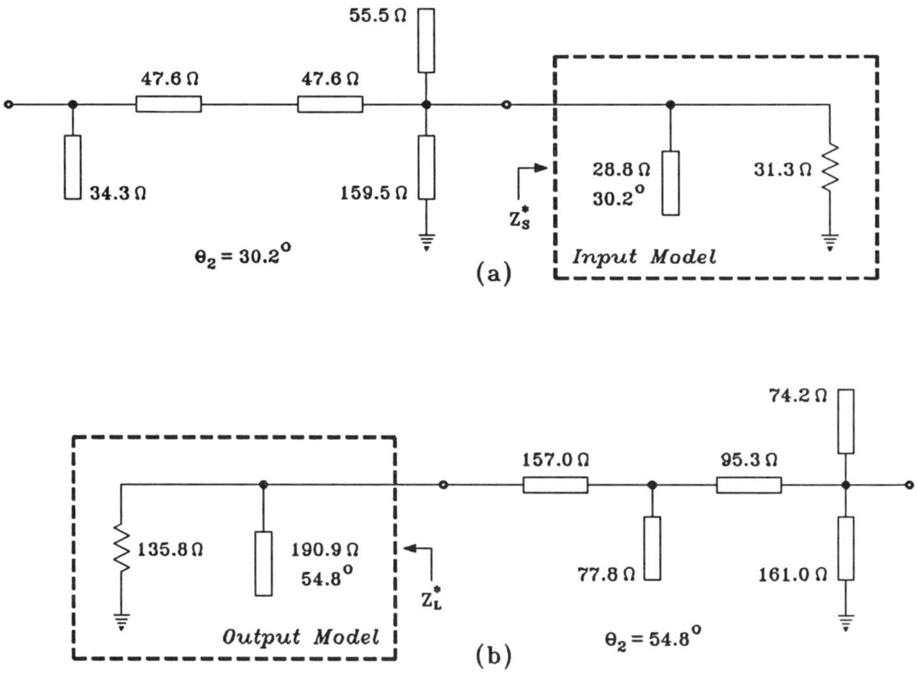

Figure 6.30 Matching networks and models: (a) input; (b) output.

However, we can increase the ripple to 0.02651 dB and set the MIL to 0, which results in excess absorption capability, but does have a 50Ω termination. Figure 6.30(a) shows the synthesized input matching network, where a portion of the open stub remaining after removing the parasitic value has been used to provide a double-length TLE. The matching network for the output load shown in Figure 6.30(b) is designed in the same manner as the input. The optimum loss is 0.03 dB using $n = 2$, $q = 1$, and $m = 1$. The output termination also presents a problem transforming to 50Ω, and increasing the ripple and MIL results in inconvenient element values. Therefore, we change the elements to $n = 2$, $q = 3$, and $m = 1$, compute the optimum loss as 0.007 dB, and then synthesize the matching network shown in Figure 6.30(b).

We can verify the accuracy of the design thus far by replacing the ideal matching networks in Figure 6.28 with the actual networks shown in Figure 6.30. The darker plot in Figure 6.31 shows that the amplifier gain is 10.0 ± 0.5 dB, without any optimization other than the negative-image models. The lighter traces represent the insertion loss of the matching networks operating into the

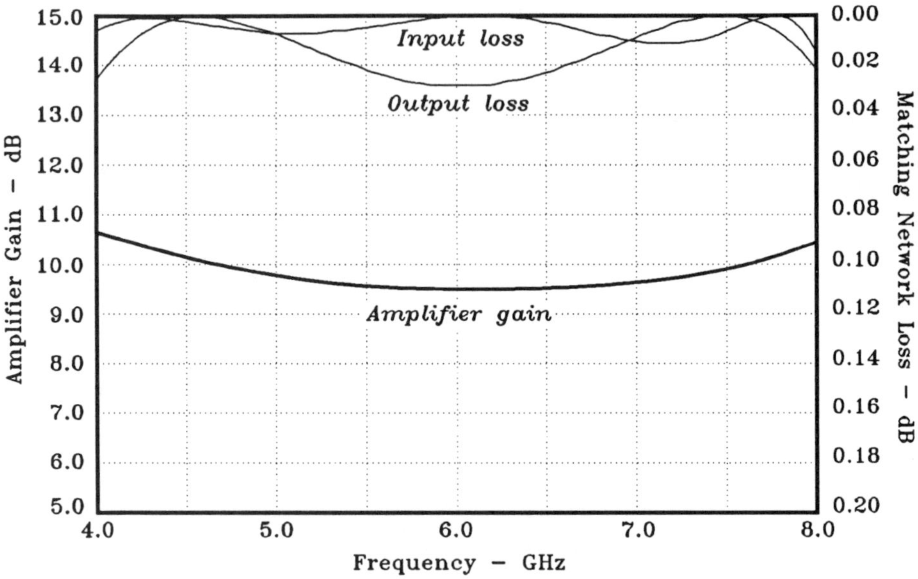

Figure 6.31 Amplifier and synthesized matching network response.

input and output loads of Figure 6.30. The scale on the right shows that the maximum loss is approximately 0.03 dB across the 4.0- to 8.0-GHz frequency range, and the minimum input and output matching network loss occurs at 4.5 and 7.5 GHz, where the corresponding amplifier gain is very close to 10.0 dB. Unfortunately, we cannot use the matching networks without modification. The input network has a shunt shorted stub that is too high in impedance, and the output matching network has a TLE and shorted shunt stub whose impedances are also too high. However, one of the advantages of the negative-image technique is that we need only be concerned with matching into the appropriate loads. Thus, we can constrain the element values and individually minimize the insertion loss of the two networks in Figure 6.30 when terminated with the positive load models. Essentially, the optimization problem for the overall amplifier is reduced by one-half and the task is greatly simplified.

Constraining the element values between 40Ω and 100Ω and then optimizing the individual matching networks in Figure 6.30 result in a maximum insertion loss of 0.03 dB for the input network and 0.07 dB for the output network. The lighter curve in Figure 6.32 shows the amplifier response using the constrained matching networks. The gain ripples between 9.0 and 10.6 dB across the 4.0- to 8.0-GHz frequency range. The ripple is primarily due to the insertion loss of the output network and is somewhat higher than the 0.5-dB design specification. However, we can optimize the complete amplifier and use the interaction between the input and output matching networks to minimize the ripple. The darker plot in Figure 6.32 shows the results of the final optimization. The amplifier gain is 10.00 $\pm$ 0.01 dB from 4.0 to 8.0 GHz, and 10.00 $\pm$ 0.08 dB from 4.0 to 9.0 GHz! The complete amplifier is shown in Figure 6.33, and an analysis reveals that all specifications have been met except $|s_{11}|$, which is slightly greater than one below 2.3 GHz. The reason for this problem is apparent from Figure 6.27, which shows that the output plane stability circles enclose more of the positive real region at lower frequencies. However, we can correct this situation by terminating the input shorted shunt stub with a parallel R-C, as shown in Figure 6.33. At the operating frequency, the 10-pF capacitor approximates a short, and at lower frequencies, the resistor dominates and reduces $|s_{11}|$ below one. The figure also shows that the output shorted stub is RF grounded through a bypass capacitor and the drain bias injected at the low impedance point.

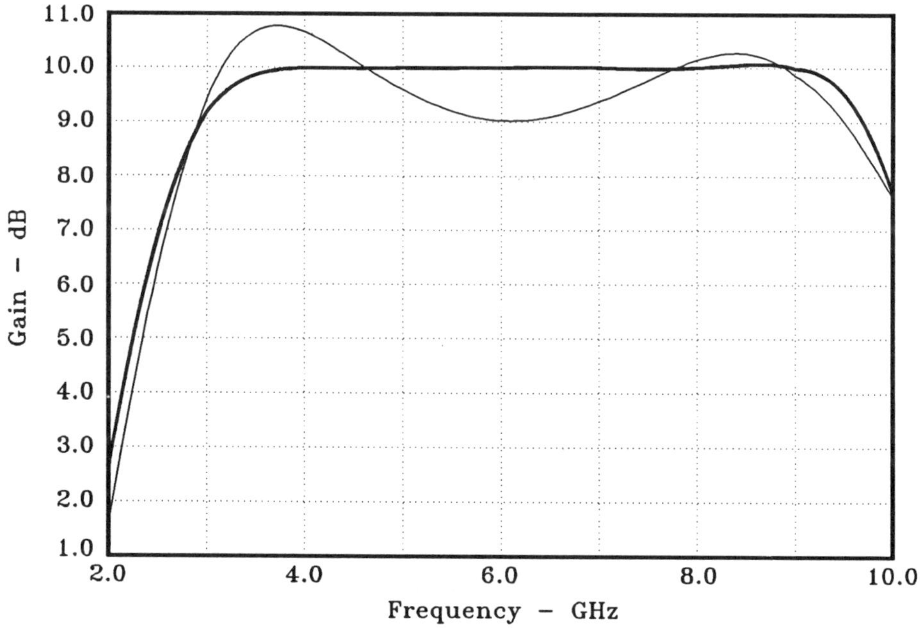

Figure 6.32 Final 4.0- to 8.0-GHz amplifier response.

This example demonstrates the advantages of using the negative-image design approach in conjunction with matching network synthesis. We can analyze various device parameters, assume a topology and initial element values for ideal matching networks, and then optimize a simplified circuit using all performance specifications, as well as goals and constraints affecting the realizability of the actual matching networks. We immediately know the predicted circuit performance and have a good idea of the requirements for the matching networks. The method is completely general, can be used whenever a matching network is required, and is simple to apply. In fact, we can easily try all combinations of matching network topologies and then select the optimum. Once the ideal matching networks are defined, we automatically have the required synthesis load models. Matching networks can be synthesized, constrained, and optimized using only the load models. If necessary, we can even transform to physical parameters, add parasitics and bias networks, and then analyze or optimize, still using the load models. Once the individual matching networks are finalized, we can model the complete circuit. The response of the complete circuit depends on the behavior of the individual matching networks. When each accurately matches the corresponding load, the response will be identical to that of the original model using ideal matching networks.

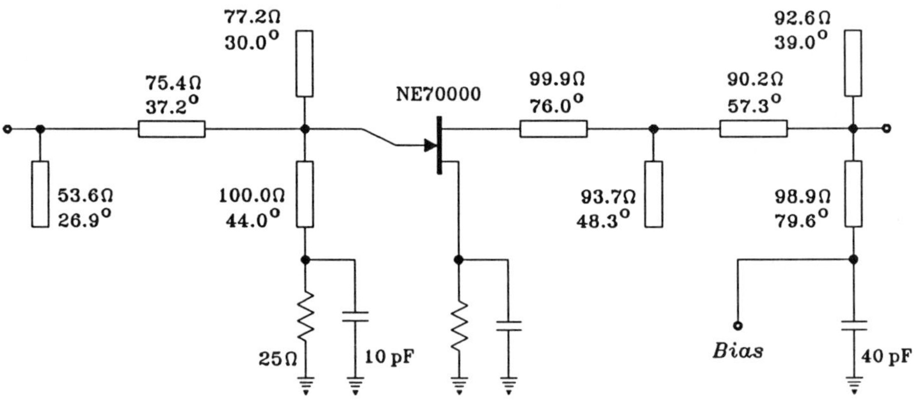

Figure 6.33 Final 4.0- to 8.0- GHz amplifier.

6.3 DOUBLY TERMINATED MATCHING NETWORKS

The most common matching problem is that of transforming a real impedance Z_O into a complex impedance Z. However, some applications require complex impedances at both ends of the matching network (i.e., a match between a complex source and complex load). The interstage matching network between the output of one amplifier stage and the input of the next is one example. One approach is to match both complex loads into a common real impedance and then combine the networks. When the ripple and MIL of both networks approach zero, the real output impedance of one network provides a constant input impedance for the other network and the combination accurately matches the two loads. However, as the loss of either network increases, the common termination is not constant, and significant errors are introduced when the networks are cascaded; that is, the termination is not the synthesized value. The real impedance can be determined by synthesizing both networks without regard for the output termination and then transforming either network into the impedance of the other. When the Norton or Kuroda transformations are not sufficient, we can use the maximum available transformation from one network, then adjust the other. In some cases it may be necessary to increase the loss or number of elements in one or both networks to provide equal terminations.

We can eliminate the errors associated with cascading two networks by directly synthesizing a matching network with complex loads at both ends. This approach usually results in a design having fewer elements and higher efficiency. Figure 6.34 shows a matching network terminated with a complex load at both ports. We can approach this problem in exactly the same manner as the matching network for a single complex termination. That is, we design a filter structure with the appropriate type of input and output elements and then adjust the ripple and MIL for exact or excess absorption. However, there are several factors that complicate this procedure when both terminations are complex. Perhaps the most important consideration is the type of elements at the input and output ports. Figure 6.1(b) shows the topology for an $n = 3$ Chebyshev bandpass filter. In this case, the parasitic elements are either series inductors or capacitors at both ports. We can also use $-s_{11}$ to compute the driving point impedance, which results in the dual circuit, parallel inductors and capacitors at both ports. However, we cannot use this approximation to match a series element at one port with a parallel element at the other port; both elements must be either series or parallel. The bandpass filter can be designed from a low-pass prototype using the transformations shown in Figure 4.19. The low-pass series inductor is transformed into a bandpass series L-C and the shunt capacitor into a parallel L-C. Since the low-pass filter consists of alternating series L and shunt C, the basic bandpass topology is one of alternating series and shunt L-C elements. Therefore, odd orders result in identical element types at both ports, and even orders provide series elements at one port and shunt elements at the other. The Norton network transformations alters the topology; however, equation (4.12) shows that $s_{11}(s)$ and $s_{22}(s)$ are functions of the poles and zeros, not the topology. Thus, the input and output reflection coefficient is the same for all topologies, and the type of elements at each port is restricted. For example, the $n = 3$ filter in Figure 6.1(a) has a series L-C at each port, and both s_{11} and s_{22} are $+1$ (open circuit) at zero and infinity. It is not possible to transform the basic circuit shown in the figure such that the last reactive element is a shunt L or C (i.e., short circuit at zero or infinity). However, the output elements can be used in a Norton transformation to adjust the termination, and we must be careful not to alter the parasitic element unless there is sufficient absorption capability.

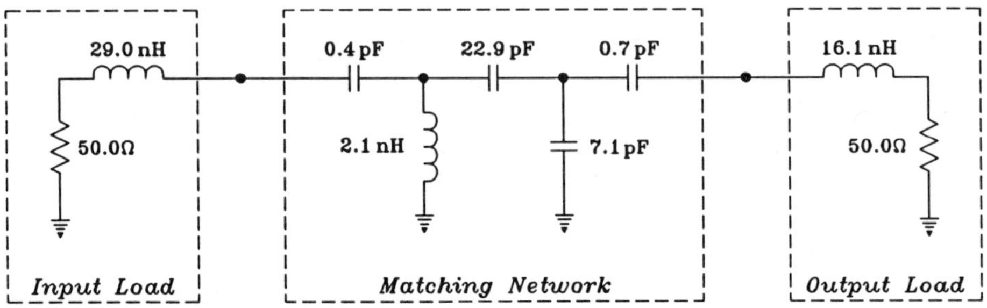

Figure 6.34 Matching network with complex input and output loads.

Once the number of elements or order of the approximation is resolved by comparing the topology of the loads with s_{11} and s_{22} of the matching network, we must determine the insertion loss that provides optimum efficiency. The GBW limitations given by (6.1) and (6.6) can readily be evaluated for the input and output loads. We use s_{11} for the input load in the normal manner, and then substitute s_{22} in place of s_{11} to evaluate the output load constraint. From the perspective of the output load, s_{22} is the input reflection coefficient of the network. However, the network is lossless, and from the analysis in Section 2.3.1, we know that $|s_{11}|^2 = |s_{22}|^2$. Therefore, the integral expressions in (6.1) and (6.6) are identical for both ends of the network. The output reflection coefficient is formed from the negative zeros of s_{11}, as illustrated in (4.12). Hence, as we increase the absorption capability at the input of the network by placing the zeros in the left half-plane, the output absorption capability decreases, because the corresponding zeros of s_{22} are in the right half-plane, and subtracts from the load constant. Thus, in addition to an optimum value for the ripple and MIL, we must also determine the zero combination that provides the lowest total loss. Usually the GBW requirement for the load at each end of the network will be different, and, in fact, we may also have resonant loads with different requirements for two elements at each port. In most instances the optimum solution will be exact absorption for the load at one end and excess absorption for the other load.

Figure 6.4 shows the value of the series input inductance versus ripple and zero locations for the $n = 3$ Chebyshev filter discussed in Section 6.1. The output element is also a series inductor, as illustrated in Figure 6.1(b), and since the left half-plane zeros of s_{11} are the right half-plane zeros of s_{22}, the curves represent the value of the output inductor when the L and R labels are interchanged. For example, at 0.5 dB ripple, the input inductor is 20 nH for an L-R-R zero combination (i.e., left-right-right). The output inductor can be determined from the *R-L-L* curve and is approximately 34 nH at 0.5 dB ripple. Figure 6.4 can also be applied to solve the complex matching problem shown in Figure 6.34. When all three zero pairs are placed in the left half-plane, the ripple is 0.01 dB for a 29-nH input inductor and the corresponding output inductor is 7 nH from the *R-R-R* curve. The output absorption capability is insufficient, and we must increase the ripple to 0.5 dB for exact absorption. Using the *R-L-L* curve, 0.25 dB of ripple provides a 29-nH input inductor, and the corresponding output inductor is 16 nH from the *L-R-R* curve. Both loads can be exactly absorbed, and the ripple is minimized for 0.5 dB of MIL. However, this combination of ripple, MIL, and zero location is not necessarily optimum. We must change the MIL, recompute the minimum ripple, and then iterate until the lowest total loss is found. Reducing the MIL to 0.25 dB, the minimum ripple is 0.31 dB when all zeros are placed in the left half-plane. This combination results in exact absorption for the output element and excess absorption for the input element. The optimum is all left half-plane zeros, 0.25 dB of ripple, and 0.13 dB of MIL, which provides exact absorption for both the input and output elements.

Amplifier Design Using Interstage Matching Networks

A common requirement for many RF and microwave systems is a *gain block*; that is, a general purpose amplifier that can be used to increase the signal levels following a low noise stage or preceding a power amplifier. We can use the doubly terminated matching network for the design of a gain block with 20 ± 1 dB gain, maximum VSWR of 2:1, and unconditional stability across an operating frequency range of 4.0 to 8.0 GHz. A good choice for this application is the low cost NEC NE70000 MESFET chip listed in Table 6.7. The device has a G_{MSG} ranging from 16 to 13 dB across the operating frequency, as shown in Table 6.8, and two stages will provide the required 20-dB gain. We can stabilize the device and then match the input of the first stage and output of the second stage to achieve the 2:1 VSWR specification. The interstage matching network between the output of the first stage and input of the second stage will be designed to provide a mismatch that corrects for the gain slope versus frequency such that the overall gain is 20 ± 1 dB across the 4.0- to 8.0-GHz operating range.

Figure 6.26 shows the input plane stability circles and Figure 6.27 the output plane circles for this device. The output plane stability circles are almost concentric, with an increasing radius at lower frequencies. Therefore, unconditional stability requires either a large series or a small shunt resistance. In fact, it may not be possible to stabilize this device at very low frequencies using the output plane. However, the input plane stability circles shown in Figure 6.26 intersect the positive real region at approximately the same distance from the origin and rotate counterclockwise with increasing frequency. The tendency for the input plane stability circles to rotate toward 0° at lower frequencies requires higher series resistance values to ensure stability. However, a parallel resistance of approximately 175Ω is sufficient to stabilize the device at all frequencies. The characteristics of the 175Ω shunt resistor and active device are shown in Table 6.9.

Table 6.9 Stabilized NE70000 Chip

Frequency (GHz)	*K*	G_{MAX} *(dB)*	Γ_{MS}	Γ_{ML}
4.0	1.313	12.588	$0.719\angle 67.74°$	$0.810\angle 32.28°$
6.0	1.077	12.712	$0.844\angle 87.03°$	$0.884\angle 43.79°$
8.0	1.078	11.360	$0.836\angle 102.65°$	$0.859\angle 53.20°$

Evaluating the necessary and sufficient conditions for unconditional stability as discussed in Section 2.3.2 reveals that the stabilized device is unconditionally stable from 2.0 to 18.0 GHz, the range of measured data provided by the manufacturer. Thus, we can design the matching networks without regard for stability; that is, the device is stable for all positive real input and output impedances. We intend to match the input of the first stage, the output of the second stage, and provide a mismatch between stages. One approach is to construct the output plane constant gain circles for the first stage and input plane gain circles for the second stage. We then choose a first and second stage gain such that the total adds up to 20 dB. The division of gain between stages is normally selected to provide convenient impedance values or is based on other considerations, such as stability, cascaded noise figure, or output power. The gain circles assume a conjugate match in the opposite plane and the amplifier is matched at the input and output ports. However, we must be able to synthesize practical matching, networks which usually requires a tradeoff between the impedances presented to each device and the overall circuit performance. We can constrain the negative-image models to predict the performance from realizable matching networks; however, we must place additional restrictions on the interstage models.

The matching network for a single complex load essentially transforms Z_O into the complex conjugate impedance of the load or, at least, a best fit to the load impedance. Z_O is fixed and always provides a known termination at one port of the matching network. When the matching network has a complex termination at both ports, a complex termination at either port is transformed into the complex conjugate of the load at the other port. However, in the case of an interstage matching network, the loads used for synthesis *represent a mismatch and not the actual network terminations*. The actual matched impedance is given by Γ_{MS} and

Γ_{ML}, and the corresponding gain is G_{MAX}. We can provide the proper termination at one port of the interstage matching network by requiring that either negative--image model match the impedance at that port. That is, we transform the actual output impedance of the first stage into a mismatch with the input of the second stage, or vice versa. The amplifier will be realized using distributed elements and the interstage negative-image models must be commensurate (i.e., the same electrical length) in order for the parasitic elements to be absorbed from both ends of the synthesized network. Thus, the electrical length of the parasitic elements in the models at each end of the interstage network must be equal.

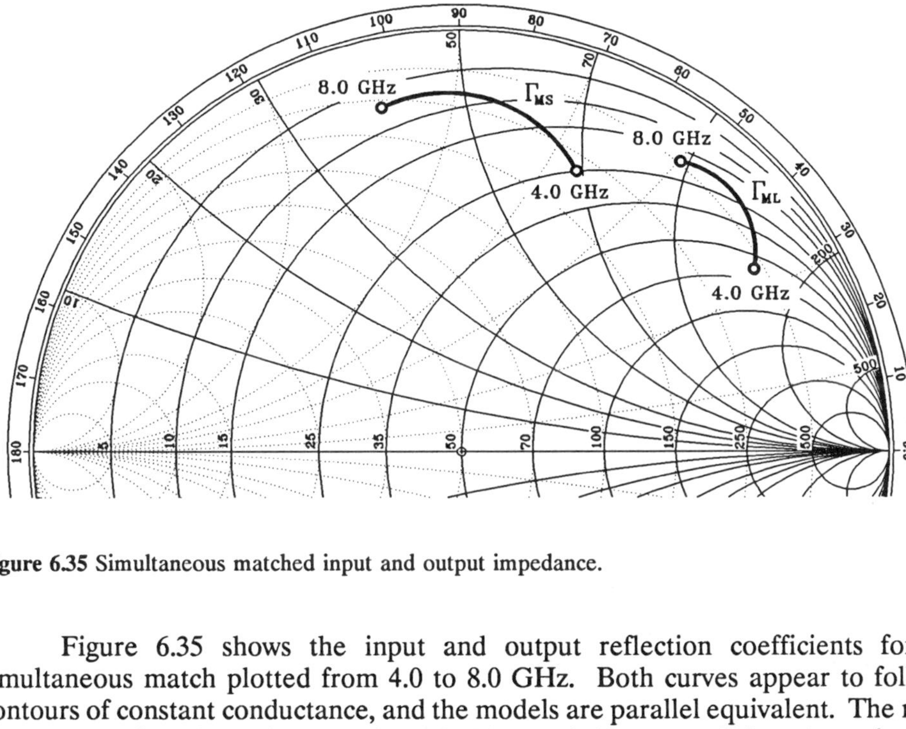

Figure 6.35 Simultaneous matched input and output impedance.

Figure 6.35 shows the input and output reflection coefficients for a simultaneous match plotted from 4.0 to 8.0 GHz. Both curves appear to follow contours of constant conductance, and the models are parallel equivalent. The real component of the input is approximately 250Ω and the output 500Ω. Assuming an electrical length of 45° at 8.0 GHz, we can compute the initial element values from the reactance at the center frequency of 6.0 GHz. One end of the interstage matching network must be mismatched to provide flat 20-dB gain, and we can determine the model from the appropriate constant gain circles. However, the interstage mismatch affects the input and output match; that is, Γ_{MS} and Γ_{ML} are no longer applicable. Thus, we must select an impedance from the constant gain circles, recompute the input and output impedances, and then determine the corresponding models. This procedure is complicated by the interaction between the input and output impedances of the two devices and is best accomplished using optimization. The Γ_{MS} and Γ_{ML} models can be used to provide initial values for the optimization, as shown in Figure 6.36. The gain will be higher than desired (twice G_{MAX}), but at least the input and output will be matched.

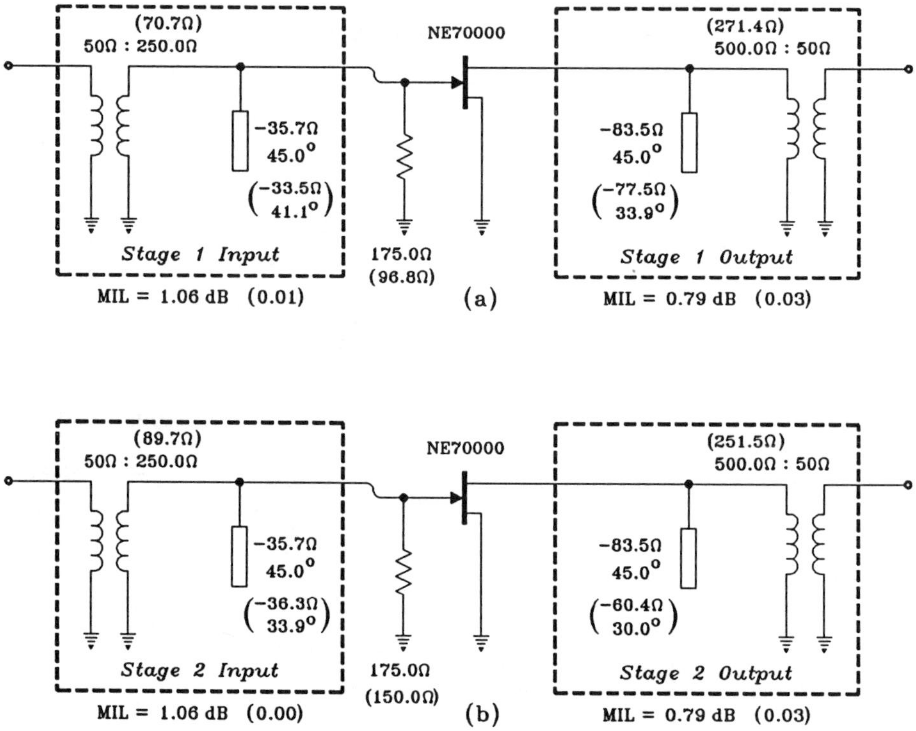

Figure 6.36 Two-stage amplifier models: (a) first stage; (b) second stage.

The lighter traces in Figure 6.37 illustrates the gain, input, and output return loss using the initial values. As expected, the 6.0-GHz gain is slightly over 25 dB and the input and output are well matched. We observe from the figure that there is a significant reduction in gain and return loss above 6.0 GHz, which indicates an error in the initial element values. The optimizer should be able to improve the performance using these models. However, we must apply the proper constraints to ensure that we will be able to synthesize practical matching networks. The real component of all four models is higher than desired, 250Ω for the input and 500Ω for the output. In addition, the ideal MIL shown in Figure 6.36 is 1.06 dB for the input models and 0.79 dB for the output models, which will result in significant impedance errors from the synthesized matching networks. Since R is in the denominator of (6.8b), a reduction will result in lower MIL, as well as more convenient distributed element values. The value of the shunt gate resistor was set equal to 175Ω, which is the maximum for unconditional stability. We can allow the optimizer to reduce these values, which will improve stability, reduce the maximum available gain, and, most importantly, lower the real component of impedance. Therefore, we constrain the maximum resistance of the shunt resistors to 175Ω, the input models to 250Ω, and the output models to 300Ω. The electrical length of all four parasitic elements are also constrained between 30° and 60°, and we add an optimization goal of 0.05 dB or less for the ideal MIL of each model. The goals for the complete two-stage amplifier are set at 20-dB gain, and both $|s_{11}|$ and $|s_{22}|$ less than -20 dB (i.e., input and output return loss greater than 20 dB). The interstage matching network will be terminated at the input by the output load in Figure 6.36(a), and at the output by the input load shown in Figure 6.36(b). Although either end can be used, we elect to transform the output impedance of the first stage and mismatch the input of the second stage. Therefore, we add an optimization goal specifying that $|s_{22}|$ of the first stage should be less than -15 dB, which will provide a best fit model to the output impedance of the first stage. We also equate the electrical lengths of both interstage models and then optimize. The optimized response is shown in Figure 6.37 as the darker curves, and the final element values are enclosed by parentheses in Figure 6.36.

The optimized performance is acceptable, gain is 20.0 ± 0.8 dB, and input and output return loss are greater than 19 dB. The two-stage amplifier is unconditionally stable since each stage is stabilized by the gate resistor. The models are also acceptable, the highest ideal MIL is 0.03 dB, and all element values are reasonable, although lower resistive values for the two output models would be preferred. We can now synthesize all three matching networks, replace the ideal networks with the actual circuits, and then analyze and optimize. However, it is usually advantageous to synthesize the interstage network first and then reoptimize the amplifier using the negative-image input and output models, which will compensate for errors introduced by the interstage network. As shown in Figure 6.36, the parasitic element at both ends of the interstage matching network is a shunt open stub. Therefore, both $s_{11}(S)$ and $s_{22}(S)$ must be -1 at $S = \infty$; that is, a short circuit at infinite frequency or 90° electrical length. The electrical lengths of the parasitic elements are less than 90° at 8.0 GHz, and we can use the generalized distributed approximation from Section 5.3.4 to synthesize a network with low-pass, high-pass, and transmission line elements. Since both parasitics are low-pass elements, q must be at least 2. We can try an approximation with $n = 2$, $m = 1$, and $q = 3$, which will result in a matching network with two TLEs, one low-pass, and one high-pass element; that is, two of the low-pass elements are absorbed by the loads. The network may actually have two low-pass elements if either end has excess absorption capability. We can easily verify that

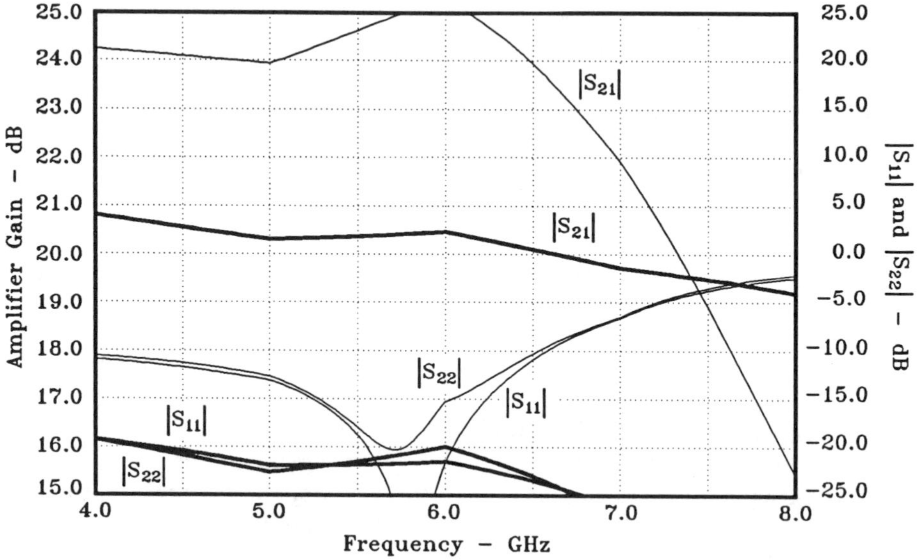

Figure 6.37 Amplifier gain, input and output match.

this combination of elements is acceptable by assuming a ripple and MIL value, computing $s_{11}(S)$ from the poles and zeros, and then either computing $s_{22}(S)$ or simply synthesizing any topology and examining the output element or reflection coefficient at $\theta = 90°$. This combination does provide the required short circuit at both ends, and we must determine the optimum zero location, ripple, and MIL for minimum loss. The ideal MIL calculated from (6.8b) for the input load shown in Figure 6.36 is 0.033 dB, and the output is 0.004 dB. Since the input requirement is an order of magnitude higher than the output, we place all zeros in the left half-plane to maximize the absorption capability and then compute the optimum ripple as 0.05 dB and MIL as 0.066 dB, or a total loss of 0.125 dB. The synthesized matching network topology is shown in Figure 6.38. However, after using the shunt shorted stub to transform the output termination to 89.7Ω, we find that the output parasitic is 45.5Ω, which is insufficient. We could change the zero locations, which would decrease the absorption capability at the input and increase the corresponding capability at the output. However, we can also retain the left half-plane zeros and change the ripple and MIL, which will increase the total loss, but can sometimes be used to provide exact absorption at both ends of the network. In this

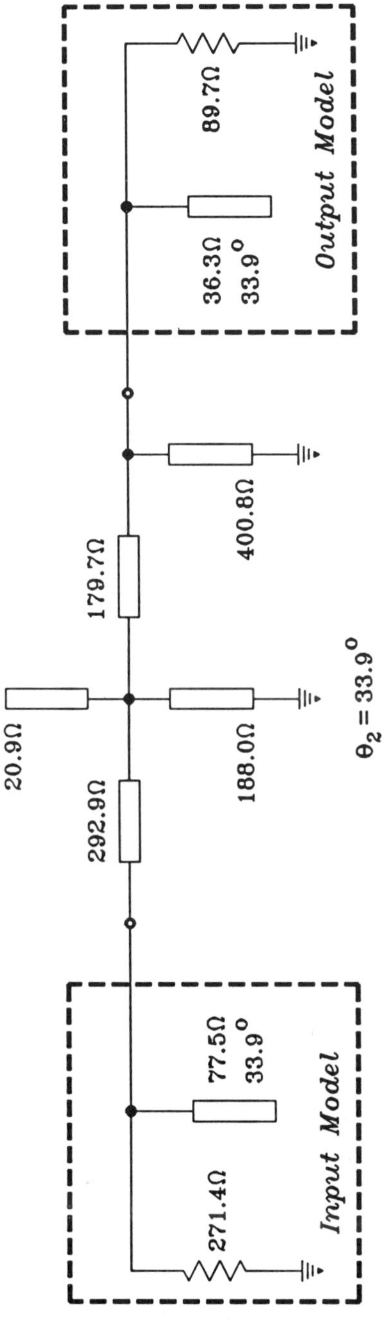

Figure 6.38 Amplifier response using the synthesized interstage network.

case, 0.120 dB of ripple and 0.025 dB of MIL provide exact absorption at both ends. The total loss is 0.145 dB, only 0.020 dB higher than the optimum. The synthesized matching network is shown in Figure 6.38.

The lighter traces in Figure 6.39 show the response of the amplifier with the synthesized interstage matching network and negative-image input and output models. Comparing this response with the darker curves in Figure 6.37, which illustrate the original response using negative-image interstage models, reveals similar characteristics. The gain is approximately 20.0 ± 1.0 dB, and the minimum return loss at both ports is 15 dB. Unfortunately, we cannot realize any of the interstage elements and must modify the network. The impedance of shunt stubs can usually be adjusted by changing the electrical length. In fact, the reactance is exact at a single frequency as, illustrated in Chapter 5. However, the TLEs are more difficult to change. The input TLE is almost 300Ω and will present the greatest difficulty. However, the NEC NE70000 FET is a chip that requires a bond wire to connect from the drain pad to the matching network. We can replace the high impedance TLE with an inductor that consists of the appropriate length of bond wire. The maximum length of the bond wire must be limited to avoid mechanical and radiation problems. Assuming a maximum length of 150 mils (3.81 mm), a 1-mil (0.025-mm) gold wire has approximately 4 nH of inductance [15]. Since we need only be concerned with the performance of the interstage matching network when terminated with the loads in Figure 6.38, we can replace the TLE with a series inductor, constrain the inductance between 0 and 4 nH and the distributed elements between 40Ω and 100Ω, and then optimize for minimum insertion loss. The results are 3.6 nH for the inductor, 40Ω for the open shunt stub, and 100Ω for the other elements. The electrical lengths are between 42° and 78°, and the maximum insertion loss is 0.19 dB, 0.04 dB higher than the synthesized value. The response of the amplifier with the modified interstage network is not substantially different from that of the synthesized network, shown by the lighter curves in Figure 6.39.

We now have an amplifier circuit that consists of negative-image input and output models and the interstage matching network, as shown by the topology in Figure 6.40. We can take advantage of the simplified circuit model by reoptimizing in this form to compensate for errors introduced by replacing the ideal interstage network with the actual elements. The optimized element values are shown in Figure 6.40, and the corresponding performance by the darker traces in Figure 6.39. The most significant difference is a slight improvement in input and output return loss. At this point the procedure is similar to that of the single-stage design in Section 6.2.3. We synthesize input and output matching networks for the load models in Figure 6.40, constrain the elements to realizable values, optimize to minimize the insertion loss when terminated with the appropriate load, and then replace the ideal matching networks with the optimized ones. Using the generalized approximation with $n = 2$, $m = 1$, and $q = 3$ requires 0.077 dB of ripple and 0.099 dB of MIL to exactly absorb the input parasitic and transform to

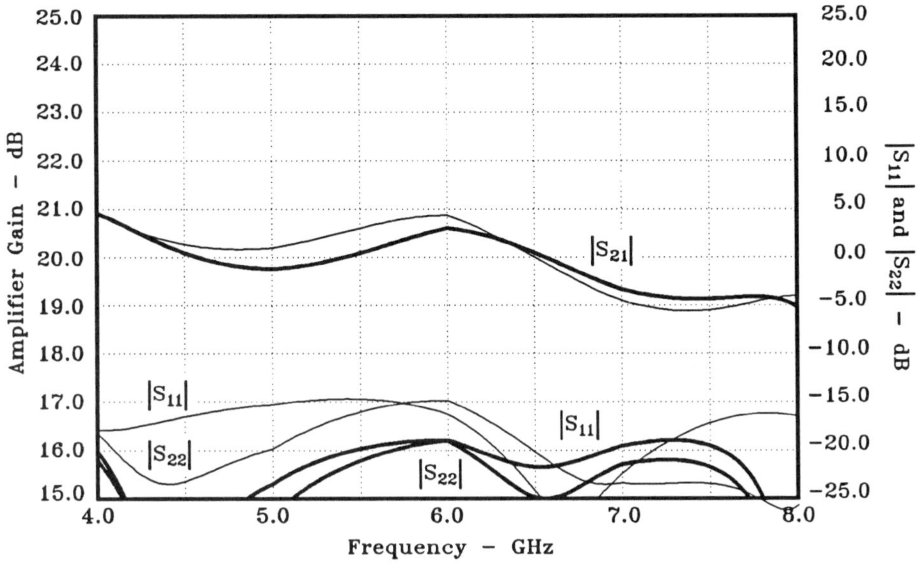

Figure 6.39 Amplifier model using negative-image and interstage matching networks.

50Ω. The output network uses the same approximation, with optimum values of 0.077 dB ripple and 0.104 dB MIL. However, this network has an input termination of 300Ω, which results in a topology and element values similar to that of the interstage network shown in Figure 6.38, and we can use the same technique to realize this circuit. That is, the high impedance input TLE is replaced with an inductor, the element values constrained, and the network is optimized to minimize the insertion loss. Figure 6.41 shows the complete amplifier topology after the ideal matching networks are replaced by the synthesized matching networks.

The lighter traces in Figure 6.42 show the response of the two-stage amplifier using the element values of the input and output matching networks as optimized into the respective loads. The gain is 19.77 ± 1.09 dB and the worst-case return loss is 11.04 dB or approximately 1.78:1 VSWR. We can now optimize the complete amplifier, which results in the element values shown in Figure 6.41 and the response depicted by the darker curves in Figure 6.42. The gain is 20.00 ± 0.51 dB, and the minimum return loss is 11.50 dB or 1.73:1 VSWR across the 4.0- to 8.0-GHz frequency range. The amplifier meets the design specifications, and we have some margin for component tolerance in both gain and VSWR. However, there is a tradeoff between the gain flatness and match. Depending on the design requirements, we could increase the weight on either specification and then reoptimize to improve one parameter at the expense of the other. For example,

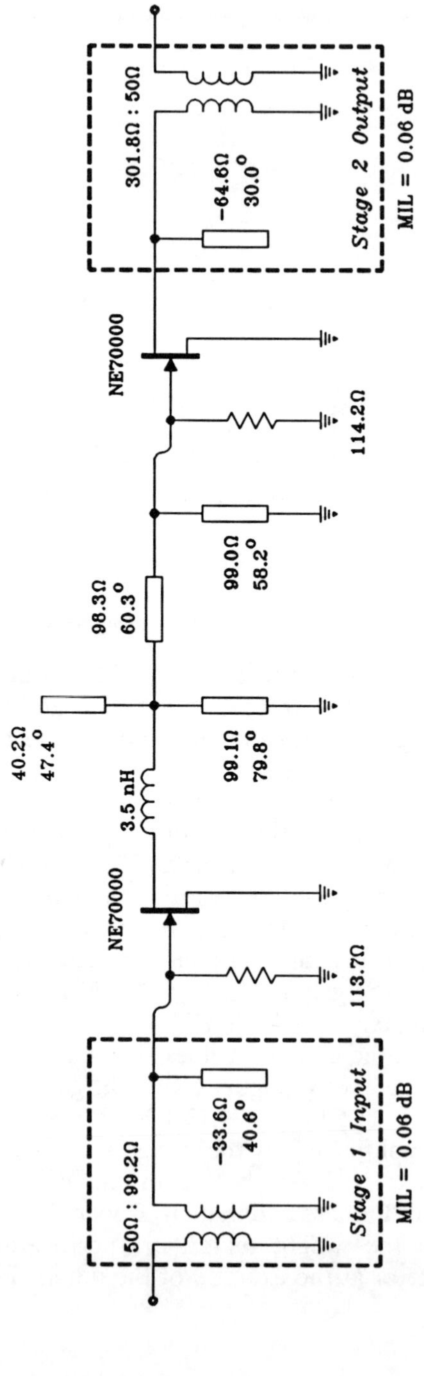

Figure 6.40 Amplifier model using negative-image and interstage matching networks.

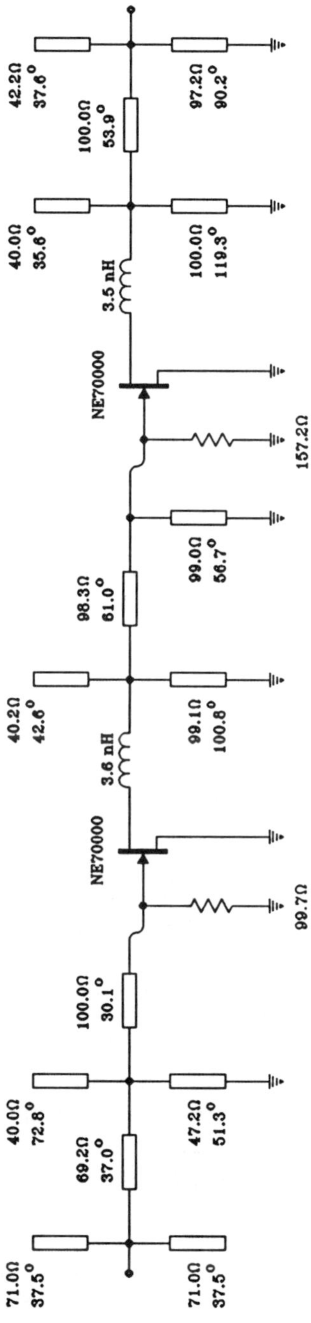

Figure 6.41 Complete 4.0- to 8.0-GHz amplifier.

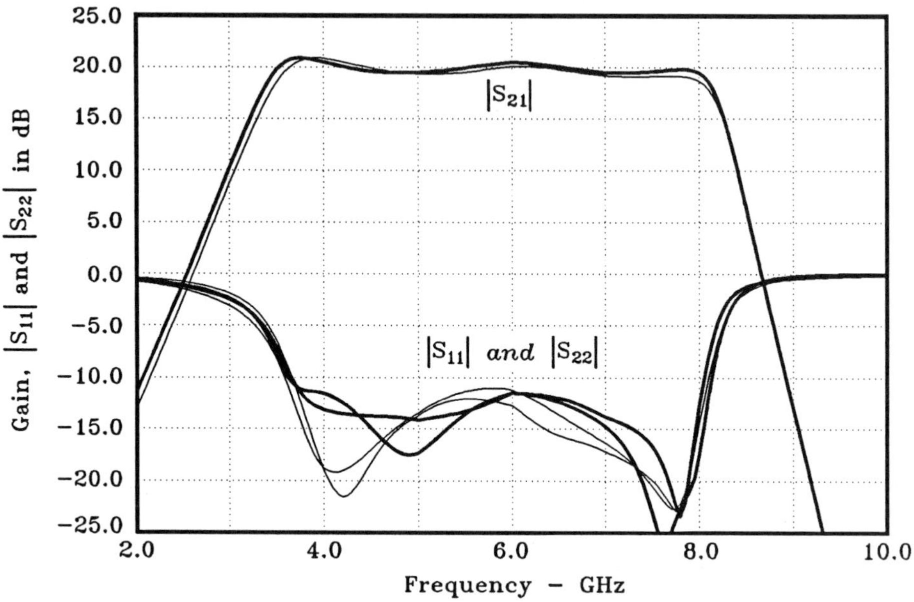

Figure 6.42 Complete 4.0- to 8.0-GHz amplifier response.

Figure 6.43 shows the results when we optimize for flat 20-dB gain without regard for input or output match. The gain is 20.00 $\pm$ 0.03 dB from 4.0 to 8.0 GHz, and the minimum return loss is 6.0 dB or 3.0:1 VSWR. The bias networks are not shown in Figure 6.41 to emphasize the form of the matching networks. However, the topologies of the synthesized matching networks are convenient for injecting bias without affecting the RF characteristics. Both gates are grounded by a shorted shunt stub, and we can add a bypassed source resistor as shown in Figure 6.33 to provide self bias. The shorted shunt stubs adjacent to the drain inductors can be grounded through a capacitor and the bias voltage connected to this low impedance point, which is also illustrated in Figure 6.33. A coupling capacitor must also be inserted between the two interstage shorted stubs to avoid short-circuiting the drain bias. One method is to split the connecting transmission line and insert the capacitor somewhere in the middle. The impedance is a function of distance along the transmission lines, and we can usually find one point where the series capacitor has minimal effect.

We were able to successfully design this two-stage amplifier by using the negative-image matching networks to determine impedance as a function of frequency at each terminal of the two active devices. The ideal networks not only predict the circuit performance before designing any circuitry, but also provide

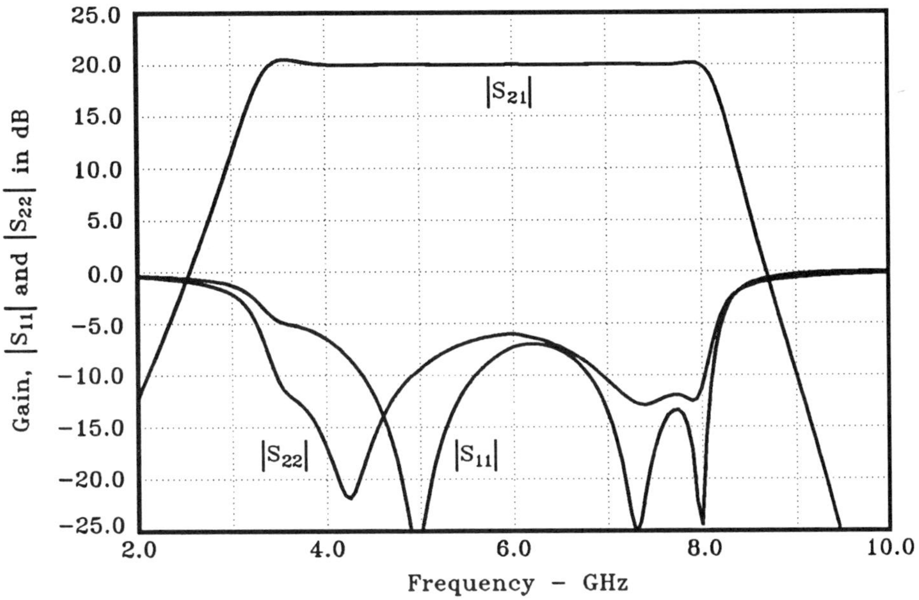

Figure 6.43 Complete 4.0- to 8.0-GHz amplifier optimized for flat gain.

accurate load models for matching network synthesis and a measure of the efficiency of each matching network. Once the negative-image elements are defined, the problem is essentially one of designing three completely independent passive matching networks. We can modify and optimize the networks as needed and then replace the ideal circuits with practical, realizable networks. If the synthesized networks accurately match the defined loads, the response of the final design will be the same as the original model, assuming lossless elements. We can include the effects of component loss by increasing the gain slightly in the original ideal model. The negative-image models were all parallel equivalent for this example. However, we have 15 other combinations that can be investigated (i.e., series or parallel configuration for four different models). In addition, we elected to match the output of the first stage and mismatch the input of the second stage. We could also try matching the input of the second stage and mismatching the output of the first stage. Once the interstage network is defined, we might also try different types of negative-image models for the amplifier input and output, or at least plot s_{11} and s_{22} to ensure that the models are correct.

6.4 SLOPED APPROXIMATIONS

The unilateral assumption discussed in Section 6.2.3 results in an input and output impedance invariant to terminations in the opposite plane; that is, s_{12} is zero in (2.40a) and (2.40b). We can model s_{11} and s_{22} independently and then design the appropriate matching networks. Assuming the magnitude of s_{12} is small and the device is unconditionally stable, this approach results in an amplifier gain equal to G_{MAX}. The maximum available gain usually decreases with increasing frequency and the result is a gain slope. However, we can compensate for the gain slope by designing a matching network that exhibits the opposite characteristics as a function of frequency [16-19]. That is, the load is matched at the highest frequency and mismatched at lower frequencies. Ideally, the slope of the matching network loss is exactly opposite G_{MAX}, which results in flat gain when cascaded with the active device.

Mellor [16] shows that we can introduce a gain slope of 6S dB per octave by multiplying the transducer gain by ω^{2S}. For example, we can introduce a 6-dB gain slope in the response of the 1.25- to 1.75-GHz bandpass filter in Section 4.6.2. Multiplying $G_t(\omega^2)$ by ω^2 yields

$$\frac{\omega^8}{2941.225\,\omega^{12} - 12965.4\,\omega^{10} + 23549.4\,\omega^8 - 22554.75\,\omega^6 + 12015\,\omega^4 - 3375\,\omega^2 + 390.625}$$

Then, following the same procedure as was used in Section 4.6.2, we compute the driving-point impedance as

$$\frac{s^6 + 0.301391s^5 + 2.252541s^4 + 0.450008s^3 + 1.610239s^2 + 0.157729s + 0.364431}{0.077986s^5 + 0.023504s^4 + 0.110398s^3 + 0.015508s^2 + 0.035831s}$$

The original filter is shown in Figure 4.21. We can extract the elements in the same order to synthesize a similar topology. However, after removing all but the last element, we have an input impedance of

$$Z'(s) = \frac{0.153856s}{s + 2.310495}$$

Referring to Figure 4.1, we see that the only choice is a shunt inductor. We cannot synthesize the same topology. After removing the shunt inductor, we find an output termination of 0.153856, and we can scale the element values to 50Ω and 1.75 GHz, as shown in Figure 6.44. Comparing the sloped filter with the original in Figure 4.21 reveals that one of the low-pass elements has been transformed into a high-pass element. The sloped filter has the same number of elements, but two are low-pass and four are high-pass.

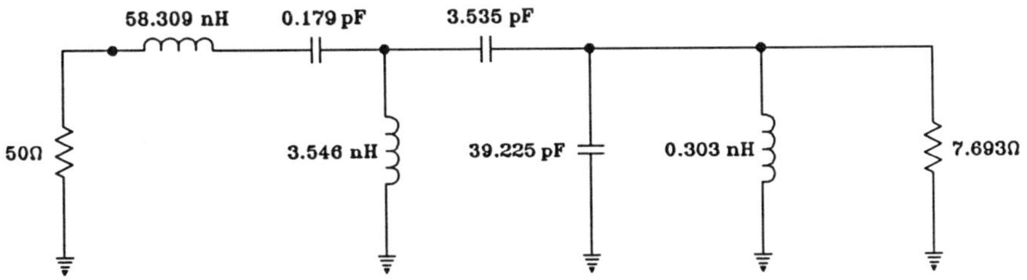

Figure 6.44 Sloped Chebyshev bandpass filter.

Figure 6.45 shows the response of the original and modified filters. The broken line represents a linear slope of 6 dB per octave, and we see that the response of the sloped approximation is exact at the band edges, but appears to exhibit a small error at the two minimum ripple points in the passband. However, the slope is a function of ω^2 and appears as a straight line when plotted using a logarithmic frequency scale. Mellor shows that we can realize a sloped response that is not an integer multiple of 6 dB per octave by combining two sloped approximations. The two approximations are computed such that each has a slope that is an integer multiple of 6 dB per octave, but one is less than the desired slope while the other is greater. For example, we can synthesize a structure with 3.0-dB slope by combining a flat approximation (0-dB slope) with a 6-dB sloped approximation. However, the flat approximation from Section 4.6.2 has an equal number of high-pass and low-pass elements, while the sloped function has two low-pass and four high-pass elements. Therefore, we must use a different form for the bandpass approximation [20], which allows arbitrary numbers of low-pass and high-pass elements; that is, the topologies of the two networks must be the same. The noninteger approximation is not exact across the entire passband. However, Mellor provides expressions for the combining constants such that the insertion loss is correct at each cutoff frequency and the maximum error is easily predicted.

Ku [18] and Ku and Petersen [19] consider the problem of designing sloped matching networks using distributed approximations. However, sloped matching networks, whether lumped or distributed, have some disadvantages when used for the design of active circuits. One problem is obtaining accurate load models. When the device is unconditionally stable, we can use the unilateral assumption, an equivalent circuit model for the device, or compute Γ_{MS} and Γ_{ML} and then design for a slope opposite that of G_{MAX}. Most high-performance microwave transistors are potentially unstable, and the impedance for simultaneous match is not defined. When the design specifications preclude stabilization (i.e., low noise or high output power), we can use the maximum stable gain and stability circles to define an acceptable load impedance and then synthesize a matching network with a slope opposite that of G_{MSG}. However, we have no control over the input impedance of the matching network. The matching network must provide the complex conjugate load impedance at the upper end, where the mismatch is zero. But at lower frequencies, where the mismatch is nonzero, the impedance can be anywhere on the appropriate mismatch circle because the approximation is based on insertion loss rather than on impedance. The negative-image models discussed in Section 6.2.3 provide more control over the input impedance as illustrated by the single-stage amplifier design example, where we designed matching networks to specifically avoid certain impedance regions. The load models are exact, and the required slope or mismatch is automatically incorporated and distributed to improve the efficiency of each matching network. However, we can also use this modeling technique to synthesize sloped matching networks.

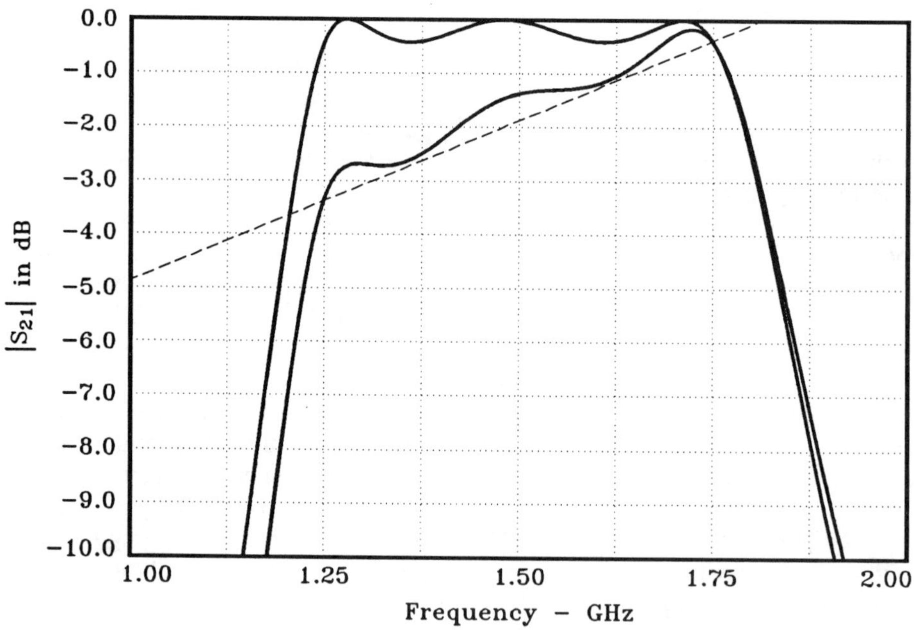

Figure 6.45 Original and sloped filter response.

Sloped Matching Metworks Using Negative-Image Models

Rather than modify the approximation to provide a sloped response, we can use the load model to provide the required characteristics. That is, the load model is exact at the highest frequency and mismatched at lower frequencies. We can examine the impedance of the model, add constraints as necessary to avoid undesired regions, and then synthesize a matching network using standard lumped or distributed approximations. When the matching network is terminated with the actual impedance, the response will have the same characteristics as those exhibited by the load model. The slope of the matching network may show some errors due to the combination of a mismatch with the load and nonzero ripple and MIL. However, optimization can be used to correct minor variations. As an example, the network in Figure 6.44 matches a load consisting of the 50Ω input termination and series L-C with a slope of 6 dB per octave. We will design a matching network with similar characteristics for the same load using negative-image models.

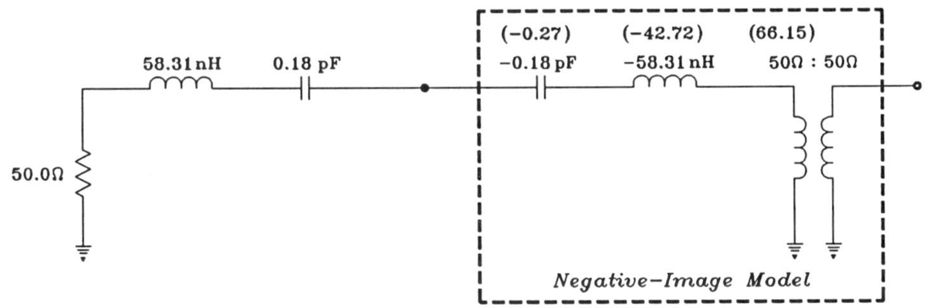

Figure 6.46 Sloped negative-image model.

Figure 6.46 shows an ideal matching network connected to the load. The topology of the model is the mirror image of the load, and the reactive elements are initially set equal to the negative values of the corresponding load elements. The ideal matching network matches the load across the entire bandwidth with no loss. However, the original matching network has a 3-dB slope from 1.25 to 1.75 GHz, and a similar response can be realized by optimizing the element values. We can set the optimization goals equal to the desired insertion loss at a number of discrete frequencies and then use minimax optimization to find a best fit to the slope. We also have the option of optimizing to a linear or logarithmic function of frequency describing the desired slope. However, for the purposes of this example, we specify 3 dB of loss at 1.25 GHz and 0 dB of loss at 1.25 GHz, and then optimize. The final element values are enclosed in parentheses in Figure 6.46, and the corresponding insertion loss versus frequency response is shown in Figure 6.47.

Figure 6.47 shows a comparison between the optimized ideal matching network plotted as the darker trace, the original matching network which is the lighter trace, and the ideal 6 dB per octave slope shown as a dashed line. The ideal response has been shifted down by 0.327 dB to coincide with the original network at 1.75 GHz. The ideal matching network has 0.087-dB loss at 1.75 GHz and the original 0.414-dB loss. We see that the mismatched load model has an error of approximately 1 dB in the center of the band when compared with the ideal 6 dB per octave slope. We could increase the MIL to 0.5 dB, which would change the error to ±0.5 dB, or reoptimize with additional frequency specifications. However, we accept this response and synthesize a matching network for the optimized load model shown in Figure 6.46. We can realize the same topology as that of the original filter in Figure 4.21 by using the $n = 3$ Chebyshev bandpass approximation. In this case, we specify a minimum insertion loss of 0.25 dB, which will improve the fit to the ideal slope, and then compute 0.221 dB of ripple for exact absorption of the optimized load model in Figure 6.46. The synthesized matching network is shown in Figure 6.48 and has an output termination of 0.855Ω. The two capacitors form a Norton transformation which can be used to adjust the termination to 50Ω. However, we will leave the topology in minimum form for comparison with the sloped network in Figure 6.44.

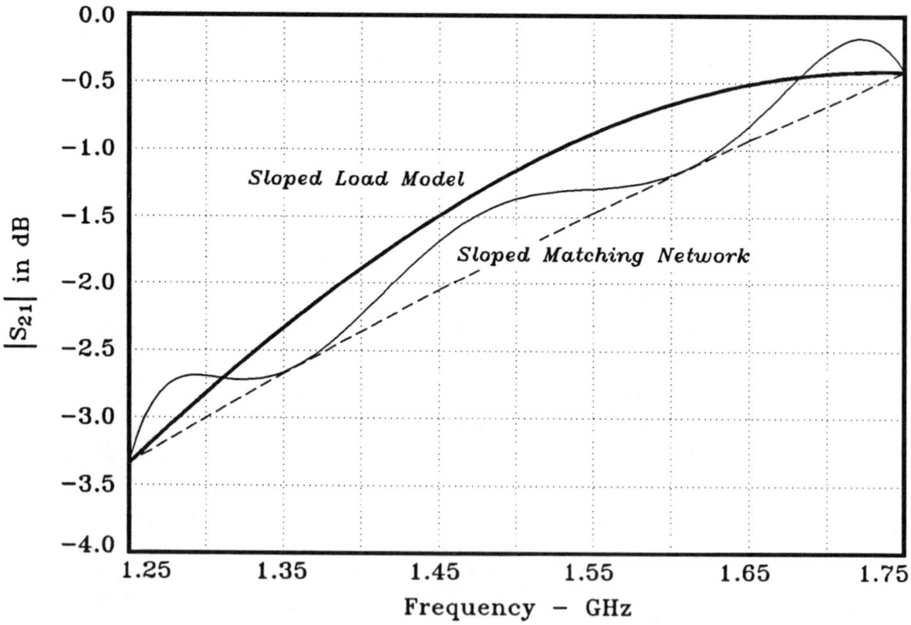

Figure 6.47 Sloped matching network and load.

Figure 6.49 shows the response of the sloped network as the broken curve and the synthesized matching network as the lighter trace. Between 1.30 and 1.75

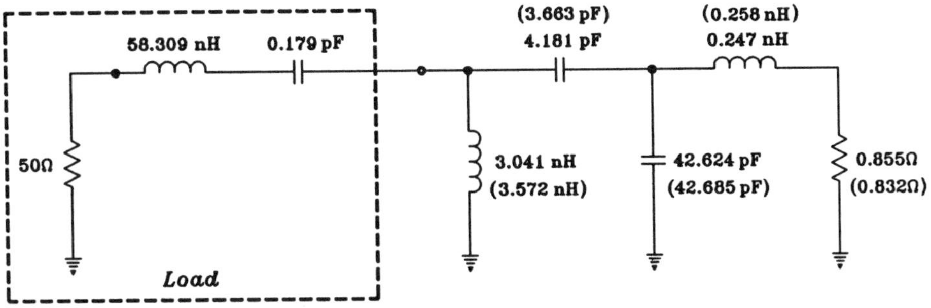

Figure 6.48 Synthesized matching network.

GHz, the maximum difference is approximately 0.5 dB. However, the synthesized matching network appears to have a lower cutoff frequency at 1.30 GHz due to the combination of load mismatch, ripple, and MIL. We can synthesize another network with increased bandwidth which might improve the low frequency performance or optimize the matching network. We elect to optimize the network and define goals which specify insertion loss equal to that of the sloped network at five evenly distributed frequency points across the passband. Values enclosed within the parentheses in Figure 6.48 denote the optimized element values, and the response is the darker curve shown in Figure 6.49. The element values are changed only slightly, but the maximum difference has been reduced to 0.16 dB across the entire passband. The figure shows almost identical ripple characteristics in the passband, but the rejection of the optimized network is superior above 1.75 GHz. The topology of the matching network using the mismatched load model contains two high-pass and low-pass elements, while the network designed by modifying the approximation contains three high-pass elements and one low-pass element.

6.5 THE REAL FREQUENCY MATCHING TECHNIQUE

A significant portion of the material in this chapter deals with the problem of determining a load model for matching network synthesis. This step is the most important aspect of the design process because the efficiency and topology of the matching networks are dependent on the loads. However, in 1977 Carlin [21] introduced a method for synthesizing matching networks which eliminates the problem of load characterization. The synthesis is based solely on the characteristics of the real component of load impedance or admittance and is known as the *Real Frequency technique*. Subsequent development of the Real Frequency technique [22-30] has resulted in a useful design procedure which provides an attractive alternative to load characterization.

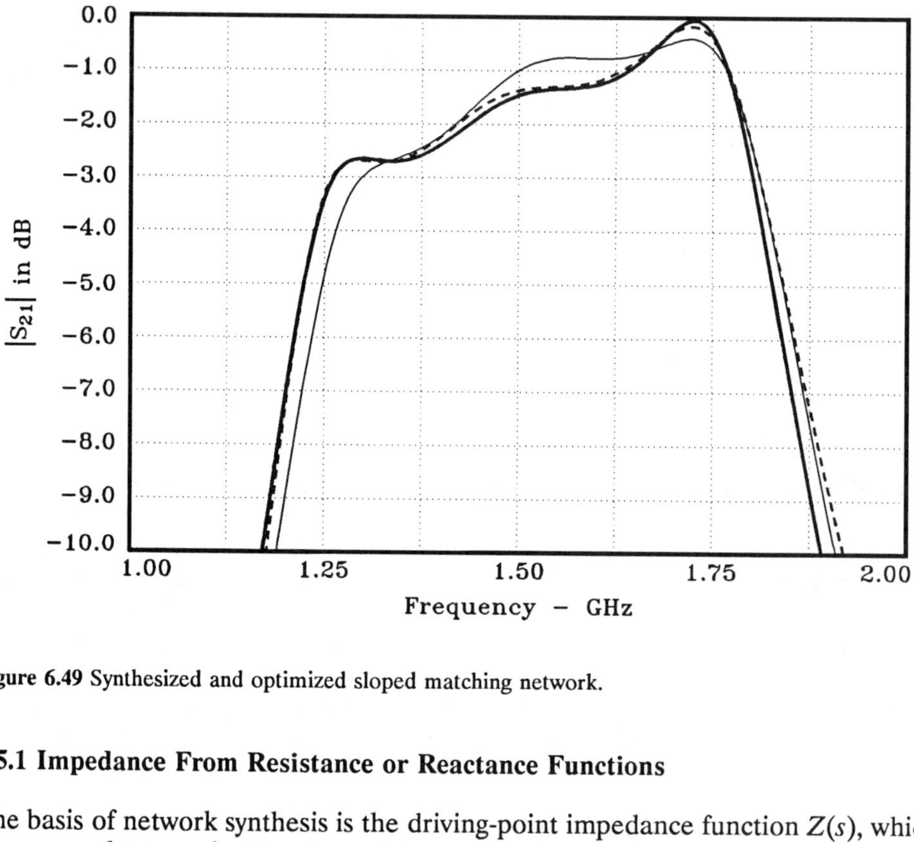

Figure 6.49 Synthesized and optimized sloped matching network.

6.5.1 Impedance From Resistance or Reactance Functions

The basis of network synthesis is the driving-point impedance function $Z(s)$, which is expressed as a rational polynomial representing the input or output impedance of a network. Once $Z(s)$ is defined, it is a simple matter to extract the elements as illustrated in Sections 4.1 and 4.2. The lumped and distributed insertion loss approximations discussed in Chapters 4 and 5 provide $Z(s)$ functions which result in a specified gain response. However, the synthesis of singly terminated networks discussed in Section 4.10 provides an impedance function, where the real part of the impedance or admittance exhibits a prescribed behavior over frequency, as illustrated in Figure 4.35. In fact, a method due to O. Brune [31] and C.M. Gewertz [32] allows $Z(s)$ to be determined from either $R(\omega)$ or $X(\omega)$ to within an arbitrary resistance or reactance function. For example, $R(\omega)$ remains unchanged when a series reactive element is added to the input of a network. Similarly, $X(\omega)$ is independent of additional series resistive elements. However, $Z(s)$ is explicit when $R(\omega)$ or $X(\omega)$ represents a minimum reactance or resistance network; that is, left half-plane transfer function zeros or, equivalently, no imaginary axis poles or zeros of $Z(s)$.

We can write the impedance function as

$$Z(s) = \frac{E_n + O_n}{E_d + O_d} \tag{6.15}$$

where E_n and O_n represents the even and odd parts of the numerator, respectively, and E_d and O_d the even and odd parts of the denominator. The resistance function $R(\omega)$ is Re $Z(j\omega)$ or the even part of $Z(s)$:

$$R(\omega) = \text{Ev}\, Z(s) = \frac{E_n E_d - O_n O_d}{E_d^2 - O_d^2} \tag{6.16a}$$

Replacing $j\omega$ with s in (6.16a), then expanding, we observe that $R(\omega)$ is a rational polynomial in ω^2 or $-s^2$; that is,

$$R(\omega) = \frac{A(\omega^2)}{B(\omega^2)} = \frac{A(-s^2)}{B(-s^2)} \tag{6.16b}$$

Van Valkenburg [33] discusses properties of the Ev $Z(s)$ poles, which are:

- For $Z(s)$ positive real, Ev $Z(s)$ has no poles on the imaginary axis.
- The complex poles of Ev $Z(s)$ have quadrantal symmetry; that is, the poles appear in pairs and one is always the negative of the other.
- If the denominator of $Z(s)$ is of degree t, then $B(-s^2)$ is of degree $2t$, and the degree $2r$ of $A(-s^2)$ is bounded by $0 \leq 2r \leq 2t$.
- Poles of Ev $Z(s)$ may be either simple or of higher multiplicity.

The zeros are only restricted to even multiplicity, which guarantees positive $R(\omega)$. Rewriting (6.16b) in terms of the ratio of two polynomials in ω^2 yields

$$R(\omega) = \frac{A_0 + A_2\omega^2 + A_4\omega^4 + \cdots + A_{2r}\omega^{2r}}{B_0 + B_2\omega^2 + B_4\omega^4 + \cdots + B_{2t}\omega^{2t}} \tag{6.17a}$$

where $r \leq t$. The driving-point impedance can be written as

$$Z(s) = \frac{a_0 + a_1 s + a_2 s^2 + \cdots + a_t s^t}{b_0 + b_1 s + b_2 s^2 + \cdots + b_t s^t} \tag{6.17b}$$

Realizability theory discussed in Section 4.1 states that the degree of the numerator and denominator polynomials of $Z(s)$ can differ at most by one. When $Z(s)$ is restricted to minimum resistance or reactance, the Brune-Gewertz method provides a simple two-step procedure for determining $Z(s)$ from a resistance function in the form of (6.17a), or a similar reactance function. Using (6.16b) in (6.16a), we can write the denominator as

$$B(-s^2) = E_d^2 - O_d^2 = (E_d + O_d)(E_d - O_d) \tag{6.18}$$

The roots of $B(-s^2)$ have quadrantal symmetry and are not located on the imaginary axis. The left half-plane roots are identified with the Hurwitz polynomial $E_d + O_d$, which is also the denominator of $Z(s)$. The right half-plane roots provide the anti-Hurwitz polynomial $E_d - O_d$. The first step in the Brune-Gewertz procedure is to compute the roots of $B(-s^2)$, select those in the left half-plane as the poles of $Z(s)$, and then expand to form the denominator of (6.17b). Next, we substitute $j\omega$ for s and apply (6.16a) to the $Z(j\omega)$ polynomials in (6.17b), which defines $R(\omega)$ in terms of the a and b coefficients. The numerator terms are

$$\begin{aligned} A(\omega^2) = {} & \left(a_0 - a_2\omega^2 + a_4\omega^4 - \cdots\right)\left(b_0 - b_2\omega^2 + b_4\omega^4 - \cdots\right) \\ & + \left(a_1\omega - a_3\omega^3 + a_5\omega^5 - \cdots\right)\left(b_1\omega - b_3\omega^3 + a_5\omega^5 - \cdots\right) \end{aligned} \tag{6.19}$$

which is equal to the numerator of (6.17a). Expanding (6.19) and equating coefficients with like powers of ω provide a system of simultaneous equations in the unknown a coefficients:

$$\begin{bmatrix} b_0 & 0 & 0 & 0 & \cdots \\ -b_2 & b_1 & -b_0 & 0 & \cdots \\ b_4 & -b_3 & b_2 & -b_1 & \cdots \\ -b_6 & b_5 & -b_4 & b_3 & \cdots \\ \vdots & \vdots & \vdots & \vdots & \vdots \end{bmatrix} \begin{bmatrix} a_0 \\ a_1 \\ a_2 \\ a_3 \\ \vdots \end{bmatrix} = \begin{bmatrix} A_0 \\ A_2 \\ A_4 \\ A_6 \\ \vdots \end{bmatrix} \tag{6.20}$$

and the solution provides the numerator of $Z(s)$.

Most practical problems are specified in terms of measured or computed data (i.e., input reflection coefficient or impedance versus frequency). We are seldom given a driving-point impedance polynomial and asked to synthesize the corresponding network. Therefore, a rational polynomial is required that fits the measured data and can be used for synthesis. One approach is to assume the degree and coefficient values of the numerator and denominator polynomials and then use optimization to adjust the values for a best fit to the measured data. We can formulate an error function in the usual manner; that is,

$$E = \sum_{k=0}^{m} \left[W_k \varepsilon_k \right]^2 \tag{6.21a}$$

W_k is the weight at each frequency, m is one less than the total number of frequency points, and ϵ_k is the error component defined as

$$\varepsilon_k = F(\omega_k) - Z(\omega_k) \tag{6.21b}$$

where $F(\omega_k)$ is the measured data and $Z(\omega_k)$ the value of the rational polynomial evaluated at frequency ω_k. Cuthbert [28] points out that the polynomials are ill conditioned, and most optimizers have difficulty finding the minimum; however, a method due to E.C. Levy [34] is usually successful. We can separate a rational polynomial in the form of (6.17b) into its real and imaginary components and then write

$$Z(s) = \frac{\alpha + j\omega\beta}{\sigma + j\omega\tau} = \frac{N(\omega)}{D(\omega)} \tag{6.22}$$

Substituting into (6.21b),

$$\varepsilon(\omega) = F(\omega) - \frac{N(\omega)}{D(\omega)} \tag{6.23a}$$

Then, multiplying both sides by $D(\omega)$ and squaring,

$$[D(\omega)\varepsilon(\omega)]^2 = [D(\omega)F(\omega) - N(\omega)]^2 \tag{6.23b}$$

and we see that $D(\omega)$ corresponds to the weight W_k in (6.21a). Writing $F(\omega)$ as

$$F(\omega) = R(\omega) + jX(\omega)$$

we can expand (6.23b) in terms of the real and imaginary components of the rational polynomial and target data; then the error function in (6.21a) becomes

$$E = \sum_{k=0}^{m} \left[(R_k \sigma_k - \omega_k \tau_k X_k - \alpha_k)^2 + (\omega_k \tau_k R_k + \sigma_k X_k - \omega_k \beta_k)^2 \right] \tag{6.24}$$

When the error function is minimized, the partial derivatives with respect to the polynomial coefficients are equal to zero; that is,

$$\frac{\partial E}{\partial a_i} = 0$$

$$\frac{\partial E}{\partial b_j} = 0$$

for all i and j. The real and imaginary values of the rational polynomial are functions of the coefficients, and the partial derivatives can be written explicitly from (6.24) and then set equal to zero, which forms a system of simultaneous equations. The solution provides the unknown numerator and denominator coefficients of $Z(s)$. Cuthbert gives a detailed description of Levy's method and provides Basic language computer programs which can be used to define the system of equations and to solve for the unknown coefficients.

Values of resistance versus frequency are shown in Table 6.10 and can be used to illustrate the method of synthesizing a network from a resistance function. Assuming a degree of 6, we apply Levy's method, which provides coefficients of a rational polynomial in s:

$$Z(s) = \frac{0s^6 + 0s^5 + 0s^4 + 0s^3 + 0s^2 + 0s + 1}{-4s^6 + 0s^5 - 4s^4 + 0s^3 + 0s^2 + 0s + 1} \tag{6.25a}$$

The nonzero coefficients are associated with even powers of s, and the polynomial is real for all ω. Hence, we can substitute $j\omega$ for s:

$$R(\omega) = \frac{1}{4\omega^6 - 4\omega^4 + 1} \tag{6.25b}$$

and then use the Brune-Gewertz method to compute $Z(s)$. The roots of the denominator of (6.25a) are

$\pm\ 0.647799$
$\pm\ (0.176101 \pm j0.860717)$

Selecting the three left half-plane roots which are the poles of $Z(s)$, we have

$$Z(s) = \frac{a_3 s^3 + a_2 s^2 + a_1 s + a_0}{s^3 + s^2 + s + 0.5} \tag{6.26}$$

The real part of (6.26) is the same as (6.25b), and we can apply (6.16a) to (6.26), substitute jω for s, and then equate the numerators:

$$\left(-a_2\omega^2 + a_0\right)\left(-\omega^2 + 0.5\right) + \left(-a_3\omega^3 + a_1\omega\right)\left(-\omega^3 + \omega\right) = 1 \tag{6.27a}$$

Expanding (6.27a) and equating like powers of ω on either side of the equal sign, we have

$$\begin{bmatrix} 0.5 & 0 & 0 & 0 \\ -1 & 1 & -0.5 & 0 \\ 0 & -1 & 1 & -1 \\ 0 & 0 & 0 & 1 \end{bmatrix} \begin{bmatrix} a_0 \\ a_1 \\ a_2 \\ a_3 \end{bmatrix} = \begin{bmatrix} 1 \\ 0 \\ 0 \\ 0 \end{bmatrix} \tag{6.27b}$$

which has the solution $a_0 = 0.5$, $a_1 = 1$, $a_2 = 1$, and $a_3 = 0$. Thus, the driving-point impedance for the data in Table 6.10 is

$$Z(s) = \frac{s^2 + s + 0.5}{s^3 + s^2 + s + 0.5} \tag{6.28}$$

and we need only extract the elements to synthesize an appropriate network. Comparing this impedance function with $Z(s)$, given in Section 4.4 for the low-pass Butterworth filter, reveals that (6.28) is the inverse and the circuit is the dual of the network in Figure 4.6 (i.e., shunt C, series L, and shunt C). In fact, the data in Table 6.10 was created using the inverse of the Butterworth $Z(s)$ function. The input resistance of the network shown in the figure is independent of L_1, and we cannot explicitly determine $Z(s)$; that is, the circuit is not a minimum reactance network. However, we could use the input conductance and follow a similar procedure, or simply invert the driving-point impedance, as in this example.

Table 6.10 Resistance Versus Radian Frequency

ω	$R(\omega)$	ω	$R(\omega)$
0.00	1.000000	3.50	$1.480645 \cdot 10^{-4}$
0.25	1.014866	4.00	$6.509993 \cdot 10^{-5}$
0.50	1.230769	4.50	$3.166981 \cdot 10^{-5}$
0.75	2.240700	5.00	$1.666639 \cdot 10^{-5}$
1.00	1.000000	6.00	$5.511433 \cdot 10^{-6}$
1.50	$3.800475 \cdot 10^{-2}$	7.00	$2.169230 \cdot 10^{-6}$
2.00	$5.181347 \cdot 10^{-3}$	8.00	$9.688110 \cdot 10^{-7}$
2.50	$1.217563 \cdot 10^{-3}$	9.00	$4.762991 \cdot 10^{-7}$
3.00	$3.856537 \cdot 10^{-4}$	10.00	$2.525252 \cdot 10^{-7}$

6.5.2 Computing Reactance From Discrete Resistance Functions

Implicit in the Brune-Gewertz method of determining $Z(s)$ from the minimum reactance $R(\omega)$ function or the minimum resistance $X(\omega)$ function is a relationship between the real and imaginary parts of the impedance. That is, once we know the resistance or reactance function, the other component is determined and the impedance defined as the sum of the real and imaginary components. $R(\omega)$ and $X(\omega)$ are related by the *Hilbert transform* [1,28,35,36], which is

$$X(\omega) = \frac{1}{\pi}\int_0^{\infty} \frac{dR}{dy} \ln\left|\frac{y+\omega}{y-\omega}\right| dy \tag{6.29}$$

in the form expressed by Bode [1] for a linear frequency scale. We usually work with discrete data points, and (6.29) cannot be solved in closed form. However, Carlin's approach [21] is to approximate $R(\omega)$ with a number of straight line segments and then integrate (6.29) using the piecewise linear interpolation functions. The $R(\omega)$ function can be expressed as a linear combination of the total resistance excursion r_k in each segment; that is,

$$R(\omega) = r_0 + \sum_{k=1}^{n} a_k(\omega) r_k \tag{6.30a}$$

where $r_0 = R(0)$, the resistance at zero frequency, and $a_k(\omega)$ is the interpolation function:

$$a_k(\omega) = \begin{cases} 0, & \omega <= \omega_{k-1} \\ \dfrac{\omega - \omega_{k-1}}{\omega_k - \omega_{k-1}}, & \omega_{k-1} <= \omega <= \omega_k \\ 1, & \omega >= \omega_k \end{cases} \tag{6.30b}$$

Figure 6.50 shows five data points connected by straight line segments. At zero frequency, all the a_k values are zero and $R(0) = r_0$. Between 0 and ω_1, a_1 linearly increases from 0 to 1, and the resistance contribution from the first segment is $a_1 \bullet r_1$. At and above ω_1, a_1 is equal to 1 and the contribution is r_1. Between ω_1 and ω_2, $a_2 \bullet r_2$ linearly increases from 0 to r_2 and $R(\omega) = r_0 + r_1 + a_2 \bullet r_2$. Thus, the total resistance at any frequency ω is the excursion interpolated from the line segment containing ω plus the excursions from all segments lower in frequency. For example, ω, shown in Figure 6.50, is midway between ω_2 and ω_3. $a_3 = 0.5$, r_3 is (0.6-0.4) or 0.2, and the interpolated excursion is equal to (0.2)(0.5), or 0.1. $R(\omega)$ is equal to $(r_0 + r_1 + r_2 + 0.1) = (0.1 + 0.2 + 0.1 + 0.1)$, or 0.5. The function in (6.29) is to be integrated from zero to infinity; hence, $R(\omega)$ must be zero above some frequency in order for the integral to have a finite value. Cuthbert [28] suggests making r_n dependent; that is,

$$r_n = -\sum_{k=0}^{n-1} r_k \tag{6.30c}$$

which forces $R(\omega_n)$ to equal zero. We must also constrain $R(\omega)$ to be positive if the network is to be realized using passive elements. A negative resistance requires at least one active element in the network.

The consequence of representing $R(\omega)$ as straight line segments is that the reactance can also be expressed in terms of the resistance excursions, r_k; that is,

$$X(\omega) = \sum_{k=1}^{n} b_k(\omega) r_k \tag{6.31a}$$

Applying (6.29), we have

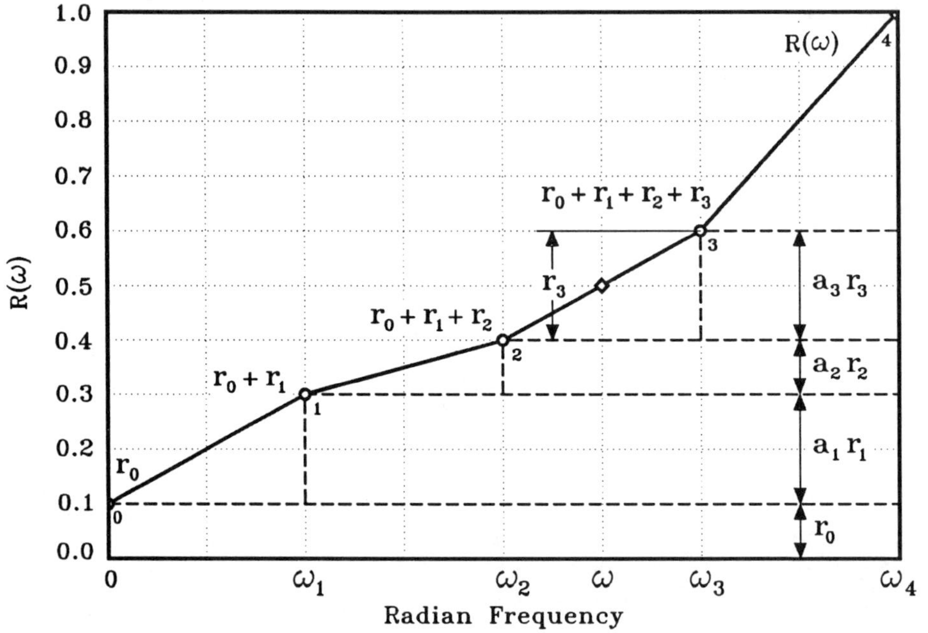

Figure 6.50 Linear interpolation functions.

$$b_k(\omega) = \frac{1}{\pi(\omega_k - \omega_{k-1})} \int_{\omega_{k-1}}^{\omega_k} \ln\left|\frac{y + \omega}{y - \omega}\right| dy \tag{6.31b}$$

which is limited to a single line segment due to the zero slope of the resistance contribution above and below. Bode [1] solved (6.31b) in closed form, and Cuthbert gives the expression for b_k, which is

$$b_k = \frac{B(\omega, \omega_k) - B(\omega, \omega_{k-1})}{\pi(\omega_k - \omega_{k-1})} \tag{6.31c}$$

$b_0 = 0$, and the function $B(\omega,\omega')$ is defined as

$$B(\omega, \omega') = \omega'[(X + 1)\ln(X + 1) + (X - 1)\ln|X - 1| - 2X\ln(X)] \tag{6.31d}$$

where

$$X = \frac{\omega}{\omega'}$$

$B(\omega,0) = 0$ and X is not equal to 0 or 1. Given a data set representing resistance at discrete frequency points, we can compute the impedance at any frequency by first approximating $R(\omega)$ with straight line segments. Next, we compute the r_k excursions for each segment, evaluate $a_k(\omega)$ using (6.30b), and evaluate $b_k(\omega)$ using (6.31c) and (6.31d). Then the impedance is computed using

$$Z = \sum_{k=0}^{n} r_k(a_k + jb_k) \tag{6.32}$$

As an example, we will use the first seven data points in Table 6.10 to compute the impedance at $\omega = 1.25$. Figure 6.51 shows the seven resistance values connected by straight line segments. The first excursion, r_0, is equal to $R(0)$ or 1.0, and we compute r_1 through r_5 as the difference between $R(\omega)$ at points k and k - 1; that is, $R(\omega_k) - R(\omega_{k-1})$. The last excursion, r_6, is the negative sum of all other excursions, which ensures that $R(2.0)$ is zero. Table 6.11 shows the excursions, and the next step is to compute the a_k terms. The frequency of interest is contained in the fifth line segment, as shown in the figure. Therefore, a_0 through a_4 are equal to one, a_6 is zero, and we apply (6.30b) to compute a_5 as 0.5. The b_k terms are computed using (6.31c) and (6.31d), where b_0 is defined as zero. Substituting into equation (6.31c) for $k = 1$,

$$b_1 = \frac{B(1.25, 0.25) - B(1.25, 0.00)}{\pi(0.25 - 0.00)}$$

and we can use (6.31d) to evaluate $B(1.25,0.25)$ and $B(1.25,0.00)$. For $\omega = 1.25$ and $\omega' = 0.25$, $X = (1.25/0.25) = 5$, and

$$B(1.25, 0.25) = 0.25\,[6 \ln(6) + 4 \ln(4) - 10 \ln(5)] = 0.050339$$

$B(1.25,0.00)$ is defined as zero and we have

$$b_1 = \frac{0.050339 - 0.000000}{\pi(0.25)} = 0.064093$$

The r_k, a_k, and b_k values are tabulated in Table 6.11. Using (6.32), the sum of $r_k \bullet a_k$ provides the resistance, the sum of $r_k \bullet b_k$ the reactance, and the impedance at $\omega = 1.25$ is

$Z(1.25) = 0.519003 - j1.326608$

which is labeled as $R(1.25)$ and $X(1.25)$ in Figure 6.51.

The solid traces in Figure 6.51 show the resistance and reactance of the rational polynomial used to create the data in Table 6.10. The broken traces in the figure represent $Z(\omega)$, computed using all 18 resistance values from Table 6.10. The figure shows that the impedance computed from the discrete data points has the same general shape as but does not accurately characterize $Z(\omega)$. The straight line approximations between points 3, 4, and 5 exhibit a significant deviation from the actual $R(\omega)$ function, causing errors in the computed resistance and reactance. We can, however, accurately compute $Z(\omega)$ from $R(\omega)$ by increasing the number of data points between $\omega = 0.25$ and $\omega = 1.5$, providing a better approximation of the resistance function.

Table 6.11 Data for $Z(1.25)$ Calculations

k	ω	$R(\omega)$	r_k	$a_k(1.25)$	$b_k(1.25)$
0	0.00	1.000000	1.000000	1.000000	0.000000
1	0.25	1.014866	0.014866	1.000000	0.064093
2	0.50	1.230769	0.215903	1.000000	0.197821
3	0.75	2.240700	1.009931	1.000000	0.351611
4	1.00	1.000000	-1.240700	1.000000	0.558059
5	1.50	0.038005	-0.961995	0.500000	1.050713
6	2.00	0.005181	-0.038005	0.000000	0.584362

6.5.3 Impedance-Matching Using Resistance Functions

Given a load impedance $Z_L(\omega)$, we can design an appropriate matching network by synthesizing a network with a specified input impedance $Z(\omega)$. A perfect match requires $Z(\omega)$ equal to the complex conjugate of $Z_L(\omega)$ at all frequencies in the passband. However, in most cases a perfect match is not possible using a finite number of elements and, in fact, we may require a controlled mismatch to provide a sloped response. Carlin [21] describes the transducer gain of the matching network and load as

$$T(\omega^2) = \frac{4\,R(\omega)\,R_L(\omega)}{\left[R(\omega) + R_L(\omega)\right]^2 + \left[X(\omega) + X_L(\omega)\right]^2} \qquad (6.33)$$

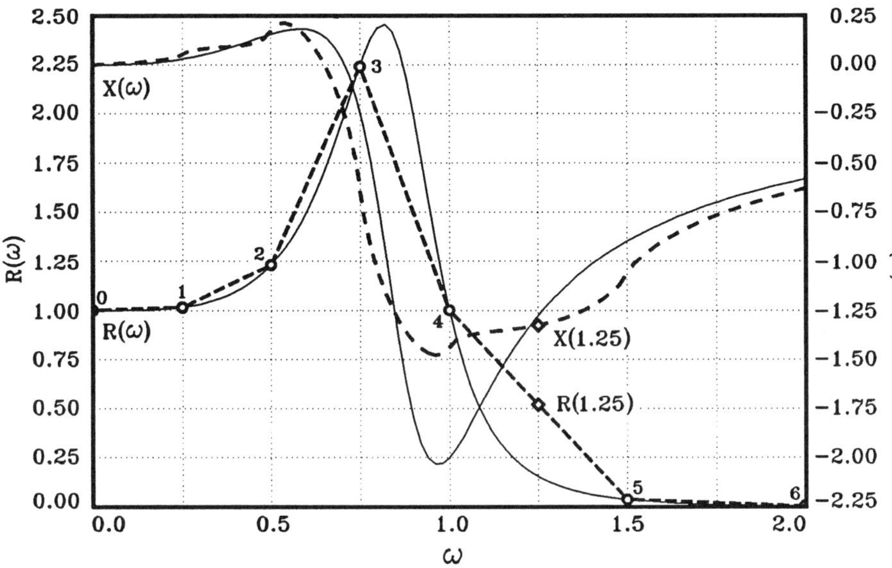

Figure 6.51 Real and imaginary components of *Z(s)*.

where the load impedance is $Z_L = R_L + jX_L$ and the matching network input $Z = R + jX$. The analysis in Section 6.5.2 reveals that the impedance is completely described by the resistance excursion vector **r** when $R(\omega)$ is approximated in a piecewise linear fashion. Therefore, the transducer gain is a function of frequency and the resistance excursions (i.e., $T(\mathbf{r},\omega^2)$), and we need only find the appropriate values of r_k, which provide the required frequency response. Carlin and Komiak [22] suggest minimizing an error function given by

$$E = \sum_{j=1}^{m} e_j^2 = \sum_{j=1}^{m} \left[\frac{T\left(r, \omega_j^2\right)}{g_t\left(\omega_j\right)} - 1 \right]^2 \tag{6.34}$$

where m is the number of frequencies for the gain specification and g_t is the target gain expressed as a ratio. This result illustrates one advantage of the Real Frequency technique for designing matching networks. The optimization variables are real numbers, R, X and $T(\mathbf{r},\omega^2)$ can readily be computed from simple expressions, and the error function is well behaved. However, Cuthbert [28] derives an equally important result. The efficient gradient optimizers require partial derivatives of the error function with respect to the optimization variables, r_k in this case. Numerical differentiation can introduce errors causing convergence problems, as illustrated in Chapter 3. The partial derivative of the error function with respect to r_k is

$$\frac{\partial E}{\partial r_k} = \sum_{j=1}^{m} 2e_j \frac{\partial e_j}{\partial r_k} \tag{6.35a}$$

The transducer gain $T(\omega^2)$ is a function of R and X; hence, the chain rule for partial derivatives applies, and

$$\frac{\partial e_j}{\partial r_k} = \frac{1}{g_t}\left(\frac{\partial T}{\partial R}\frac{\partial R}{\partial r_k} + \frac{\partial T}{\partial X}\frac{\partial X}{\partial r_k}\right) \tag{6.35b}$$

Cuthbert provides an analytical expression for each partial derivative in (6.35b):

$$\frac{\partial T}{\partial R} = 4R_L \frac{\Delta - 2R(R + R_L)}{\Delta^2} \tag{6.35c}$$

$$\frac{\partial T}{\partial X} = 4R_L \Delta^2 \left[-2R(X + X_L)\right] \tag{6.35d}$$

$$\frac{\partial R}{\partial r_k} = a_k - a_n \tag{6.35e}$$

and

$$\frac{\partial X}{\partial r_k} = b_k - b_n \tag{6.35f}$$

where Δ is the denominator of the transducer gain given by (6.33). Thus, the optimization procedure is particularly efficient and accurate. Most optimizers will rapidly converge to a minimum. In fact, Cuthbert lists a short Basic language computer program incorporating a simple gradient optimizer to find the resistance excursions that provide a specified transducer gain when $Z(\omega)$ is terminated with a given load $Z_L(\omega)$. The Real Frequency technique is applied to the design of impedance-matching networks by first defining a load impedance $Z_L(\omega)$ and target gain response $g_t(\omega)$. Next, we approximate $R(\omega)$ using straight line segments, provide initial values for the **r** vector, and then minimize the error function given in (6.34), which results in a transducer gain as close as possible to the desired response. Once the resistance function is defined, the methods of Section 6.5.1 are used to find an appropriate rational polynomial that fits the discrete $R(\omega)$ data points, and then $Z(s)$ is computed and the network synthesized.

Given the load data tabulated in Table 6.12, we can design a low-pass matching network that provides less than 0.1-dB insertion loss from 1.0 to 2.0 GHz. The load data, normalized to 50Ω and 2.0 GHz, is shown in Figure 6.52. Since the matching network is a low-pass structure, we can use the impedance function given by (6.28) to initialize $R(\omega)$. The initial resistance function is shown in Figures 6.51 and 6.52 as the lighter traces. The error function in (6.34) is modified to set each error component to zero when the transducer gain is greater than -0.1 dB. $R(\omega)$ is specified at 32 frequencies between $\omega = 0$ and $\omega = 3$ to ensure a reasonably good piecewise approximation, and then r_1 through r_{30} are optimized. r_{31} is dependent to ensure that $R(3.0) = 0$, and r_0 is the resistance at zero frequency or the output termination of the low-pass network which should equal 1.0. The optimization is successful; the final values are tabulated in Table 6.13 and plotted in Figure 6.52, along with $X(\omega)$ from $\omega = 0.5$ to $\omega = 1.0$ (i.e., the passband).

Table 6.12 Load Data

Frequency (GHz)	$R_L(\omega)$	$X_L(\omega)$	*Frequency (GHz)*	$R_L(\omega)$	$X_L(\omega)$
1.00	21.583	8.588	1.60	23.544	5.855
1.10	22.107	7.997	1.70	23.702	5.547
1.20	22.524	7.468	1.80	23.835	5.269
1.30	22.859	6.996	1.90	23.950	5.015
1.40	23.132	6.574	2.00	24.048	4.784
1.50	23.357	6.196			

The next step is to use Levy's method described in Section 6.5.1 to approximate the $R(\omega)$ data in Table 6.13 with a rational polynomial. $R(\omega)$ must be real for all ω, i.e., even powers of ω. Cuthbert suggests providing $R(\omega)$ data at positive and negative frequencies distributed symmetrically about zero and setting all $X(\omega)$ values to zero. That is, $R(\omega_i) = R(-\omega_i)$ and $X(\omega_i) = 0$, which is the case when the resistance function is written in the form of (6.16b) or (6.17a). Applying Levy's method with 63 positive and negative frequency values from Table 6.13, we have

$$R(\omega) = \frac{0.997689}{-2.792556\,\omega^6 - 8.820153\,\omega^4 - 6.822446\,\omega^2 + 1} \tag{6.36a}$$

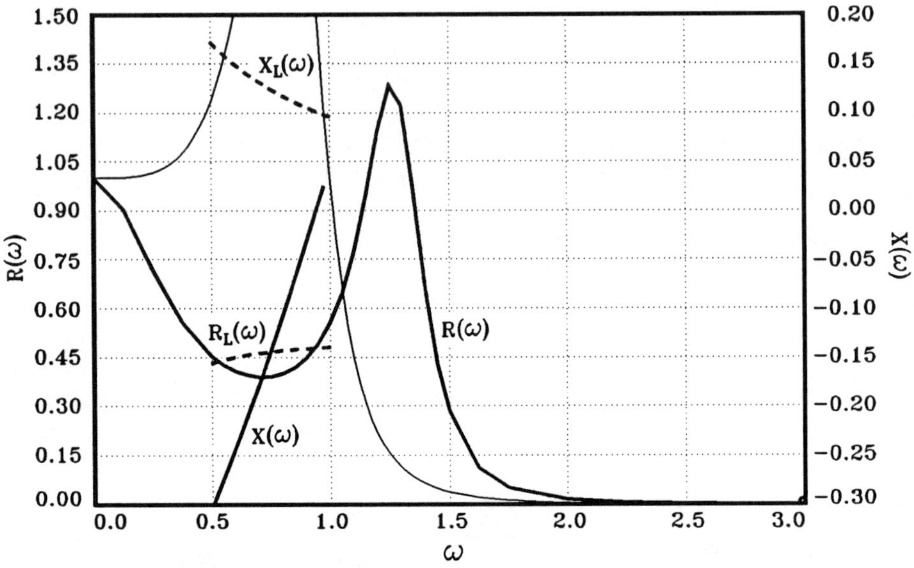

Figure 6.52 Load impedance and optimized network impedance.

Table 6.13 Optimized $R(\omega)$ Values

ω	$R(\omega)$	ω	$R(\omega)$	ω	$R(\omega)$
0.000	1.000000	0.850	0.419120	1.400	0.659320
0.125	0.905040	0.900	0.448300	1.450	0.432480
0.250	0.717180	0.950	0.491980	1.500	0.284600
0.375	0.556700	1.000	0.555340	1.625	0.111700
0.500	0.453820	1.050	0.645980	1.750	0.051460
0.550	0.427320	1.100	0.773580	2.000	0.015140
0.600	0.408140	1.150	0.944800	2.250	0.005800
0.650	0.395880	1.200	1.142420	2.500	0.002620
0.700	0.390400	1.250	1.281320	2.750	0.001320
0.750	0.391880	1.300	1.221620	3.000	0.000000
0.800	0.401040	1.350	0.958660		

for the rational approximating polynomial. We use the Brune-Gewertz method to determine $Z(s)$ by substituting $s^2 = -\omega^2$ in the denominator of $R(\omega)$, computing the roots, and then expanding the left half-plane roots to form the denominator of $Z(s)$. The numerator of Ev $Z(s)$ is equated with the numerator of $R(\omega)$, and the unknown coefficients are determined from the system of simultaneous equations. The driving-point impedance is

$$Z(s) = \frac{0.665832s^2 + 0.441523s + 0.597028}{s^3 + 0.663115s^2 + 1.799087s + 0.598410} \tag{6.36b}$$

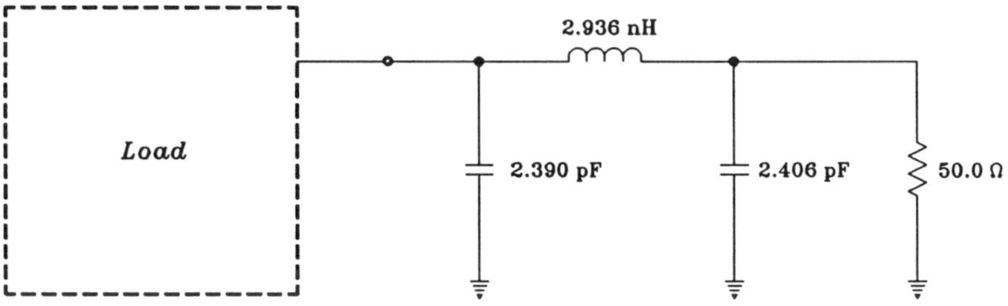

Figure 6.53 Real frequency matching network.

and we can extract the elements, which provides the matching network for the load in Table 6.12. Figure 6.53 shows the topology and element values scaled to 50Ω and 2 GHz, and Figure 6.54 shows the response from 0 to 4.0 GHz.

A simple circuit consisting of a 25Ω resistor in parallel with a 10-nH inductor was used to create the load data in Table 6.12 and to terminate the matching network in Figure 6.53. The shunt inductor accounts for the low-frequency response in Figure 6.54. However, the Real Frequency matching network provides less than 0.1 dB of insertion loss in the passband extending from 1.0 to 2.0 GHz, has the characteristic low-pass response above 2.0 GHz, and is terminated with 50Ω. In this case, we could not design the matching network using classical methods since the high-pass load element cannot be absorbed from the synthesized low-pass network. The Real Frequency technique has certain limitations because of the minimum reactance requirement. For example, we cannot design a matching network with a series input inductor using $R(\omega)$ or adjust the transmission zeros for a complex load at both ends of the network. However, references 21 through 30 discuss various methods that can be used to overcome these limitations and provide a useful, general purpose impedance-matching technique.

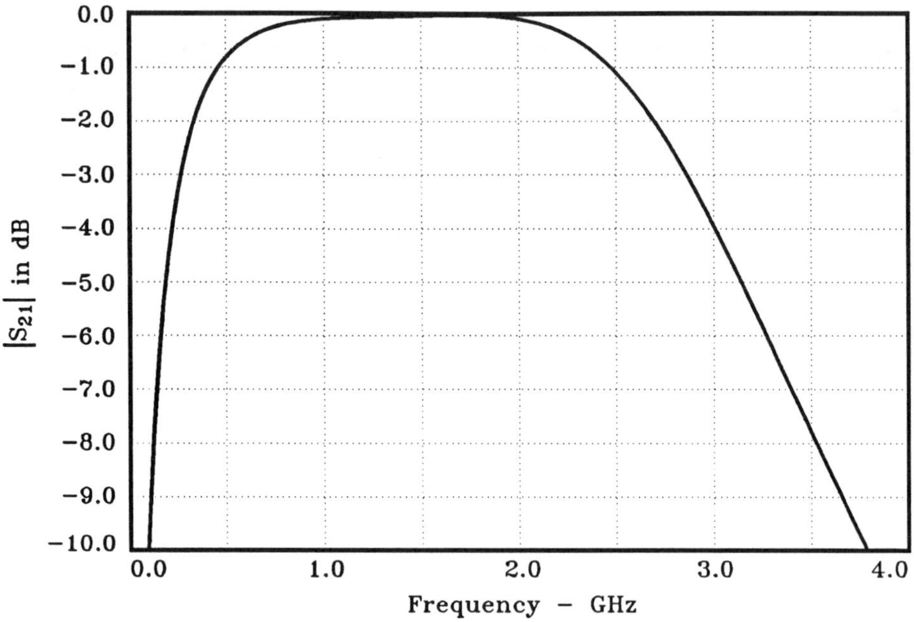

Figure 6.54 Response of the real frequency matching network.

6.6 SUMMARY

The design of broadband matching networks requires a systematic approach to achieve optimum performance. Although we can use graphical or *cut-and-try* methods, the results are not usually acceptable for modern circuits. Network synthesis is a powerful design tool that has several advantages when applied to the design of broadband matching networks. The most important advantage is the number of topologies with identical matching characteristics. We can usually identify one or more networks that have attractive element values or topologies useful for bias injection or other purposes. Another advantage is the efficiency of the matching networks, which, in most cases, will be optimum as synthesized.

The classical method of matching network synthesis is based on insertion loss approximations, and simply consists of defining a network with minimum total loss that has an input termination and reactive element equal to the load. The load elements are removed or absorbed and the remainder is the matching network. Using this approach, the load must be represented as a simple combination of a resistor and reactive element. However, the negative-image modeling technique provides a method that accurately predicts the performance of the circuit, as well as the efficiency of individual matching networks.

The Real Frequency matching technique is fundamentally different from the classical method in that the load does not need to be characterized. Instead, the real component of the load data is used to define the driving-point impedance of the matching network; hence the topology and element values. The Real Frequency technique requires optimization to determine the appropriate resistance excursions of the network. However, the classical method also requires optimization to classify the load model and determine the optimum ripple and MIL. Direct comparison between the two methods is somewhat limited in the literature and there is insufficient information to define the advantages of one technique over the other. It is possible to find examples where each method provides superior results. However, it may turn out that the best approach for the design of impedance-matching networks is some combination of the classical and Real Frequency techniques.

REFERENCES

[1] H.W. Bode, *Network Analysis and Feedback Amplifier Design*, D. Van Nostrand, New York, 1945.

[2] R.M. Fano, "Theoretical Limitations of the Broadband Matching of Arbitrary Impedances," *Journal of the Franklin Institute*, Vol. 249, January 1950, pp. 57-84, and February 1950, pp. 139-154.

[3] D.C. Youla, "A New Theory of Broad-Band Matching," *IEEE Trans. on Circuit Theory*, Vol. CT-11, March 1964, pp. 30-50.

[4] R.E. Collin, *Foundations for Microwave Engineering*, McGraw-Hill Book Company, New York, 1966.

[5] G.L. Matthaei, L. Young, and E.M.T. Jones, *Microwave Filters, Impedance-Matching Networks, and Coupling Structures*, Artech House, Inc., Norwood, Massachusetts.

[6] E. Kreyszig, *Advanced Engineering Mathematics*, John Wiley & Sons, Inc., New York, 1979.

[7] J. Curtis, "Modeling Transistors For Improved Performance," *Microwaves*, February 1975, pp. 33-43.

[8] G.D. Vendelin and M. Omori, "Try CAD for Accurate GaAs MESFET Models," *Microwaves*, June 1975, pp. 58-70.

[9] R.S. Tucker, "Gain-Bandwidth Limitations of Microwave Transistor Amplifiers," *IEEE Trans. on Microwave Theory and Techniques*, Vol. MTT-21, No. 5, May 1973, pp. 322-327.

[10] W.H. Ku and W.C. Petersen, "Optimum Gain-Bandwidth Limitations of Transistor Amplifiers as Reactively Constrained Active Two-Port Networks," *IEEE Trans. on Circuits and Systems*, Vol. CAS-22, No. 6, June 1975, pp. 523-533.

[11] D.J. Mellor, "Improved Computer-Aided Synthesis Tools for the Design of Matching Networks for Wide-Band Microwave Amplifiers," *IEEE Trans. on Microwave Theory and Techniques*, Vol. MTT-34, No. 12, December 1986, pp. 1276-1281.

[12] M.W. Medley, Jr., "Stretch FET Amp Design Beyond Octave Bandwidths," *Microwaves*, Vol. 16, No. 5, May 1977, pp. 54-59.

[13] M.W. Medley, Jr., and J.L. Allen, "Improved Device Modeling for Matching Network Synthesis," *1979 IEEE MTT-S International Microwave Symposium Digest*, April 1979, pp. 350-352.

[14] M.W. Medley, Jr., and J.L. Allen, "Broad-Band GaAs FET Amplifier Design Using Negative-Image Device Models," *IEEE Trans. on Microwave Theory and Techniques*, Vol. MTT-27, No. 9, September 1979, pp. 784-788.

[15] K.C. Gupta, R. Garg, and R. Chadha, *Computer-Aided Design of Microwave Circuits*, Artech House, Inc., Dedham, Massachusetts, 1981, pp. 206-213.

[16] D.J. Mellor and J.G. Linvill, "Synthesis of Interstage Networks of Prescribed Gain Versus Frequency Slopes," *IEEE Trans. on Microwave Theory and Techniques*, Vol. MTT-23, No. 12, December 1975, pp. 1013-1020.

[17] O. Pitzalis, Jr., and R.A. Gilson, "Tables of Impedance Matching Networks Which Approximate Prescribed Attenuation Versus Frequency Slopes," *IEEE Trans. on Microwave Theory and Techniques*, Vol. MTT-19, No. 4, April 1971, pp. 381-386.

[18] W.H. Ku, "Exact Synthesis of Interstage Matching Networks for Broadband Microwave GaAs FET Amplifiers," *1977 IEEE MTT-S International Microwave Symposium Digest*, June 1977, pp. 312-315.

[19] W.H. Ku and W.C. Petersen, "Optimum Gain-Bandwidth Limitations of Transistor Amplifiers as Reactively Constrained Active Two-Port Networks," *IEEE Trans. on Circuits and Systems*, Vol. CAS-22, No. 6, June 1975, pp. 523-533.

[20] G. Szentirmai, "A Filter Synthesis Program," *Systems Analysis by Digital Computer*, F.F. Kuo and J.F. Kaiser, eds., John Wiley & Sons, Inc., New York, 1966.

[21] H.J. Carlin, "A New Approach to Gain-Bandwidth Problems," *IEEE Trans. on Circuits and Systems*, Vol. CAS-24, No. 4, April 1977, pp. 170-175.

[22] H.J. Carlin and J.J. Komiak, "A New Method of Broad-Band Equalization Applied to Microwave Amplifiers," *IEEE Trans. on Microwave Theory and Techniques*, Vol. MTT-27, No. 2, February 1979, pp. 93-99.

[23] H.J. Carlin and P. Amstutz, "On Optimum Broad-Band Matching," *IEEE Trans. on Circuits and Systems*, Vol. CAS-28, No. 5, May 1981, pp. 401-405.

[24] B.S. Yarman and H.J. Carlin, "A Simplified Real Frequency Technique Appliable to Broadband Multistage Microwave Amplifiers," *1982 IEEE MTT-S International Microwave Symposium Digest*, June 1982, pp. 529-531.

[25] B.S. Yarman and H.J. Carlin, "A Simplified Real Frequency Technique Applied to Broad-Band Multistage Microwave Amplifiers," *IEEE Trans. on Microwave Theory and Techniques*, Vol. MTT-30, No. 12, December 1982, pp. 2216-2222.

[26] B.S. Yarman, "Real Frequency Broadband Matching Using Linear Programming," *RCA Review*, Vol. 43, December 1982, pp. 626-654.

[27] H.J. Carlin and B.S. Yarman, "The Double Matching Problem: Analytic and Real Frequency Solutions," *IEEE Trans. on Circuits and Systems*, Vol. CAS-30, No. 1, January 1983, pp. 15-28.

[28] T.R. Cuthbert, Jr., *Circuit Design Using Personal Computers*, John Wiley & Sons, Inc., New York, 1983.

[29] L. Zhu, B. Wu, and C. Sheng, "Real Frequency Technique Applied to the Synthesis of Lumped Broad-Band Matching Networks with Arbitrary Nonuniform Losses for MMIC's," *IEEE Trans. on Microwave Theory and Techniques*, Vol. MTT-36, No. 12, December 1988, pp. 1614-1620.

[30] L. Zhu, "Computer-Aided Synthesis of a Lossy Commensurate Line Network and Its Application in MMIC's," *IEEE Trans. on Microwave Theory and Techniques*, Vol. MTT-39, No. 4, April 1991, pp. 654-659.

[31] O. Brune, "Synthesis of a Finite Two-Terminal Network Whose Driving-Point Impedance Is a Prescribed Function of Frequency," *Journal of Math and Physics*, Vol. 10, No. 3, 1931, pp. 191-236.

[32] C.M. Gewertz, *Network Synthesis*, The Williams and Wilkins Co., Baltimore, 1933.

[33] M.E. Van Valkenburg, *Introduction to Modern Network Synthesis*, John Wiley & Sons, Inc., New York, 1960.

[34] E.C. Levy, "Complex Curve Fitting," *IRE Trans. on Automated Control*, May 1959, pp. 37-43.

[35] E.A. Guillemin, *The Mathematics of Circuit Analysis*, The MIT Press, Cambridge, Massachusetts, 1969.

[36] H.J. Blinchikoff and A.I. Zverev, *Filtering in the Time and Frequency Domains*, John Wiley & Sons, Inc., New York, 1976.

Chapter 7

Practical Applications

The numerical examples in the preceding chapters were created using a programmable calculator, general purpose mathematics software, or simple Basic and C language programs. For the most part, the circuits consisted of simple ideal elements with no loss or parasitics, and explicit expressions were easy to write and evaluate. However, setting design goals and synthesizing the ideal network is only the first step in the design process. We must model the circuit in as much detail as possible to ensure an accurate representation of the hardware. After all, the intent of most design efforts is to produce a competitive product that meets specifications, rather than simply as an academic exercise. When the computer model is accurate, the design should perform as expected the first time power is applied. However, when the model is inaccurate or incomplete, the measured performance will not be that predicted, and the circuit will require tuning and one or more design iterations. Tuning microwave integrated circuits (MIC) is difficult, and tuning monolithic microwave integrated circuits (MMIC) is almost impossible. Additional design iterations are both expensive and time consuming. Although we might occasionally exceed a budget or delivery schedule on a particular project, the overall record must reflect meeting or exceeding specifications under budget and on time. If a competitor can consistently produce similar products that meet specifications, are cheaper, and are delivered on time, we will not be in business for long.

As the circuit model becomes more complex, we can no longer afford the luxury of writing special-purpose procedures for each design. Instead, we must rely on a general-purpose program that contains all the necessary circuit elements, physical models, and interconnections. The methods described in this book can be used to write such a program for in-house use. However, several RF and microwave computer-aided design (CAD) packages are commercially available. These programs represent investments of hundreds of man-years and we cannot hope to provide comparable capability without a similar effort. The software industry is highly competitive, directly benefiting the users. The programs are easy to use, accurate, and capable of analyzing almost any conceivable circuit. The companies often employ leading experts in certain fields to develop techniques that enhance the capability of the software. This means that some methods that represent the state of the art are proprietary and not published. For example, one of the most difficult modeling tasks is accurately characterizing microstrip and stripline discontinuities, such as bends, tee, or cross junctions. We can analyze these structures using numerical techniques [1]. However, these methods require considerable computing power and are too slow when a number of junctions must be analyzed at many frequencies for different physical dimensions. For this reason, most programs use equivalent circuits that converge much faster but not as accurate. Another example is active device models which provide accurate small- and large-signal analysis.

The real test of any CAD program is how well it predicts hardware performance. Numerous articles are published each year by users of the various systems. Measured data and the computer analysis is often included, which can be used to compare the accuracy of the methods used by each. However, for an accurate comparison, we must be certain that the computer model represents the circuit as assembled and tested. Manufacturer's data is seldom accurate for packaged or chip devices at higher microwave or millimeter-wave frequencies. The *s*-parameters are usually measured under specific conditions that may not be representative of our assembly methods. We can usually obtain accurate data by characterizing the device in a test fixture which simulates the circuit environment as accurately as possible. We can include bond wires, bypass capacitors, and bias resistors, which provide *s*-parameters measured at the reference plane of the input and output matching networks. However, even the computer model of a passive circuit may not include all factors that affect the measured performance. The enclosure usually affects the response to some extent, and the transition from a coaxial connector to microstrip has associated discontinuities, as does the connector itself. There may also be a small gap between the connector and substrate which results in a ground path substantially longer than the gap spacing. Time domain analysis and modeling can be an invaluable asset in characterizing these discontinuities to ensure that the simulated circuit accurately represents the hardware.

Historically, it has been very difficult for the designer to justify software tools to upper management. The cost of a CAD system is comparable with that of a complete test station, and it is sometimes easier to procure the hardware and manpower necessary to accommodate extensive tuning and multiple design iterations. However, as a result of the development of inexpensive personal computers and workstations, as well as competitive pressures, most designers now have access to some form of CAD tool. This chapter is intended to illustrate the applications of methods covered in Chapters 1 through 6. The vast majority of new products are designed using CAD systems and the examples in this chapter reflect this trend. The particular software used for the examples includes ACADEMY™, Touchstone™, Libra™, E-Syn™, and LineCalc™[1]. ACADEMY™ is a schematic and layout editor, Touchstone™ is a linear circuit simulator, and Libra™ is a nonlinear circuit simulator which includes the linear simulator as a subset. E-Syn™ is a network synthesis program used for the design of filters and impedance-matching networks, and LineCalc™ is used to compute physical dimensions from electrical parameters.

The amplifier represents one of the most common and complex design tasks undertaken by engineers. The examples in Chapters 2, 3, and 6 illustrate some of the methods that can be used for the design of either narrow or broadband amplifiers. However, most engineers do not exclusively design amplifiers. We are often called upon to design oscillators, filters, mixers, multipliers, switches, phase shifters, attenuators, detectors, and other components. The design of most components is essentially that of providing the appropriate matching network. Since an example of each type of component would require more space than is available in a book this size, we will use the amplifier to illustrate typical CAD techniques. The reader can easily adapt these methods to the design of other components.

7.1 DIRECTIONAL COUPLERS

A story, which is plausible but may or may not be true, relates the experience of engineers in the early days of long-distance telephone service. One side of a conversation between two stations in a telephone network could be heard at a third station even though there was no direct connection. The problem was finally traced to a phenomenon whereby a portion of a signal traveling in one direction down a transmission line is coupled into an adjacent transmission line and sent in the opposite direction. This observation, whether first noticed by telephone engineers or not, leads to one of the more useful circuit elements, which is known by several names: *coupled transmission lines*, *backward-wave coupler*, *directional coupler*, or simply a *coupler*. Coupled transmission lines are used for a number of applications, including power splitting and combining, sampling or signal injection, filters, and the balanced amplifier discussed in Section 7.2. There is a considerable amount of reference material available for the coupled line structure, far too much to include in this book. References 2 through 23 are a small sample of some of this material and can be used as a starting point for further study.

---

[1] ACADEMY, Touchstone, Libra, E-Syn, and LineCalc are registered trademarks of EEsof, Inc., 5601 Lindero Canyon Road, Westlake Village, California, 91362.

We can determine *s*-parameters of the coupled line structure by applying the methods discussed in Section 1.3. We first connect a $Z_O\Omega$ voltage source at port 1, terminate ports 2, 3, and 4 with $Z_O\Omega$ loads, compute the voltage and current at each port, and then compute the incident and reflected waves. The first column of the **S** matrix is computed from a_i and b_i (i.e., s_{11}, s_{21}, s_{31}, and s_{41}). The remaining columns are filled in by moving the source to the other three ports. Figure 7.1 shows a $2V_S$ voltage source connected to port 1 of a symmetrical coupler. Both lines are assumed identical in this example. That is, they have the same impedance and electrical length when analyzed as isolated lines. The figure does not show the common ground plane, which is necessary to form a pair of coupled transmission lines. In the case of microstrip, the ground plane is below the strips, while stripline has a ground plane both above and below. Numerous other physical layouts are possible, such as two isolated conductors inside a coaxial shield. However, the analysis is the same. We need the voltage and currents at the four ports and can use the principle of superposition, which is the basis of a useful technique known as *even and odd mode analysis* [4].

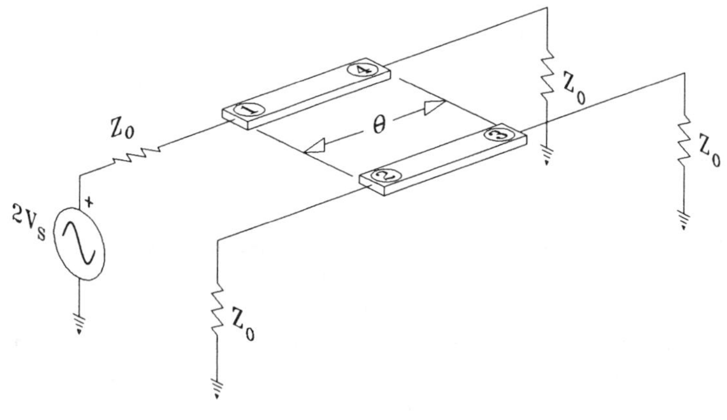

Figure 7.1 Coupled transmission lines excited at port 1.

Figure 7.2(a) shows two in-phase V_S voltage generators connected at ports 1 and 2, which provide an even-mode excitation, or simply the even mode. Figure 7.2(b) shows the odd mode where the generator at port 2 is out of phase. When we add the even and odd mode analysis, the voltage at port 1 adds to $2V_S$, and the positive and negative voltage cancels at port 2, which is the original configuration shown in Figure 7.1. In the even mode, both strips are at the same potential, and there is no coupling between them. Under these conditions, we have

$$V_{1e} = V_{2e} \qquad I_{1e} = I_{2e}$$
$$V_{3e} = V_{4e} \qquad I_{3e} = I_{4e} \tag{7.1a}$$

where the numeric subscript refers to the port number and *e* denotes even mode. The strips are at opposite potentials when excited in the odd mode, and we can write

$$V_{1o} = -V_{2o} \qquad I_{1o} = -I_{2o}$$
$$V_{3o} = -V_{4o} \qquad I_{3o} = -I_{4o} \tag{7.1b}$$

using the subscript *o* to symbolize the odd mode. Essentially, the plane of symmetry parallel to the lines represents an open circuit in the even mode and a short circuit in the odd mode, or an electric and magnetic wall between the two transmission lines, respectively. Therefore, we can treat this problem as two identical isolated transmission lines with a characteristic impedance of Z_{oe} when excited in the even mode, and Z_{oo} when excited in the odd mode. We can use (5.4) to write the input impedance at each port, and then the equalities in (7.1) to compute the four sets of port voltages and currents. This exercise provides *s*-parameters in terms of Z_{oe}, Z_{oo}, and Z_o. However, when we compare the incident and reflected waves with those of an isolated transmission line discussed in Section 2.1.2, the *s*-parameters of the coupled line structure turn out to be simple combinations of the *s*-parameters of the $Z_{oe}\Omega$ and $Z_{oo}\Omega$ transmission lines; that is,

$$s_{11} = s_{22} = s_{33} = s_{44} = \frac{1}{2}\cdot\left[s_{11}(Z_{oe}) + s_{11}(Z_{oo})\right] \tag{7.2a}$$

$$s_{12} = s_{21} = s_{34} = s_{43} = \frac{1}{2}\cdot\left[s_{11}(Z_{oe}) - s_{11}(Z_{oo})\right] \tag{7.2b}$$

$$s_{13} = s_{31} = s_{24} = s_{42} = \frac{1}{2}\cdot\left[s_{12}(Z_{oe}) - s_{12}(Z_{oo})\right] \tag{7.2c}$$

and

$$s_{14} = s_{41} = s_{23} = s_{32} = \frac{1}{2}\cdot\left[s_{12}(Z_{oe}) + s_{12}(Z_{oo})\right] \tag{7.2d}$$

The *s*-parameter symmetry of the $Z_{oe}\Omega$ and $Z_{oo}\Omega$ transmission line elements (TLE) is used to simplify (7.2); that is, $s_{11} = s_{22}$ and $s_{12} = s_{21}$ for both elements, and there are only four distinct s-parameters as a result of the symmetry of the coupled line structure. The factor of one-half is caused by the superposition of two V_S sources into one $2V_S$ source, as shown in Figure 7.1. The negative signs in (7.2b) and (7.2c) reflect the out-of-phase odd-mode voltages.

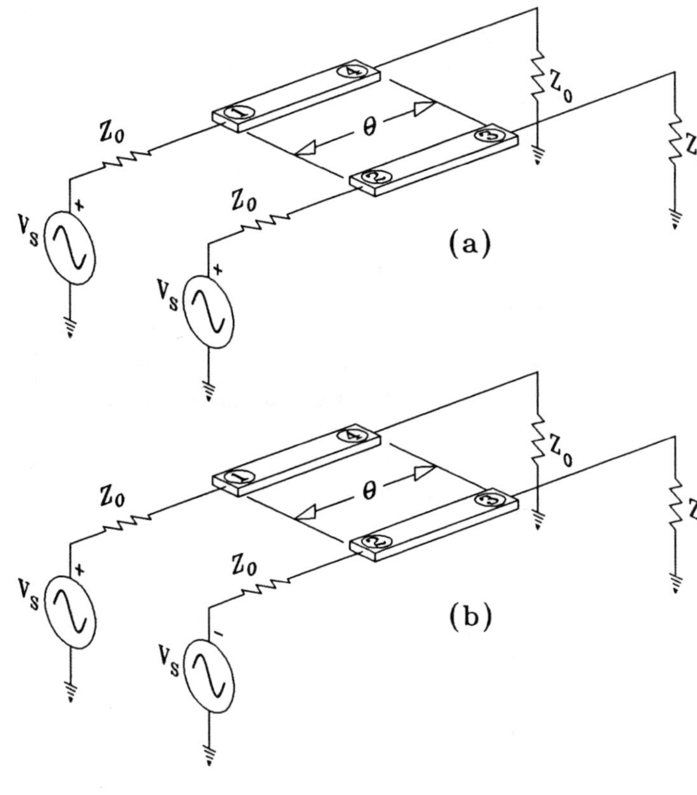

Figure 7.2 Excitation: (a) even-mode; (b) odd-mode.

We normally use the coupler in a $Z_o\Omega$ system and require the input reflection coefficient at all four ports to equal zero (i.e., the matched condition). Setting (7.2a) to zero we have,

$$s_{11}(Z_{oe}) = -s_{11}(Z_{oo})$$

Then, substituting $s_{11}(Z_{oe})$ and $s_{11}(Z_{oo})$ from (2.8), we find that

$$Z_o = \sqrt{Z_{oe} Z_{oo}} \tag{7.3a}$$

where Z_o is known as the *characteristic impedance of the coupler.* When the coupler is matched at all ports, the voltage at port 1 is V_S and, using superposition, the voltage at port 2 is $V_2 = V_{2e} - V_{2o}$. Applying standard transmission line equations [4], we can write the port 2 voltage as

$$V_2 = \frac{jV_s \sin\theta \left[\sqrt{\dfrac{Z_{oe}}{Z_{oo}}} - \sqrt{\dfrac{Z_{oo}}{Z_{oe}}} \right]}{2\cos\theta + j\sin\theta \left[\sqrt{\dfrac{Z_{oe}}{Z_{oo}}} + \sqrt{\dfrac{Z_{oo}}{Z_{oe}}} \right]}$$

The voltage is a maximum when the coupler is a quarter-wavelength long or an odd multiple of a quarter-wavelength (i.e., Sinθ = ± 1 and Cosθ = 0). Thus, we can define a *maximum voltage coupling coefficient* as

$$\left| \frac{V_2}{V_s} \right| = k = \frac{Z_{oe} - Z_{oo}}{Z_{oe} + Z_{oo}} \tag{7.3b}$$

Usually Z_o and the coupling coefficient are design parameters, and we must determine the values for Z_{oe} and Z_{oo}. Rearranging (7.3b) and then substituting into (7.3a), we have

$$Z_{oe} = Z_o \sqrt{\frac{1 + k}{1 - k}} \tag{7.3c}$$

$$Z_{oo} = Z_o \sqrt{\frac{1 - k}{1 + k}} \tag{7.3d}$$

which provide the necessary expressions.

Using (2.8) and (7.3) in (7.2), we can write all 16 *s*-parameters of the $Z_o\Omega$ coupler:

$$s_{11} = s_{22} = s_{33} = s_{44} = 0 \tag{7.4a}$$

$$s_{12} = s_{21} = s_{34} = s_{43} = \frac{jk\sin\theta}{\sqrt{1 - k^2}\cos\theta + j\sin\theta} \tag{7.4b}$$

$$s_{13} = s_{31} = s_{24} = s_{42} = 0 \tag{7.4c}$$

and

$$s_{14} = s_{41} = s_{23} = s_{32} = \frac{\sqrt{1 - k^2}}{\sqrt{1 - k^2}\cos\theta + j\sin\theta} \tag{7.4d}$$

At all frequencies, the input reflection coefficient at the four ports is zero and the coupler is matched. Similarly, s_{13}, s_{31}, s_{24}, and s_{42} is zero, and we see that there is total isolation between diagonal ports when numbered as shown in Figure 7.1. For example, we usually consider port 1 as the *input port*. Since $s_{31} = 0$, no power is coupled to port 3 and it is known as the *isolated port*. At the quarter-wavelength frequency, s_{12}, s_{21}, s_{24}, and s_{42} are equal to k, the maximum voltage coupling coefficient. Thus, power entering any port is coupled to the adjacent port, and that port is known as the *coupled port*. Evaluating (7.4d) at $\theta = 90°$, we see that the power emerging from the opposite or *direct port* is the difference between the incident power and that reflected from the coupled port (i.e., square root of (1 - k^2)). We also observe that there is a phase shift of -90° due to the j factor in the denominator of (7.4d). Thus, the coupled port is in phase with the input and the direct port is 90° out of phase. This type of structure is sometimes called a *quadature coupler.*

It is apparent from the analysis that the coupler is useful for coupling power from one signal path to another, as the name implies. When power is applied to port 1 in Figure 7.1, some in-phase power is coupled to port 2, the remainder is reflected from port 4, and port 3 is isolated. Similarly, if we apply power to port 4, port 3 becomes the coupled port, port 1 the direct port, and port 2 the isolated port. Thus, if ports 1 and 4 are inserted in a signal path, we can sample the forward power at port 2 and the reflected power at port 3. In fact, the directional coupler is sometimes used for this purpose in network analyzers or to monitor power delivered to and reflected from an antenna system. However, further analysis of this structure reveals other, equally important properties. Figure 7.3 shows the signal flowgraph of a $Z_o\Omega$ coupler, where the coupled and direct ports (i.e., ports 2 and 4) are terminated with an arbitrary reflection coefficient. The flowgraph does not show the zero coefficient paths which provide a simplified diagram. In fact, there are no loops. We are interested in the input reflection coefficient and the power delivered to port 3. We have only one path to a_1, and the value is 1. There are two paths to b_1: a path from a_1 to b_4, b_4 to a_4, a_4 to b_1, and a second path from a_1 to b_2, b_2 to a_2, then a_2 to b_1. Using the methods of Section 1.3.3, we can write b_1/b_s as

$$\frac{b_1}{b_s} = s_{12} \cdot \Gamma_C \cdot s_{21} + s_{41} \cdot \Gamma_D \cdot s_{14}$$

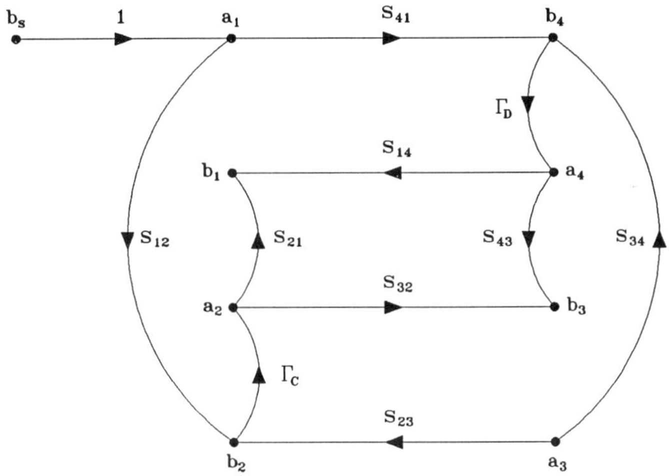

Figure 7.3 Signal flowgraph of terminated coupler.

Similarly, there are two paths from b_s to b_3, and

$$\frac{b_3}{b_s} = s_{12} \cdot \Gamma_C \cdot s_{32} + s_{41} \cdot \Gamma_D \cdot s_{43}$$

Dividing the appropriate quantities and then using (7.4),

$$s'_{11} = s_{12}^2 \Gamma_C + s_{14}^2 \Gamma_D \tag{7.5a}$$

and

$$s'_{31} = s_{12} s_{32} (\Gamma_C + \Gamma_D) \tag{7.5b}$$

where the prime is used to denote the combined *s*-parameters. The special case where $k^2 = 0.5$ (i.e., -3 dB or equal power split between the coupled and direct ports) proves to be a useful condition. Evaluating (7.4) at the quarter-wavelength frequency for $k^2 = 0.5$ and then substituting into (7.5), we have

$$s'_{11} = \frac{1}{2}(\Gamma_C - \Gamma_D) \tag{7.6a}$$

and

$$s'_{31} = \frac{-j}{2}(\Gamma_C + \Gamma_D) \tag{7.6b}$$

Thus, the input reflection coefficient is one-half the difference of the two terminations, and the transfer coefficient from port 1 to port 3 is one-half the sum shifted in phase by -90°.

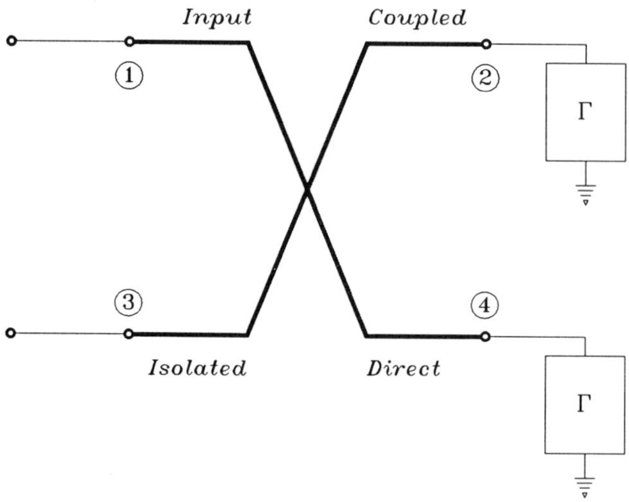

Figure 7.4 Coupler terminated at coupled and direct ports.

Figure 7.4 shows a coupler which has identical terminations connected at the direct and coupled ports. The figure shows the structure in a configuration where the coupled lines cross over, resulting in the coupled and direct ports on the same side. It turns out that this configuration is convenient for schematic representation as well as physical implementation. Since both terminations are equal, (7.6a) reveals that the input reflection coefficient is equal to zero for all values of Γ. Equation (7.6b) also shows that $|s_{31}| = |\Gamma|$, the phase of s_{31} is the phase of Γ shifted by -90°, and, since the coupler is symmetric, we have the same conditions for s_{33} and s_{13}. Hence, the two-port is matched under all conditions, and we can control the magnitude and phase of s_{13} and s_{31} by adjusting Γ. If we terminate the coupled and direct ports with a variable resistance (i.e., a PIN diode), we have maximum power transfer between ports 1 and 3 when R is very small or very large. When R = 50Ω, all power is absorbed by the terminations, and $s_{13} = s_{31} = 0$. Thus, we can make a variable attenuator using a relatively simple circuit. When the terminations are realized as a pure reactance, variable capacitance, or a varactor diode, for example, the magnitude of Γ is one for all values, and maximum power is transferred between ports 1 and 3. However, the phase is controlled by the phase of Γ. Small capacitance values approximate an open circuit, $1\angle 0°$, and $s_{31} = 1\angle 0°\text{-}90°$ or $1\angle -90°$. Large capacitive reactance forms a short circuit, $1\angle 180°$, $s_{31} = 1\angle 180°\text{-}90°$ or $1\angle 90°$, and we have a matched phase-shifter capable of almost 180° of continuous phase shift.

Coupled transmission lines also find widespread use in filter structures [8]. Parallel-coupled, interdigital, and comb-line are all common filter types and can easily be realized using microstrip, stripline, suspended substrate, or mechanical fabrication methods. In many instances the coupled lines are asymmetric; that is, the two lines are not identical when analyzed as isolated transmission lines. The even- and odd-mode analysis does not apply and we must resort to other means to characterize the structure [6,8,14-16]. The coupled lines may also be part of an array of coupled lines where there is coupling to the nearest neighbor on either side, as well as second order coupling to all other lines in the array. When two ports of a coupler are terminated using combinations of opens and shorts, we often find that there is a simple equivalent circuit consisting of TLEs, open and shorted stubs, and sometimes ideal transformers [8,21]. The equivalent circuits are useful for realizing a synthesized distributed-element filter in coupled line form. However, some topologies contain series open stubs and can be used to realize this element on microstrip or stripline.

7.2 BALANCED AMPLIFIERS

Figure 7.4 suggests a possible solution to a problem often encountered in the design of amplifiers: mismatched input or output resulting from a requirement for low noise or high output power. If we replace the one-port terminations with identical two-port amplifiers, the reflection coefficients at the coupled and direct ports are equal (i.e., s'_{11} of each amplifier), and the input is matched at the quarter-wavelength frequency. However, the input power is split between the two amplifiers and we must recombine the output. Figure 7.5 shows two identical amplifiers with gains of $A\angle\theta$ cascaded between a pair of -3-dB couplers. We can apply an incident wave of $1\angle 0°$ and then compute the wave reflected from the output port. Essentially, Figure 7.5 is a pictorial flowgraph of the forward paths: a_1/b_s is $1\angle 0°$, and we can use the diagram to compute b_2/b_s and s_{21}. The input wave emerges from the direct port of the input coupler with a value of -j0.707 or $0.707\angle -90°$ and enters amplifier a where it is amplified and emerges with a value of $0.707A\angle \theta - 90°$. Similarly, the input wave emerges from the coupled port with a value of $0.707\angle 0^\circ$, enters amplifier b, is amplified, and emerges with a value of $0.707A\angle\theta$. From the perspective of amplifier a, the output shown in the figure is the coupled port, and the reflected wave has a value of $(0.707\angle 0^\circ)(0.707A\angle\theta - 90^\circ)$ or $0.5A\angle\theta - 90^\circ$. With respect to amplifier b, the output is connected to the direct port and the wave emerges with a value of $(0.707\angle -90^\circ)(0.707A\angle\theta)$ or $0.5A\angle\theta - 90^\circ$,

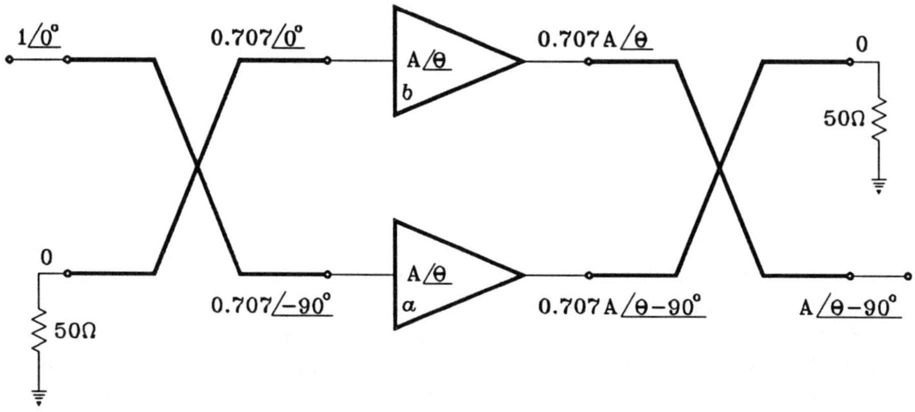

Figure 7.5 Balanced amplifier gain analysis.

and the total output is the sum or $A\angle\theta - 90^\circ$. Analyzing the waves incident on the terminated output port shown in Figure 7.5, we have $(0.707\angle -90^\circ)(0.707A\angle\theta - 90^\circ)$ from amplifier a and $(0.707\angle 0^\circ)(0.707A\angle\theta)$ from amplifier b. The sum is zero and we see that no power is lost in the coupler terminations. However, the terminations do absorb the power, which is reflected from the input and output ports of the amplifier. We can use a similar analysis to show that the reflected power cancels at the input and output ports and adds at the terminated ports.

The configuration shown in Figure 7.5 was introduced by Engelbrecht and Kurokawa [24] in 1965 and is known as a *balanced amplifier*. Although we can use isolators and circulators to realize a matched input and output, the balanced amplifier has several characteristics that make it one of the most common amplifier topologies in use today. One obvious advantage is the equal power split between the two amplifiers. Since each active device operates at one-half the power level, we can double the output power of a single device. Another advantage is reliability. If one amplifier fails, the gain is reduced by approximately 6 dB, which, depending on the system requirements, may or may not be classified as a critical failure. In fact, we often remove power from one side when troubleshooting a balanced amplifier. If the gain drops by 6 dB, we assume the amplifier is good and look at the other side. Kurokawa [25] has rigorously analyzed the balanced amplifier and shows that the noise figure is the same as that of the individual amplifiers (assuming no loss in the couplers), and the third-order intermodulation is improved by an average of 9 dB. He also investigated the effects of imperfect terminations, nonideal couplers, and an imbalance between the two amplifiers. The analysis is rather involved only because there are so many paths in the network: 16 for each coupler and 4 for each amplifier. However, we can take advantage of modern CAD tools and analyze this circuit with little effort.

Thus far we have only considered the coupler operating at an electrical length of 90° where (7.4) is reduced to simple expressions in k. However, we normally operate over broader bandwidths and must consider the effects of electrical lengths different from 90°. Figure 7.6 shows the schematic of an ideal balanced amplifier. That is, two 0-dB gain amplifiers with matched input and

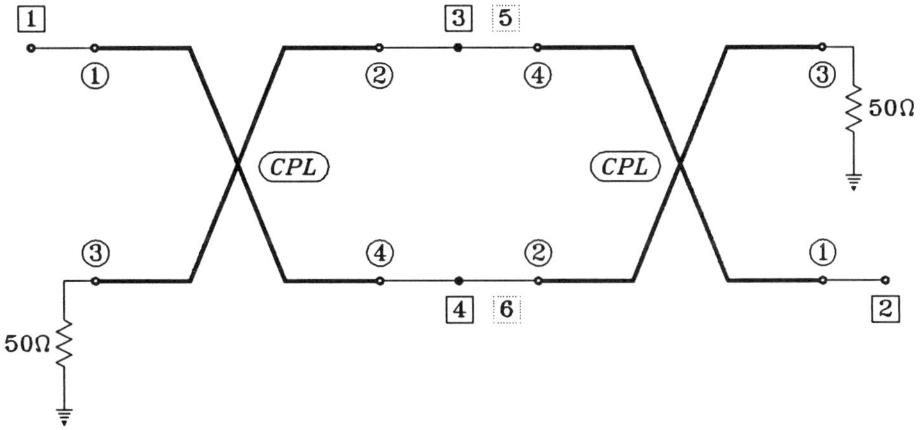

Figure 7.6 Ideal balanced amplifier analysis.

output, which we simulate by directly cascading two couplers. We will model this circuit in two steps. First we will create the four-port s-parameters of one coupler, terminate the isolated port, and then save the three-port s-parameters as a subnetwork called *CPL*. Then we will cascade two CPL networks to form the two-port balanced amplifier, which we will call *BAMP*. The figure shows two sets of node numbers. The circled numbers represent node numbers of the coupler, and the numbers enclosed in boxes are nodes of the balanced amplifier. Once we terminate node 3 of the coupler, the four-port becomes a three-port and node 4 is renumbered to node 3 in subnetwork CPL. Nodes 3 and 4 of the balanced amplifier are internal and used to connect the two couplers. This approach has two advantages. By saving the terminated coupler, the program computes the parameters of the coupler and resistor once, reduces the network to a three-port, and then recalls the data from memory when used again, thus reducing the execution time. We can also plot or list the parameters of the terminated coupler to examine the behavior of this portion of the circuit.

We will design the couplers to operate in a 50Ω system and a frequency range extending from 4.0 to 8.0 GHz. Equal power split requires $k^2 = 0.5$. Substituting into (7.3c), we have Z_{oe} = 120.710678Ω, and from (7.3d), Z_{oo} = 20.710678Ω. Figure 7.7 shows the Libra™ circuit file corresponding to the network in Figure 7.6. Although the documentation adequately describes the syntax, it is

```
1     ! Chapter 7, Example 1
2     ! Analysis of a coupler and ideal balanced amplifier
3
4     CKT
5        CLIN 1 2 3 4 ZE=120.711 ZO=20.711 E=90 F=6
6        RES  3 0     R=50
7     DEF3P    1 2 4    CPL                        ! Save CPL network
8
9        CPL  1 3 4                                ! Input coupler
10       CPL  2 4 3                                ! Output coupler
11    DEF2P    1 2      BAMP                       ! Save balanced amp
12
13    OUT
14       CPL     DB[S21] GR1                       ! Coupled port
15       CPL    ANG[S21] GR2
16       CPL     DB[S31] GR1                       ! Direct port
17       CPL    ANG[S31] GR2
18       BAMP    DB[S21] GR3                       ! Balanced amp gain
19
20    FREQ
21       SWEEP  4.0  8.0  0.05                     ! Frequency
22
23    GRID
24       RANGE  4.0     8.0  0.5                   ! Frequency axis
25       GR1   -2.0    -4.0  0.2                   ! CPL gain
26       GR2   30.0  -120.0  10.0                  ! CPL phase
27       GR3    0.0    -0.2  0.02                  ! BAMP gain
```

Figure 7.7 Ideal balanced amplifier circuit file.

anticipated that not all readers will have access to this particular program and a brief explanation of the format may be useful. The line numbers shown in the figure are used to aid in the discussion and are not part of the circuit file. All text following an exclamation point is assumed to be a comment and the remainder of the line is ignored. Similarly, blank lines can be used to improve readability and are also ignored. The circuit description begins with the block labeled *CKT* on line 4. Line 5 is the definition of an ideal four-port coupler. *ZE* is the even-mode impedance, *ZO* is the odd-mode impedance, *E* is the electrical length, and *F* is the corresponding frequency. The node numbers are shown in Figure 7.6, and we have specified an electrical length of 90° at a frequency of 6.0 GHz (i.e., the center frequency as defined in Ch. 5 for distributed elements). Line 6 defines a 50Ω resistor which is connected from node 3 to node 0, the common ground. Line 7 defines a three-port network called *CPL* with ports created from nodes 1, 2, and 4. Each node is paired with the ground node to form a port. The sequence of node numbers corresponds with the port numbers of the saved network in numerical order; that is, the first number becomes port 1 of the saved network, and so forth. In this case, nodes 1, 2, and 4 of the coupler form ports 1, 2, and 3 of CPL, respectively. Once a network is defined, the circuit is cleared and we can model another network using the same node numbers. Line 9 connects the input coupler to nodes of the balanced amplifier, line 10 connects the output coupler, and line 11 defines the node numbers and name of the saved two-port. The circuit description is complete and we must specify the output parameters, which is the purpose of the *OUT* block beginning on line 13. Each line has a format consisting of a circuit name, parameter, and output specification. In this instance, we have specified that the gain in decibels from the input to the coupled and direct ports of the coupler should be plotted on *GR1*, rectangular grid number 1, and the associated phase is plotted on *GR2*. Line 18 designates the gain of the ideal balanced amplifier to be plotted on *GR3*. Following the output section, we have the frequency specifications on lines 20 and 21. Line 21 results in a frequency sweep from 4.0 to 8.0 GHz in 50-MHz steps. The *GRID* block beginning on line 23 defines the x and y axes for GR1, GR2, and GR3 in the form of an upper, lower, and step size value. We also have the option of defining units in a *DIM* block. In this case, the default units of ohms, gigahertz, and degrees are acceptable.

Figure 7.8 is the GR1 plot of the ideal coupler, CPL, where the response of the coupled port is marked with box symbols, and the direct port is identified by plus symbols. We observe an equal power split at the center frequency, but at higher and lower frequencies less power emerges from the coupled port. In fact, Equation (7.4b) predicts no coupling at 0 and 12.0 GHz since $\sin\theta$ is in the numerator and equal to zero for $\theta = 0°$ and 180°. Figure 7.9 is the GR2 plot of the insertion phase, and the traces are identified in the same manner as Figure 7.8. The figure shows a negative slope with increasing frequency and phase angles of 0° and -90° for the coupled and direct ports at the center frequency of 6.0 GHz. However, the phase difference of 90° is preserved at all frequencies. Figure 7.10 shows the response of the ideal balanced amplifier, and we see that the insertion loss of the lossless network is less than 0.01 dB from 5.0 to 7.0 GHz, but increases rapidly at the band edges. The response is due to the unequal power split of the ideal couplers shown in Figure 7.8. In this case, the maximum insertion loss is less than 0.1 dB and would be acceptable for most applications. However, multioctave designs usually require some modifications to reduce the maximum loss and achieve an equal-ripple response. Matthaei, Young, and Jones [8] provide methods that can be used to design multisection couplers achieving multioctave performance. References 22 and 23 also discuss broadbanding techniques. However, within limits, we can modify the design procedure and achieve similar results with a single coupler.

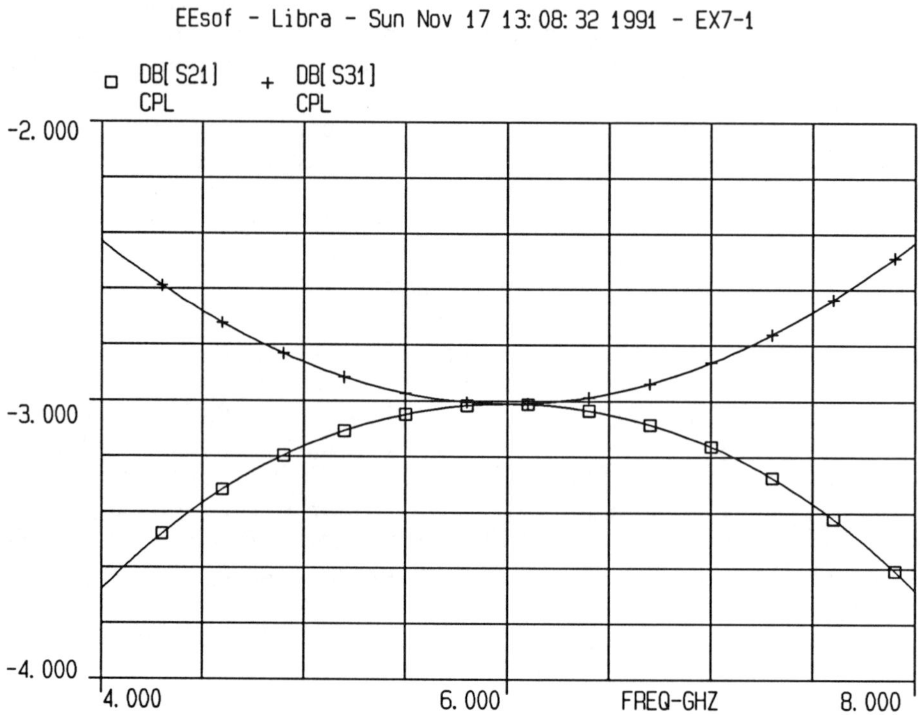

Figure 7.8 S_{21} and S_{31} in decibels for the ideal coupler.

The only option we have for reducing the power imbalance shown in Figure 7.8 is to change the coupling coefficient such that the coupled response shifts up and the direct response shifts down. The result will be an overcoupling at the center frequency and a reduction of the maximum imbalance across the passband. The transmission coefficient of the coupled port, s_{21}, is given by Equation (7.4b) and is equal to k at the center frequency. The response at the direct port, s_{41}, is given by (7.4d), and the magnitude at f_1 and f_2 must equal k for an equal-ripple response. Thus, choosing f_2, we have

$$k^2 = \left| \frac{\sqrt{1 - k^2}}{\sqrt{1 - k^2}\cos\theta_2 + j\sin\theta_2} \right|^2$$

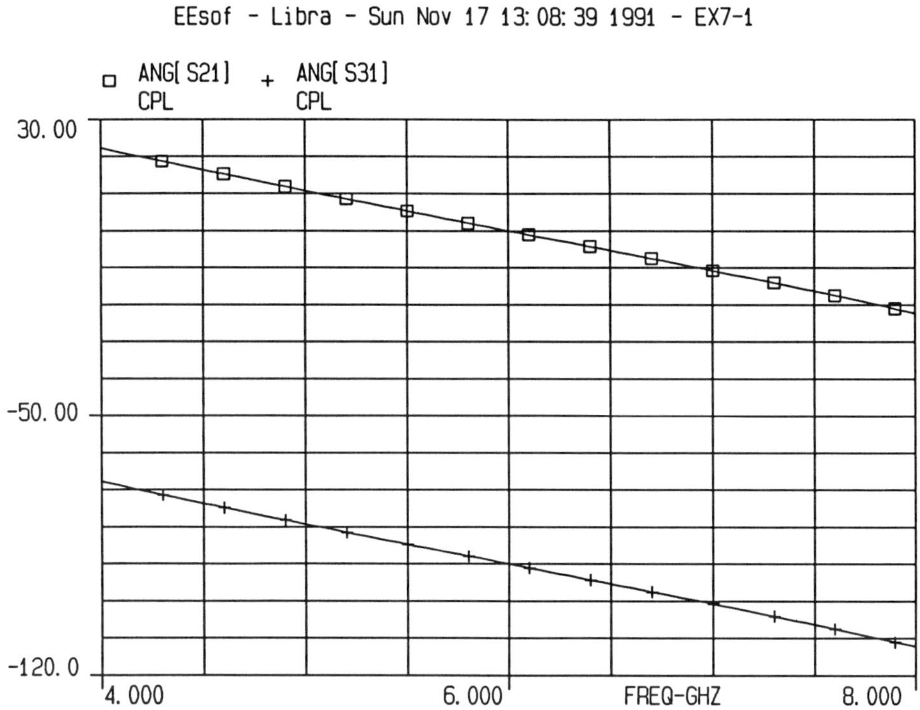

Figure 7.9 Angle of s_{21} and s_{31} in degrees for the ideal coupler.

where the magnitude squared is used for convenience. Expanding and using a trigonometric identity for $\text{Sin}^2\theta$, the expression becomes

$$k^4 \cos^2\theta_2 - 2k^2 + 1 = 0$$

which has the solution

$$k^2 = \frac{1}{1 - \sin\theta_2} ; \quad \frac{1}{1 + \sin\theta_2}$$

Since $\text{Sin}\theta$ is positive for angles less than 180° and we require k to be less than one,

$$k^2 = \frac{1}{1 + \sin\theta_2} \tag{7.7}$$

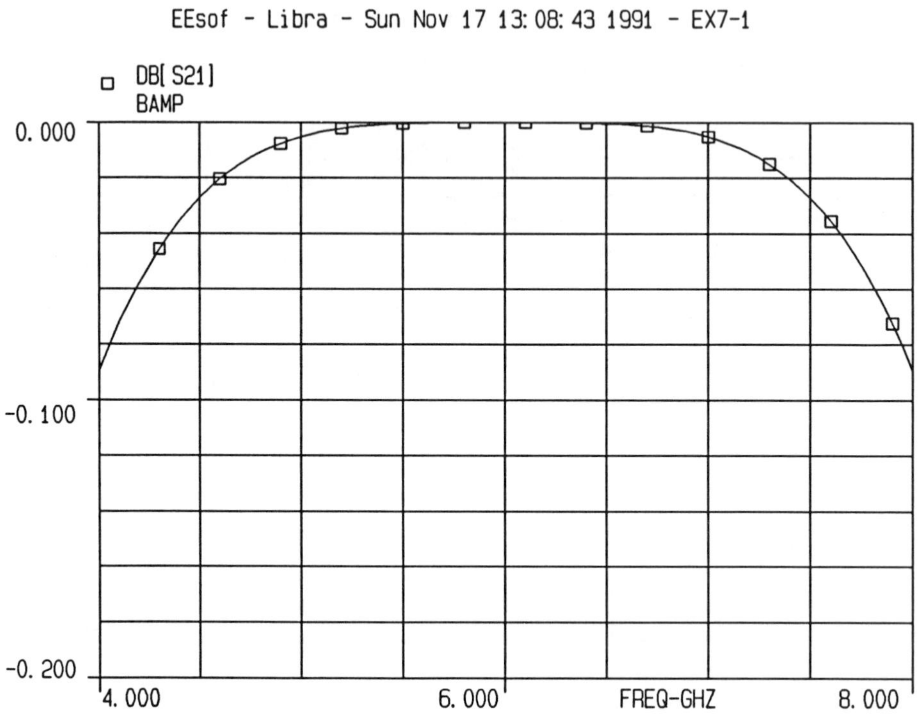

Figure 7.10 S_{21} in decibels for the ideal balanced amplifier.

In this case, f_2 = 8.0 GHz and θ_2 = 120°. Substituting into (7.7) provides the coupling coefficient; k^2 = 0.535898 or approximately 2.709-dB coupling at 6.0 GHz. Using (7.3), we have Z_{oe} = 127.122988Ω and Z_{oo} = 19.665995Ω, and we can change line 5 in Figure 7.7 and then reanalyze the equal-ripple response. Figure 7.11 is a plot of the coupled and direct ports of the ideal coupler, which illustrates the modified performance. The imbalance is equal ripple and the response of the balanced amplifier shown in Figure 7.12 has similar characteristics. The maximum ripple of the ideal amplifier has been reduced from 0.09 to 0.02 dB across an octave bandwidth.

7.2.1 A 4.0- to 8.0-GHz Balanced Amplifier

We designed a single-stage 4.0- to 8.0-GHz amplifier in Section 6.2.3 which can be combined with the equal ripple couplers to form a balanced amplifier. The amplifier represents a typical design. It is not matched because of low noise or high power requirements and is potentially unstable. However, it does exhibit flat gain. The amplifier topology is shown in Figure 6.33, and we can modify the circuit file of Figure 7.7 by inserting the circuit definition between lines 7 and 9. That is, we first model the three-port terminated coupler model, the two-port amplifier stage, and then connect two couplers and two amplifiers to form a model of the complete balanced amplifier. Referring to Figure 7.6, we break the circuit at nodes 3 and 4 where the couplers are joined and insert the two amplifier stages between nodes 3-5 and 4-6 as shown. Figure 7.13 shows the portion of the circuit file used to define the individual amplifier stage and complete balanced amplifier. The amplifier model is simplified somewhat from that shown in Figure 6.33. The bias networks and stabilizing circuit adjacent to the field-effect transistor (FET) gate are not included. The first four lines define the input matching network, the next line defines s-parameters for the active device, and the following five lines describe the output matching network. The *TLOC* element is an open stub, *TLSC* is a shorted stub, and *TLIN* represents a transmission line element. The *S2PA* keyword designates a two-port s-parameter disk file called *N70000A.S2P*, with the three terminals connected to the specified circuit nodes.

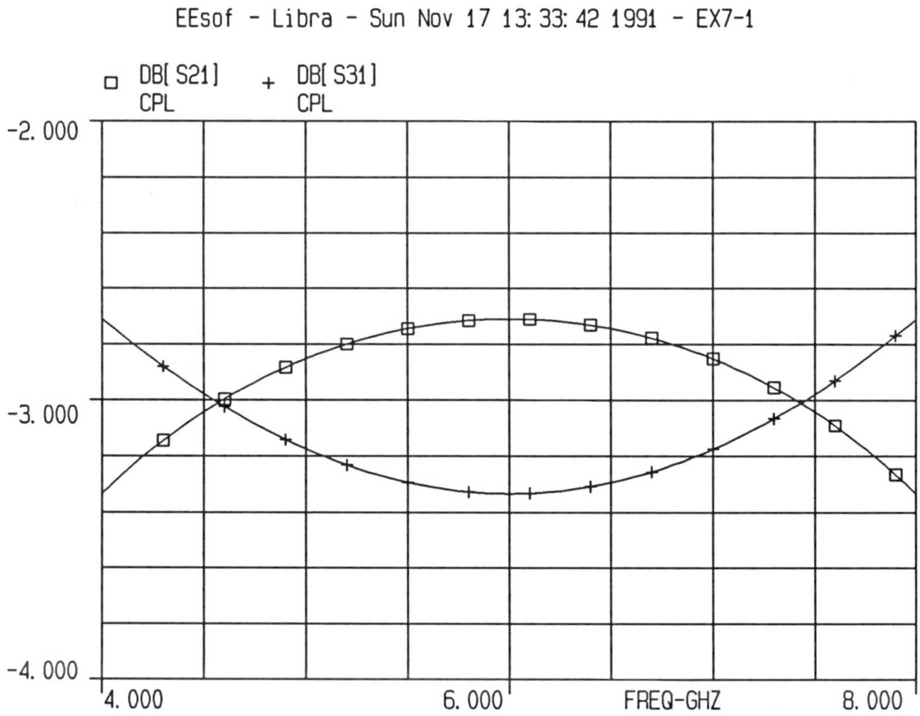

Figure 7.11 Response of the ideal equal-ripple coupler.

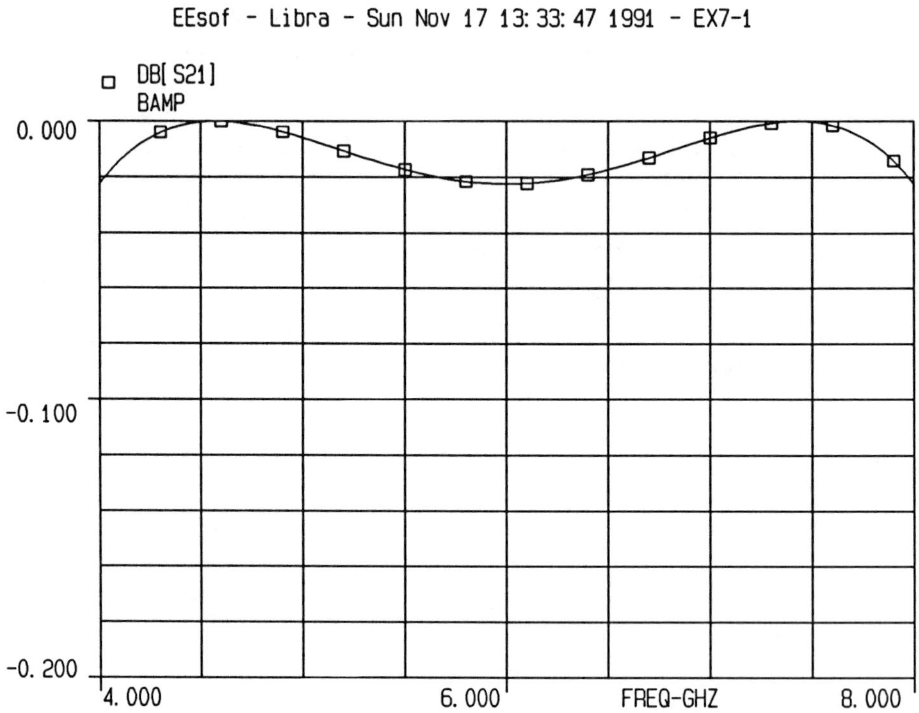

Figure 7.12 Response of the ideal equal-ripple balanced amplifier.

```
  TLOC 1 0     Z=53.59561 E=26.93698 F=8     ! Input match
  TLIN 1 2     Z=75.44379 E=37.15846 F=8
  TLOC 2 0     Z=77.24053 E=30.00000 F=8
  TLSC 2 0     Z=100.0000 E=43.96482 F=8
  S2PA 2 3 0   N70000A.S2P                   ! Device data
  TLIN 3 4     Z=99.87277 E=76.03526 F=8     ! Output match
  TLOC 4 0     Z=93.74686 E=48.29771 F=8
  TLIN 4 5     Z=90.16510 E=57.26069 F=8
  TLOC 5 0     Z=92.64652 E=39.00423 F=8
  TLSC 5 0     Z=98.91803 E=79.58394 F=8
DEF2P   1 5    AMP                           ! Save amp

  CPL   1 3 4                         ! Input coupler
  AMP   3 5                           ! Side B
  AMP   4 6                           ! Side A
  CPL   2 6 5                         ! Output coupler
DEF2P   1 2    BAMP                   ! Save balanced amp
```

Figure 7.13 Partial circuit file for the 4.0 to 8.0 GHz balanced amplifier.

Figure 7.14 shows the Libra™ analysis of the gain, $|s_{11}|$, and $|s_{22}|$ in decibels. The scale for gain is shown on the left, and the scale for reflection coefficients is on the right. The curves are identified by the symbols at the top of the figure. The gain response shown for the balanced amplifier in Figure 7.14 is

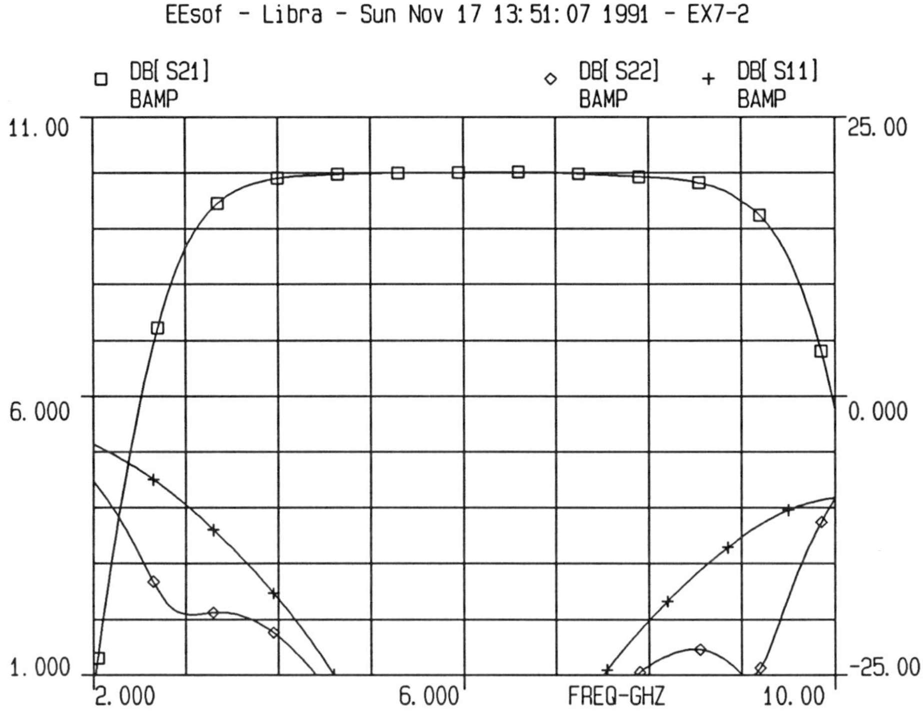

Figure 7.14 4.0- to 8.0- GHz balanced amplifier analysis.

not much different from the original in Figure 6.32. However, the couplers add an additional 0.1-dB loss at the band edges, and the gain is between 9.9 and 10.0 dB in the passband, compared with 10.00 $\pm$ 0.01 dB for the original amplifier. The coupler response also adds 1.5 dB of rejection at 2.0 and 10.0 GHz. The worst-case passband return loss is 18 dB for $|s_{11}|$ and 20 dB for $|s_{22}|$. This amplifier was designed directly from the manufacturer's data, and the performance is acceptable for most applications: flat gain, low input and output voltage standing wave ratio (VSWR), noise figure equal to that of the individual amplifiers, and twice the output power. However, the original design is potentially unstable, and we must consider the stability of the balanced amplifier as well. An analysis reveals that the minimum stability factor is 1.23 across the full 2.0- to 18.0-GHz frequency range where data is provided. Further analysis confirms that the necessary and sufficient conditions for unconditional stability given in Section 2.3.2 have been met. As it turns out, an added benefit of the balanced amplifier is improved stability.

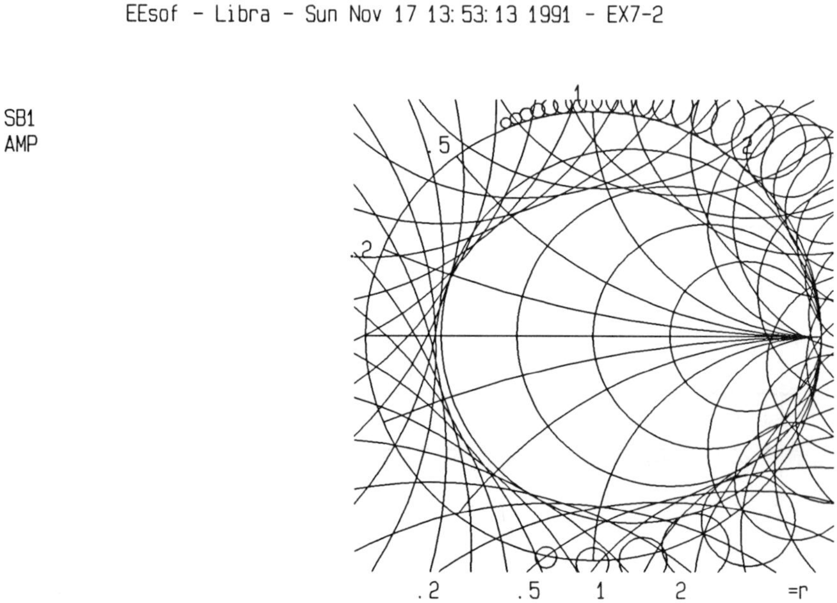

Figure 7.15 Input plane stability circles for the original amplifier.

Analysis of the input reflection coefficient of a coupler terminated with arbitrary loads at the coupled and direct ports is given by (7.6a) and is simply one-half the difference of the reflection coefficients of the two terminations. Referring to the balanced amplifier shown in Figure 7.5, we see that the reflection coefficients at the input and output ports of each amplifier are one-half the difference between the input or output termination and the resistor. For example, with respect to the input of side *a*, the balanced amplifier input is the direct port, and the resistor is connected to the coupled port. Since the reflection coefficient of the resistor is zero, the input of each amplifier sees $\pm\Gamma_S/2$ and the output, $\pm\Gamma_L/2$. Thus, the maximum magnitude of the reflection coefficient is 0.5. In fact, if we map the two coupler ports, the result is a Smith chart reduced by one-half and possibly rotated by 180°, depending on whether the resistor is connected to the direct or coupled port (i.e., $(\Gamma - 0)/2$ or $(0 - \Gamma)/2$). Examining Figure 6.26, which shows the input plane stability circles for the active device, we find that all reflection coefficients less than 0.5 fall completely within the stable region. The device stability circles are not directly applicable since the matching networks transform the impedances, but only within the positive real region. However, Figure 7.15 shows the input plane stability circles referenced to the input of the amplifier and plotted from 2.0 to 18.0 GHz in steps of 0.25 GHz. We see that reflection coefficients less than 0.5 are, in fact, completely within the stable regions and this balanced amplifier is unconditionally stable.

7.3 MICROSTRIP DESIGN

The balanced amplifier in Section 7.2.1 was modeled using ideal components. Unfortunately, ideal elements only exist somewhere in computer memory. We must decide on a method of construction, compute the physical parameters, and then model a circuit that represents the actual hardware. The key to a successful product is accurate physical modeling. CAD tools provide the means to easily analyze and optimize a specified circuit model. However, the correlation between the measured and predicted performance is a direct result of the accuracy of the physical model. The axiom of GIOG (garbage in, garbage out) is certainly true when it comes to computer modeling. RF and microwave circuit design may once have been an art. However, it is now more of a science and success depends directly on the knowledge of the engineer. It is somehow comforting to know that after thousands of man-hours of collective effort to develop CAD tools, the designer is still the most important factor.

7.3.1 Microstrip Couplers

The introduction in Chapter 5 discussed the quasi-TEM propagation mode and dispersive characteristics of microstrip, which cause the impedance and effective dielectric constant to be functions of frequency. Although the change is only a few percent across normal operating frequencies, it can affect the performance of broadband components. These properties also affect the performance of the backward-wave coupler. Figure 7.2 shows the even- and odd-mode excitation for a pair of coupled lines. The fields are considerably different when the lines are at equal and opposite potentials. The percentage of fields contained in air and the dielectric media are also different, which leads to dissimilar effective dielectric constants for the two modes or unequal even- and odd-mode phase velocities. The coupler *s*-parameters are computed from combinations of Z_{oe} and Z_{oo}. Since both modes have equal physical length but differing phase velocities, the *s*-parameters do not cancel as in the ideal case. In fact, each mode has a different quarter-wavelength frequency, and we cannot even define the center frequency without considerable effort. The most pronounced effect is a reduction in isolation and a corresponding degradation of *directivity* (i.e., the difference between the isolation and coupling in decibels). Typically, a 3-dB octave band microstrip coupler exhibits 15 to 20 dB of isolation. March [26] discusses several methods for compensating the unequal microstrip phase velocities. One technique that can be used to improve the directivity [27] is to meander the coupled region in much the same manner as a saw blade. This method effectively increases the physical length of the coupling region and alters the odd-mode length without affecting the even-mode length too much. Another technique is to add lumped capacitors between the lines [28]. However, placing a cover one substrate height above the microstrip [20] equalizes the even and odd modes and is sometimes easier than the other methods. When high directivity is a design goal, stripline is a better choice.

The first balanced amplifiers [24] used overlay couplers. That is, the coupled lines were placed one on top of the other and separated by a thin dielectric material. The physical dimensions of a 3-dB coupler are realizable and the configuration is convenient because the coupled and direct ports can easily be made to exit the same side providing the required symmetry. The simplest microstrip couplers are planar (i.e., two parallel lines separated by a gap). However, the spacing between lines for 3-dB coupling is too small to be consistently realized using practical fabrication methods. It is also difficult to cross the coupled lines without degrading the performance. Fortunately, in 1969 Lange [29] described an interdigitated coupler which solved both of these problems. In fact, the structure is in widespread use today and is known as a *Lange coupler*. Figure 7.16 shows the conductor pattern of a microstrip Lange coupler from the automatic layout facility of ACADEMY™. The four ports are located as noted and bond wires attach the unconnected end of each line to alternating lines. Essentially, the coupler consists of four interdigitated fingers and three coupling regions. The bond wires force equal voltage conditions on alternate lines, and the result is tight coupling and practical dimensions. The structure is not perfect. There is some phase shift between the fingers attached to the isolated and coupled ports in the figure, as well as finite reactance of the bond wires and discontinuities caused by the transition from a wide to narrow linewidth and fringing off the open ends of the fingers. However, it does perform very well into millimeter wavelengths, although the directivity is degraded due to the unequal microstrip phase velocities. We can change the number of fingers(even integers) as necessary to obtain practical dimensions for a number of different substrate materials and coupling factors. The four- and six-finger couplers are the most practical configurations.

7.3.2 Microstrip Discontinuities

The most significant sources of error in the computer model for microstrip and stripline circuits are the discontinuities caused by intersecting or otherwise changing the width or direction of a TLE or stub. Considerable effort has been expended

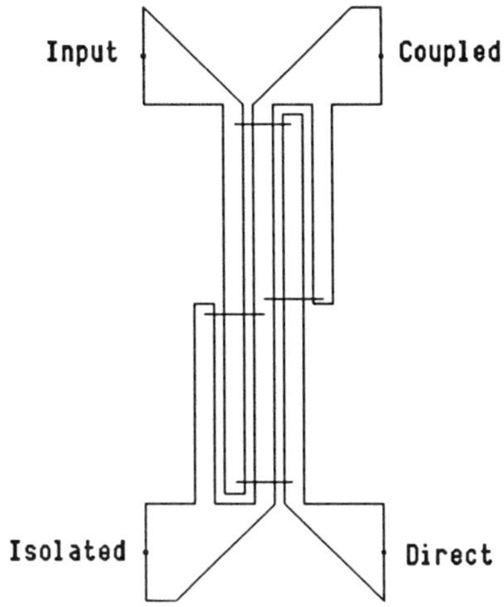

Figure 7.16 Conductor pattern of a microstrip Lange coupler.

to accurately characterize microstrip and stripline discontinuities [10-13]. Three of the most common are shown in Figure 7.17. The *tee junction* results when we connect a stub, and the *cross junction* is formed by two stubs on opposite sides of a transmission line. The *bend* is caused by a change in direction, which is often necessary to efficiently utilize a given area. One problem with these junctions is determining the physical reference plane. For example, if we attach a stub to a TLE, is the length measured from the center of the transmission line, the edge, or some other point? We can measure these discontinuities as two-, three-, or four-port networks and then recall the data when necessary. However, this approach is not practical since it is valid only for one specific set of linewidths. Most programs use equivalent circuits valid for a broad range of substrate parameters and linewidths. The models provide the necessary accuracy within the constraints of reasonable execution times. In fact, software companies often use the accuracy of these computer models as selling points and as a basis for comparison with competing products.

There are numerous discontinuities or physical models in addition to the tee, cross, and bend, such as changes in linewidth due to cascaded TLEs, fringing off the end of open stubs, via holes used for grounding, tapered lines, radial stubs, bonding wires, and flat ribbons. The best source of information on available models is the Element Catalog which accompanies the software. Usually, if the developers have taken time to implement a particular model, it is worthwhile including it in the circuit when applicable. The complete circuit model is dependent on fabrication techniques, materials, and components. Thus, the models will vary from company to company depending on operating procedures and, in

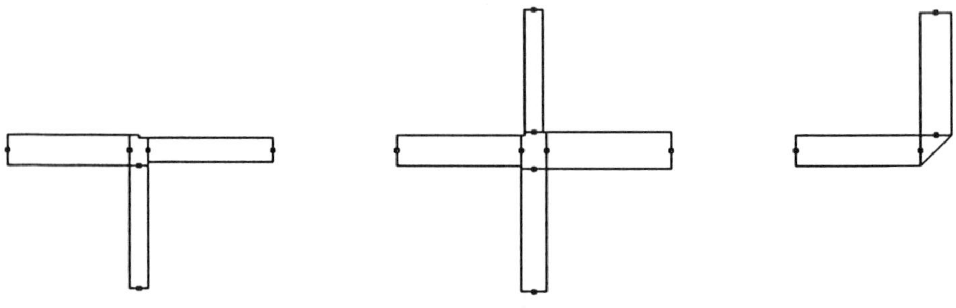

Figure 7.17 Microstrip discontinuities: (a) tee; (b) cross; (c) bend.

fact, may be different for engineers within the same design group. When creating the physical model of a particular circuit, it usually helps to draw a picture of the assembled parts. For example, we designed a two-stage 4.0- to 8.0-GHz amplifier in Section 6.3. Figure 6.41 shows the topology, which requires a blocking capacitor between the two shorted stubs in the output matching network to avoid shorting the drain bias. Figure 7.18(a) shows a profile view of one possible method for attaching the capacitor. The transmission line is fabricated in two sections separated by a small gap. A parallel-plate chip capacitor is attached to one side and a wire is bonded from the top plate to the other section of transmission line. Figure 7.18(b) shows an equivalent circuit model. The microstrip gap is modeled as three capacitors. The two capacitors labeled C_f represent fringing off the open ends of the transmission lines, and C_g accounts for coupling across the gap. The bond wire is modeled [30] as the inductor labeled L, and the capacitor is represented by the *R-L-C* elements enclosed within the broken box. Manufacturer's sometimes provide the Q and resonant frequency of capacitors. The resistor accounts for the loss, and a small inductor is included to model series resonance. However, the capacitor may also have a parallel resonant frequency which requires a parallel inductor. We must provide a model representing the capacitor over a broad frequency range, especially if the series and parallel resonant frequencies fall within the operating range. Some power is lost in the bond wire and fringing off the open ends of the TLEs, which, if significant, can be accounted for by placing resistors at the appropriate points. However, a close examination of Figure 7.18(a) reveals that the capacitor and bond wire have not been placed exactly at the end of each TLE. Thus, a more accurate representation would be open shunt stubs instead of the fringing capacitors. In fact, if manual assembly methods are used, the length of the open stubs and TLEs, as well as the bond wire will vary from circuit to circuit, depending on the exact placement of the capacitor. Another consideration is the reference plane of the capacitor. The dimensions of the component may be an appreciable fraction of a wavelength, and we must decide exactly where physical lengths should be measured. Does the TLE extend to the left side of the capacitor, center, or the right side? A comparable MMIC component might consist of a dielectric layer deposited over one transmission line and then the conductor pattern of the second TLE deposited directly on top of the dielectric. In this case, the model would be quite different. The end of the top transmission line forms one edge of the capacitor, and the fringing capacitance is included with the total. The connection from the top of the capacitor is more accurately modeled as a TLE, although the impedance might change somewhat because of the dielectric and added height above the ground plane.

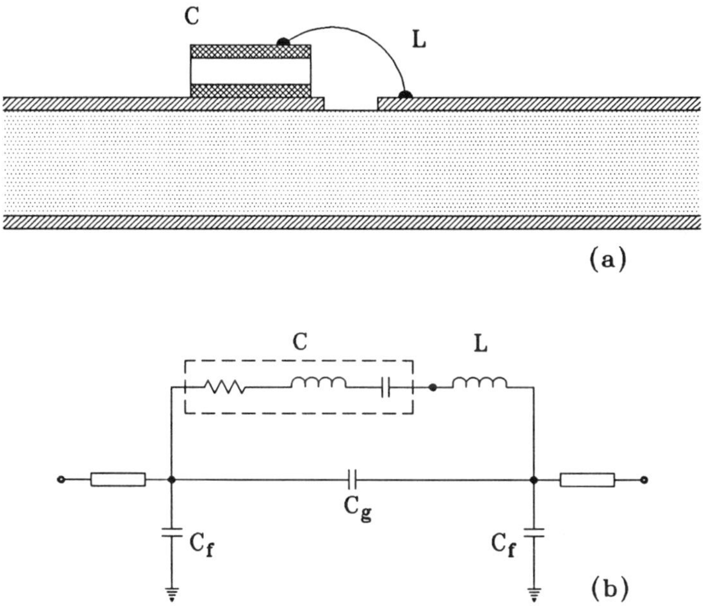

Figure 7.18 Series capacitor: (a) physical; (b) equivalent.

It may seem as if modeling a single component is complicated and the task almost impossible when we consider the number of possibilities in a typical circuit. However, the latest generation of CAD tools includes electromagnetic simulators which can be used to analyze complex physical structures. We also have a better alternative of using measured data to create accurate models for components mounted in common configurations. For example, we usually use capacitors at microwave and millimeter-wave frequencies for bypass or blocking functions. Distributed elements are better controlled and we can usually synthesize effective matching networks using only TLEs and stubs. We can measure several capacitor values in series and shunt configurations using various-width transmission lines and then develop models that are functions of capacitance and linewidth. Time-domain analysis and modeling are often useful since we can see the individual parasitic elements as a function of distance, although the resolution is sometimes inadequate for smaller components. The equivalent circuits can often be simplified by using existing models. For example, the capacitor in Figure 7.18 might be a combination of the microstrip gap and bond wire models, a TLE to account for the reference plane shift, and *R-L-C* elements or a built-in capacitor model to simulate the parallel-plate capacitor. The CAD software sometimes incorporates a means of adding user-defined elements, and we can include an accurate representation of the element as mounted in our circuit. The effort to measure and model components can be time consuming. However, the benefit of accurate physical models more than compensates in short order. Hardware performs as expected, thus eliminating redesigns and reducing design costs and delivery time. The models may be valid for several years, at least until we change the assembly process or the manufacturer makes a change in the component.

7.3.3 A Microstrip Balanced Amplifier

As an example of the steps involved in a typical microstrip design, we can use the 4.0- to 8.0-GHz amplifier originally designed in Section 6.2.3 and analyzed as a balanced amplifier in Section 7.2.1. The most effective procedure is to determine the negative-image models, synthesize ideal matching networks as shown in Figure 6.30, and then separately convert each network into a microstrip equivalent. This approach has the advantage that the only optimization goal is to minimize the insertion loss of each network when terminated with the appropriate load. We can easily compute the physical dimensions and then add discontinuities and bias networks as necessary to accurately model the circuits. However, the entire design is based on the *s*-parameters of the active device. We plotted the input and output plane in Figures 6.26 and 6.27, defined ideal matching networks as shown in Figure 6.28, and then optimized to achieve the specified performance. Manufacturer's data for the NE70000 FET chip does not include bond wires or other parasitics. Hence, the circuit in Figure 6.33 does not represent the parameters of the FET as assembled. In this instance we will add the parasitics after the fact; that is, we will add bonding wires and the self-bias network and then reoptimize in an attempt to compensate. Ideally, we would begin the design with measured data which includes bond wires and bias networks, thus providing an accurate representation of the device characteristics throughout all phases of the design.

Before beginning the conversion we must define the substrate material. In this case, we will use 25-mil (0.635-mm) gold-plated alumina with a relative dielectric constant of 10, conductor thickness of 150 μin (3.81 μm), and surface finish of 2 μin (0.0508 μm) RMS. The substrate loss tangent is 0.001, and we assume a cover height of 500 mils (1.27 cm). After adding the device parasitics and reoptimizing, we will convert the electrical parameters of the input matching network into physical dimensions, add discontinuities, and then optimize again. Next, we will do the same for the output matching network, optimize the complete model, and add Lange couplers to form a balanced amplifier. However, the Lange couplers add loss and we must increase the gain to compensate. Using LineCalc™, we compute a linewidth of 1.67 mils (0.042 mm), spacing equal to 1.56 mils (0.040 mm), and a length of 199.97 mils (5.08 mm) for the 50Ω, 2.709-dB Lange coupler centered at 6.0 GHz. We also compute a 50Ω linewidth of 23.59 mils (0.60 mm).

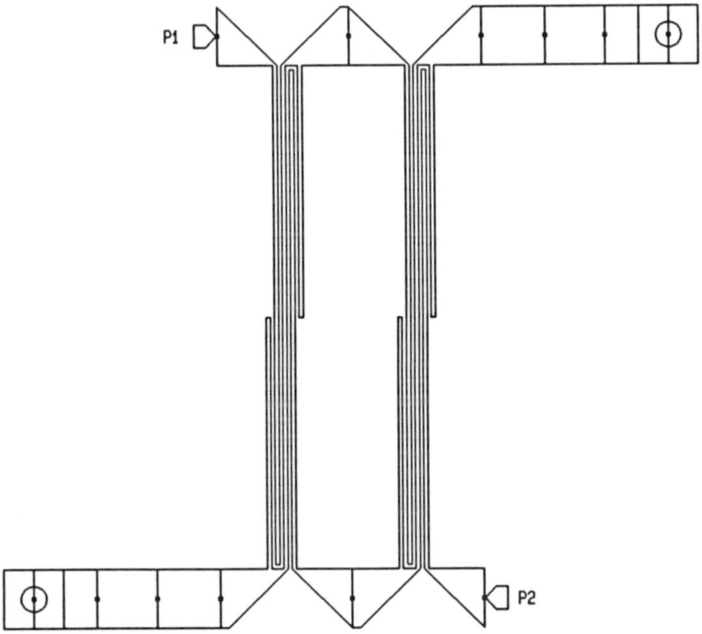

Figure 7.19 Cascaded microstrip Lange couplers with 50Ω terminations.

Figure 7.19 shows the physical topology corresponding to the ideal balanced amplifier in Figure 7.6, as created by the ACADEMY™ layout editor. Two Lange couplers are cascaded and each isolated port is connected to a short section of 50Ω transmission line, a 50Ω thin-film resistor, another short section of 50Ω transmission line, and a metalized via hole to ground. The analysis in Figure 7.20 shows the characteristic equal-ripple response. However, the insertion loss increases with frequency and is approximately 0.2 dB at 8.0 GHz. Microstrip loss is a function of frequency. Therefore, we will design the amplifier for 10.2-dB gain, including losses in the input and output matching networks, and then connect the couplers and perform a final optimization to compensate.

The parasitics associated with attaching the FET include the bonding wires, bypass capacitors, and the bias resistor. Figure 7.21(a) shows one construction method. The FET is attached to a metal rib with a bypass capacitor mounted on

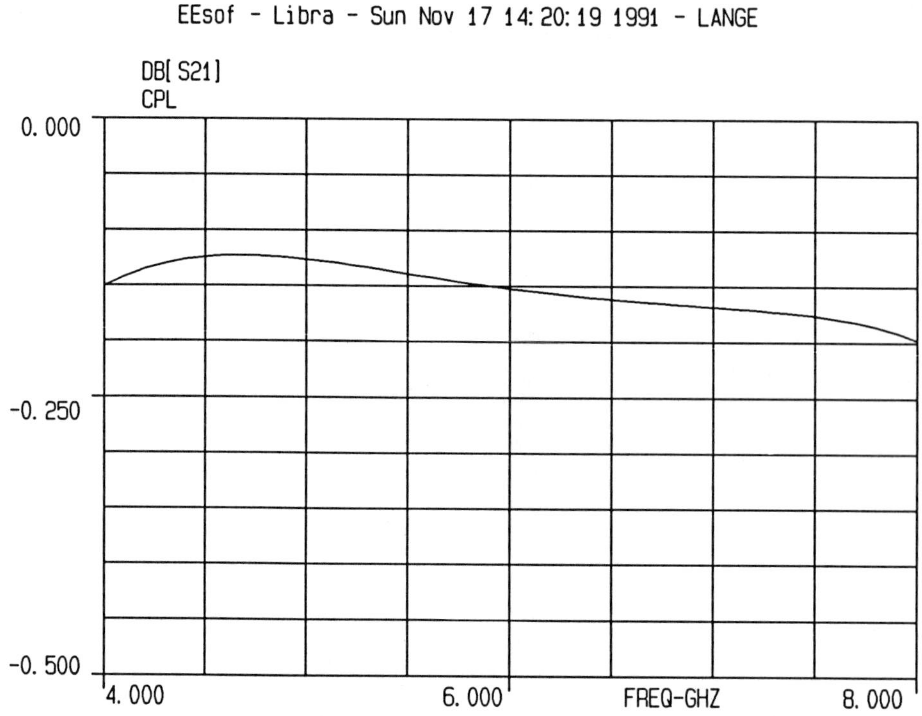

Figure 7.20 Microstrip Lange coupler analysis.

either side and a chip resistor also mounted beside one of the capacitors. The input and output matching networks are constructed on separate substrates which are mounted next to the metal rib. The FET is approximately 16 by 18 mils (0.41 by 0.46 mm), and two bonding pads are provided for the gate and drain connections. The source is a large metalized area surrounding the gate, and we (or a skilled assembler) can easily attach two bond wires to each side, as shown in the figure. The parallel-plate capacitors are approximately the same physical size as the FET and we elect to use the same value on either side. The series resonant frequency for a 20-pF capacitor of this size is approximately 6 GHz, which will provide zero reactance at the center frequency. The associated inductance is 0.035 nH, and we include a 0.5Ω resistor in the model to account for loss. In some cases we might use different values on either side, such as a 20-pF capacitor on one side for high-frequency bypass and a 100-pF on the other side for low-frequency bypass. A typical resistor value for a 10-mA drain current is 200Ω, and we assume the element is a pure resistance. Figure 7.21(b) shows the circuit model for this arrangement. Each gold bond wire is 1 mil (0.025 mm) in diameter and we assume identical lengths of 15 mils (0.38 mm).

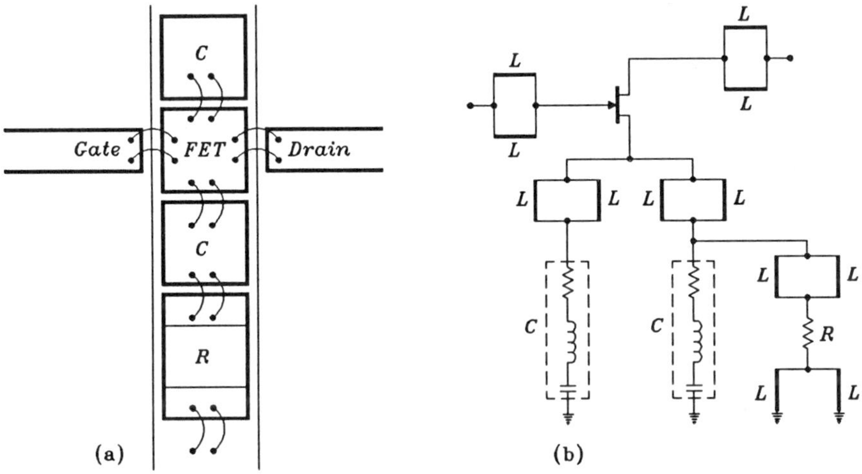

Figure 7.21 FET model: (a) physical; (b) electrical.

We first modify the original amplifier, shown in Figure 6.33, to include the source parasitics shown in Figure 7.21(b), and then analyze. Figure 7.22 shows the initial analysis, identified as *INITIAL*. The maximum gain has dropped below 10 dB, and there is almost a 0.5-dB slope across the 4- to 8-GHz frequency range. The response illustrates the error caused by ignoring the parasitics in the original design. Most of the degradation is due to the bond wires. Although we used four parallel bond wires which are only 15 mils long, the effect is significant. In fact, series feedback can have a pronounced effect on the device characteristics, as demonstrated in Chapter 3. After considerable optimization, the gain is restored to 10.0 ± 0.1 dB, and we add the gate and drain bond wires, change the gain specification to 10.2 dB, and then optimize again. The final results are labeled *AMP* in Figure 7.22. In this case, we were successful. The gain is

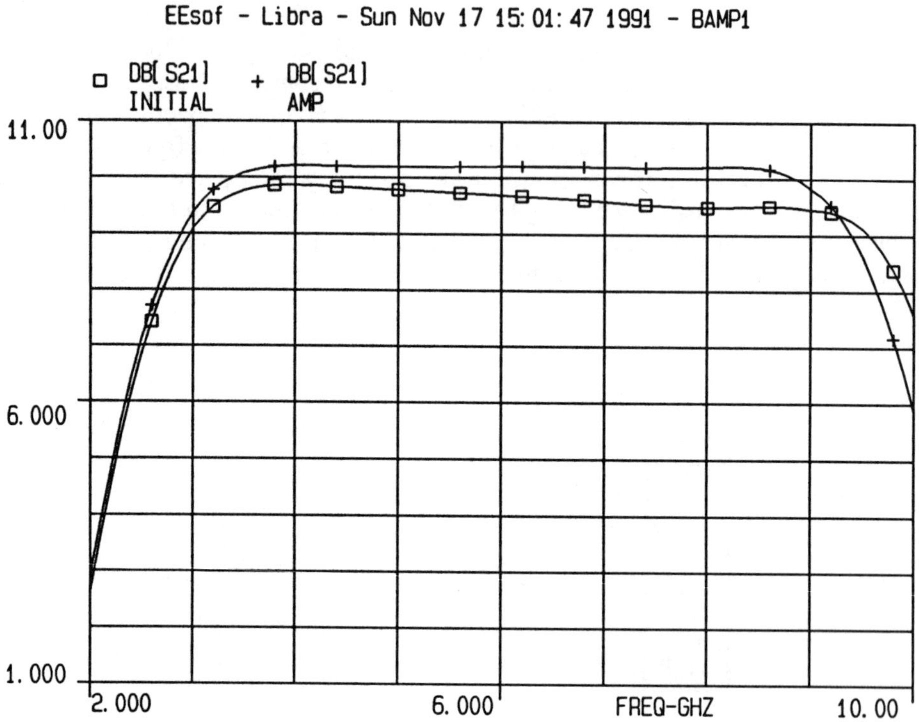

Figure 7.22 Initial and optimized response with FET parasitics.

10.20 ± 0.01 dB, and we can begin the conversion to physical dimensions. Table 7.1 shows the final optimized element values and corresponding physical dimensions from LineCalc™. With reference to the amplifier schematic shown in Figure 6.33, the elements listed in the table are in order from input to output. *OP* represents an open shunt stub, *SP* a shorted shunt stub, and *TLE* a transmission line element.

We will model the microstrip amplifier using a number of subcircuits, which offers several advantages. The networks can be used repeatedly simply by referring to the name, and we can analyze, optimize, or otherwise change a subnetwork, which automatically updates all references in the complete assembly. The first, or lowest level, network is *BWIRE*, which consists of two parallel 1-mil (0.025 mm) bond wires, each 15 mils (0.381 mm) long. This step is not necessary; however, the model of the FET and bias network is simplified somewhat because there are a total of 12 bond wires or 6 *BWIRE* connections. The next level in the

Table 7.1 Physical and Electrical Amplifier Parameters

Element	$Z_o \Omega$	*E at 8 GHz*	*W (mils)*	*L (mils)*
OP	53.29	27.50°	W1 = 20.71	43.04
TLE	79.91	35.11°	W2 = 7.04	56.93
OP	78.68	30.00°	W3 = 7.40	48.59
SP	100.0	44.98°	W4 = 3.11	74.06
TLE	99.91	74.74°	W5 = 3.12	123.05
OP	96.59	52.03°	W6 = 3.58	85.48
TLE	92.46	56.25°	W7 = 4.24	92.15
OP	92.39	39.75°	W8 = 4.25	65.12
SP	100.0	79.94°	W9 = 3.11	131.62

amplifier hierarchy contains three separate circuits. *DEVICE* is the FET and bias network shown in Figure 7.21(b), *IMN* is a model of the input matching network, and *OMN* is the output matching network. The gate and drain bond wires are modeled as part of the input and output matching networks, although we could just as easily include the wires as part of the device model. On the next highest level we have *CPL*, which is a single terminated Lange coupler, as shown in Figure 7.19, and a model for the amplifier called *AMP*, consisting of the input matching network, active device, and output matching network, as shown in Figure 7.23. The top assembly is *BAMP*, which is shown in Figure 7.24. BAMP contains two amplifiers and couplers along with an input and output transmission line to form the complete microstrip balanced amplifier. Both figures show the schematic representation directly from ACADEMY™. The wedge-shaped symbols at either end of the networks denote the input and output ports, and each block is labeled with the element name, followed by an internal reference tag.

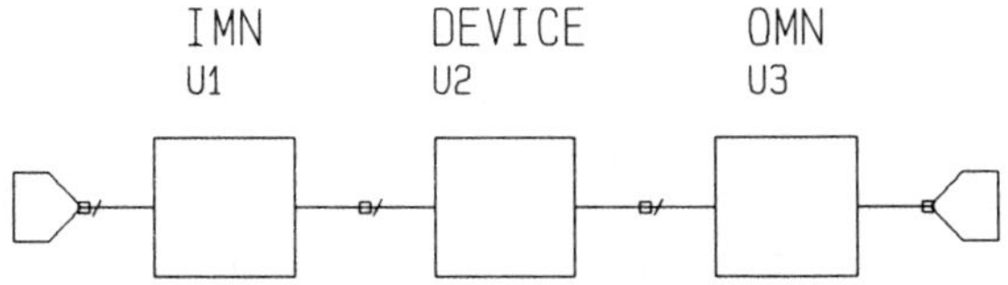

Figure 7.23 AMP: Amplifier model using IMN, DEVICE, and OMN.

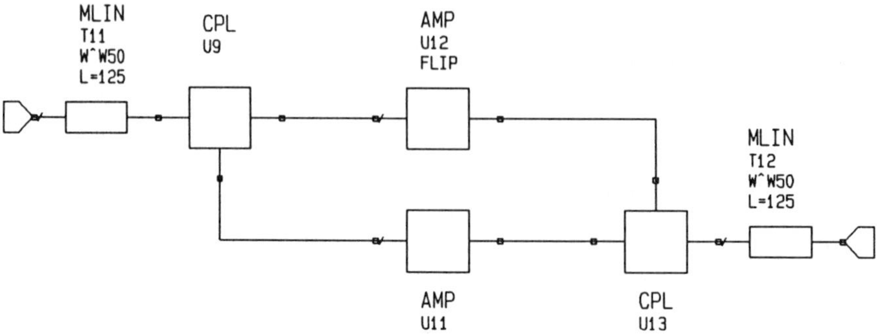

Figure 7.24 BAMP: Complete microstrip balanced amplifier model.

The input matching network, IMN, is modeled using the schematic in Figure 6.33 as a guide for the topology, and the optimized dimensions listed in Table 7.1 as the element values. The output network, OMN, is modeled in the same manner. An analysis of AMP reveals that the gain is approximately 10.2 dB, which confirms that the amplifier model is correct and we have not made any serious errors. We must now begin the process of generating a physical layout and start with the input matching network. The circuit model and physical layout are closely related. In fact, we can either work with a physical topology, then allow the program to automatically create a circuit model, or create a circuit which is automatically translated into a layout pattern. We elect to use the circuit model because our design is in this form. However, the purpose of this exercise is to create artwork that can be used to build the balanced amplifier, and we will continually switch between the layout and circuit. Therefore, starting with the simplified model in Figure 6.33, we add elements necessary for physical realization, examine the layout, analyze and optimize if necessary, and then reexamine the layout. Thus, the schematic and layout must be considered together. We can best illustrate this procedure by starting with the final results and working backwards. Figure 7.25 shows the schematic representation of the final version of the input matching network, and Figure 7.26 shows the corresponding layout pattern. Although the equivalent circuit appears complex, we make the conversion one step at a time to ensure that the amplifier response does not deviate too far from the optimum.

The input matching network will be cascaded with the Lange coupler, and we first add an input transmission line to provide some spacing. Figure 7.25 shows the TLE identified as *MLIN T8*, where MLIN identifies a microstrip transmission line and T8 is the element ID or tag. The physical dimensions are specified as a width and length, *W* and *L*, respectively. The caret(^) associates the parameter with a variable name defined in the *VAR* block. In this case, the width is equal to W50, which is defined as the 50Ω linewidth, 23.59 mils (0.60 mm), in the VAR

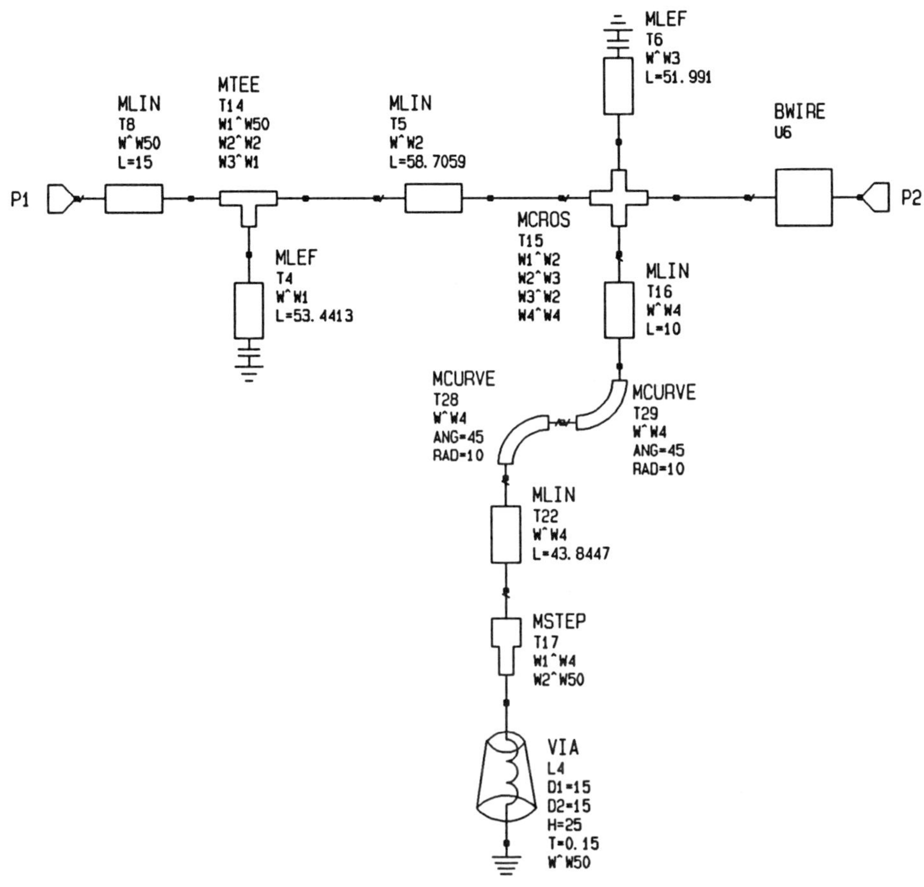

Figure 7.25 Schematic representation of the input matching network.

block. The length is specified as 15 mils (0.38 mm), and the corresponding conductor pattern is shown as the left most rectangle in Figure 7.26. It is not necessary to reanalyze the amplifier since we have only added a short 50Ω TLE, which will have minimal effect on the response. Working from input to output, the first significant discontinuity we encounter is the junction of the 50Ω input line, open shunt stub, and another TLE. This arrangement forms a tee junction shown in the schematic as the three-port labeled *MTEE*. The corresponding conductor pattern is shown in Figure 7.26, which provides a physical reference for each connecting line. The computer model for this discontinuity provides an electrical reference plane and accounts for the interaction between the three transmission

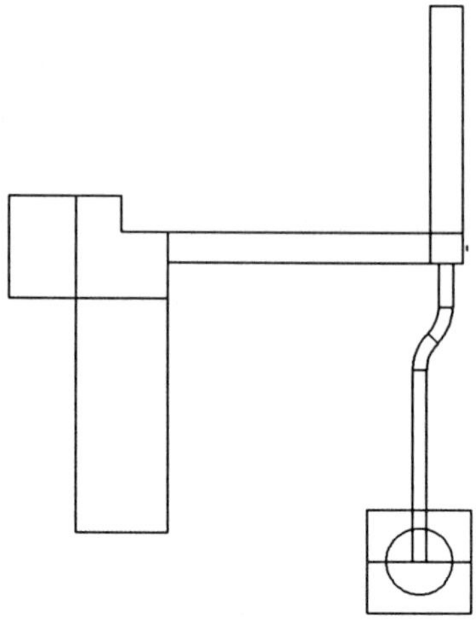

Figure 7.26 Layout pattern of the input matching network.

lines. The *s*-parameters of the tee junction are functions of the linewidth at all three ports, as well as the frequency. When we consider the tee junction without a shunt stub, the input and output TLEs meet at the center point where there is a change in linewidth and a second step at the output. Thus, the equivalent model consists of an input TLE with a length one-half the width of the shunt stub, a shunt capacitor to account for different linewidths, an output TLE the same length as the input, and another shunt capacitor to account for the step discontinuity at the output port. When the impedance of the two transmission lines are equal, there is no change in linewidth, and both capacitors are zero. As the shunt stub is lengthened, we no longer have a simple connection of two TLEs and two capacitors, and the equivalent circuit becomes more complex. Port definitions for the MTEE element are numbered in a clockwise direction beginning with the input port, which is connected to the 50Ω line. The schematic shows the linewidth at all three ports specified in terms of variable names defined in the VAR block. The linewidths W1 through W9 are listed in Table 7.1, and this approach allows us to reference the tee junction to the width of connecting lines, which can then be optimized or changed as desired while maintaining an accurate model. We have another discontinuity associated with the open shunt stub attached to the tee. The effect of the open end of the stub is to produce a fringing field which can be modeled as a shunt capacitor and shunt resistor to account for losses. The net result is a reduced length of the physical stub to compensate for the excess capacitance. The *MLEF* element shown in the schematic is a model for a microstrip open stub which includes the end effects.

At this point we have made two significant changes to the ideal circuit by adding a microstrip tee and changing the open stub to a physical model. The impact of these changes is to alter the electrical lengths of the three lines connected to the tee. The total length of each includes the original element length plus the distance to the electrical reference plane in the tee junction. Since the length of the 50Ω TLE does not affect the performance, we optimize only the lengths of the open stub MLEF T4 and the output line MLIN T5. After only a few iterations the response is restored to 10.2 ± 0.1 dB, and we can continue with the conversion. The next discontinuity is located at the output where the TLE is connected to an open and shorted shunt stub. This junction is also a tee. However, for the purposes of the analysis, we intend to connect bond wires from this junction to the FET gate, which requires the four-port *MCROS* shown in both figures. After adding the MCROS and changing the open stub to a physical model, we successfully optimize the lengths of the three connecting elements. The input model now represents a potential conductor pattern for the input matching network. We have added a short section of 50Ω line to provide some spacing from the coupler, replaced the ideal junctions with physical ones, and corrected for the end effect of the open stubs. However, the shunt shorted stub is terminated with an ideal short circuit which we cannot realize on microstrip. We have several options for providing the required ground. We can bend the stub in the direction of the metal rib where the FET and bias network are mounted, and then connect bond wires from the end of the stub to the rib, assuming a capacitor or resistor is not in the way. Substrate material can be purchased metalized on all six sides, and we can bring the stub to one edge of the substrate and then use a wrap-around ground. Another possibility is to drill a hole in the substrate and deposit metalization which bonds to the interior walls of the hole and forms a path from a top pad to the bottom ground plane, called a *VIA*. We elect to use the VIA model shown in Figure 7.25 where the hole diameter is specified as 15 mils (0.381 mm), the substrate height as 25 mils (0.635 mm), metalization thickness as 150 μin (3.81 μm), and pad size as 23.59 mils (0.599 mm) square. Details of the model are not provided in the documentation, but for the purposes of this example we assume connecting a TLE to one side of the pad creates a step discontinuity, which is modeled by the *MSTEP* element shown in Figure 7.25. The shorted stub is converted into a cascade of the MLIN, MSTEP, and VIA models; then the length is optimized and the layout generated. Examining the layout in Figure 7.26 reveals that a portion of the VIA pad extends off the edge of the substrate, denoted by the small line adjacent to the cross junction, when the stub extends straight down from the junction. We correct this problem by breaking the TLE into a 10-mil (0.254mm) section, followed by two 45° curves and the remaining length and termination. The layout is now acceptable and we have the final form shown in both figures. At this point we perform a final optimization on all element lengths corresponding to the original elements in Figure 6.33 and turn our attention to the output matching network shown in Figures 7.27 and 7.28.

We approach the realization of the output matching network in the same manner as the input. That is, we work from input to output adding parasitics and

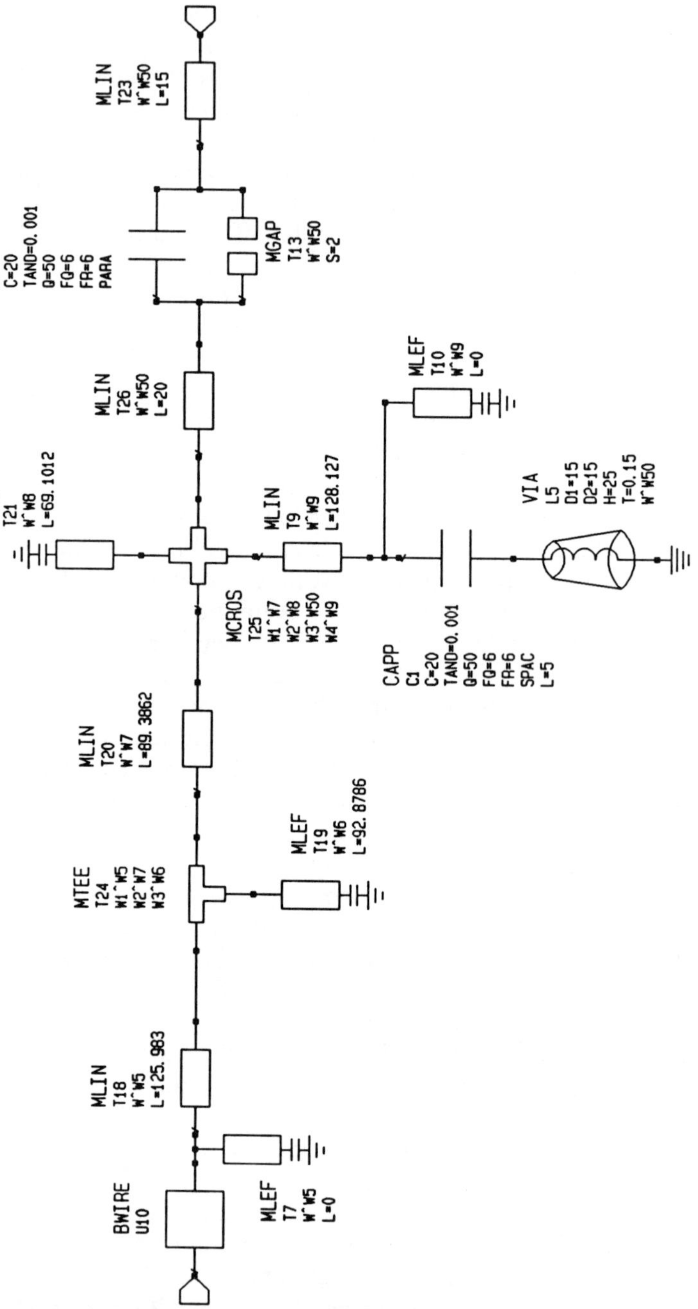

Figure 7.27 Schematic representation of the output matching network.

occasionally optimizing the affected elements to ensure that the electrical characteristics remain close to those of the ideal model. The first discontinuity we encounter is the TLE, which is connected to the drain. Although bond wires are connected to the end, the conductor abruptly terminates, which causes the same fringing associated with an open stub. In fact, we can use the MLEF element to model the end effects by specifying a zero length. The next discontinuity is the tee junction, where an open stub is connected to the two TLEs.

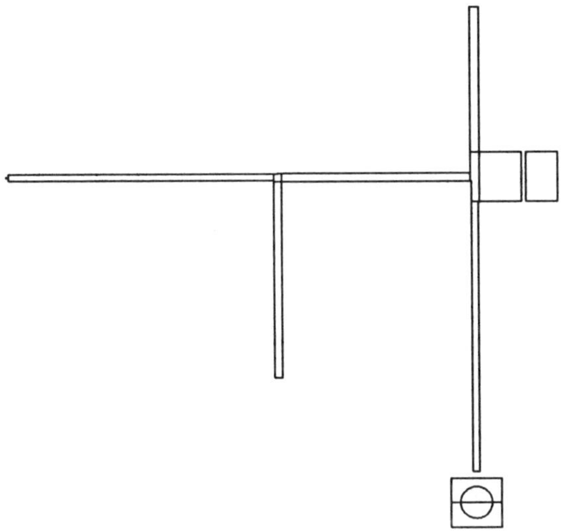

Figure 7.28 Layout pattern of the output matching network.

We add the MTEE, change the open stub to an MLEF element, and optimize the lengths of the three connecting lines. We have another junction at the output port where the open and shorted stubs are connected. We will add a short section of 50Ω line to provide some spacing from the coupler, and the junction becomes a cross, as shown in Figures 7.27 and 7.28. We alter the model to include the MCROS, change the open stub to a physical model, and then reoptimize, which provides the topology and element values for everything but the grounded end of the shorted stub. In this case, we must provide drain bias which requires an RF ground without shorting the bias voltage. We can accomplish this by adding a VIA, mounting a capacitor on the pad, and connecting to the end of the stub. The bias wire can either be connected directly to the capacitor or soldered to a larger pad, which, in turn, is connected to the capacitor using bond wires. If significant, we can model the bias circuitry, including additional wires and pads. We can easily investigate the effects of additional circuitry by adding elements at the junction of the capacitor and TLE. The amplifier is relatively insensitive to the bias network and we elect to omit the additional elements. However, we do include the fringing at the capacitor end of the TLE, and we also use a built-in physical capacitor model, as shown in Figure 7.27. We must also add a blocking capacitor to avoid a DC path through the coupler to the 50Ω termination or the output load. A 2-mil (0.051-mm) gap is added to the output line, followed by another 15-mil (0.381-mm) section of 50Ω transmission line. A capacitor will be mounted across the space, and we model the discontinuity using a microstrip gap *MGAP* in parallel with a physical capacitor model *CAPP*. Optimizing lengths of the five original elements in the output matching network results in an amplifier gain of 10.20 ± 0.01 dB. Final element values are shown in Figure 7.27, and the corresponding conductor pattern in Figure 7.28. There is not much point in further optimization since the response of this design is almost perfectly flat. However, we can often improve the response by simultaneously optimizing both the input and output networks. We may also find it necessary to optimize the linewidth as well as length to realize the desired level of performance. In this case, the linewidths are the original calculated values listed in Table 7.1.

The final step in the design is to use the CPL and AMP subcircuits to model the balanced amplifier shown in Figure 7.24, and then automatically generate the complete layout shown in Figure 7.29. We examine the layout to ensure there are no problems caused by elements too close to one another or other modifications that might be necessary or desirable. The arrangement shown in the figure is acceptable. However, had we not included the 50Ω output lines, the open stubs would be too close to the coupler. Also, if the length had increased much more, it would be necessary to bend the open stubs to avoid coupling between the two opposing open stubs. Depending on any size constraints, we also have the option of bending the output shorted stubs to reduce the substrate width and folding the two TLEs in the output matching network to reduce the total length. We can also bend the 50Ω input and output lines such that the two connections are on the center line, which simplifies cascading multiple stages. Once we incorporate any changes in the amplifier model, then reoptimize if necessary, we are ready to analyze the complete design and generate assembly masks. Final artwork is similar to the layout shown in Figure 7.29, except that the connecting lines are removed and there may be more than one mask layer. For example, one mask may define the conductor pattern and another the resistor pattern.

Analysis of the complete microstrip balanced amplifier shows that the gain is 10.0 ± 0.3 dB. The input and output return loss is greater than 15 dB. The correction we made for the coupler loss is correct since the gain is centered about 10 dB. However, the ripple of the balanced configuration is significantly higher than that of the single-ended amplifier. We attribute the ripple to a phase imbalance in the microstrip couplers which have unequal even- and odd-mode phase velocities. The balanced amplifier performance can be improved somewhat by optimizing the physical dimensions of the coupler. That is, we allow the

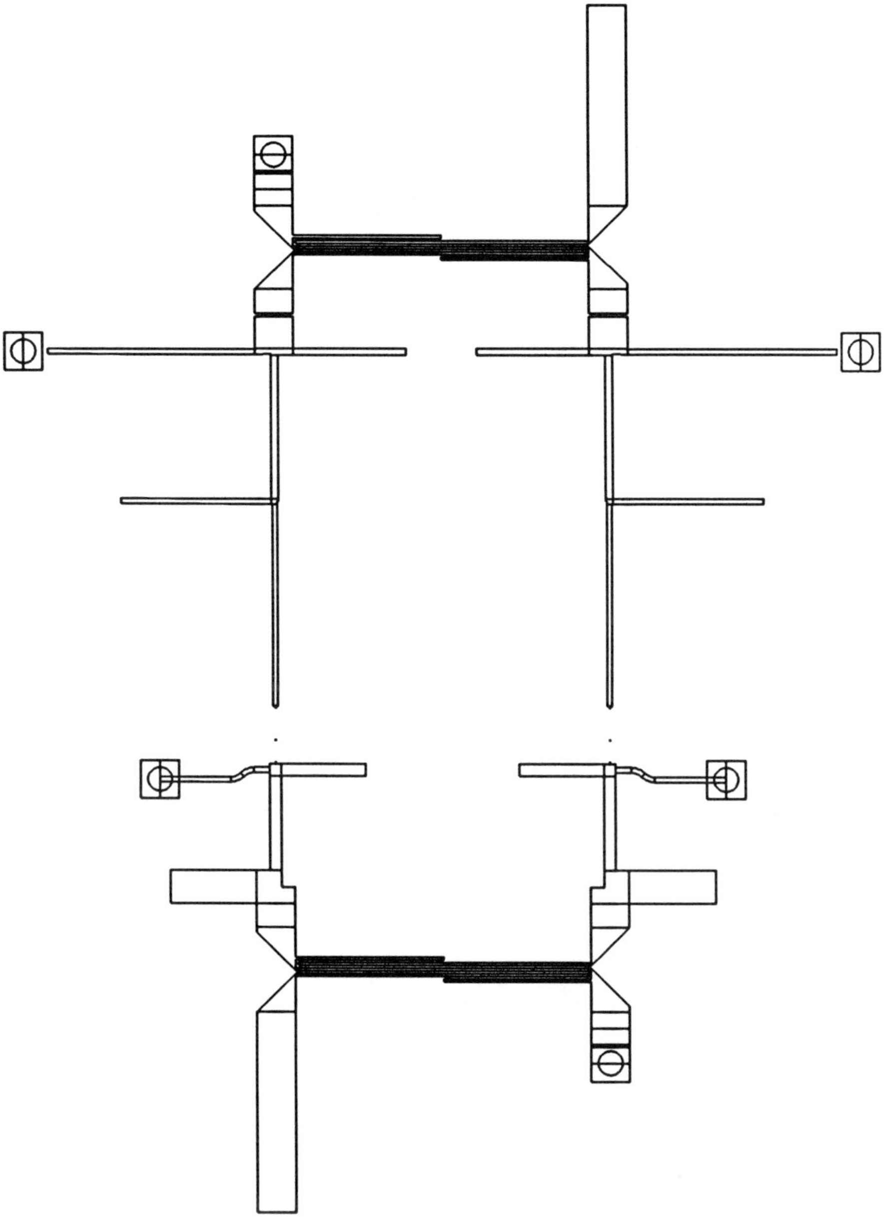

Figure 7.29 Microstrip 4.0- to 8.0-GHz balanced amplifier.

optimizer to vary the center frequency and coupling in an effort to reduce the ripple. The optimized width remains the same, but the gap is increased slightly from 1.56 to 1.68 mils (0.040 to 0.043 mm), and the length is reduced from 199.97 to 180.02 mils (5.079 to 4.573 mm), or an increase in the center frequency of approximately 10%. Figure 7.30 shows the response of the complete balanced amplifier, as well as the single-ended amplifier for comparison. The balanced amplifier has a gain of 10.000 ± 0.126 dB, and the out-of-band rejection is improved because of the coupler response. Figure 7.31 shows that the worst-case input and output return loss is 18 dB or a VSWR of 1.29:1 across an octave bandwidth. It may be possible to slightly modify one side of the balanced amplifier to reduce the ripple even further. For example, we might try changing the length of the gate and drain bond wires or the open stubs.

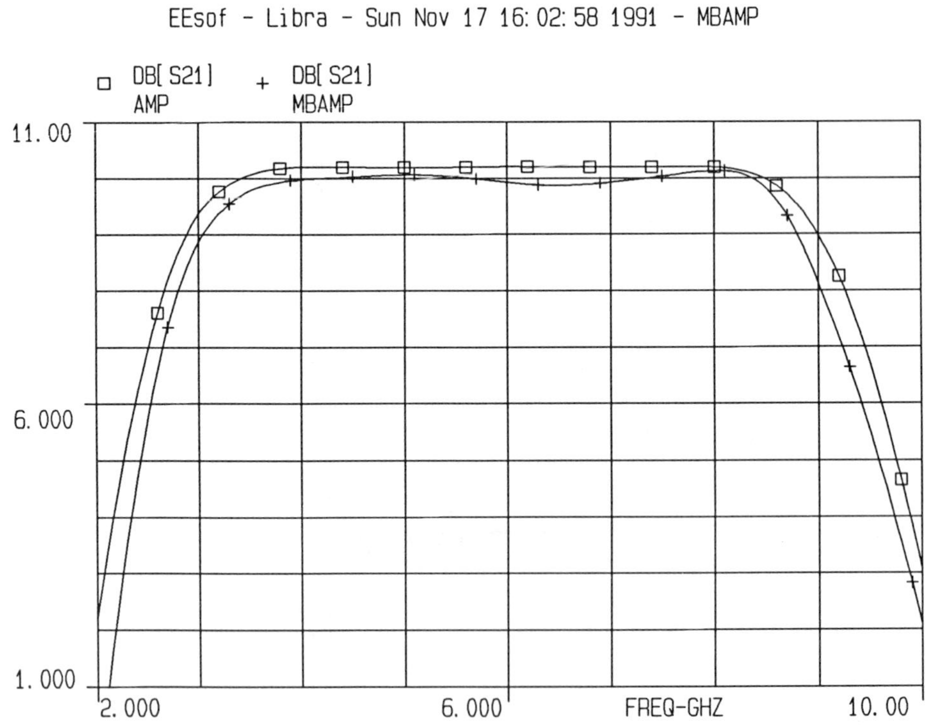

Figure 7.30 Gain response of the balanced and single-ended amplifiers.

Although most discontinuities are well defined, some are open for discussion and can lead to heated debates on exactly how a particular region should be modeled. For example, the tee, cross, curve, and bend are obvious, and we have only one choice, although the equivalent discontinuity model may be debatable. However, other connections, such as the capacitor shown in Figure 7.18(a), are less obvious. Is the equivalent shown in Figure 7.18(b) sufficient, or should we also

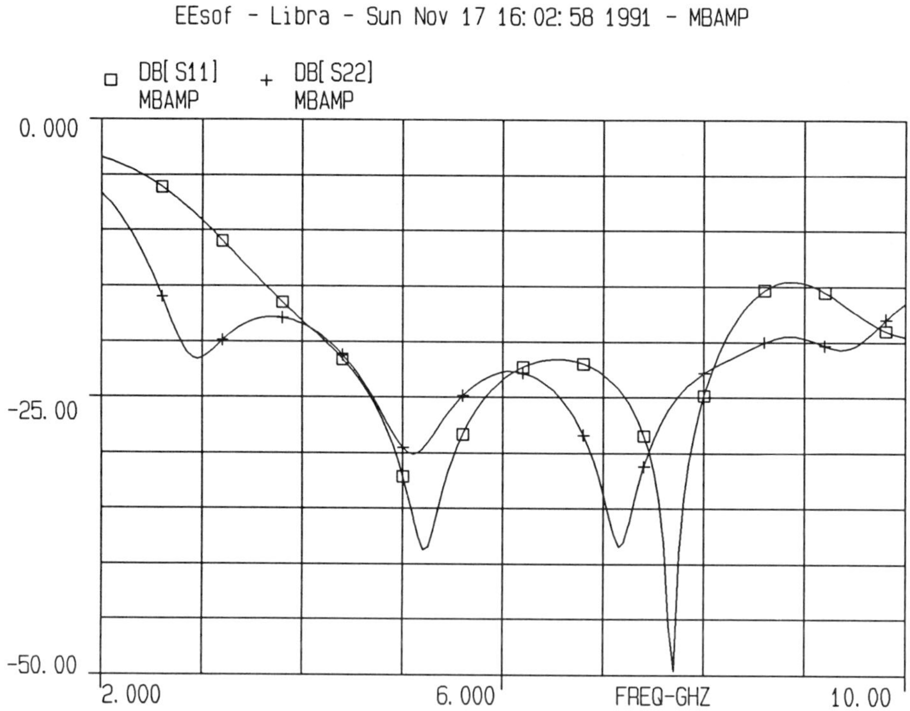

Figure 7.31 Input and output return loss of the balanced amplifier.

add short open stubs or TLEs from the capacitor and wire to the gap? Perhaps the model should also include transmission lines to account for the phase shift from one end of the capacitor to the other, as well as the reference plane beneath the component. The equivalent model is highly dependent on the assembly methods as well as upon the physical component. As we move higher in frequency, where the lumped elements begin to act as distributed elements and small physical errors represent significant electrical lengths, these questions become increasingly important. Electromagnetic simulators can answer some of these questions. However, accurate measurements are often the only way to completely resolve these questions.

7.4 MMIC DESIGN

MMIC technology has advanced significantly in recent years as more effort is applied in an effort to realize the full potential of this fabrication method. All active and passive elements, as well as the interconnections, are formed on a semi-insulating semiconductor substrate. This approach eliminates the wire bonds and component assembly associated with MICs and leads to high performance, low cost, small size, and reproducible circuits. However, the manufacturing tolerances are somewhat higher than those of a MIC and the design cost is significantly higher. Thus, it is essential that the computer model be accurate and the design centered to maximize yields.

In some respects the design of MMICs is easier than MICs. We can use distributed elements in the same manner as microwave integrated circuits. However, we also have lumped elements that are useful up to 20 GHz or higher due to the small physical size. Foundries often furnish accurate models for the lumped elements, active devices, and discontinuities. Many parasitics are significantly reduced. For example, short lengths of metalization replaces long bond wires, and the distance for a via or wrap-around ground is considerably reduced. However, thinner metalization and substrate thickness results in increased loss and parasitic capacitance to the ground plane. Lower Q components make it more difficult to realize high-performance filters, low-noise amplifiers, and other circuits affected by lossy elements. Added shunt parasitic capacitance can be a problem, as is the relatively narrow range of lumped and distributed element values. However, network synthesis offers substantial advantages when applied to the design of MMICs. Synthesis usually results in numerous circuits with identical electrical performance. We can often identify one or more with desirable MMIC characteristics, such as a narrow range of element values or topologies which are useful for absorbing parasitic elements.

A typical range for the characteristic impedance of distributed elements is between 20Ω and 100Ω [31]. Capacitor values between 0 and 30 pF and inductor values between 0.5 and 10 nH are also easily fabricated using MMIC technology. Inductor and capacitor Q factors on the order of 50 have been measured at X-band. The range of element values as well as the equivalent circuit models are dependent on the wafer processes and differs for each foundry. However, for the purposes of examples in this section we will assume Z_o between 20Ω and 100Ω, capacitance less than 30 pF, and inductor values between 0.5 and 10 nH. Q factors for lumped elements will also be assumed constant and equal to 50.

7.4.1 Feedback Amplifiers

The bandwidth of the reactive mismatch amplifier is limited by the gain-bandwidth of the required matching networks, as well as by the bandwidth of the isolator or hybrid coupler, which must be used to provide a reasonable input and output match. Microwave amplifiers can routinely be built to operate over several octaves using the methods discussed in Chapter 6. However, decade bandwidths require a reduction in gain, and perhaps noise figure, in order to realize the matching networks with a finite number of elements, if they can at all. The physical dimensions of the couplers become unrealizable for very broadband operation, and multisection couplers must be used. The diplexer discussed in Chapter 4 can sometimes be used to overcome these problems. We design several amplifiers to operate over separate portions of the frequency range and then combine them into a single broadband network. However, the amplifier becomes complicated and expensive. Few system requirements are stringent enough to justify this approach.

One method that can be used for the design of very broadband components is the feedback amplifier. Figure 7.32 shows an active device where an admittance Y_F is connected from input to output, and another admittance Y_E is connected from the common terminal to the ground node. Y_F forms a parallel feedback path and Y_E forms series feedback. The simplest method of analyzing this circuit is nodal analysis, discussed in Section 2.2.2. We can write the definite nodal admittance matrix by inspection; that is,

$$Y = \begin{bmatrix} (y_{11} + Y_F) & (y_{12} - Y_F) & -(y_{11} + y_{12}) \\ (y_{21} - Y_F) & (y_{22} + Y_F) & -(y_{21} + y_{22}) \\ -(y_{11} + y_{21}) & -(y_{12} + y_{22}) & (y_{11} + y_{12} + y_{21} + y_{22} + Y_E) \end{bmatrix} \tag{7.8a}$$

Node 3 is of no interest since it is not connected to the input or output, and we can use pivotal condensation to suppress it. Substituting into (2.30), we have

$$Y = \begin{bmatrix} \dfrac{y_{11} Y_E}{\Sigma} + \alpha & \dfrac{y_{12} Y_E}{\Sigma} - \alpha \\ \dfrac{y_{21} Y_E}{\Sigma} - \alpha & \dfrac{y_{22} Y_E}{\Sigma} + \alpha \end{bmatrix} \tag{7.8b}$$

where

$$\Sigma = y_{11} + y_{12} + y_{21} + y_{22} + Y_E \tag{7.8c}$$

and

$$\alpha = \frac{y_{11}y_{22} - y_{12}y_{21}}{\Sigma} + Y_F \tag{7.8d}$$

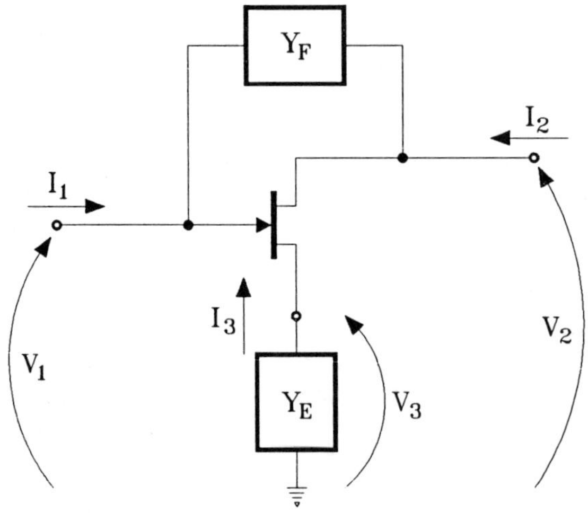

Figure 7.32 Feedback amplifier.

The parameters of most interest for the feedback amplifier are the input impedance, output impedance, and the gain. These parameters can be computed from the following two-port relationships:

$$Z_{\text{in}} = \frac{y_{22} + Y_L}{\Delta_y + y_{11}Y_L} \tag{7.9a}$$

$$Z_{\text{out}} = \frac{y_{11} + Y_S}{\Delta_y + y_{22}Y_S} \tag{7.9b}$$

and

$$A_V = \frac{-y_{21}}{y_{22} + Y_L} \tag{7.9c}$$

where y_{ij} is a component of the two-port admittance matrix, (7.8b) in this case, Δ_y is the determinant, Y_S is the source admittance, and Y_L is the load admittance. Substituting (7.8) into (7.9), we have

$$Z_{\text{in}} = \frac{(Y_F + Y_L)(\beta + Y_E) + (Y_E y_{22} + \Delta_y)}{(Y_E + Y_L)(\Delta_y + \beta Y_F) + Y_E Y_L (Y_F + y_{11})} \tag{7.10a}$$

$$Z_{\text{out}} = \frac{(Y_F + Y_S)(\beta + Y_E) + (Y_E y_{11} + \Delta_y)}{(Y_E + Y_S)(\Delta_y + \beta Y_F) + Y_E Y_S (Y_F + y_{22})} \tag{7.10b}$$

and

$$A_V = \frac{Y_F(\beta + Y_E) + (\Delta_y - Y_E y_{21})}{(Y_E + \beta)(Y_F + Y_L) + (Y_E y_{22} + \Delta_y)} \tag{7.10c}$$

where

$$\beta = y_{11} + y_{12} + y_{21} + y_{22} \tag{7.10d}$$

and

$$\Delta_y = y_{11} y_{22} - y_{12} y_{21} \tag{7.10e}$$

Equation (7.10) reveals that the input impedance, output impedance, and voltage gain can be controlled by the proper selection of the feedback admittances, Y_E and Y_F. Unfortunately, all three cannot be specified simultaneously. In fact, the input and output impedance cannot be equal except in one special case, rarely encountered in practice. A close examination of (7.10a) and (7.10b) shows that both have the same form. The only difference is that Y_S and Y_L are interchanged, as is y_{11} and y_{22}. Thus, if Y_S is equal to Y_L, the only way Z_{in} can equal Z_{out} is if y_{11} and y_{22} are identical. Actually, we can add an admittance at either port to force y_{11} and y_{22} equal. However, it is often difficult to find a topology acceptable for multioctave performance, and the improvement usually does not justify the additional circuit complexity.

The most useful condition for the feedback amplifier is when $Z_O = Z_{\text{in}} = Z_{\text{out}} = 1/Y$; that is, the amplifier is matched in a $Z_O \Omega$ system. Under these conditions, s_{11} and Γ_L both equal zero, and the voltage gain in (2.68c) reduces to $A_V = s_{21}$. We can solve for Y_E and Y_F by combining equations (7.10a) and (7.10c). However, this provides an exact solution for Z_{in} and A_V, but not Z_{out}. An alternative is to replace y_{11} and y_{22} with y'_{11}, which is the average value of y_{11} and

y_{22}. This substitution increases the error in input impedance but reduces the error in output impedance, thus distributing the error between the input and output. Substituting y'_{11} for y_{11} and y_{22} in (7.10a) and (7.10c) and then solving for the feedback admittance yields

$$Y_F^2[\beta(A_V - 1)] + Y_F[2\Delta_y(A_V - 1) + A_V\beta Y] + \left[Y^2(A_V y'_{11} + y_{21}) + \frac{\Delta_y}{\beta}(A_V - 1)(\Delta_y - Y^2) + A_V Y \Delta_y\right] = 0 \tag{7.11a}$$

and

$$Y_E = \frac{\beta Y^2}{(\Delta_y - Y^2) + \beta Y_F} \tag{7.11b}$$

The two solutions for Y_F are determined from (7.11a) by applying the quadratic formula. Then we select the root that provides a positive real feedback element and substitute into (7.11b) to compute Y_E. When the design specifications are not feasible (i.e., the open loop gain is not sufficient to meet the feedback requirements), one or both elements will be negative real and the specifications must be changed or a different device selected.

As an example of the feedback amplifier, we can compute the feedback elements for 6-dB gain using a device with 50Ω s-parameters given by

$$S = \begin{bmatrix} 0.42\angle{-40.0^\circ} & 0.005\angle{80.0^\circ} \\ 35.0\angle{170.0^\circ} & 0.90\angle{-10.0^\circ} \end{bmatrix} \tag{7.12a}$$

The voltage gain is equal to s_{21} for the matched amplifier, and we have A_V = $1.995\angle{180^\circ}$. Using Figure 1.11, we convert to normalized y-parameters,

$$Y = \begin{bmatrix} 0.4995\angle{23.102^\circ} & 0.0039\angle{-87.615^\circ} \\ 27.2297\angle{2.385^\circ} & 0.0518\angle{17.313^\circ} \end{bmatrix} \tag{7.12b}$$

and then compute y'_{11}, the average of y_{11} and y_{22}, as $0.2755\angle{22.559^\circ}$. Next we evaluate (7.10d),

$$\beta = 27.7476\angle{2.769^\circ}$$

and (7.10e),

$$\Delta_y = 0.1231\angle 84.934°$$

We then substitute into (7.11a) and solve for the two roots, which are

$$Y_F = 0.3240\angle -1.353° ; \qquad 0.9912\angle -179.934°$$

The first root is positive real and Z_F = 154.294Ω + j3.6440Ω. Evaluating (7.11b), we have

$$Y_E = 3.466\angle 0.301°$$

which is also positive real, and Z_E = 14.4242Ω - j0.0756Ω. Using (7.10) to check the results, we find that $A_V = 2.0300\angle -179.611°$, $Z_{in} = 48.3158\angle -0.765°$, and $Z_{out} = 51.7430\angle 0.765°$Ω. The input and output impedances are equally balanced above and below 50Ω, and the VSWRs are 1.04:1. The voltage gain is 6.15 dB, slightly high due to the substitution of y'_{11}. However, since s_{11} is not zero, A_V is not equal to s_{21} in (2.68c). In this case, $|s_{21}|$ = 6.00 dB, which is the desired value.

When the open loop gain is sufficiently higher than the closed loop gain, the feedback elements will be resistive. However, as we increase the closed loop gain, the resistive and reactive components of Y_F and Y_E change until a point is finally reached where one or both are negative real and the design is not feasible using passive elements. For example, if we use the parameters in (7.12) to design a 25-dB gain amplifier, $A_V = 17.7828\angle 180°$, Z_F = 1121.5Ω + j198.4Ω, and Z_E = 0.4Ω - j0.08Ω. Evaluating (7.10), we have $A_V = 20.5551\angle -176.462°$, $Z_{in} = 38.0185\angle -6.228°$Ω, and $Z_{out} = 65.7574\angle 6.228°$Ω. Again, the error is not too large, A_V is 26.26 dB, and the input and output VSWRs are 1.34:1. The transducer gain $|s_{21}|$ is 24.989 dB, which is even closer. The error is directly attributable to unequal y_{11} and y_{22}. If we set y_{22} equal to y_{11} and then recompute, the parallel impedance becomes 1300.4Ω + j67.7Ω and the series impedance 0.06Ω - j0.2Ω. A check shows that A_V = 25.0 dB and both Z_{in} and Z_{out} are equal to 50Ω. However, as we increase the gain beyond 25 dB, the parallel resistance increases while the series resistance decreases. Since the series resistance is already close to zero, a further reduction will result in a negative element.

The open loop gain of active devices decreases with increasing frequency, resulting in significant reactive feedback components at higher frequencies. We can usually find a feedback topology that provides the necessary impedance to operate over very wide bandwidths. For example, if we realize the parallel feedback as a series *R-L*, the inductive reactance is reduced at lower frequencies and we can choose a value for the resistance that equalizes the error between the upper and lower ends of the frequency range. Similarly, we can realize series feedback as a parallel *R-C*, which has the required frequency characteristics. We can also extend the frequency range by using input and output matching networks. The low-pass topology is ideal since it has minimal effect at low frequencies where the amplifier is matched. We also have the option of synthesizing networks that have the required driving-point impedance for the series and parallel feedback paths. However, in some instances the addition of a single element can significantly improve the performance. For example, the 50Ω *s*-parameters of an active device measured at 2.0 GHz are

$$S = \begin{bmatrix} 0.59\underline{/-149.0^\circ} & 0.042\underline{/26.0^\circ} \\ 3.15\underline{/71.0^\circ} & 0.75\underline{/-43.0^\circ} \end{bmatrix}$$

and for 6-dB gain, Z_F = 71.4060Ω + j53.4581Ω and Z_E = 8.7600Ω - j15.8858Ω. Substituting into (7.10a) and (7.10b), Z_{in} = 50.7410$\underline{/-42.737^\circ}$Ω and Z_{out} = 49.2698 $\underline{/-42.737^\circ}$Ω. The transducer gain is 6.00 dB, but the input and output VSWRs are 2.29:1. This device has an insertion gain of 9.966 dB. The excess gain, or difference between the open and closed loop gains, is only 4 dB. However, an analysis shows that the stability factor is 0.99996 and the maximum stable gain (MSG) is 18.751 dB. Thus, it is possible to provide almost 13 dB of excess gain at 2.0 GHz by stabilizing the device and then matching the input and output. Calculating the input and output impedances of the device from s_{11} and s_{22} reveals that Z_{in} = 13.8Ω - j12.9Ω and Z_{out} = 47.0Ω - j109.9Ω. Rather than match the input and output, we can simply add a series +j109.9Ω reactance at the output, which will improve the match and increase the transducer gain. Connecting a 8.74-nH inductor to the output terminal results in an insertion gain of 4.76, or 13.55 dB, and the excess gain has been increased to 7.5 dB. Recomputing, we have Z_F = 73.0264Ω + j16.5298Ω and Z_E = 23.1420Ω - j4.5272Ω, which provides 6.00-dB gain and input and output VSWRs of 1.40:1. The performance is significantly improved by the addition of a single element, and we have the final topology shown in Figure 7.33.

An analysis of the feedback amplifier shown in the figure reveals that the stability factor is 0.965. Although reactive elements cascaded with a two-port network do not affect the stability factor, the same is not true of series and shunt feedback. The mapping examples in Chapter 3 illustrate that some values of feedback impedance improve the stability while other values have the opposite effect. We could add a series or shunt admittance at the input or output ports of Figure 7.32, include the stability factor in equation (7.10), and then solve for Y_F, Y_E, and a third admittance which provides unconditional stability. However, we can just as easily modify the design in the usual manner and then use optimization; that is, we examine the input and output plane stability circles and select the stabilization element which causes minimal degradation. In this case, a 910Ω shunt resistor

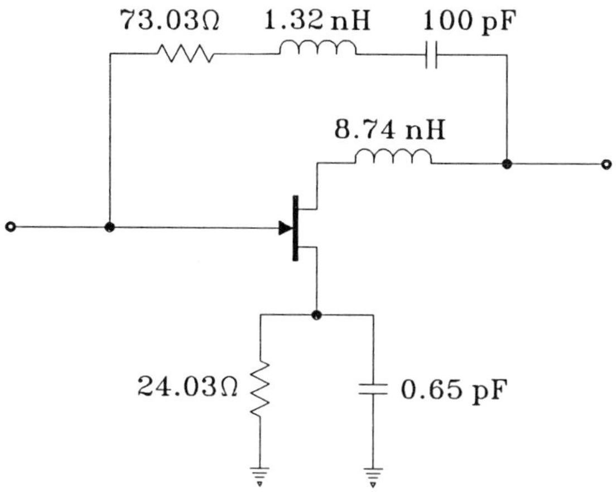

Figure 7.33 2.0-GHz amplifier.

connected at the input provides a stability factor of one and reduces the gain slightly to 5.722 dB. However, after only a few optimization iterations, the original response is restored and the stability factor is greater than one.

Equation (7.11) is a complicated expression which provides an exact solution for the feedback elements. However, when the reference impedance is real and the excess gain high, Y_F and Y_E are both real and we can derive simple first-order approximations for the resistive elements R_F and R_E. We can replace the device in Figure 7.32 with a simple model [32,33]; that is, the hybrid-π model which contains a dependent current generator $V_1 \bullet g_m$, where g_m is the transconductance. If $R_S = R_L = R$ and $g_m \bullet R_E >> 1$, the input and output resistance can be written as

$$R_{in} \approx R_F \parallel R_F R_E \left(\frac{1}{R} + \frac{1}{R_F} \right) \tag{7.13a}$$

$$R_{out} \approx \frac{R_F + R}{1 + R/R_E} \tag{7.13b}$$

When the parallel resistance is much greater than R and the series resistance is much less than R (i.e., $R_F >> R$ and $R_E << R$),

$$R_{in} \approx \frac{R_F R_E}{R} \tag{7.13c}$$

$$R_{out} \simeq \frac{R_F R_E}{R} \tag{7.13d}$$

and when $g_m \bullet R_E >> 1$, we can write the transducer gain as

$$s_{21} \simeq -\frac{1}{R_E} \cdot \frac{1}{1/R_F + 1/R} \tag{7.13e}$$

Since $R_{in} = R_{out} = R$, we can write (7.13c) or (7.13d) as

$$R \simeq \sqrt{R_F R_E} \tag{7.14a}$$

and then solve (7.13e) and (7.14a) for the feedback elements:

$$R_F \simeq \frac{R}{2}\left\{-s_{21} \pm \sqrt{s_{21}^2 - 4s_{21}}\right\} \tag{7.14b}$$

and

$$R_E \simeq \frac{R^2}{R_F} \tag{7.14c}$$

However, if $R_F >> R$, then (7.13e) reduces to

$$s_{21} \simeq -\frac{R}{R_E} \tag{7.15a}$$

and (7.14b) becomes

$$R_F \simeq R_E s_{21}^2 \tag{7.15b}$$

As an example of the simplified expressions in (7.14), we can recompute the feedback elements for the 6-dB gain amplifier using the device parameters given by (7.12). Substituting into (7.14b), we have R_F = 136.3473Ω (154.3Ω), and from (7.14c), R_E = 18.3355Ω (14.4Ω). The numbers in parenthesis are the original resistive values computed from (7.11). Equation (7.10) reveals that these values result in a voltage gain of 3.57 dB, Z_{in} = 51.8968$\angle$-1.281°Ω, and Z_{out} = 54.8797 $\angle$-0.092°Ω. The input and output impedances are close to the specified values; however, the voltage gain exhibits a significant error. If we use the simpler expression given by (7.15b), R_F = 99.7631Ω, R_E = 25.0594Ω, and the corresponding voltage gain is -1.11 dB. A large error indeed. The simplified expressions can be used when the assumptions are valid; that is, $R_F >> R$, $R_E << R$, and $g_m \bullet R_E >> 1$. However, in most cases the exact expressions in (7.11) give better results at RF and microwave frequencies and, additionally, provide some insight into the form of the feedback networks.

We have alternatives for feedback other than the simple network shown in Figure 7.32. For example, we can apply feedback around two or more amplifier stages or, in fact, between any two nodes in the network. Another option is to connect two active devices together and apply feedback to the combination. Figure 2.30 shows different device configurations which have characteristics desirable for certain applications. When contemplating a feedback design, it is usually worthwhile investigating the various alternatives in an effort to find the topology that provides maximum performance or the least expensive product.

7.4.2 Resistive Compensated Amplifiers

In Section 3.2.1 we used mapping techniques to apply parallel feedback to a device that simultaneously increased the MSG from 18.8 to 31.2 dB, and the stability factor from 0.13 to 0.65. We then added an 18350Ω resistor from the gate to ground, which resulted in unconditional stability and G_{MAX} equal to 31.07 dB. The design goal was a 25-dB gain matched amplifier, and we simply reduced the value of the shunt resistor until G_{MAX} equaled 25 dB and then matched the input and output. Essentially, we used resistive compensation, which resulted in the single-stage design shown in Figure 3.15. Had we used reactive matching networks, the amplifier would require two stages similar to the design in Section 6.3; that is, the MSG is less than 25.0 dB and any gain compensation must be between stages if the input and output is to be matched. The feedback amplifier approach discussed in Section 7.4.1 would also require two stages due to the low value of MSG. In many instances we cannot synthesize the feedback impedance necessary to improve the MSG and stability across multioctave bandwidths. However, resistive compensation is a useful technique and can often be used over multioctave or even decade bandwidths.

The stability factor of most RF and microwave devices increases with frequency, and we often find that the device is potentially unstable over the lower portion of its useful frequency range and unconditionally stable at higher frequencies. The corresponding MSG or G_{MAX} decreases in a like manner. However, if we can modify the characteristics such that G_{MAX} is constant over the frequency range and the device is unconditionally stable, then the input and output can be matched, resulting in flat gain and low VSWRs. Figure 7.34 shows three networks that are useful for this purpose. The subscript i refers to the input, o the output, and f refers to the feedback network. The capacitor C_f is a blocking capacitor used to avoid a DC connection between the gate and drain. At

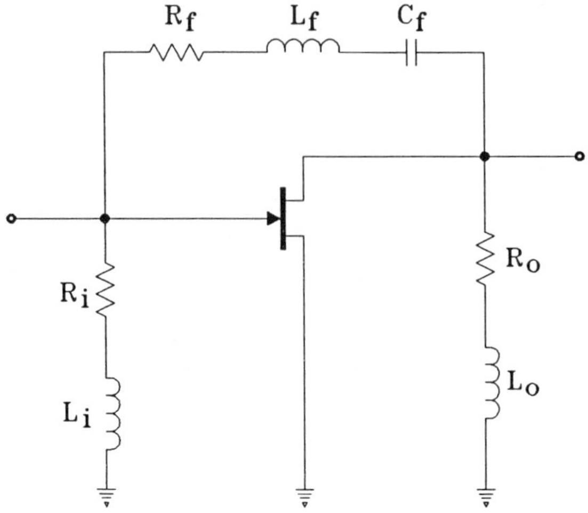

Figure 7.34 Resistive compensation networks.

low frequencies, all three networks are resistive, which causes a reduction in G_{MAX} and a corresponding improvement in stability. At higher frequencies, the inductive reactance dominates and the predominate effect of the input and output networks is to change the port impedance but not G_{MAX} or K. Similarly, the feedback network is also inductive, but the effect on G_{MAX} and K is dependent on the device parameters. There are other networks that could be used as well. For example, a parallel *R-C* connected from source to ground forming series feedback, or parallel *R-C* networks in series with the gate or drain. In this case, the networks are resistive at low frequencies, but at higher frequencies the capacitor acts as a short circuit and the device parameters are unchanged. We can also use shorted shunt stubs in place of L_i and L_o or a high impedance TLE in place of L_f. The distributed elements often provide an advantage since we can change both the electrical length and characteristic impedance. The electrical lengths of the shorted shunt stubs can be set equal to one-quarter wavelength at the upper cutoff frequency, which provides an open circuit termination for R_i and R_o.

The best choice for the compensation network depends on a number of factors. One consideration is the amplifier stability. The mapping of source and load plane stability circles shows that, in most cases, we have four choices: series or shunt resistance connected at either the input or output terminal. However, the output plane circles shown in Figure 6.27 for the NEC NE70000 GaAs FET are almost concentric, and a large series or small shunt resistor is required to stabilize the device at the output. The input plane circles shown in Figure 7.15 reveals that a small series resistor or large shunt resistor will provide unconditional stability and have less effect on the maximum available gain. The resistive compensation technique requires that we synthesize input and output matching networks. In fact, once the device is compensated, the design is no different than the approach discussed in Chapter 6. Usually the input of a GaAs FET is more difficult to match than the output. However, resistive compensation at the input is often beneficial in providing an input impedance easily matched over broad frequency ranges. Unfortunately, resistive loss anywhere in the circuit has a detrimental effect on the noise figure. The noise figure is particularly sensitive to loss at the input. Similarly, power is absorbed by resistive loss in the output network and the result is a reduction in output power. When the specifications require low noise figure or significant output power, we will not be able to realize the design using either the feedback or resistive compensation approach.

The design of a resistively compensated amplifier is best accomplished using the ideal negative-image matching networks discussed in Chapter 6. We model the device and one or more compensation networks and then optimize for unconditional stability and constant G_{MAX}. Next, we plot the reflection coefficients for simultaneous input and output match to determine the form of the matching networks. The ideal matching networks are connected to the compensated device and then another optimization is performed to minimize the matching requirements. At this point, the ideal minimum insertion loss (MIL) of the input and output matching networks, as well as the performance, is defined. If all parameters are acceptable, we can proceed by synthesizing the matching networks and modeling the physical circuit. However, we also have the option of changing the specifications, using different compensation networks, or changing the form of the negative-image models. For example, a reduction in G_{MAX} usually results in improved gain ripple and gain-bandwidth (GBW) requirement for the matching networks. Reference 34 illustrates this design approach. The NE70000 GaAs FET was compensated using the input network shown in Figure 7.34. R_i and L_i were adjusted to provide G_{MAX} equal to 7.5 ± 0.5 dB from 50 MHz to 18 GHz. An analysis of the matching requirements revealed that the input could be matched, but not the output. Therefore, since the output plane is difficult to stabilize, feedback compensation was tried. A combination of R_f and L_f was found that provided similar G_{MAX} characteristics, but now the input could not be matched with a reasonable number of elements. However, the feedback and input compensation networks used together provided a G_{MAX} equal to 5.00 ± 0.25 dB and reasonable matching requirements at both ports. The reference shows a MMIC amplifier with 5.00 ± 0.25-dB gain from 2.0 to 18.0 Ghz, return loss greater than 15 dB, and unconditional stability. A second design is shown that uses distributed elements and has 5.0 ± 0.4-dB gain from 50 MHz to 18 GHz, greater than 11-dB return loss, and is also unconditionally stable. This design has a bandwidth of almost two and one-half decades!

7.4.3 MMIC Resistive Compensated Amplifier

As an example of resistive compensation, we will design a MMIC amplifier to operate over a frequency range from 6.0 to 12.0 GHz with at least 10-dB gain, ripple less than ±0.5 dB, VSWRs of 1.5:1 or less, and unconditional stability. Although this approach is not the best choice for low noise, we will design for the minimum possible noise figure consistent with the other specifications. Rather than base the design on a specific MMIC device, we will use the NEC NE20200 chip data, which is readily available from the manufacturer or a library provided with most CAD systems. The NE20200 is a low-noise heterojunction GaAs FET with a gate length of 0.3 μm. *S*-parameter and noise data is provided from 1 to 26.5 GHz for bias conditions of 2V drain to source and 10-mA drain current. Analysis reveals that the device is potentially unstable up to 22 GHz, and the MSG ranges from 14.3 to 17.1 dB in the 6.0- to 12.0-GHz range. The minimum noise figure is 1.0 dB at 12.0 GHz.

The first step is to determine the best choice for the compensation network. Since noise figure is a consideration, we will avoid adding loss at the input, which would significantly increase the noise figure. The input and output plane stability circles show characteristics similar to those of the NE70000 plotted in Figures 6.26, 6.27, and 7.15. Thus, stabilization at the output will require a large series resistor or small shunt resistor, and G_{MAX} may be reduced below 10 dB. However, feedback compensation shown in Figure 7.34 is a potential candidate. The device and series *R-L-C* feedback circuit is modeled and goals are specified for 10-dB maximum available gain and 2-dB maximum noise figure from 6.0 to 12.0 GHz. A stability factor greater than 1 from 1 to 26 GHz is also specified and then the network is optimized. The result is G_{MAX} = 10.0 ± 0.4 dB, noise figure slightly greater than 2.0 dB from 6.0 to 12.0 GHz, and unconditional stability across the entire frequency range. Noise parameters are affected by series and shunt feedback, as discussed in Section 2.3.11, and it is worthwhile investigating series feedback as a means of reducing the noise figure of this design. Adding a series *R-L-C* network from the source to ground and then reoptimizing provides somewhat better performance. The noise figure is between 1.54 and 1.75 dB from 6.0 to 12.0 GHz, and the maximum available gain is 10.000 ± 0.125 dB. The series feedback network is a 0.1-nH inductor; that is, the optimizer set the resistance to zero and the capacitance to a large value. Unfortunately, the small inductor has a detrimental effect on the stability factor. *K* is less than 1 above 20 GHz where the inductive reactance is significant. However, the output plane stability circles for the combination of series and shunt feedback encloses a small region of the Smith chart, and a 20Ω resistor in series with the output is sufficient to provide unconditional stability. The final form of the feedback and stabilization networks is shown in Figure 7.35.

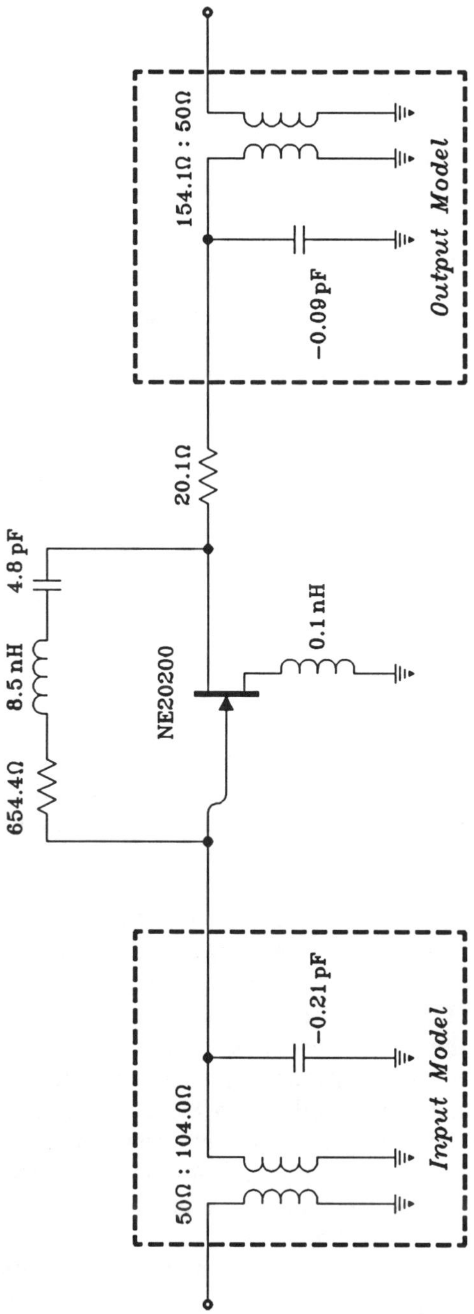

Figure 7.35 Resistive compensated 6.0- to 12.0-GHz amplifier model.

The maximum available gain is 10.000 ± 0.125 dB, and, if we can synthesize efficient input and output matching networks, the amplifier will achieve G_{MAX} and low VSWRs. The noise figure is dependent on the source impedance and the input matching network will affect the noise figure. In fact, since the optimum noise impedance does not coincide with that for a matched input, we have conflicting requirements. However, the optimum noise impedance is also affected by series and shunt feedback. Thus, the task is to find a combination of source, load, and feedback impedances that provide acceptable performance and can easily be synthesized. The input and output reflection coefficients for the optimized device and feedback networks follow contours of constant conductance in the capacitive half of the Smith chart, which results in the negative-image models shown in Figure 7.35. We will use Libra™ to optimize the circuit shown in the figure for flat 10-dB gain, input and output match, low noise, and unconditional stability. The ideal MIL of both negative-image models must be low in order to synthesize efficient matching networks. Libra™ does not provide the necessary models. Nevertheless, we can use several built-in features to analyze the matching requirements of the source and load models. Figure 7.36 shows a portion of the circuit file corresponding to the ideal amplifier in Figure 7.35 and illustrates one approach to modeling the negative-image networks. We first define the resistor and *positive* capacitor for each network in the *VAR* or variable block. *RI-CI* corresponds to the input model, and *RO-CO* the output model. The number sign (#) inserted between the variable name and element values denotes constrained optimization in the form of a minimum, initial, and maximum value. The portion of the circuit file beginning with *EQN* is used to specify variables in terms of equations. The first two lines defines negative values for the capacitors and the next two lines define the transformer turns ratios as the square root of the input resistance divided by the output resistance. The variables labeled *KI* and *KO* evaluate equation (6.6b) for the input and output models, respectively. The factor of 1000 in the numerator is due to the default units of gigahertz and picofarad. Equation (6.3) is used to define *AI* and *AO* as the ideal MIL in decibels for the input and output networks, respectively. The constant 4.343 is used to convert the natural logarithm to base 10. The block labeled *CKT* contains the actual circuit description. *NAIN* is the input negative-image model, *NA2P* is the active device, *NAPAR* is the parallel feedback network, *NASER* is the series feedback, and *NAOUT* is the output negative-image model. The five subcircuit names have special meaning. When each is combined to form the complete amplifier in Figure 7.35, the total noise contribution will be computed, including any loss and transformation of noise parameters. We cannot realize the series feedback inductor using MMIC technology; it is too small. Therefore, the inductive reactance is implemented as a shorted stub, forming the NASER subcircuit.

We intend to minimize the ideal MIL of both matching networks, and can do so by modeling each as an ideal gain block with the gain equal to the computed

```
VAR       ! =================== VARIABLES ==========================
   RI # 10 103.98198 500                         ! Input Rs
   CI # 0  0.21045  10                           ! Input Cp

   RO # 10 154.07637 250                         ! Output Rl
   CO # 0  0.08872  10                           ! Output Cp

EQN       ! =================== EQUATIONS ==========================
   CIN = -CI                                     ! Negative C
   CON = -CO                                     ! Negative C

   NI = (50.0 / RI)**(0.5)                       ! Input transformer
   NO = (RO / 50.0)**(0.5)                       ! Output transformer

   KI = 1000.0 / (2.0 * RI * CI *  6.0)          ! Input K
   KO = 1000.0 / (2.0 * RO * CO *  6.0)          ! Output K

   AI = 4.342945 * LN(1.0 - EXP(-2.0 * KI))      ! Input MIL
   AO = 4.342945 * LN(1.0 - EXP(-2.0 * KO))      ! Output MIL

CKT       ! ============== CIRCUIT DESCRIPTION ====================
   XFER   1 2 0 0  N^NI                   ! Rs
   CAP    2 0      C^CIN                  ! Cp
DEF2P     1 2      NAIN                   ! Input matching network

   S2PA   1 2 3    N20200B.S2P
DEF3P     1 2 3    NA2P                   ! Active device

   SRLC   1 2      R#0 654.38275 5000  L#0 8.47371 10  C#0 4.75524 30
DEF2P     1 2      NAPAR                  ! Parallel feedback

   TLSC   1 0      Z#20 89.14434 100  E#0 10.14569 45  F=26
DEF1P     1        NASER                  ! Series feedback

   RES    1 2      R#0 20.12017 50        ! Stabilization
   CAP    2 0      C^CON                  ! Cp
   XFER   2 3 0 0  N^NO                   ! Rl
DEF2P     1 3      NAOUT                  ! Output matching network

   GAIN   1 2      A^AI  S=0  F=1
DEF2P     1 2      IMN                    ! Input match MIL

   GAIN   1 2      A^AO  S=0  F=1
DEF2P     1 2      OMN                    ! Output match MIL

OPT       ! ============== OPTIMIZATION GOALS =====================
   AMP  K          > 1.0 10               ! K from 1 to 26 GHz
RANGE 6.0  12.0                           ! Frequency 6 to 12 GHz
   AMP  DB[S21]  = 10.0                   ! Gain = 10
   AMP  DB[NF]  < 1.75                    ! Noise less than 1.75
   AMP  DB[S11]  < -15                    ! Input return loss
   AMP  DB[S22]  < -15                    ! Output return loss
   IMN  DB[S21]  > -0.010 100             ! Input match MIL
   OMN  DB[S21]  > -0.010 100             ! Output match MIL
```

Figure 7.36 Circuit file for negative-image amplifier model.

MIL. The *IMN* subcircuit in Figure 7.36 is the input model and *OMN* is the output model. Optimization goals are specified in the *OPT* block. The analysis range is from 1 to 26 GHz, and the first line specifies that the stability factor should be greater than 1 at all frequencies. A weight of 10 is associated with *K* to ensure a significant contribution to the error function for values less than 1. The *RANGE* statement restricts the contribution of each of the following goals to the 6.0- to

12.0-GHz range. The goals following the *RANGE* statement are ignored outside this frequency band. The amplifier goals are specified as 10-dB gain, noise figure less than 1.75 dB, and input and output return loss greater than 15 dB. The last two lines define targets for the ideal MIL of each negative-image model. In this case, we have specified the MIL as less than 0.01 dB and used a weight of 100 to compensate for the small value; that is, an error of 0.1 dB will contribute a value of 1 to the error function. The results of the optimization are shown in Figure 7.37, and the final element values are shown in Figures 7.35 and 7.36. Figure 7.37 shows that the gain is 10.000 ± 0.125 dB and the noise figure is between 1.54 and 1.75 dB from 6.0 to 12.0 GHz, essentially the same performance obtained without the matching networks. However, the input and output return loss is 15 dB minimum, and the amplifier is unconditionally stable from 1 to 26 GHz. We need only synthesize matching networks to replace the negative-image models to determine a realizable circuit topology for the resistively compensated MMIC amplifier.

Since the ideal MIL of each model is less than or equal to 0.01 dB, the insertion loss of the synthesized matching networks should be low. In fact, E-Syn™ computes a total loss of 0.073 dB for the input network and 0.015 dB for

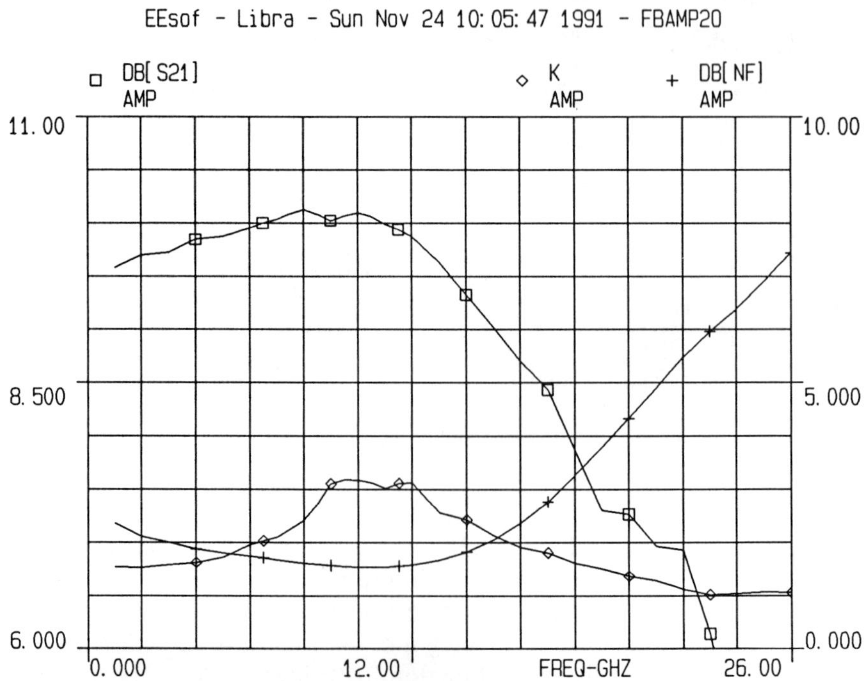

Figure 7.37 Optimization results for the negative-image amplifier model.

the output network using an $n = 2$ Chebyshev lumped-element approximation. However, assuming full absorption, this leaves only three elements in each matching network, which does not provide an assortment of topologies. Therefore, $n = 3$ is specified and E-Syn™ is allowed to automatically synthesize all possible matching networks. The input model results in a total loss of 0.025 dB and six topologies. Lumped-element values are constrained between the realizable MMIC limits defined in Section 7.4 and then each network is optimized. Four of the networks provide a maximum loss of 0.024 dB, and the other two 0.029 dB. A topology is selected from the group of input matching networks that has lowest loss and is convenient for grounding the gate, as shown in Figure 7.41. We also have six choices for the output matching network, but the insertion loss is even better at 0.006 dB. However, when constrained, two networks provide 0.003-dB maximum loss, two have 0.007-dB loss, and the other two have 0.018-dB loss. One of the two lowest loss output matching networks has a convenient topology for drain bias and is shown in Figure 7.41. An analysis of the complete amplifier with the synthesized input and output matching networks shows only minor differences when compared with the original negative-image network. The gain is reduced slightly to 9.936 $\pm$ 0.116 dB, and the noise figure is between 1.52 and 1.76 dB. The input return loss is greater than 13 dB and the output greater than 14 dB. Optimization increases the gain to 9.984 $\pm$ 0.061 dB and reduces the return loss below 15 dB at both ports. However, the noise figure has increased slightly and is between 1.58 and 1.81 dB in the 6.0- to 12.0-GHz frequency range. The gain and noise figure are plotted in Figure 7.38.

The amplifier meets all the design goals, although the model is incomplete since the loss of the MMIC lumped elements is not considered. Assuming a constant Q factor of 50, we change the model for each inductor and capacitor to more accurately reflect the loss and then perform another analysis. This time we observe a considerable change. The gain has been reduced by almost 1 dB to 9.015 $\pm$ 0.216 dB, and the noise figure has increased by 0.5 dB and is between 2.08 and 2.48 dB. However, the stability factor has increased to a minimum of 2.5 and the return loss is greater than 16 dB at both ports. A final optimization compensates for the additional loss as much as possible. Figure 7.39 shows that the gain is 9.969 $\pm$ 0.147 dB, and the noise figure is between 1.88 and 2.24 dB. Figure 7.40 shows the magnitude of the input and output reflection coefficients in decibels which are less than or equal to -14.87 dB, or a maximum VSWR of 1.44:1 at both ports. The stability factor is greater than 2.0 from 1 to 26 GHz, and the amplifier is unconditionally stable.

Figure 7.41 shows the topology and element values of the complete 6.0- to 12.0-GHz MMIC amplifier. When we added loss to the matching networks, the optimizer set the output stabilization resistor shown in Figure 7.35 to zero. The combined loss of the output matching elements is more than sufficient to stabilize the device as illustrated by a minimum stability factor of 2. Similarly, the two capacitors shown in dotted outlines were also set to zero and are not necessary in

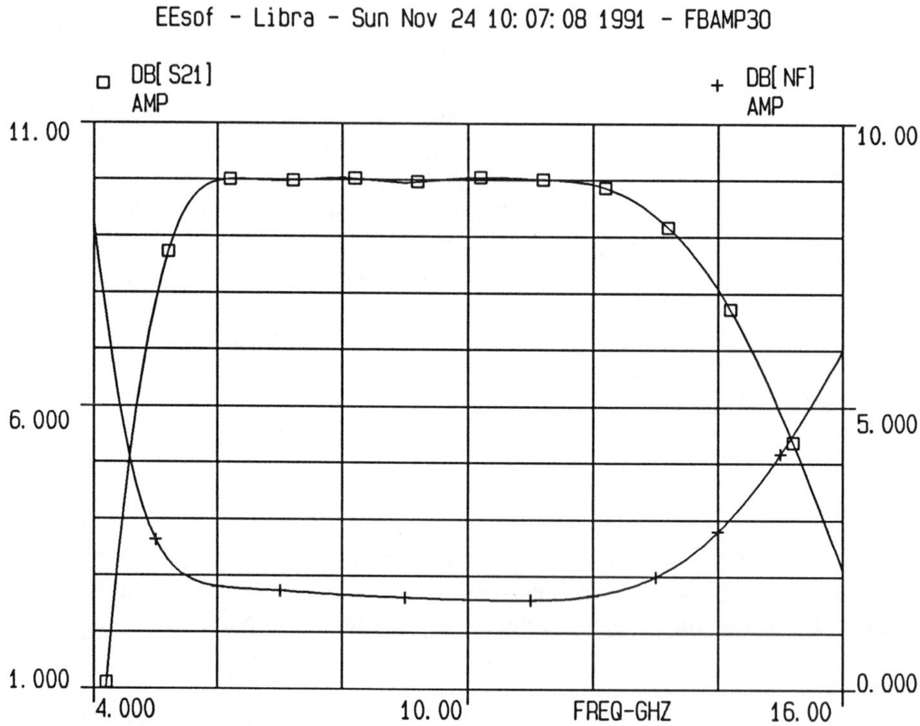

Figure 7.38 Optimized performance of the lossless amplifier.

the final network. The circuit more than meets the design goals. The gain is 9.97 ± 0.15 dB, input and output VSWR are less than 1.45:1, and the noise figure is less than 2.25 dB, approximately 1.25 dB above the optimum. Inductor values range between 0.76 and 8.64 nH, while the capacitance range falls between 0.31 and 5.00 pF. The design also requires an 88Ω shorted stub with an electrical length of 10.6° at 26 GHz. We selected the input and output matching networks based on topologies convenient for bias, as well as a range of realizable element values. However, when designing for a specific fabrication process, we have well-defined equivalent models for the lumped elements and can sometimes choose a topology where the parasitic elements can be absorbed by the matching network. For example, if we specify $n = 4$, there are 16 choices for the input network and 32 for the output network. However, additional elements adds to the total loss, and we must weigh the advantages of parasitic absorption against the disadvantage of increased loss. This usually means that we begin with the minimum order, $n = 2$ in this example, select matching networks, accurately model all elements, and then

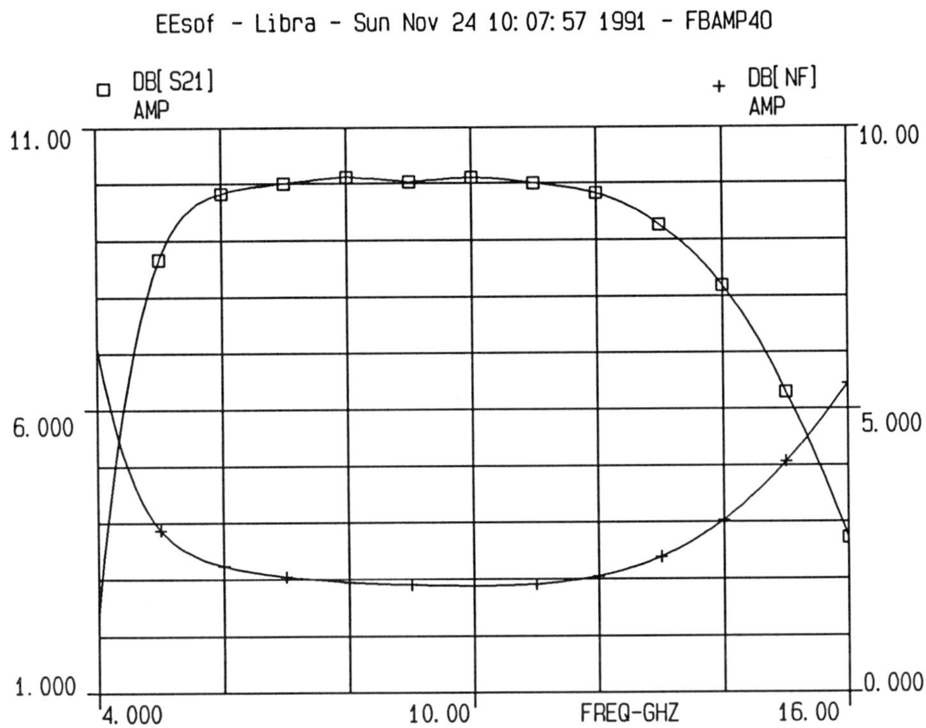

Figure 7.39 Gain and noise figure of the 6.0- to 12.0-GHz MMIC amplifier.

optimize. We then increase the order and repeat the process until the optimum combination of input and output topologies is found. In this case, a lower order input network may result in lower noise figures due to reduced input loss. However, the input VSWR may also be higher. In fact, we can reduce the input VSWR specification for the design shown in Figure 7.41 and achieve slightly lower noise figure and gain ripple.

The data for the NE20200 device extends from 1.0 to 26.5 GHz, adequate for this design. The stability factor is almost flat below 4.0 GHz, as shown in Figure 7.37, and the design should be stable below 1.0 GHz. However, in some cases we require data above or below the measured range. For example, one design discussed in reference 34 used the NE70000 chip for a 50-MHz to 18-GHz amplifier, although data is only provided between 2.0 and 18.0 GHz. A technique that can be used successfully is to fit a built-in device model to the measured data, and then analyze the model to obtain the necessary data. That is, we constrain elements of the device model to reasonable values and optimize in an effort to set the four measured *s*-parameters equal to the *s*-parameters of the model. We can use this approach to obtain the low-frequency parameters of the NE20200 device and possibly design a DC to 26-GHz MMIC amplifier using low-pass matching networks. The FET has more than 10 dB maximum available gain at 26 GHz, and we can eliminate the blocking capacitor in the feedback network of Figure 7.41 to provide compensation from DC to 26 GHz. The gate is grounded and the voltage across the resistor is the sum of the drain-to-source and source-to-ground voltages, approximately 3V. Thus, the resistor will dissipate less than 15 mW of power, which is a small price to pay for a design capable of DC to 26 GHz operation.

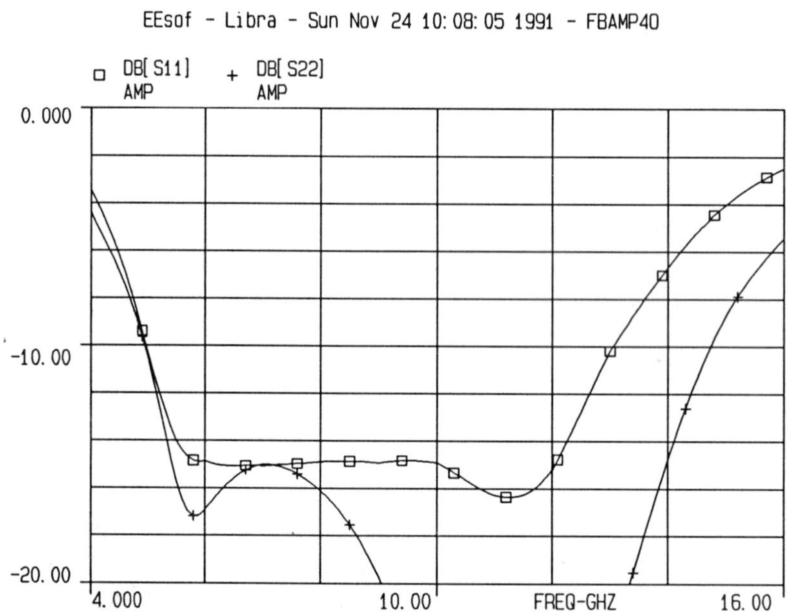

Figure 7.40 $|s_{11}|$ and $|s_{22}|$ of the 6.0- to 12.0-GHz MMIC amplifier.

7.5 SUMMARY

The examples in this chapter illustrate some of the applications of CAD tools to the design of RF and microwave circuits. We can easily investigate various approaches to a particular design problem, synthesize matching networks, analyze, and optimize as necessary. Once the design is acceptable in ideal form, we can alter the model to include the effects of fabrication parasitics and nonideal components, which provides an accurate representation of the physical circuit. The physical model can be analyzed to determine the effects of component and

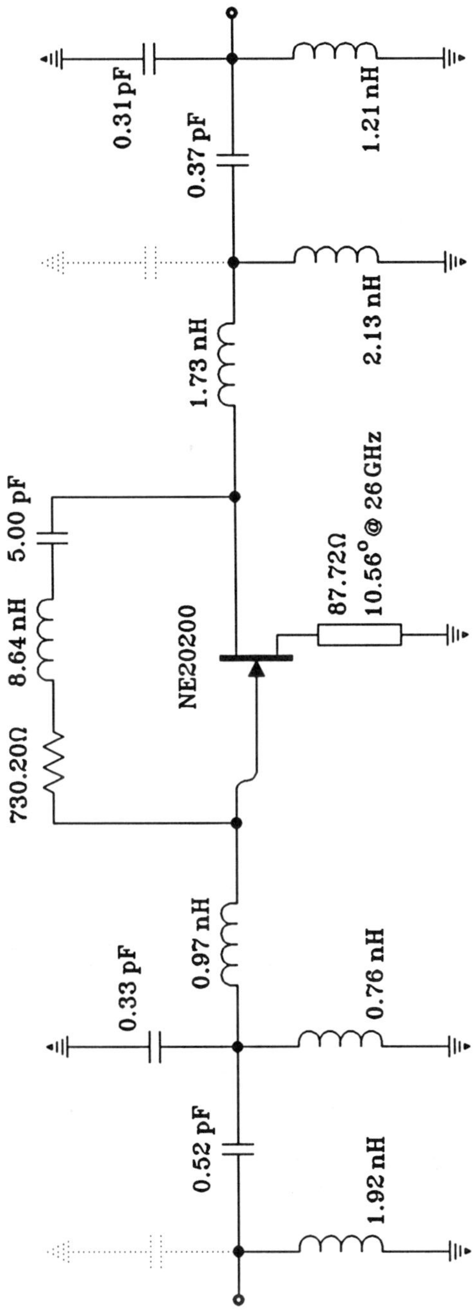

Figure 7.41 Resistive compensated 6.0- to 12.0-GHz amplifier circuit.

assembly tolerances. Then the design can be centered to improve yields and reduce production costs.

Modern CAD tools provide the capability of schematic capture, which virtually eliminates modeling errors. We simply draw the circuit on the screen, and the software automatically creates the necessary mathematical model. Interaction between the circuit model and the physical layout is also a valuable asset. We can either create a physical topology which is automatically translated into a circuit model, or define the model and generate an equivalent layout. Layout changes are often necessary to accommodate physical constraints, and either the physical or electrical model can easily be modified. Changes are automatically reflected in both models. Personal computers and inexpensive workstations have resulted in powerful design tools which are available to most designers. As a rule, the software is easy to use and often includes features such as extensive built-in help and multiple windows. However, perhaps the most beneficial aspect of these programs is a hidden one. Leading experts are often employed to develop circuit models and techniques, improving the correlation between a computer prediction and measured data. In effect, we have access to a great deal of expert knowledge every time we use a CAD program to design a circuit. As artificial intelligence is applied to future generations of CAD products, the knowledge base will be extended to other aspects of circuit design, resulting in truly remarkable design tools for both the experienced and inexperienced engineer.

REFERENCES

[1] T. Itoh, ed., *Numerical Techniques for Microwave and Millimeter-Wave Passive Structures*, John Wiley & Sons, Inc., New York, 1989.

[2] B.M. Oliver, "Directional Electromagnetic Couplers," *IRE Proc.*, Vol. 42, No. 11, November 1954, pp. 1686-1692.

[3] G.D. Monteath, "Coupled Transmission Lines as Symmetrical Directional Couplers," *IEE Proc.*, Part B, No. 3, May 1955, pp. 383-392.

[4] E.M.T. Jones and J.T. Bolljahn, "Coupled-Strip Transmission-Line Filters and Directional Couplers," *IRE Trans. on Microwave Theory and Techniques*, Vol. MTT-4, No. 2, April 1956, pp. 75-81.

[5] L. Young, "The Analytical Equivalence of TEM-Mode Directional Couplers and Transmission--Line Stepped-Impedance Filters," *IEE Proc.*, Vol 110, No. 2, February 1963, pp. 275-281.

[6] E.G. Cristal, "Coupled-Transmission-Line Directional Couplers with Coupled Lines of Unequal Characteristic Impedances," *IEEE Trans. on Microwave Theory and Techniques*, Vol. MTT-14, No. 7, July 1966, pp. 337-346.

[7] C.B. Sharpe, "An Equivalent Principle for Nonuniform Transmission-Line Directional Couplers," *IEEE Trans. on Microwave Theory and Techniques*, Vol. MTT-15, No. 7, July 1967, pp. 398-405.

[8] G.L. Matthaei, L. Young, and E.M.T. Jones, *Microwave Filters, Impedance-Matching Networks, and Coupling Structures*, Artech House, Inc., Dedham, Massachusetts, 1964.

[9] R.E. Collin, *Foundations for Microwave Engineering*, McGraw-Hill Book Company, New York, 1966.

[10] K.C. Gupta, R. Garg, and I.J. Bahl, *Microstrip Lines and Slotlines*, Artech House, Inc., Dedham, Massachusetts, 1979.

[11] K.C. Gupta, R. Garg, and R. Chadha, *Computer-Aided Design of Microwave Circuits*, Artech House, Inc., Dedham, Massachusetts, 1981.

[12] T.C. Edwards, *Foundations for Microstrip Circuit Design*, John Wiley & Sons, Inc., New York, 1981.

[13] H. Howe, Jr., *Stripline Circuit Design*, Artech House, Inc., Dedham, Massachusetts, 1974.

[14] V.K. Tripathi, "Asymmetric Coupled Transmission Lines in an Inhomogeneous Medium," *IEEE Trans. on Microwave Theory and Techniques*, Vol. MTT-23, No. 9, September 1975, pp. 734-739.

[15] V.K. Tripathi, "Equivalent Circuits and Characteristics of Inhomogeneous Nonsymmetrical Coupled-Line Two-Port Circuits," *IEEE Trans. on Microwave Theory and Techniques*, Vol. MTT-25, No. 2, February 1977, pp. 140-142.

[16] J.L. Allen, "Non-Symmetrical Coupled Lines in an Inhomogeneous Dielectric Medium," *Int. J. Electronics*, Vol. 38, No. 3, March 1975, pp. 337-347.

[17] L. Young, ed., *Parallel Coupled Lines and Directional Couplers*, Artech House, Inc., Dedham, Massachusetts, 1972.

[18] L. Young, ed., *Microwave Filters Using Parallel Coupled Lines*, Artech House, Inc., Dedham, Massachusetts, 1972.

[19] T.G. Bryant and J.A. Weiss, "Parameters of Microstrip Transmission Lines and of Coupled Pairs of Microstrip Lines," *IEEE Trans. on Microwave Theory and Techniques*, Vol. MTT-16, No. 12, December 1968, pp. 1021-1027.

[20] J.I. Smith, "The Even and Odd-Odd Mode Capacitance Parameters for Coupled Lines in Suspended Substrate," *IEEE Trans. on Microwave Theory and Techniques*, Vol. MTT-19, No. 5, May 1971, pp. 424-431.

[21] R.J. Wenzel, "Exact Design of TEM Microwave Networks Using Quarter-Wave Lines," *IEEE Trans. on Microwave Theory and Techniques*, Vol. MTT-12, No. 1, January 1964, pp. 94-111.

[22] J.P. Shelton and J.A. Mosko, "Synthesis and Design of Wideband Equal Ripple TEM Directional Couplers and Fixed Phase Shifters," *IEEE Trans. on Microwave Theory and Techniques*, Vol. MTT-14, No. 10, October 1966, pp. 462-473.

[23] J.P. Shelton and R.C. Van Wagoner, "Tandem Couplers and Phase Shifters," *Microwaves*, Vol. 4, No. 4, April 1965, pp. 14-19.

[24] R.S. Engelbrecht and K. Kurokawa, "A Wide-Band Low Noise L-Band Balanced Transistor Amplifier," *IEEE Proc.*, Vol. 53, No. 3, March 1965, pp. 237-247.

[25] K. Kurokawa, *Design Theory of Balanced Transistor Amplifiers*, Bell Telephone Systems, Monograph No. 5070, 1965, and *The Bell System Technical Journal*, Vol. 44, No. 3, October 1965, pp. 1675-1698.

[26] S.L. March, "Phase Velocity Compensation in Parallel-Coupled Microstrip," 1982 *IEEE MTT-S International Microwave Symposium Digest*, June 1982, pp. 410-412.

[27] A. Podell, "A High Directivity Microstrip Coupler Technique," *1970 IEEE MTT-S International Microwave Symposium Digest*, May 1970, pp. 33-36.

[28] H.E. Brenner, "Perturbations of the Critical Paramaters of Quarter-Wave Directional Couplers," *IEEE Trans. on Microwave Theory and Techniques*, Vol. MTT-15, No. 6, June 1967, pp. 384-385.

[29] J. Lange, "Interdigitated Stripline Quadrature Hybrid," *IEEE Trans. on Microwave Theory and Techniques*, Vol. MTT-17, No. 12, December 1969, pp. 1150-1151.

[30] S.L. March, "Simple Equations Characterize Bond Wires," *Microwaves and RF*, Vol. 30, No. 11, November 1991, pp. 105-110.

[31] R.A. Pucel, ed., *Monolithic Microwave Integrated Circuits*, IEEE Press, New York, 1985.

[32] J.B. Coughlin, R.J. Gelsing, P.J. Jochems, and H.J.M. Van der Laak, "A Monolithic Silicon Wide-Band Amplifier from DC to 1 GHz," *IEEE Journal of Solid-State Circuits*, Vol. SC-8, No. 6, December 1973, pp. 414-419.

[33] R.G. Meyer, R. Eschenbach, and R. Chin, "A Wide-Band Ultralinear Amplifier from 3 to 300 MHz," *IEEE Journal of Solid-State Circuits*, Vol. SC-9, No. 4, August 1979, pp. 167-175.

[34] M.W. Medley, Jr., "Network Synthesis CAD Program Applied to RF and Microwave Circuit Design," *Microwave System News & Communications Technology*, Vol. 17, No. 1, January 1987, pp. 97-107; Vol. 17, No. 3, March 1987, pp. 88-93; Vol. 17, No. 4, April 1987, pp. 90-96.

Index

The Artech House Microwave Library

Acoustic Charge Transport: Device Technology and Applications, R. Miller, C. Nothnick, and D. Bailey

Advanced Automated Smith Chart Software and User's Manual, Version 2.0, Leonard M. Schwab

Algorithms for Computer-Aided Design of Linear Microwave Circuits, Stanislaw Rosloniec

Analysis, Design, and Applications of Fin Lines, Bharathi Bhat and Shiban K. Koul

Analysis Methods for Electromagnetic Wave Problems, Eikichi Yamashita, editor

Automated Smith Chart Software and User's Manual, Leonard M. Schwab

C/NL2 for Windows: Linear and Nonlinear Microwave Circuit Analysis and Optimization, Software and User's Manual, Stephen A. Maas and Arthur Nichols

Capacitance, Inductance, and Crosstalk Analysis, Charles S. Walker

Design of Impedance-Matching Networks for RF and Microwave Amplifiers, Pieter L. D. Abrie

Dielectric Materials and Applications, Arthur von Hippel, editor

Dielectrics and Waves, Arthur von Hippel

Digital Microwave Receivers, James B. Tsui

Electric Filters, Martin Hasler and Jacques Neirynck

Electrical and Thermal Characterization of MESFETs, HEMTs, and HBTs, Robert Anholt

E-Plane Integrated Circuits, P. Bhartia and P. Pramanick, editors

Feedback Maximization, Boris J. Lurie

Filters with Helical and Folded Helical Resonators, Peter Vizmuller

Frequency Synthesizer Design Toolkit Software and User's Manual, Version 1.0, James A. Crawford

Fundamentals of Distributed Amplification, Thomas T. Y. Wong

GaAs MESFET Circuit Design, Robert A. Soares, editor

GASMAP: Gallium Arsenide Model Analysis Program, J. Michael Golio *et al.*

Handbook of Microwave Integrated Circuits, Reinmut K. Hoffmann

Handbook for the Mechanical Tolerancing of Waveguide Components, W. B. W. Alison

HELENA: HEMT Electrical Properties and Noise Analysis Software and User's Manual, Henri Happy and Alain Cappy

HEMTs and HBTs: Devices, Fabrication, and Circuits, Fazal Ali, Aditya Gupta, and Inder Bahl, editors

High-Power Microwave Sources, Victor Granatstein and Igor Alexeff, edtiors

High-Power GaAs FET Amplifiers, John Walker, editor

High-Power Microwaves, James Benford and John Swegle

Introduction to Microwaves, Fred E. Gardiol

Introduction to Computer Methods for Microwave Circuit Analysis and Design, Janusz A. Dobrowolski

Introduction to the Uniform Geometrical Theory of Diffraction, D. A. McNamara, C. W. I. Pistorius and J. A. G. Malherbe

LOSLIN: Lossy Line Calculation Software and User's Manual, Fred E. Gardiol

Lossy Transmission Lines, Fred E. Gardiol

Low-Angle Microwave Propagation: Physics and Modeling, Adolf Giger

Low Phase Noise Microwave Oscillator Design, Robert G. Rogers

MATCHNET: Microwave Matching Networks Synthesis, Stephen V. Sussman-Fort

Matrix Parameters for Multiconductor Transmission Lines: Software and User's Manual, A. R. Djordjevic *et al.*

Microelectronic Reliability, Volume I: Reliability, Test, and Diagnostics, Edward B. Hakim, editor

Microelectronic Reliability, Volume II: Integrity Assessment and Assurance, Emiliano Pollino, editor

Microwave and RF Circuits: Analysis, Synthesis, and Design, Max Medley

Microwave and RF Component and Subsystem Manufacturing Technology, Heriot-Watt University

Microwave Circulator Design, Douglas K. Linkhart

Microwave Engineers' Handbook, 2 Volumes, Theodore Saad, editor

Microwave Materials and Fabrication Techniques, Second Edition, Thomas S. Laverghetta

Microwave MESFETs and HEMTs, J. Michael Golio *et al.*

Microwave and Millimeter Wave Heterostructure Transistors and Applicatons, F. Ali, editor

Microwave and Millimeter Wave Phase Shifters, Volume I: Dielectric and Ferrite Phase Shifters, S. Koul and B. Bhat

Microwave and Millimeter Wave Phase Shifters, Volume II: Semiconductor and Delay Line Phase Shifters, S. Koul and B. Bhat

Microwave Mixers, Second Edition, Stephen Maas

Microwave Transmission Design Data, Theodore Moreno

Microwave Transition Design, Jamal S. Izadian and Shahin M. Izadian

Microwave Transmission Line Couplers, J. A. G. Malherbe

Microwave Tubes, A. S. Gilmour, Jr.

Microwaves: Industrial, Scientific, and Medical Applications, J. Thuery

Microwaves Made Simple: Principles and Applicatons, Stephen W. Cheung, Frederick H. Levien *et al.*

MMIC Design: GaAs FETs and HEMTs, Peter H. Ladbrooke

Modern GaAs Processing Techniques, Ralph Williams

Modern Microwave Measurements and Techniques, Thomas S. Laverghetta

Monolithic Microwave Integrated Circuits: Technology and Design, Ravender Goyal *et al.*

Nonuniform Line Microstrip Directional Couplers, Sener Uysal

PC Filter: Electronic Filter Design Software and User's Guide, Michael G. Ellis, Sr.

PLL: Linear Phase-Locked Loop Control Systems Analysis Software and User's Manual, Eric L. Unruh

RF Design Guide: Systems, Circuits, and Equations, Peter Vizmuller

Scattering Parameters of Microwave Networks with Multiconductor Transmission Lines: Software & User's Manual, A. R. Djordjevic *et al.*

Solid-State Microwave Power Oscillator Design, Eric Holzman and Ralston Robertson

Terrestrial Digital Microwave Communications, Ferdo Ivanek *et al.*

Time-Domain Response of Multiconductor Transmission Lines: Software and User's Manual, A. R. Djordjevic *et al.*

Transmission Line Design Handbook, Brian C. Waddell

Yield and Reliability in Microwave Circuit and System Design, Michael Meehan and John Purviance

For further information on these and other Artech House titles, contact:

Artech House
685 Canton Street
Norwood, MA 02062
617-769-9750
Fax: 617-769-6334
Telex: 951-659
email: artech@world.std.com

Artech House
Portland House, Stag Place
London SW1E 5XA England
+44 (0) 171-973-8077
Fax: +44 (0) 171-630-0166
Telex: 951-659
bookco@artech.demon.co.uk